Lecture Notes in Computer Science 10020

Commenced Publication in 1973
Founding and Former Series Editors:
Gerhard Goos, Juris Hartmanis, and Jan van Leeuwen

More information about this series at http://www.springer.com/series/7151

James F. Peters · Andrzej Skowron (Eds.)

Transactions on Rough Sets XX

Springer

ISSN 0302-9743 ISSN 1611-3349 (electronic)
Lecture Notes in Computer Science
ISSN 1861-2059 ISSN 1861-2067 (electronic)
Transactions on Rough Sets
ISBN 978-3-662-53610-0 ISBN 978-3-662-53611-7 (eBook)
DOI 10.1007/978-3-662-53611-7

Library of Congress Control Number: 2016954935

Printed on acid-free paper

This Springer imprint is published by Springer Nature
The registered company is Springer-Verlag GmbH Germany
The registered company address is: Heidelberger Platz 3, 14197 Berlin, Germany

Preface

Volume XX of the *Transactions on Rough Sets* (TRS) is a continuation of a number of research streams that have grown out of the seminal work of Zdzisław Pawlak[1] during the first decade of the twenty-first century.

The paper co-authored by Javad Rahimipour Anaraki, Saeed Samet, Wolfgang Banzhaf, and Mahdi Eftekhari introduces a new hybrid merit based on a conjunction of correlation feature selection and fuzzy-rough feature selection methods. The new merit selects fewer redundant features and finds the most relevant features resulting in reasonable classification accuracy. The paper co-authored by Mohammad Azad, Mikhail Moshkov, and Beata Zielosko presents a study of a greedy algorithm for construction of approximate decision rules. This algorithm has polynomial time complexity for binary decision tables with many-valued decisions. The proposed greedy algorithm constructs relatively short α-decision rules. The paper by Mani presents algebraic semantics of proto-transitive rough sets. Proto-transitivity, according to the author, can be considered as a possible generalization of transitivity that happens often in the context of applications. The paper by Piero Pagliani presents a uniform approach to previously introduced covering-based approximation operators from the point of view of pointless topology. The monograph authored by Mohammad Aquil Khan is devoted to the study of multiple-source approximation systems, evolving information systems, and corresponding logics based on rough sets.

The editors would like to express their gratitude to the authors of all submitted papers. Special thanks are due to the following reviewers: Jan Bazan, Chris Cornelis, Davide Cuicci, Ivo Düntsch, Soma Dutta, Jouni Järvinen, Richard Jensen, Pradipta Maji, Sheela Ramanna, Zbigniew Suraj, and Marcin Wolski.

The editors and authors of this volume extend their gratitude to Alfred Hofmann, Christine Reiss, and the LNCS staff at Springer for their support in making this volume of TRS possible.

The Editors-in-Chief were supported by the Polish National Science Centre (NCN) grants DEC-2012/05/B/ST6/06981 and DEC-2013/09/B/ST6/01568, the Polish National Centre for Research and Development (NCBiR) DZP/RID-I-44/8/NCBR/2016, as well as the Natural Sciences and Engineering Research Council of Canada (NSERC) discovery grant 185986.

August 2016 James F. Peters
Andrzej Skowron

---

[1] See, *e.g.*, Pawlak, Z., A Treatise on Rough Sets, *Transactions on Rough Sets* IV, (2006), 1–17. See, also, Pawlak, Z., Skowron, A.: Rudiments of rough sets, *Information Sciences* 177 (2007) 3–27; Pawlak, Z., Skowron, A.: Rough sets: Some extensions, *Information Sciences* 177 (2007) 28–40; Pawlak, Z., Skowron, A.: Rough sets and Boolean reasoning, *Information Sciences* 177 (2007) 41–73.

LNCS Transactions on Rough Sets

The *Transactions on Rough Sets* series has as its principal aim the fostering of professional exchanges between scientists and practitioners who are interested in the foundations and applications of rough sets. Topics include foundations and applications of rough sets as well as foundations and applications of hybrid methods combining rough sets with other approaches important for the development of intelligent systems. The journal includes high-quality research articles accepted for publication on the basis of thorough peer reviews. Dissertations and monographs of up to 250 pages that include new research results can also be considered as regular papers. Extended and revised versions of selected papers from conferences can also be included in regular or special issues of the journal.

Contents

A New Fuzzy-Rough Hybrid Merit to Feature Selection

Javad Rahimipour Anaraki[1(✉)], Saeed Samet[2], Wolfgang Banzhaf[3], and Mahdi Eftekhari[4]

[1] Department of Computer Science, Memorial University of Newfoundland, St. John's, Nl A1B 3X5, Canada
jra066@mun.ca

[2] Faculty of Medicine, Memorial University of Newfoundland, St. John's, Nl A1B 3V6, Canada
saeed.samet@med.mun.ca

[3] Department of Computer Science, Memorial University of Newfoundland, St. John's, Nl A1B 3X5, Canada
banzhaf@mun.ca

[4] Department of Computer Engineering, Shahid Bahonar University of Kerman, 7616914111 Kerman, Iran
m.eftekhari@uk.ac.ir

Abstract. Feature selecting is considered as one of the most important pre-process methods in machine learning, data mining and bioinformatics. By applying pre-process techniques, we can defy the curse of dimensionality by reducing computational and storage costs, facilitate data understanding and visualization, and diminish training and testing times, leading to overall performance improvement, especially when dealing with large datasets. Correlation feature selection method uses a conventional merit to evaluate different feature subsets. In this paper, we propose a new merit by adapting and employing of correlation feature selection in conjunction with fuzzy-rough feature selection, to improve the effectiveness and quality of the conventional methods. It also outperforms the newly introduced gradient boosted feature selection, by selecting more relevant and less redundant features. The two-step experimental results show the applicability and efficiency of our proposed method over some well known and mostly used datasets, as well as newly introduced ones, especially from the UCI collection with various sizes from small to large numbers of features and samples.

Keywords: Feature selection · Fuzzy-rough dependency degree · Correlation merit

1 Introduction

Each year the amount of generated data increases dramatically. This expansion needs to be handled to minimize the time and space complexities as well as the

J.F. Peters and A. Skowron (Eds.): TRS XX, LNCS 10020, pp. 1–23, 2016.
DOI: 10.1007/978-3-662-53611-7_1

comprehensibility challenges inherent in big datasets. Machine learning methods tend to sacrifice some accuracy to decrease running time, and to increase the clarity of the results [1].

Datasets may contain hundreds of thousand of samples with thousands of features that make further processing on data a tedious job. Reduction can be done on either features or on samples. However, due to the high cost of sample gathering and their undoubted utility, such as in bioinformatics and health systems, data owners usually prefer to keep only the useful and informative features and remove the rest, by applying Feature Selection (FS) techniques that are usually considered as a preprocessing step to further processing (such as classification). These methods lead to less classification errors or at least to minimal diminishing of performance [2].

In terms of data usability, each dataset contains three types of features: 1- informative, 2- redundant, and 3- irrelevant. Informative features are those that contain enough information on the classification outcome. In other words, they are non-redundant, relevant features. Redundant features contain identical information compared to other features, whereas irrelevant features have no information about the outcome. The ideal goal of FS methods is to remove the last two types of features [1].

FS methods can generally be divided into two main categories [3]. One approach is *wrapper* based, in which a learning algorithm estimates the accuracy of the subset of features. This approach is computationally intensive and slow due to the large number of executions over selected subsets of features, that make it impractical for large datasets. The second approach is *filter* based, in which features are selected based on their quality regardless of the results of learning algorithm. As a result, it is fast but less accurate. Also, a combinational approach of both methods called *embedded* has been proposed to accurately handle big datasets [4]. In the methods based on this approach, feature subset selection is done while classifier structure is being built.

One of the very first feature selection methods for binary classification datasets is Relief [5]. This method constructs and updates a weight vector of a feature, based on the nearest feature vector of the same and different classes using Euclidean distance. After a predefined number of iterations l, relevant vector is calculated by dividing the weight vector by l, and the features with relevancy higher than a specific threshold will be selected. Hall [1] has proposed a merit based on the average intra-correlation of features and inter-correlation of features and the outcome. Those features with higher correlation to the outcome and lower correlation to other features are selected.

Jensen et al. [6] have introduced a novel feature selection method based on lower approximation of the fuzzy-rough set, in which features and outcome dependencies are calculated using a merit called Dependency Degree (DD). In [7], two modifications of the fuzzy-rough feature selection have been introduced to improve the performance of the conventional method: 1- Encompassing the selection process in *equal* situations, where more than one feature result in an identical fitness value by using correlation merit [1] and 2- Combining the first improvement with the stopping criterion [8]. Qian et al. [9], have proposed an accelerator to perform sample and feature selection simultaneously in order to improve the time complexity of fuzzy-rough feature selection. Jensen et al. [10] have designed a new version of fuzzy-rough feature selection to deal with semi-supervised datasets, in which class feature is partially labelled. Shang et al. [11] have introduced a hybrid system for Mars images based on conjunction of fuzzy-rough feature selection and support vector machines. The behaviour of k-nearest neighbours classifier has been improved by Derrac et al. [12], using fuzzy rough feature selection and steady-state genetic algorithm for both feature and prototype selection. Dai et al. [13], have designed a system using fuzzy information gain ratio based on fuzzy rough feature selection structure to classify tumor data in gene expression.

Xu et al. [14] have proposed a non-linear feature selection method based on gradient boosting of limited depth trees. This method combines classification and feature selection processes into one by using gradient boosted regression trees resulting from the greedy CART algorithm.

In this paper, we propose a new merit, which is not only capable of effectively removing redundant features, selecting relevant ones, and enhancing the classification accuracy, but it also outperforms when applied to large datasets, compared to the other existing methods.

In Sect. 2, background and preliminaries of correlation based and fuzzy-rough feature selection methods are described in detail. Our proposed method is discussed in Sect. 3. Section 4 is dedicated to experimental results and discussion on the performance and effectiveness of the new approach comparing with previously introduced methods. Conclusions and future directions are explained in Sect. 5.

2 Preliminaries

In this section, the idea and explanation of the Correlation-based Feature Selection (CFS) method will be presented in Sect. 2.1. Subsection 2.2 illustrates the rough set theory and the rough set based feature selection approach.

2.1 Correlation Based Feature Selection (CFS)

In the feature selection process, selecting those features that are highly correlated with the class attribute while loosely correlated with the rest of the features, is the ultimate goal. One of the most successful feature selection methods based on this is CFS [1]. The evaluation measure of CFS is designed in such a way that it selects predictive and low level inter-correlated features on the class and other features, respectively. Equation 1 shows the merit.

$$Merit_S = \frac{k\overline{r}_{cf}}{\sqrt{k + k(k-1)\overline{r}_{ff}}}, \tag{1}$$

where S is a subset of features, k is the number of selected features, $\overline{r}_{cf}$ is the mean of the correlations of the selected features to the class attribute, and $\overline{r}_{ff}$ is the average of inter-correlations of features. The enumerator calculates how much the selected subset is correlated with the class, and the denominator controls the redundancy of selected features within the subsets. At the heart of the merit, correlation undeniably plays the most important role. Therefore, maximizing merit requires the most relevant features (to maximize the numerator) and the least redundant ones (to minimize the denominator) to be included in the subset. The relevancy and non-redundancy are two important factors in feature selection that are handled in CFS. However, correlation is only capable of measuring linear relationships of two vectors [15]; therefore, in the case of non-linear relationships, the result will be inaccurate.

2.2 Rough Set Feature Selection

The rough set theory has been proposed by Pawlak that is a mathematical tool to handle vagueness in effective way [16]. Suppose U and A to be the universe of discourse and a nonempty set of attributes, respectively, and the information system is presented by $I = (U, A)$. Consider X as a subset of U, and P and Q as subsets of A; approximating a subset in rough set theory is done through the lower and upper approximations. The lower approximation of X, $(\underline{P}X)$ involves those objects which are surely classified in X with regarding to attributes in P. Whereas, upper approximation of X, $(\overline{P}X)$ accommodates those objects which can possibly classified in X considering attributes of P. By defining the lower and upper approximations, a rough set is shown using an ordered pair $(\underline{P}X, \overline{P}X)$. Based on these approximations, different regions in rough set theory is illustrated by Eqs. 2, 3 and 4.

The union of all objects in different regions of $\mathbb{U}$ partitioned by Q with regarding to P is called positive region $POS_P(Q)$.

$$POS_P(Q) = \bigcup_{X \in \mathbb{U}/Q} \underline{P}X \tag{2}$$

The negative region is collection of object that are in $\mathbb{U}$ but not in $POS_P(Q)$, and is shown by $NEG_P(Q)$ [17].

$$NEG_P(Q) = \mathbb{U} - \bigcup_{X \in \mathbb{U}/Q} \overline{P}X \tag{3}$$

The boundary region has determinative role in specifying the type of a set. If the region is a non-empty set, it is called a rough set, otherwise, it is a crisp set.

$$BND_P(Q) = \bigcup_{X \in \mathbb{U}/Q} \overline{P}X - \bigcup_{X \in \mathbb{U}/Q} \underline{P}X \tag{4}$$

The rough set theory can be used to measure the magnitude of dependency between attributes. The dependency of attributes in Q on attribute(s) in P is shown by $P \Rightarrow_k Q$, in which k equals to $\gamma_P(Q)$ and it is labeled Dependency

Degree (DD) [17]. If $0 < k < 1$, then Q partially depends on P, otherwise if $k = 1$ then Q completely depends on P. Equation 5 calculates the DD of Q on P.

$$k = \gamma_P(Q) = \frac{|POS_P(Q)|}{|\mathbb{U}|}, \tag{5}$$

where notation $|.|$ is number of objects in a set.

The reduct set is a subset of features which has identical DD as considering all features. The members of the reduct set are the most informative features which feature outcome is highly dependent on them, while non-members are irrelevant and/or redundant ones.

The most important drawback of rough set based FS methods is their incapability of handing continuous data. One way to govern this imperfection is to discretize continuous data in advance that is necessary but not enough, as long as the amount of similarity between discretized data is unspecified. The ultimate way to handle continuous data using rough set theory is fuzzy-rough set. To begin with, the definition of the X-lower and X-upper approximations and the degree of fuzzy similarity [6] are given by Eqs. 6 to 8, respectively

$$\mu_{\underline{P}X}(x) = \inf_{y \in \mathbb{U}} I\{\mu_{R_P}(x,y), \mu_X(y)\}, \tag{6}$$

$$\mu_{\overline{P}X}(x) = \sup_{y \in \mathbb{U}} T\{\mu_{R_P}(x,y), \mu_X(y)\}, \tag{7}$$

$$\mu_{R_P}(x,y) = \bigcap_{a \in P} \{\mu_{R_a}(x,y)\}, \tag{8}$$

where I is a Łukasiewicz fuzzy *implicator*, which is defined by $min(1 - x + y, 1)$, and T is a Łukasiewicz fuzzy t-norm, which is defined by $max(x+y-1, 0)$. In [18], three classes of fuzzy-rough sets based on three different classes of implicators, namely S-, R-, and QL-implicators, and their properties have been investigated. Here, R_P is the fuzzy similarity relation considering the set of features in P, and $\mu_{R_P}(x,y)$ is the degree of similarity between objects x and y over all features in P. Also, $\mu_X(y)$ is the membership degree of y to X. One of the best fuzzy similarity relations as suggested in [6] is given by Eq. 9.

$$\mu_{R_a}(x,y) = max\left\{min\left\{\frac{(a(y) - (a(x) - \sigma_a))}{\sigma_a}, \frac{((a(x) + \sigma_a) - a(y))}{\sigma_a}\right\}, 0\right\} \tag{9}$$

where σ_a is variance of feature a. Definitions of fuzzy lower and upper approximations are the same as rough lower and upper approximations, except the fact that fuzzy approximations deal with fuzzy values, operators, and output; however, rough approximations deal with discrete and categorical values.

The positive region in the rough set theory is defined as a union of lower approximations. By referring to the extension principle [6], the membership of object x to a fuzzy positive region is given by Eq. 10.

$$\mu_{POS_P(Q)}(x) = \sup_{X \in \mathbb{U}/Q} \mu_{\underline{P}X}(x) \tag{10}$$

where supremum of lower approximations of all partitions resulting from U/Q construct positive region.

If the equivalence class that includes x does not belong to a positive region, clearly x will not be part of a positive region. Using the definition of positive region, the FRDD function is defined as:

$$\gamma'_P(Q) = \frac{|\mu_{POS_{P(Q)}}(x)|}{|\mathbb{U}|} = \frac{\sum_{x \in \mathbb{U}} \mu_{POS_{P(Q)}}(x)}{|\mathbb{U}|} \tag{11}$$

where notation $|.|$ is number of objects in a set; however, in numerator we are dealing with fuzzy values and cardinality can be calculated using summation. For denominator $|\mathbb{U}|$ is size of samples in dataset.

The Lower approximation Fuzzy-Rough Feature Selection (L-FRFS) as shown in Algorithm 1 is based on FRDD as shown in Eq. 11, and greedy forward search algorithm, which is capable of being applied to real-valued datasets. The L-FRFS algorithm finds a reduct set without finding all the subsets [6]. It begins with an empty set and each time selects the feature that causes the greatest increase in the FRDD. The algorithm stops when adding more features does not increase the FRDD. Since it employs a greedy algorithm, it does not guarantee that the minimal reduct set will be found. For this reason, a new feature selection merit presented in this section.

Algorithm 1. Lower approximation Fuzzy-Rough Feature Selection

C, the set of all conditional attributes
D, the set of decision attributes
$R \leftarrow \{\}; \gamma'_{best} = 0; \gamma'_{prev} = 0$
do
  $T \leftarrow R$
  $\gamma'_{prev} \leftarrow \gamma'_{best}$
  foreach $x \in (C - R)$
    if $\gamma'_{R\cup\{x\}}(D) > \gamma'_T(D)$
      $T \leftarrow R \cup \{x\}$
      $\gamma'_{best} \leftarrow \gamma'_T(D)$
  $R \leftarrow T$
until $\gamma'_{best} = \gamma'_{prev}$
return R

3 Proposed Method

On the one hand, FRDD is capable and effective in uncovering the dependency of a feature to another, and the feature selection method based on the merit has shown remarkable performance on resulting classification accuracies [6]. The L-FRFS algorithm evaluates every feature to find the one with the highest dependency, and continues the search by considering every features combination to asset the most dependent features subset to the outcome. However, tracking and finding highly dependent features to the class might end in selecting redundant features.

On the other hand, CFS merit, as shown in Eq. 1, has the potentiality of selecting less redundant features due to the structure of the denominator, in which the square root of the mean of the correlation of the features to each other has a positive impact on the number of redundant features being selected.

By considering capabilities of CFS merit, substituting the correlation with Fuzzy-Rough Dependency Degree (FRDD) that is fuzzy version of DD could take advantage of both criteria to construct a more powerful merit. In this section, the proposed approach is defined based on the two main concepts of feature selection: 1- Evaluation measure, and 2- Search method. The evaluation measure is the new hybrid merit and the search method is hill-climbing.

3.1 A New Hybrid Merit

Based on the concepts of the FRDD and CFS, we have developed a new hybrid merit by substituting the correlation in CFS with FRDD to benefit from both merits. Equation 12 shows the proposed merit.

$$\delta = \frac{\sum_{i=1}^{k} \gamma_i'(c)}{\sqrt{k \times \left(1 + \sum_{j=1}^{k-1} \gamma_j'(f)\right)}}, \tag{12}$$

where $\gamma_i'(c)$ is the FRDD of already selected feature i to the class c, and $\gamma_j'(f)$ is the FRDD of selected feature j to the new under consideration candidate feature f. The numerator is summation of the FRDD of already selected $k-1$ features as well as newly selected k's feature to the outcome, while the summation in denominator is aggregation of the FRDD of all features except currently under consideration one k's, to itself. It is worth to mention that k in denominator controls the number of selected features. We call the feature selection method based on our proposed merit, Delta Feature Selection (DFS). The numerator can vary from zero to one for each k (since $\gamma_i' \in [0,1]$), so we have interval of $[0,k]$ in the numerator. However, summation in the denominator varies from zero to $k-1$ for each k, and the whole portion changes in interval of $[\sqrt{k}, k]$ since k is always positive.

The search algorithm of our proposed, that is a greedy forward search method shown in Algorithm 2. The QuickReduct algorithm starts from an empty subset and each time selects one feature to be added to the subset, if the selected feature causes the highest increase in δ; therefore, it will be added to the subset, otherwise, the algorithm evaluates next feature. This process will be continued until no more feature can improve the δ.

To evaluate the applicability of the proposed merit to different types of datasets, a series of criteria have been considered as follows [19]:

Algorithm 2. Delta QuickReduct (DQR)

```
/* S_f: best subset of features
   δ'_curr: current DFS
   δ'_prev: previous DFS
   nF: number of features
   bF: best feature
*/
S_f = {};
δ_curr, δ_prev = 0;
do
{
  δ_prev = δ_curr;
  for i = 1 to i ≤ nF
  {
    if ((f_i ∉ S_f) AND (δ_{S_f ∪ {f_i}} > δ_prev))
    {
      δ_curr = δ_{S_f ∪ {f_i}};
      bF = f_i;
    }
  }
  S_f = S_f ∪ bF;
} while (δ_curr != δ_prev)
return S_f;
```

1. Correlated and redundant features
2. Non-linearity of data
3. Noisy input
4. Noisy target
5. Small ratio of samples/features
6. Complex datasets

Based on each criteria, thirteen datasets have been adopted from different papers as mentioned in [19] to examine the appropriateness of DFS. Datasts are shown in Table 1. The last column depicts corresponding criterion to the current dataset.

CorrAL dataset has six features, and features one to four are relevant and they generate the outcome by calculating $(f_1 \wedge f_2) \vee (f_3 \wedge f_4)$, feature five is irrelevant, and feature six has 75 % of correlation to the outcome. CorrAL-100 has 99 features that the first six are exactly the same as CorrAL, and the rest are irrelevant and randomly assigned. For both datasets, DFS was able to uncover all four relevant features and also the correlated one.

XOR-100 dataset is a non-linear dataset with two relevant features that compute the output by calculating $(f_1 \oplus f_2)$, and the other 97 features are irrelevant. Again, DFS was able to find two relevant features.

Led-25 dataset is composed of seven relevant features and 17 irrelevant ones. Each dataset, contains the amount of noise (i.e. replacing zero with one or vice

Table 1. Sample datasets to probe different capabilities of a feature selection method

Dataset	#Relevant	#Irrelevant	#Correlated	Criteria
CorrAL [20]	4	1	6	1
CorrAL-100 [21]	4	94	1	1
XOR-100 [21]	2	97	–	2
Led-25 [22] (2 %)	7	17	–	3
Led-25 [22] (6 %)	7	17	–	3
Led-25 [22] (10 %)	7	17	–	3
Led-25 [22] (15 %)	7	17	–	3
Led-25 [22] (20 %)	7	17	–	3
Monk3 [23]	3	3	–	4
SD1 [24]	FCR = 20	4000	–	5
SD2 [24]	FCR = 10, PCR = 30	4000	–	5
SD3 [24]	PCR = 60	4000	–	5
Madelon	5	480	15	6

versa) that is mentioned in parenthesis in front of dataset. Based on the resulting subsets containing two relevant features for all cases, of applying DFS it can be understood that DFS cannot perform well for datasets with noisy inputs.

Monk3 dataset has 5 % of misclassification values as a dataset with noisy target. The DFS has selected features one and five that are irrelevant and relevant, respectively. Therefore, DFS was not able to find all relevant features and also has been misled by noisy target.

SD1, SD2 and SD3 datasets each has three classes, and 75 samples, containing both relevant and irrelevant features. Relevant ones are generated based on a normal distribution, and irrelevant features have been generated based on two distributions namely, normal distribution with mean zero and variance one, and uniform distribution in interval of $[-1, 1]$, each 2000 features. All cancer types can be distinguished by using some genes (or features) called full class relevant (FCR). However, the other genes that are helpful in contrasting some portion of cancer types are called partial class relevant (PCR). Table 2 shows the optimal subset for each dataset, in which nine features out of 10 are redundant features.

Table 2. Optimal features and subsets of SD1, SD2, and SD3

Dataset	#Optimal features/subset	Optimal subsets
SD1 [24]	2	{1–10} {11–20}
SD2 [24]	4	{1–10} {11–20} {21–30} {31–40}
SD3 [24]	6	{1–10} {11–20} {21–30}
		{31–40} {41–50} {51–60}

The DFS has selected 2, 11, and 2 features for SD1, SD2, and SD3, respectively. For SD1, the DFS has selected one feature from the second optimal subset, and one feature from 4000 irrelevant features. For SD2, 11 features have been selected, in which, 10 of them are from the second optimal subset and one feature from 4000 irrelevant features. Finally, two features have been selected from SD3 that one of them is from the third optimal subset and the other one is from irrelevant features.

Madelon dataset has five relevant, 15 linearly correlated to relevant features, and 480 distractor features that are noisy, flipped and shifted [19]. The DFS was able to find five features, in which none of them were among relevant features.

Based on the resulting subsets, our proposed method is capable of dealing with datasets having characteristics mentioned in Table 3.

Table 3. DFS capabilities

Dataset	DFS capability
Correlated and redundant features	✓
Nonlinearity of data	✓
Noisy input	depends on data
Noisy target	depends on data
Small ratio of samples/features	✓
Complex datasets	×

For datasets with noisy input and target, the DFS was capable of finding a subset of relevant features; however, for complex datasets such as Madelon, finding relevant features is very challenging for DFS and many state-of-art feature selection methods [19].

3.2 Performance Measures

In order to evaluate the applicability and performance of FS methods, we define three *Performance* measures to underline classification accuracy and/or reducibility power. The *Reduction* ratio is the value of reduction of total number of features resulting from applying a feature selection method to a datasets, and it is shown in Eq. 13.

$$Reduction = \frac{all\_F - sel\_F}{all\_F}, \tag{13}$$

where all_F is the number of all features, and sel_F is the number of selected features using a feature selection algorithm.

The *Performance* measure is a metric to evaluate the effectiveness of a feature selection algorithm in selecting the smallest subset of features as well as improving the classification accuracy, and is shown by Eq. 14.

$$Performance = CA \times Reduction, \tag{14}$$

where CA is the classification accuracy.

Since the primary aim of FS is to select the smallest meaningful subset of features, we propose a revision of *Performance* measure that emphasizes on the *Reduction* capability of each method and it is presented by Eq. 15.

$$Performance' = CA \times e^{Reduction}, \tag{15}$$

In some cases, data owners prefer those FS methods that lead to higher accuracies; therefore, another revision of Eq. 14 with the aforementioned preference is depicted by Eq. 16.

$$Performance'' = e^{CA} \times Reduction. \tag{16}$$

4 Experimental Results

To validate the proposed method, we have conducted a number of experiments in two steps over 25 UCI [25] traditional as well as newly introduced datasets from three different size categories; Small (S), Medium (M) and Large (L) sizes, in which the number of selected features, *Reduction* ratio, classification accuracy and *Performance* measures are compared. The small size category contains datasets with model size, i.e. $|Features| \times |Samples|$, less than 5 000, the medium size category contains 5 000 to 50 000 cells, and each dataset in the large size category has more than 50 000 cells.

In our experiments the L-FRFS, CFS, and DFS use the same search method called greedy froward search algorithm, and the GBFS uses gradient decent search method.

Computational facilities are provided by ACENET, the regional high performance computing consortium for universities in Atlantic Canada. ACEnet is funded by the Canada Foundation for Innovation (CFI), the Atlantic Canada Opportunities Agency (ACOA), and the provinces of Newfoundland and Labrador, Nova Scotia, and New Brunswick.

4.1 Step One

In this step, we consider all the 25 datasets in our experiment. Table 4 shows the number of samples, features and the size category that each dataset belongs to, and it is sorted based on the model size.

Based on the number of selected features and Eq. 13, the *Reduction* ratio of each method has been calculated and illustrated in Table 5. The cells with zero indicate that the feature selection method could not remove any feature; therefore, all of the features remain untouched.

The bold, superscripted numbers specify the best method in improving the *Reduction* ratio. L-FRFS and GBFS reaches the highest reduction ratio for four datasets, CFS for five datasets, and DFS outperforms the others by gaining the highest *Reduction* values for sixteen datasets. Based on the categories and number of successes of each method, L-FRFS and GBFS result almost similar on the small size category with two and one out of 12 datasets, respectively. However,

Table 4. Datasets specifications

Dataset	Sample	Feature	Size
BLOGGER	100	5	S
Breast Tissue	122	9	S
Qualitative Bankr	250	6	S
Soybean	47	35	S
Glass	214	9	S
Wine	178	13	S
MONK1	124	6	S
MONK2	169	6	S
MONK3	122	6	S
Olitus	120	26	S
Heart	270	13	S
Cleveland	297	13	S
Pima Indian Diab	768	8	M
Breast Cancer	699	9	M
Thoracic Surgery [26]	470	17	M
Climate Model [27]	540	18	M
Ionosphere	351	33	M
Sonar	208	60	M
Wine Quality (Red) [28]	1599	11	M
LSVT Voice Rehab. [29]	126	310	M
Seismic Bumps [30]	2584	18	M
Arrhythmia	452	279	L
Molecular Biology	3190	60	L
COIL 2000 [31]	5822	85	L
Madelon	2000	500	L

DFS highly achieves the best results in both medium and large datasets, by having six out of nine best reduction ratios in medium size category compare to two out of nine for L-FRFS and GBFS methods, and one out of all for CFS. For large datasets, DFS gains 100 % domination. Table 6, shows the number of wins of each method in three categories.

Arithmetic mean has some disadvantages, such as high sensitivity to outliers and also inappropriateness in measuring central tendency of skewed distribution [32], we have conducted the Friedman test that is a non-parametric statistical analysis [33] on the results of Tables 8, 11, 14, and 17 to make the comparison fare enough.

The nine classifiers are PART, Jrip, Naïve Bayes, Bayes Net, J48, BFTree, FT, NBTree, and RBFNetwork that have been selected from different classifier

Table 5. *Reduction* ratio of L-FRFS, CFS, DFS & GBFS

Datasets	L-FRFS	CFS	GBFS	DFS	Size
BLOGGER	0.000	0.400	0.400	**0.600**$^+$	S
Breast Tissue	0.000	0.333	**0.444**$^+$	0.111	S
Qualitative Bankr.	0.500	0.333	0.167	**0.667**$^+$	S
Soybean	**0.943**$^+$	0.743	0.886	**0.943**$^+$	S
Glass	0.000	0.111	0.333	**0.556**$^+$	S
Wine	0.615	0.154	0.692	**0.846**$^+$	S
Monk1	0.500	**0.833**$^+$	0.333	0.667	S
Monk2	0.000	**0.833**$^+$	0.167	0.667	S
Monk3	0.500	**0.833**$^+$	0.333	0.667	S
Olitus	**0.808**$^+$	0.346	0.731	0.231	S
Heart	0.462	0.462	0.538	**0.846**$^+$	S
Cleveland	0.154	**0.923**$^+$	0.538	0.846	S
Pima Indian Diab.	0.000	**0.500**$^+$	0.250	**0.500**$^+$	M
Breast Cancer	0.222	0.000	**0.444**$^+$	**0.444**$^+$	M
Thoracic Surgery	0.176	0.706	0.588	**0.882**$^+$	M
ClimateModel	0.667	0.833	**0.944**$^+$	0.889	M
Ionosphere	0.788	0.576	0.818	**0.909**$^+$	M
Sonar	**0.917**$^+$	0.683	0.900	0.050	M
Wine Quality (Red)	0.000	0.636	0.636	**0.727**$^+$	M
LSVT Voice Rehab.	**0.984**$^+$	0.900	0.977	0.923	M
Seismic Bumps	0.278	0.667	0.778	**0.889**$^+$	M
Arrhythmia	0.975	0.910	0.907	**0.993**$^+$	L
Molecular Biology	0.000	0.617	0.000	**0.950**$^+$	L
COIL 2000	0.659	0.882	0.941	**0.965**$^+$	L
Madelon	0.986	0.982	**0.990**$^+$	**0.990**$^+$	L

Table 6. Number of wins in achieving the lowest *Reduction* ratio for L-FRFS, CFS, GBFS, and DFS in each category

Algorithm/Category	Small	Medium	Large	Overall
L-FRFS	2	2	0	4
CFS	4	1	0	5
GBFS	1	2	1	4
DFS	6	6	4	16

categories to evaluate the performance of each method by applying 10-fold cross validation (10CV). These classifiers have been implemented in Weka [34], and mean of resulting classification accuracies of all selected classifiers have been used through out the paper. By considering selected features for each dataset, the resulting average of classification accuracies have been shown in Table 7.

By referring to the results in Tables 5 and 7, and applying Eq. 14, the *Performance* measure of each method has been computed and shown in Table 8. The cells that contain zero are the ones with *Reduction* ratio equal to zero. Based on the results shown in Tables 5 and 8, DFS outperforms the other methods by having the best results for 10 datasets compared to that by GBFS with seven, CFS

Table 7. Mean of classification accuracies in % resulting from PART, Jrip, Naïve Bayes, Bayes Net, J48, BFTree, FT, NBTree, and RBFNetwork based on L-FRFS, CFS, GBFS, DFS performance comparing with unreduced datasets

Datasets	L-FRFS	CFS	GBFS	DFS	Unre.
BLOGGER	**74.22**$^+$	73.78	73.78	73.56	**74.22**$^+$
Breast Tissue	**66.46**$^+$	66.35	64.88	65.72	**66.46**$^+$
Qualitative Bankr.	98.44	98.04	98.31	98.40	**98.49**$^+$
Soybean	**100.00**$^+$	75.48	97.87	95.98	98.58
Glass	**67.29**$^+$	66.93	65.42	59.71	61.89
Wine	**95.63**$^+$	95.44	94.63	74.22	85.52
Monk1	**83.13**$^+$	74.07	81.94	73.53	78.32
Monk2	-	67.13	71.89	67.13	**76.62**$^+$
Monk3	98.15	76.23	**98.28**$^+$	75.62	97.92
Olitus	66.39	**75.65**$^+$	53.8	72.69	69.81
Heart	78.48	**81.48**$^+$	81.4	71.32	79.55
Cleveland	49.76	**54.88**$^+$	52.19	54.55	50.13
Pima Indian Diab.	75.00	**75.20**$^+$	**75.20**$^+$	**75.20**$^+$	75.00
Breast Cancer	**96.23**$^+$	96.18	96.23	95.31	96.18
Thoracic Surgery	83.03	84.54	83.95	**85.11**$^+$	83.10
Climate Model	93.25	90.74	91.38	91.36	**93.54**$^+$
Ionosphere	**91.39**$^+$	90.85	89.97	84.96	89.68
Sonar	69.82	**75.48**$^+$	74.89	74.36	67.47
Wine Quality (Red)	58.59	**59.22**$^+$	58.59	56.54	58.59
LSVT Voice Rehab.	**80.60**$^+$	79.37	75.84	72.57	74.69
Seismic Bumps	91.16	91.96	**92.59**$^+$	51.87	91.13
Arrhythmia	53.74	**70.48**$^+$	63.20	59.41	65.46
Molecular Biology	-	73.66	-	51.69	**94.58**$^+$
COIL 2000	92.79	93.65	93.97	**94.02**$^+$	90.61
Madelon	65.79	69.57	**71.27**$^+$	55.26	66.32

with six, and L-FRFS with only three cases. The best performance for small sized datasets has been achieved by DFS and CFS, for medium datasets by DFS and GBFS and for large datasets by DFS. Table 9 evaluates the results of Table 8, and Friedman statistic (distributed according to chi-square with 3 degrees of freedom) is 11.772, and p-value computed by Friedman Test is 0.008206. Based on the rankings, the DFS has gained the best ranking among others; however, its distinction has been examined by post-hoc experiment. The post-hoc procedure as depicted in Table 10 rejects those hypotheses with p-value ≤ 0.030983. So, as shown, DFS and GBFS perform nearly identical. Since performances of DFS and GBFS are not statistically significant, the one with the lowest reduction ratio is selected [35]. Here, based on Table 6, the DFS is ranked the best method among others.

4.2 Step Two

Since the CFS has chosen only one feature for MONK1, MONK2, MONK3 and Cleveland, and also GBFS has selected one out of 18 of Climate Model as the most important feature, further investigations is vital on these suspicious results. The Cleveland dataset has 75 features whereas 13 features out of 75 have been suggested to be used by the published experiments [36]; therefore, all of these 13 features are important from the clinical perspective. By referring to the result of CFS, feature "sex" has been selected as the only important feature due to its highest correlation with the outcome. Neither experts in medical science nor in computer science would arrive at the point that one feature (regardless of type of the feature) out of 13 is enough to predict the outcome. Although selecting "sex" results in the highest classification accuracy, the interpretability of selecting one feature is questionable. Therefore, although "sex" might be an important factor in predicting heart diseases, it is not the only one. For MONK1, MONK2, MONK3 and Climate Model datasets, the only characteristic of the selected feature is its high correlation with the outcome, and very low correlation with the other features.

By removing Cleveland, MONK1, MONK2, MONK3 and Climate Model from Table 8, we form Table 11 and Fig. 1 in which DFS gains the best performance. The GBFS works slightly better than the L-FRFS and CFS for medium datasets, but identical in small datasets. While DFS performance surpasses the GBFS, CFS, and L-FRFS for all three categories. The overall effectiveness and capability of DFS is supported by both Table 11, and the statistical analysis in Table 12. The Friedman statistic (distributed according to chi-square with 3 degrees of freedom) is 9.345, and the p-value computed by Friedman Test is 0.025039. The Li's procedure rejects those hypotheses with p-value ≤ 0.01266, and the results are shown in Table 13. The *Performance* measures resulting form Eqs. 15 and 16 are shown in Tables 14 and 17 and also in Figs. 2 and 3, respectively. The Friedman test results are shown in Tables 15 and 18. For *Performance'*, those hypotheses with p-value ≤ 0.00257 are rejected based on Li's procedure, and the results are depicted in Table 16. For *Performance''* as Table 19

Table 8. *Performance* measure resulting from Classification Accuracy × *Reduction*

Datasets	L-FRFS	CFS	GBFS	DFS
BLOGGER	0.000	0.295	0.295	**0.441**$^+$
Breast Tissue	0.000	0.221	**0.288**$^+$	0.073
Qualitative Bankr.	0.492	0.327	0.164	**0.656**$^+$
Soybean	**0.943**$^+$	0.561	0.867	0.905
Glass	0.000	0.074	0.218	**0.332**$^+$
Wine	0.588	0.147	**0.655**$^+$	0.628
Monk1	0.416	**0.617**$^+$	0.273	0.490
Monk2	0.000	**0.559**$^+$	0.120	0.448
Monk3	0.491	**0.635**$^+$	0.328	0.504
Olitus	**0.536**$^+$	0.262	0.393	0.168
Heart	0.362	0.376	0.438	**0.603**$^+$
Cleveland	0.077	**0.507**$^+$	0.281	0.462
Pima Indian Diab.	0.000	**0.376**$^+$	0.188	**0.376**$^+$
Breast Cancer	0.214	0.000	**0.428**$^+$	0.424
Thoracic Surgery	0.147	0.597	0.494	**0.751**$^+$
ClimateModel	0.622	0.756	**0.863**$^+$	0.812
Ionosphere	0.720	0.523	0.736	**0.772**$^+$
Sonar	0.640	0.516	**0.674**$^+$	0.037
Wine Quality (Red)	0.000	0.377	0.373	**0.411**$^+$
LSVT Voice Rehab.	**0.793**$^+$	0.714	0.741	0.670
Seismic Bumps	0.253	0.613	**0.720**$^+$	0.461
Arrhythmia	0.524	**0.642**$^+$	0.573	0.590
Molecular Biology	0.000	0.454	0.000	**0.491**$^+$
COIL 2000	0.611	0.826	0.884	**0.907**$^+$
Madelon	0.649	0.683	**0.706**$^+$	0.547

Table 9. Average rankings of the algorithms based on the *Performance* measure over all datasets (Friedman)

Algorithm	Ranking
L-FRFS	3.220
CFS	2.440
GBFS	2.320
DFS	**2.020**$^+$

Table 10. Post Hoc comparison over the results of Friedman procedure of *Performance* measure

i	Algorithm	$z = (R_0 - R_i)/SE$	p	Li
3	L-FRFS	3.286335	0.001015	0.030983
2	CFS	1.150217	0.250054	0.030983
1	GBFS	0.821584	0.411314	0.05

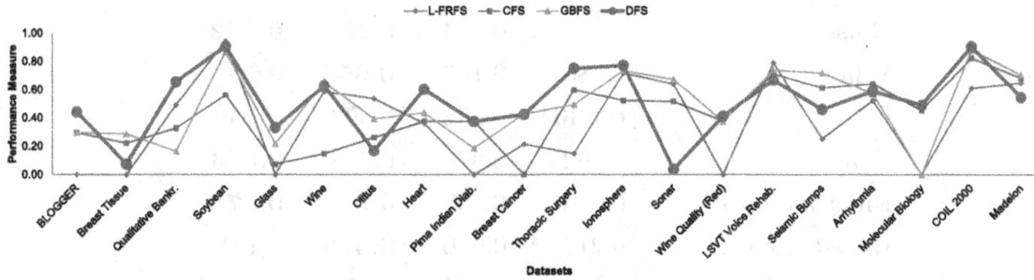

Fig. 1. *Performance* measure (Classification Accuracy × *Reduction*)

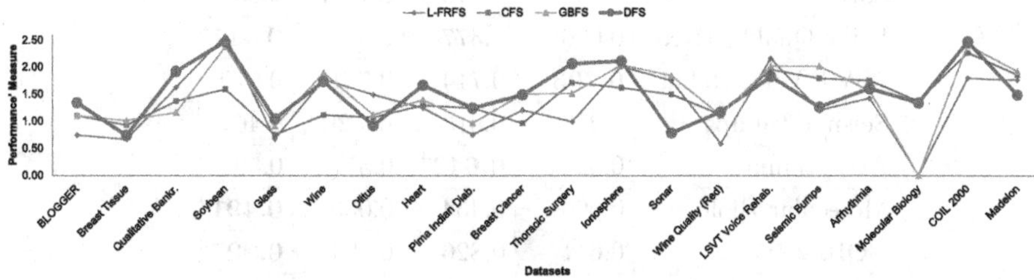

Fig. 2. *Performance'* measure (Classification Accuracy $\times e^{Reduction}$)

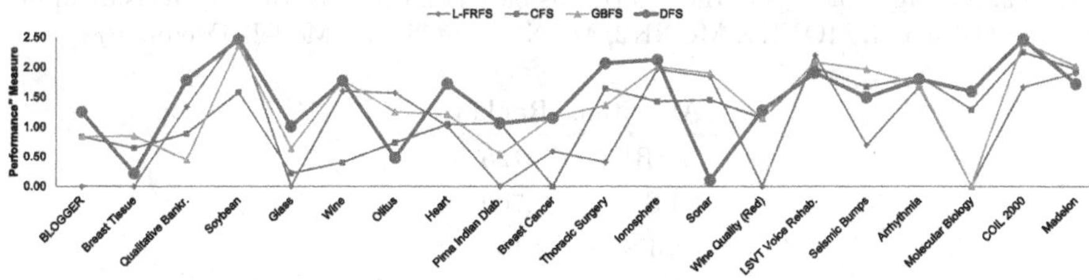

Fig. 3. *Performance''* measure ($e^{ClassificationAccuracy} \times$ *Reduction*)

shows, those hypotheses with p-value ≤ 0.01266 are rejected based on Li's procedure. Figures 1, 2 and 3 depict *Performance*, *Performance'* and *Performance''* measures values for each dataset, respectively.

Table 11. *Performance* measure resulting from classification accuracy × *reduction* after removing Cleveland, MONK1, MONK2, MONK3 & Climate Model

Datasets	L-FRFS	CFS	GBFS	DFS
BLOGGER	0.000	0.295	0.295	**0.441**$^+$
Breast Tissue	0.000	0.221	**0.288**$^+$	0.073
Qualitative Bankr.	0.492	0.327	0.164	**0.656**$^+$
Soybean	**0.943**$^+$	0.561	0.867	0.905
Glass	0.000	0.074	0.218	**0.332**$^+$
Wine	0.588	0.147	**0.655**$^+$	0.628
Olitus	**0.536**$^+$	0.262	0.393	0.168
Heart	0.362	0.376	0.438	**0.603**$^+$
Pima Indian Diab.	0.000	**0.376**$^+$	0.188	**0.376**$^+$
Breast Cancer	0.214	0.000	**0.428**$^+$	0.424
Thoracic Surgery	0.147	0.597	0.494	**0.751**$^+$
Ionosphere	0.720	0.523	0.736	**0.772**$^+$
Sonar	0.640	0.516	**0.674**$^+$	0.037
Wine Quality (Red)	0.000	0.377	0.373	**0.411**$^+$
LSVT Voice Rehab.	**0.793**$^+$	0.714	0.741	0.670
Seismic Bumps	0.253	0.613	**0.720**$^+$	0.461
Arrhythmia	0.524	**0.642**$^+$	0.573	0.590
Molecular Biology	0.000	0.454	0.000	**0.491**$^+$
COIL 2000	0.611	0.826	0.884	**0.907**$^+$
Madelon	0.649	0.683	**0.706**$^+$	0.547

Table 12. Average rankings of the algorithms based on the *Performance* measure after removing Cleveland, MONK1, MONK2, MONK3 & Climate Model (Friedman)

Algorithm	Ranking
L-FRFS	3.125
CFS	2.700
GBFS	2.150
DFS	**2.025**$^+$

Table 13. Post Hoc comparison over the results of Friedman procedure of *Performance* measure after removing Cleveland, MONK1, MONK2, MONK3 & Climate Model

i	Algorithm	$z = (R_0 - R_i)/SE$	p	Li
3	L-FRFS	2.694439	0.007051	0.01266
2	CFS	1.653406	0.098248	0.01266
1	GBFS	0.306186	0.759463	0.05

Table 14. *Performance'* measure resulting from Classification Accuracy $\times e^{Reduction}$

Datasets	L-FRFS	CFS	GBFS	DFS
BLOGGER	0.742	1.101	1.101	**1.340**$^+$
Breast Tissue	0.665	0.926	**1.012**$^+$	0.734
Qualitative Bankr.	1.623	1.368	1.161	**1.917**$^+$
Soybean	**2.567**$^+$	1.587	2.373	2.464
Glass	0.673	0.748	0.913	**1.041**$^+$
Wine	1.770	1.113	**1.891**$^+$	1.730
Olitus	**1.489**$^+$	1.069	1.117	0.916
Heart	1.245	1.293	1.395	**1.662**$^+$
Pima Indian Diab.	0.750	**1.240**$^+$	0.966	**1.240**$^+$
Breast Cancer	1.202	0.962	**1.501**$^+$	1.487
Thoracic Surgery	0.990	1.712	1.512	**2.057**$^+$
Ionosphere	2.009	1.616	2.039	**2.109**$^+$
Sonar	1.746	1.495	**1.842**$^+$	0.782
Wine Quality (Red)	0.586	1.119	1.107	**1.170**$^+$
LSVT Voice Rehab.	**2.156**$^+$	1.952	2.015	1.826
Seismic Bumps	1.204	1.791	**2.015**$^+$	1.262
Arrhythmia	1.425	**1.752**$^+$	1.565	1.604
Molecular Biology	0.000	**1.365**$^+$	0.000	1.337
COIL 2000	1.793	2.263	2.408	**2.467**$^+$
Madelon	1.763	1.857	**1.918**$^+$	1.487

Table 15. Average rankings of the algorithms based on the *Performance'* measure after removing Cleveland, MONK1, MONK2, MONK3 & Climate Model (Friedman)

Algorithm	Ranking
L-FRFS	3.075
CFS	2.650
DFS	2.150
GBFS	**2.125**$^+$

Table 16. Post Hoc comparison over the results of Friedman procedure of *Performance'* measure after removing Cleveland, MONK1, MONK2, MONK3 & Climate Model

i	Algorithm	$z = (R_0 - R_i)/SE$	p	Li
3	L-FRFS	2.327015	0.019964	0.00257
2	CFS	1.285982	0.198449	0.00257
1	GBFS	0.061237	0.95117	0.05

Table 17. *Performance''* measure resulting from $e^{ClassificationAccuracy} \times$ *Reduction*

Datasets	L-FRFS	CFS	GBFS	DFS
BLOGGER	0.000	0.837	0.837	**1.252**$^{+}$
Breast Tissue	0.000	0.647	**0.850**$^{+}$	0.214
Qualitative Bankr.	1.338	0.889	0.445	**1.783**$^{+}$
Soybean	**2.563**$^{+}$	1.580	2.357	2.462
Glass	0.000	0.217	0.641	**1.009**$^{+}$
Wine	1.601	0.400	**1.784**$^{+}$	1.777
Olitus	**1.569**$^{+}$	0.738	1.251	0.477
Heart	1.012	1.043	1.215	**1.727**$^{+}$
Pima Indian Diab.	0.000	**1.061**$^{+}$	0.530	**1.061**$^{+}$
Breast Cancer	0.582	0.000	**1.163**$^{+}$	1.153
Thoracic Surgery	0.405	1.644	1.362	**2.067**$^{+}$
Ionosphere	1.965	1.428	2.012	**2.126**$^{+}$
Sonar	1.843	1.454	**1.903**$^{+}$	0.105
Wine Quality (Red)	0.000	1.151	1.143	**1.280**$^{+}$
LSVT Voice Rehab.	**2.203**$^{+}$	1.990	2.087	1.906
Seismic Bumps	0.691	1.672	**1.963**$^{+}$	1.493
Arrhythmia	1.669	**1.842**$^{+}$	1.706	1.799
Molecular Biology	0.000	1.288	0.000	**1.593**$^{+}$
COIL 2000	1.666	2.251	2.409	**2.470**$^{+}$
Madelon	1.904	1.969	**2.019**$^{+}$	1.720

Table 18. Average rankings of the algorithms based on the *Performance''* measure after removing Cleveland, MONK1, MONK2, MONK3 & Climate Model (Friedman)

Algorithm	Ranking
L-FRFS	3.125
CFS	2.700
GBFS	2.150
DFS	**2.025**$^{+}$

Table 19. Post Hoc comparison over the results of Friedman procedure of *Performance''* measure after removing Cleveland, MONK1, MONK2, MONK3 & Climate Model

i	Algorithm	$z = (R_0 - R_i)/SE$	p	Li
3	L-FRFS	2.694439	0.007051	0.01266
2	CFS	1.653406	0.098248	0.01266
1	GBFS	0.306186	0.759463	0.05

5 Conclusions and Future Work

This paper introduces a new hybrid merit based on conjunction of correlation feature selection and fuzzy-rough feature selection. It takes advantages of both methods by integrating them into a new hybrid merit to improve the quality of the selected subsets as well as resulting reasonable classification accuracies. The new merit selects less number of redundant features, and finds the most relevant ones to the outcome.

The performance of the proposed merit is examined with a variety of different datasets with diverse number of features and samples, that have been chosen because of their predominance as well as recently introduced in the literature. The two-step experimental results show the effectiveness of our new hybrid merit over divergent UCI datasets, especially on medium and large ones. We have also proposed three measures to thoroughly figure out and compare the performance of feature selection methods.

Based on the results, we conclude that proposing a universal feature selection method might not be suitable due to the high variety of datasets and applications. Therefore, each and every newly proposed method can be "localized" to a subject and type of the data as well as the purpose of the data. In such a way, data owners can save huge amounts of processing expenses based on a set of categorized methods. As future work, we are excited to perform such categorization for the existing merits on feature selection methods. Also, we are conducting some experiments on Big Data in order to evaluate the performance of the proposed hybrid merit.

Our ongoing task is to prepare an online, web-based application for the new hybrid merit that will be available to the researchers working on datasets in various field of studies.

Acknowledgments. This work has been partially supported by the Natural Sciences and Engineering Research Council of Canada (NSERC) and the Research & Development Corporation of Newfoundland and Labrador (RDC).

References

1. Hall, M.A., Smith, L.A.: Feature subset selection: a correlation based filter approach. In: Proceedings of the 1997 International Conference on Neural Information Processing and Intelligent Information Systems, New Zealand, pp. 855–858 (1997)
2. Javed, K., Babri, H.A., Saeed, M.: Feature selection based on class-dependent densities for high-dimensional binary data. IEEE Trans. Knowl. Data Eng. **24**, 465–477 (2012)
3. Kohavi, R., John, G.H.: Wrappers for feature subset selection. Artif. Intell. **97**, 273–324 (1997)
4. Das, S.: Filters, wrappers and a boosting-based hybrid for feature selection. In: ICML, vol. 1, pp. 74–81. Citeseer (2001)
5. Kira, K., Rendell, L.A.: The feature selection problem: traditional methods and a new algorithm. In: AAAI, pp. 129–134 (1992)

6. Jensen, R., Shen, Q.: New approaches to fuzzy-rough feature selection. IEEE Trans. Fuzzy Syst. **17**, 824–838 (2009)
7. Anaraki, J.R., Eftekhari, M., Ahn, C.W.: Novel improvements on the fuzzy-rough quickreduct algorithm. IEICE Trans. Inf. Syst. **E98.D**(2), 453–456 (2015)
8. Anaraki, J.R., Eftekhari, M.: Improving fuzzy-rough quick reduct for feature selection. In: 2011 19th Iranian Conference on Electrical Engineering (ICEE), pp. 1502–1506 (2011)
9. Qian, Y., Wang, Q., Cheng, H., Liang, J., Dang, C.: Fuzzy-rough feature selection accelerator. Fuzzy Sets Syst. **258**, 61–78 (2015). Special issue: Uncertainty in Learning from Big Data
10. Jensen, R., Vluymans, S., Parthaláin, N.M., Cornelis, C., Saeys, Y.: Semi-supervised fuzzy-rough feature selection. In: Yao, Y., Hu, Q., Yu, H., Grzymala-Busse, J.W. (eds.) RSFDGrC 2015. LNCS (LNAI), vol. 9437, pp. 185–195. Springer, Heidelberg (2015). doi:10.1007/978-3-319-25783-9_17
11. Shang, C., Barnes, D.: Fuzzy-rough feature selection aided support vector machines for mars image classification. Comput. Vis. Image Underst. **117**, 202–213 (2013)
12. Derrac, J., Verbiest, N., García, S., Cornelis, C., Herrera, F.: On the use of evolutionary feature selection for improving fuzzy rough set based prototype selection. Soft Comput. **17**, 223–238 (2012)
13. Dai, J., Xu, Q.: Attribute selection based on information gain ratio in fuzzy rough set theory with application to tumor classification. Appl. Soft Comput. **13**, 211–221 (2013)
14. Xu, Z., Huang, G., Weinberger, K.Q., Zheng, A.X.: Gradient boosted feature selection. In: Proceedings of the 20th ACM SIGKDD international conference on Knowledge discovery and data mining, pp. 522–531. ACM (2014)
15. Yu, L., Liu, H.: Efficient feature selection via analysis of relevance and redundancy. J. Mach. Learn. Res. **5**, 1205–1224 (2004)
16. Pawlak, Z.: Rough sets. Int. J. Comput. Inf. Sci. **11**, 341–356 (1982)
17. Komorowski, J., Pawlak, Z., Polkowski, L., Skowron, A.: Rough sets: a tutorial. In: Pal, S.K., Skowron, A. (eds.) Rough-Fuzzy Hybridization: A New Trend in Decision Making, pp. 3–98. Springer-Verlag New York, Inc., Secaucus (1998)
18. Radzikowska, A.M., Kerre, E.E.: A comparative study of fuzzy rough sets. Fuzzy Sets Syst. **126**, 137–155 (2002)
19. Boln-Canedo, V., Snchez-Maroo, N., Alonso-Betanzos, A.: Feature Selection for High-Dimensional Data. Springer, Switzerland (2016)
20. John, G.H., Kohavi, R., Pfleger, K., et al.: Irrelevant features and the subset selection problem. In: Machine Learning: Proceedings of the Eleventh International Conference, pp. 121–129 (1994)
21. Kim, G., Kim, Y., Lim, H., Kim, H.: An mlp-based feature subset selection for HIV-1 protease cleavage site analysis. Artif. Intell. Med. **48**, 83–89 (2010). Artificial Intelligence in Biomedical Engineering and Informatics
22. Breiman, L., Friedman, J., Stone, C.J., Olshen, R.A.: Classification and regression trees. CRC Press, New York (1984)
23. Wnek, J., Michalski, R.S.: Comparing symbolic and subsymbolic learning: three studies. Mach. Learn. A Multistrategy Approach **4**, 318–362 (1994)
24. Zhu, Z., Ong, Y.S., Zurada, J.M.: Identification of full and partial class relevant genes. IEEE/ACM Trans. Comput. Biol. Bioinform. **7**, 263–277 (2010)
25. Bache, K., Lichman, M.: UCI machine learning repository (2013)
26. Zieba, M., Tomczak, J.M., Lubicz, M., Swiatek, J.: Boosted svm for extracting rules from imbalanced data in application to prediction of the post-operative life expectancy in the lung cancer patients. Appl. Soft Comput. **14**, 99–108 (2014)

27. Lucas, D.D., Klein, R., Tannahill, J., Ivanova, D., Brandon, S., Domyancic, D., Zhang, Y.: Failure analysis of parameter-induced simulation crashes in climate models. Geoscientific Model Devel. **6**, 1157–1171 (2013)
28. Cortez, P., Cerdeira, A., Almeida, F., Matos, T., Reis, J.: Modeling wine preferences by data mining from physicochemical properties. Decis. Support Syst. **47**, 547–553 (2009)
29. Tsanas, A., Little, M., Fox, C., Ramig, L.: Objective automatic assessment of rehabilitative speech treatment in parkinson's disease. IEEE Trans. Neural Syst. Rehabil. Eng. **22**, 181–190 (2014)
30. Sikora, M., Wróbel, Ł.: Application of rule induction algorithms for analysis of data collected by seismic hazard monitoring systems in coal mines. Arch. Min. Sci. **55**, 91–114 (2010)
31. Putten, P.V.D., Someren, M.V.: Coil challenge 2000: the insurance company case. Technical report 2000–2009. Leiden Institute of Advanced Computer Science, Universiteit van Leiden (2000)
32. Manikandan, S.: Measures of central tendency: the mean. J. Pharmacol. Pharmacotherapeutics **2**, 140 (2011)
33. Alcala-Fdez, J., Fernandez, A., Luengo, J., Derrac, J., Garcia, S.: Keel data-mining software tool: Data set repository, integration of algorithms and experimental analysis framework. Multiple-Valued Logic Soft Comput. **17**, 255–287 (2011)
34. Hall, M., Frank, E., Holmes, G., Pfahringer, B., Reutemann, P., Witten, I.H.: The weka data mining software: an update. SIGKDD Explor. Newsl. **11**, 10–18 (2009)
35. Guyon, I., Gunn, S., Ben-Hur, A., Dror, G.: Result analysis of the nips 2003 feature selection challenge. In: Advances in Neural Information Processing Systems, pp. 545–552 (2004)
36. Detrano, R., Janosi, A., Steinbrunn, W., Pfisterer, M., Schmid, J.J., Sandhu, S., Guppy, K.H., Lee, S., Froelicher, V.: International application of a new probability algorithm for the diagnosis of coronary artery disease. Am. J. Cardiol. **64**, 304–310 (1989)

Greedy Algorithm for the Construction of Approximate Decision Rules for Decision Tables with Many-Valued Decisions

Mohammad Azad[1(✉)], Mikhail Moshkov[1], and Beata Zielosko[2]

[1] Computer, Electrical and Mathematical Sciences and Engineering Division, King Abdullah University of Science and Technology, Thuwal 23955-6900, Saudi Arabia
{mohammad.azad,mikhail.moshkov}@kaust.edu.sa
[2] Institute of Computer Science, University of Silesia, 39, Będzińska St., 41-200 Sosnowiec, Poland
beata.zielosko@us.edu.pl

Abstract. The paper is devoted to the study of a greedy algorithm for construction of approximate decision rules. This algorithm is applicable to decision tables with many-valued decisions where each row is labeled with a set of decisions. For a given row, we should find a decision from the set attached to this row. We consider bounds on the precision of this algorithm relative to the length of rules. To illustrate proposed approach we study a problem of recognition of labels of points in the plain. This paper contains also results of experiments with modified decision tables from UCI Machine Learning Repository.

1 Introduction

In this paper, we consider one more extension of the notion of decision table – decision table with many-valued decisions. In a table with many-valued decisions, each row is labeled with a nonempty finite set of decisions, and for a given row, we should find a decision from the set of decisions attached to this row.

Such tables arise in problems of discrete optimization, pattern recognition, computational geometry, decision making etc. [10,17]. However, the main sources of decision tables with many-valued decisions are datasets filled by statistical or experimental data. In such datasets, we often have groups of objects with equal values of conditional attributes but, probably, different values of the decision attribute. Instead of a group of objects, we can consider one object given by values of conditional attributes. We attach to this object a set of decisions: either all decisions for objects from the group, or k the most frequent decisions for objects from the group, etc. As a result we obtain a decision table with many-valued decisions. In real life applications we can meet multi-label data when we study, e.g., problem of semantic annotation of images [4], music categorization into emotions [35], functional genomics [3], and text categorization [36].

In the rough set theory [22,30,31], decision tables are considered often that have equal rows labeled with different decisions. The set of decisions attached

J.F. Peters and A. Skowron (Eds.): TRS XX, LNCS 10020, pp. 24–50, 2016.
DOI: 10.1007/978-3-662-53611-7_2

to equal rows is called the *generalized decision* for that rows [23–25]. Here our aim is to find the generalized decision for a given row. In the paper, we will call this approach the generalized decision approach. However, the problem of finding an arbitrary decision or one of the most frequent decisions from the generalized decision is interesting also. Such study of decision tables with many-valued decisions can give a new tool for the rough set theory. In [2] and [18] we considered problem of construction of tests (super-reducts) and decision trees for decision tables with many-valued decisions. To choose one of the attributes we used uncertainty measure that is the number of boundary subtables.

Decision table with many-valued decisions can be considered as a decision table with an incomplete information because we don't know which decision should be chosen from the set of decisions. Incomplete information exists also in decision tables where instead of a single value of conditional attribute we have a subset of values of the attribute domain. In [13, 14] approaches to interpreting queries in a database with such incomplete information were discussed. Z. Pawlak [22] and E. Orłowska [21] proposed Non-deterministic Information System for dealing with an incomplete information. Information incompleteness is connected also with missing values of attributes or intervals on values of attributes. M. Kryszkiewicz in [11] proposed method for computing all optimal generalized rules from decision table with missing values. In [27–29] authors proposed rule generation system, based on Apriori algorithm, where incomplete information was considered as nondeterministic information.

In literature, often, problems connected with multi-label data are considered from the point of view of classification (multi-label classifications problems) [7,8,15,19,33,34,37]. Here our aim is not to deal with classification but to show that proposed approach for construction of decision rules for decision tables with many-valued decisions can be useful when we deal with knowledge representation. In various applications, we often deal with decision tables which contain noisy data. In this case, exact rules can be "over-fitted", i.e., depend essentially on the noise. So, instead of exact rules with many attributes, it is more appropriate to work with approximate rules with smaller number of attributes. Besides, classifiers based on approximate decision rules have often better accuracy than classifiers based on exact decision rules.

In the proposed approach a greedy algorithm constructs α-decision rules (α is a degree of rule uncertainty), and the number of rules for a given row is equal to the cardinality of set of decisions attached to this row. Then we choose for each row in a decision table a rule with the minimum length. The choice of shorter rules is connected with the Minimum Description Length principle [26].

The problem of construction of rules with minimum length is NP-hard. Therefore, we consider approximate polynomial algorithm for rule optimization. Based on results of U. Feige [9] it was proved in [16], for decision tables with one-valued decision, that greedy algorithm under some natural assumptions on the class NP, is close to the best polynomial approximate algorithms for partial decision rule minimization. It is natural to use these results in our approach. Note that each decision table with one-valued decision can be interpreted also as a decision table where each row is labeled with a set of decisions which has one element.

The paper, extending a conference publication [6] and some results presented in [17], is devoted to the study of a greedy algorithm for construction of approximate decision rules for decision tables with many-valued decisions. The greedy algorithm for rule construction has polynomial time complexity for the whole set of decision tables with many-valued decisions.

We discuss also a problem of recognition of labels of points in the plain which illustrates the considered approach and the obtained bounds on precision of this algorithm relative to the length of rules.

In this paper, we study only binary decision tables with many-valued decisions. However, the obtained results can be extended to the decision tables filled by numbers from the set $\{0, \ldots, k-1\}$, where $k \geq 3$. We present experimental results based on modified data sets from UCI Machine Learning Repository [12] (by removal of some conditional attributes) into the form of decision tables with many-valued decisions. Experiments are connected with length of α-decision rules, number of different rules, lower and upper bounds on the minimum length of α-decision rules and 0.5-hypothesis for α-decision rules. We also present experimental results for the generalized decision approach. It allows us to make some comparative study of length and number of different rules, based on the proposed approach and the generalized decision approach.

The paper consists of eight sections. In Sect. 2, main notions are discussed. In Sect. 3, a parameter $M(T)$ and auxiliary statement are presented. This parameter is used for analysis of a greedy algorithm. Section 4 is devoted to the consideration of a set cover problem. In Sect. 5, the greedy algorithm for construction of approximate decision rules is studied. In this section we also present a lower and upper bounds on the minimum rule length based on the information about greedy algorithm work, and 0.5-hypothesis for tables with many-valued decisions. In Sect. 6, we discuss the problem of recognition of labels of points in the plain. In Sect. 7, experimental results are presented. Section 8 contains conclusions.

2 Main Definitions

In this section, we consider definitions corresponding to decision tables with many-valued decisions.

A *(binary) decision table with many-valued* decisions is a rectangular table T filled by numbers from the set $\{0, 1\}$. Columns of this table are labeled with attributes $f_1, \ldots, f_n$. Rows of the table are pairwise different, and each row is labeled with a nonempty finite set of natural numbers (set of decisions). Note that each decision table with one-valued decisions can be interpreted also as a decision table with many-valued decisions. In such table, each row is labeled with a set of decisions which has one element. An example of decision table with many-valued decisions T_0 is presented in Table 1.

We will say that T is a *degenerate* table if either T is empty (has no rows), or the intersection of sets of decisions attached to rows of T is nonempty.

Table 1. Decision table T_0 with many-valued decisions

$T_0 =$

	f_1	f_2	f_3	
r_1	1	1	1	$\{1\}$
r_2	0	1	0	$\{1,3\}$
r_3	1	1	0	$\{2\}$
r_4	0	0	1	$\{2,3\}$
r_5	1	0	0	$\{1,2\}$

A decision which belongs to the maximum number of sets of decisions attached to rows in T is called the *most common decision for* T. If we have more than one such decision, we choose the minimum one. If T is empty then 1 is the most common decision for T.

Let $r = (b_1, \ldots, b_n)$ be a row of T labeled with a set of decisions $D(r)$ and $d \in D(r)$. By $U(T, r, d)$ we denote the set of rows r' from T for which $d \notin D(r')$. We will say that an attribute f_i separates a row $r' \in U(T, r, d)$ from the row r if the rows r and r' have different values at the intersection with the column f_i. The pair (T, r) will be called a *decision rule problem.*

Let α be a real number such that $0 \leq \alpha < 1$. A decision rule

$$f_{i_1} = b_1 \wedge \ldots \wedge f_{i_m} = b_m \rightarrow d \tag{1}$$

is called an *α-decision rule for the pair (T, r) and decision $d \in D(r)$* if attributes $f_{i_1}, \ldots, f_{i_m}$ separate from r at least $(1 - \alpha)|U(T, r, d)|$ rows from $U(T, r, d)$. The number m is called the *length* of the rule (1). For example, 0.01-decision rule means that attributes contained in the rule should separate from the row r at least 99 % of rows from $U(T, r, d)$. If α is equal to 0 we have an exact decision rule (0-decision rule) for (T, r) and d. If $U(T, r, d) = \emptyset$ then for any $f_{i_1}, \ldots, f_{i_m} \in \{f_1, \ldots, f_n\}$ the rule (1) is an α-decision rule for (T, r) and d. The rule (1) with empty left-hand side (when $m = 0$) is an α-decision rule for (T, r) and d if $U(T, r, d) = \emptyset$.

We will say that a decision rule is an *α-decision rule for the pair (T, r)* if this rule is an α-decision rule for the pair (T, r) and a decision $d \in D(r)$. We denote by $L_{\min}(\alpha, T, r, d)$ the minimum length of an α-decision rule for the pair (T, r) and decision $d \in D(r)$. We denote by $L_{\min}(\alpha, T, r)$ the minimum length of an α-decision rule for the pair (T, r). It is clear that

$$L_{\min}(\alpha, T, r) = \min\{L_{\min}(\alpha, T, r, d) : d \in D(r)\}.$$

Let α, β be real numbers such that $0 \leq \alpha \leq \beta < 1$. One can show that $L_{\min}(\alpha, T, r, d) \geq L_{\min}(\beta, T, r, d)$ and $L_{\min}(\alpha, T, r) \geq L_{\min}(\beta, T, r)$.

3 Parameter $M(T)$

In this section, we consider definition of a parameter $M(T)$ and auxiliary statement from [17]. For the completeness, we will give this statement with proof.

We will use the parameter $M(T)$ to evaluate precision of a greedy algorithm relative to the length of rules.

Let T be a decision table with many-valued decisions, which has n columns labeled with attributes $\{f_1, \ldots, f_n\}$.

Now, we define the parameter $M(T)$ of the table T. If T is a degenerate table then $M(T) = 0$. Let now T be a nondegenerate table. Let

$$\bar{\delta} = (\delta_1, \ldots, \delta_n) \in \{0, 1\}^n.$$

Then $M(T, \bar{\delta})$ is the minimum natural m such that there exist attributes $f_{i_1}, \ldots, f_{i_m} \in \{f_1, \ldots, f_n\}$ for which $T(f_{i_1}, \delta_{i_1}) \ldots (f_{i_m}, \delta_{i_m})$ is a degenerate table. Here $T(f_{i_1}, \delta_{i_1}) \ldots (f_{i_m}, \delta_{i_m})$ is a subtable of the table T consisting only rows that have numbers $\delta_{i_1}, \ldots, \delta_{i_m}$ at the intersection with the columns $f_{i_1}, \ldots, f_{i_m}$. We denote

$$M(T) = \max\{M(T, \bar{\delta}) : \bar{\delta} \in \{0, 1\}^n\}.$$

Lemma 1. *Let T be a nondegenerate decision table with many-valued decisions which have n columns labeled with attributes $f_1, \ldots, f_n$, $\bar{\delta} = (\delta_1, \ldots, \delta_n) \in \{0, 1\}^n$, and $\bar{\delta}$ be a row of T. Then*

$$L_{\min}(0, T, \bar{\delta}) \leq M(T, \bar{\delta}) \leq M(T).$$

Proof. By definition, $M(T, \bar{\delta})$ is the minimum natural m such that there exist attributes $f_{i_1}, \ldots, f_{i_m} \in \{f_1, \ldots, f_n\}$ for which subtable

$$T' = T(f_{i_1}, \delta_{i_1}) \ldots (f_{i_m}, \delta_{i_m})$$

is a degenerate table. The subtable T' is nonempty since $\bar{\delta}$ is a row of this subtable. Therefore there is a decision d which, for each row of T', belongs to the set of decisions attached to this row.

One can show that a decision rule

$$f_{i_1} = \delta_{i_1} \wedge \ldots \wedge f_{i_m} = \delta_{i_m} \rightarrow d$$

is a 0-decision rule for the pair $(T, \bar{\delta})$ and decision d. Therefore $L_{\min}(0, T, \bar{\delta}) \leq m = M(T, \bar{\delta})$. By definition, $M(T, \bar{\delta}) \leq M(T)$. □

4 Set Cover Problem

In this section, we consider a greedy algorithm for construction of an approximate cover (an α-cover).

Let α be a real number such that $0 \leq \alpha < 1$. Let A be a set containing $N > 0$ elements, and $F = \{S_1, \ldots, S_p\}$ be a family of subsets of the set A such that $A = \bigcup_{i=1}^{p} S_i$. We will say about the pair (A, F) as about a *set cover problem.* A subfamily $\{S_{i_1}, \ldots, S_{i_t}\}$ of the family F will be called an *α-cover for* (A, F) if $|\bigcup_{j=1}^{t} S_{i_j}| \geq (1 - \alpha)|A|$. The problem of searching for an α-cover with minimum cardinality for a given set cover problem (A, F) is NP-hard [20,32].

We consider now a greedy algorithm for construction of an α-cover (see Algorithm 1). At each step, this algorithm chooses a subset from F which covers the maximum number of uncovered elements from A. This algorithm stops when the constructed subfamily is an α-cover for (A, F).

Algorithm 1. Greedy algorithm for approximate set cover problem

---

Input: a set cover problem (A, F) and real number α, $0 \leq \alpha < 1$.
Output: α-cover for (A, F).

$G := \emptyset$, and $COVER := \emptyset$;
while $|G| < (1 - \alpha)|A|$ **do**
In the family F we find a set S_i with minimum index i such that

$$|S_i \cap (A \backslash G)| = \max\{|S_j \cap (A \backslash G)| : S_j \in F\}.$$

$G := G \cup S_i$ and $COVER := COVER \cup \{S_i\}$;
end while
return $COVER$

We denote by $C_{\text{greedy}}(\alpha, A, F)$ the cardinality of the constructed α-cover for (A, F), and by $C_{\min}(\alpha, A, F)$ we denote the minimum cardinality of an α-cover for (A, F).

The following statement was obtained by J. Cheriyan and R. Ravi in [5]. We present it with our own proof.

Theorem 1. *Let $0 < \alpha < 1$ and (A, F) be a set cover problem. Then*

$$C_{\text{greedy}}(\alpha, A, F) \leq C_{\min}(0, A, F)\ln(1/\alpha) + 1.$$

Proof. We denote $m = C_{\min}(0, A, F)$. If $m = 1$ then, as it is not difficult to show, $C_{\text{greedy}}(\alpha, A, F) = 1$ and the considered inequality holds. Let $m \geq 2$ and S_i be a subset of maximum cardinality in F. It is clear that $|S_i| \geq N/m$. So, after the first step we will have at most $N - N/m = N(1 - 1/m)$ uncovered elements in the set A. After the first step we have the following set cover problem: the set $A \setminus S_i$ and the family $\{S_1 \setminus S_i, \ldots, S_p \setminus S_i\}$. For this problem, the minimum cardinality of a cover is at most m. So, after the second step, when we choose a set $S_j \setminus S_i$ with maximum cardinality, the number of uncovered elements in the set A will be at most $N(1 - 1/m)^2$, etc.

Let the greedy algorithm in the process of α-cover construction make g steps and construct an α-cover of cardinality g. Then after the step number $g - 1$ more then αN elements in A are uncovered. Therefore $N(1 - 1/m)^{g-1} > \alpha N$ and $1/\alpha > (1 + 1/(m - 1))^{g-1}$. If we take the natural logarithm of both sides of this inequality we obtain $\ln 1/\alpha > (g - 1)\ln(1 + 1/(m - 1))$. It is known that for any natural p, the inequality $\ln(1 + 1/p) > 1/(p + 1)$ holds. Therefore $\ln(1/\alpha) > (g - 1)/m$ and $g < m\ln(1/\alpha) + 1$. Since $m = C_{\min}(0, A, F)$ and $g = C_{\text{greedy}}(\alpha, A, F)$, we have $C_{\text{greedy}}(\alpha, A, F) < C_{\min}(0, A, F)\ln(1/\alpha) + 1$. □

5 Greedy Algorithm for α-Decision Rule Construction

In this section, we present a greedy algorithm for α-decision rule construction, lower and upper bounds on the minimum length of α-decision rules (Sect. 5.1) and 0.5-hypothesis connected with the work of a greedy algorithm (Sect. 5.2).

We use the greedy algorithm for construction of α-covers to construct α-decision rules. Let T be a table with many-valued decisions containing n columns labeled with attributes $f_1, \ldots, f_n$, $r = (b_1, \ldots, b_n)$ be a row of T, $D(r)$ be the set of decisions attached to r, $d \in D(r)$, and α be a real number such that $0 < \alpha < 1$.

We consider a set cover problem $(A(T,r,d),\ F(T,r,d))$ where $A(T,r,d) = U(T,r,d)$ is the set of all rows r' of T such that $d \notin D(r')$ and $F(T,r,d) = \{S_1, \ldots, S_n\}$. For $i = 1, \ldots, n$, the set S_i coincides with the set of all rows from $A(T,r,d)$ which are different from r in the column f_i. One can show that the decision rule

$$f_{i_1} = b_{i_1} \wedge \ldots \wedge f_{i_m} = b_{i_m} \rightarrow d$$

is an α-decision rule for (T,r) and decision $d \in D(r)$ if and only if $\{S_{i_1}, \ldots, S_{i_m}\}$ is an α-cover for the set cover problem $(A(T,r,d), F(T,r,d))$. Evidently, for the considered set cover problem, $C_{\min}(0, A(T,r,d), F(T,r,d)) = L_{\min}(0,T,r,d)$, where $L_{\min}(0,T,r,d)$ is the minimum length of 0-decision rule for (T,r) and decision $d \in D(r)$.

Let us apply the greedy algorithm (see Algorithm 1) to the considered set cover problem. This algorithm constructs an α-cover which corresponds to an α -decision rule $rule(\alpha,T,r,d)$ for (T,r) and decision $d \in D(r)$. From Theorem 1 it follows that the length of this rule is at most

$$L_{\min}(0,T,r,d)\ln(1/\alpha) + 1.$$

We denote by $L_{\text{greedy}}(\alpha,T,r)$ the length of the rule constructed by the following polynomial time algorithm: for a given α, $0 < \alpha < 1$, decision table T, row r of T and decision $d \in D(r)$, we construct the set cover problem $(A(T,r,d), F(T,r,d))$ and then apply to this problem the greedy algorithm for construction of an α-cover. We transform the obtained α-cover into an α-decision rule $rule(\alpha,T,r,d)$. Among the α-decision rules $rule(\alpha,T,r,d)$, $d \in D(r)$, we choose a rule with the minimum length. This rule is the output of the considered algorithm. We denote by $L_{\min}(\alpha,T,r)$ the minimum length of an α-decision rule for (T,r). According to what has been said above we have the following statement.

Theorem 2. *Let T be a non-degenerate decision table with many-valued decisions, r be a row of T, and α be a real number such that $0 < \alpha < 1$. Then*

$$L_{\text{greedy}}(\alpha,T,r) \leq L_{\min}(0,T,r)\ln(1/\alpha) + 1.$$

Note that the considered algorithm is a generalization of an algorithm studied in [16].

Example 1. Let us apply the considered greedy algorithm to $\alpha = 0.1$, decision table T_0 (see Table 1) and the second row r_2 of this table.

For each $d \in D(r_2) = \{1, 3\}$ we construct the set cover problem $(A(T, r_2, d), F(T, r_2, d))$, where $A(T, r_2, d)$ is the set of all rows r' of T such that $d \notin D(r')$, $F(T, r_2, d)) = \{S_1, S_2, S_3\}$, and S_i coincides with the set of rows from $A(T, r_2, d)$ which are different from r_2 in the column f_i, $i = 1, 2, 3$. We have:

- $A(T, r_2, 1) = \{r_3, r_4\}$, $F(T, r_2, 1) = \{S_1 = \{r_3\}, S_2 = \{r_4\}, S_3 = \{r_4\}\}$,
- $A(T, r_2, 3) = \{r_1, r_3, r_5\}$, $F(T, r_2, 3) = \{S_1 = \{r_1, r_3, r_5\}, S_2 = \{r_5\}, S_3 = \{r_1\}\}$.

Now, we apply the greedy algorithm for the set cover problem (with $\alpha = 0.1$) to each of the constructed set cover problems, and transform the obtained 0.1-covers into 0.1-decision rules.

For the case $d = 1$, we obtain the 0.1-cover $\{S_1, S_2\}$ and corresponding 0.1-decision rule $f_1 = 0 \wedge f_2 = 1 \rightarrow 1$.

For the case $d = 3$, we obtain the 0.1-cover $\{S_1\}$ and corresponding 0.1-decision rule $f_1 = 0 \rightarrow 3$. We choose the shortest rule $f_1 = 0 \rightarrow 3$ which is the result of our algorithm work.

In order to show that the problem of minimization of α-decision rule length is NP-hard, let us consider a set cover problem (A, F) where $A = \{a_1, \ldots, a_N\}$ and $F = \{S_1, \ldots, S_m\}$. We define the decision table as $T(A, F)$, this table has m columns corresponding to the sets $S_1, \ldots, S_m$ respectively, and $N + 1$ rows. For $j = 1, \ldots, N$, the j-th row corresponds to the element a_j. The last $(N + 1)$-th row is filled by 0. For $j = 1, \ldots, N$ and $i = 1, \ldots, m$, at the intersection of j-th row and i-th column 1 stays if and only if $a_j \in S_i$. The set of decisions corresponding to the last row is equal to $\{2\}$. All other rows are labeled with the set of decisions $\{1\}$.

One can show that, for any α, $0 \leq \alpha < 1$, a subfamily $\{S_{i_1}, \ldots, S_{i_t}\}$ is an α-cover for (A, F) if and only if the decision rule

$$f_{i_1} = 0 \wedge \ldots \wedge f_{i_t} = 0 \rightarrow 2$$

is an α-decision rule for $T(A, F)$ and the last row of $T(A, F)$.

So, we have a polynomial time reduction of the problem of minimization of α-cover cardinality to the problem of minimization of α -decision rule length for decision tables with many-valued decisions. Since the first problem is NP-hard [20, 32], we have

Proposition 1. *For any α, $0 \leq \alpha < 1$, the problem of minimization of α-decision rule length for decision tables with many-valued decisions is NP-hard.*

5.1 Upper and Lower Bounds on $L_{\min}(\alpha, T, r)$

In this section, we present some results connected with lower and upper bounds on the minimum length of α-decision rules, based on the information obtained during the greedy algorithm work.

Let T be a decision table with many-valued decisions and r be a row of T. Let α be a real number such that $0 \le \alpha < 1$. We apply the greedy algorithm to T, r, α and each $d \in D(r)$ (really to the corresponding set cover problem) and obtain for every $d \in D(r)$ an α-decision rule for the pair (T, r) and decision d. Among these rules we choose a rule with the minimum length, and denote this length by $u(\alpha, T, r)$. It is clear that

$$L_{\min}(\alpha, T, r) \le u(\alpha, T, r).$$

Let $d \in D(r)$. We apply the greedy algorithm to T, r, α and d, and construct the α-decision rule $rule(\alpha, T, r, d)$. Let the length of this rule be equal to t, and δ_i, $i = 1, \ldots, t$, be the number of rows from $U(T, r, d)$ separated from row r at the i-th step of the greedy algorithm work. We denote

$$l(\alpha, T, r, d) = \max\left\{\left\lceil \frac{\lceil (1-\alpha)|U(T,r,d)| \rceil - (\delta_0 + \ldots + \delta_i)}{\delta_{i+1}} \right\rceil : i = 0, \ldots, t-1\right\},$$

where $\delta_0 = 0$. Let us denote

$$l(\alpha, T, r) = \min_{d \in D(r)} l(\alpha, T, r, d).$$

We can almost repeat the first part of the proof of Theorem 1.67 from [16] to obtain the following lower bound:

$$L_{\min}(\alpha, T, r, d) \ge l(\alpha, T, r, d),$$

where $L_{\min}(\alpha, T, r, d)$ is the minimum length of α-decision rule for (T, r) and d. From this inequality it follows that

$$L_{\min}(\alpha, T, r) \ge l(\alpha, T, r).$$

5.2 0.5-Hypothesis

In the book [16], the following 0.5-hypothesis was formulated for decision tables with one-valued decisions: for the most part of decision tables for each row r under the construction of decision rule, during each step the greedy algorithm chooses an attribute which separates from r at least one-half of unseparated rows with decisions other than decision attached to the row r.

Let T be a decision table with many-valued decisions and r be a row of T. We will say that 0.5-hypothesis is true for T and r if for any decision $d \in D(r)$ under the construction of decision rule for the pair (T, r) and decision d, during each step the greedy algorithm chooses an attribute which separates from r at least 50 % of unseparated rows from $U(T, r, d)$.

We will say that 0.5-hypothesis is true for T if it is true for each row of T.

Now we consider some theoretical results regarding to 0.5-hypothesis for decision tables with many-valued decisions.

A binary information system I is a table with n rows (corresponding to objects) and m columns labeled with attributes $f_1, \dots, f_m$. This table is filled by numbers from $\{0,1\}$ (values of attributes). For $j = 1, \dots, n$, we denote by r_j the j-th row of the table I.

The information system I will be called *strongly saturated* if, for any row $r_j = (b_1, \dots, b_n)$ of I, for any $k \in \{0, \dots, n-1\}$ and for any k rows with numbers different from j, there exists a column f_i which has at least $\frac{k}{2}$ numbers $\neg b_i$ (b_i is the value of the f_i column for the row r_j) at the intersection with the considered k rows.

First, we evaluate the number of strongly saturated binary information systems. After that, we study the work of the greedy algorithm on a decision table with many-valued decisions obtained from a strongly saturated binary information system by adding a set of decisions to each row. It is clear that 0.5-hypothesis holds for every such table.

Theorem 3 *[16]. Let us consider binary information systems with* n *rows and* $m \geq n + \log_2 n$ *columns labeled with attributes* $f_1, \dots, f_m$. *Then the fraction of strongly saturated information systems is at least* $1 - 1/2^{m-n-\log_2 n+1}$.

For example, if $m \geq n + \log_2 n + 6$, than at least 99 % of binary information systems are strongly saturated.

Let us consider the work of the greedy algorithm on an arbitrary decision table T with many-valued decisions obtained from the strongly saturated binary information system. Let r be an arbitrary row of table T and $d \in D(r)$. For $i = 1, 2, \dots$, after the step number i at most $|U(T,r,d)|/2^i$ rows from $U(T,r,d)$ are unseparated from r. It is not difficult to show that $L_{\text{greedy}}(\alpha, T, r) \leq \lceil \log_2(1/\alpha) \rceil$ for any real α, $0 < \alpha < 1$, where $L_{\text{greedy}}(\alpha, T, r)$ is the length of α-decision rules constructed by the greedy algorithm for (T, r). One can prove that $L_{\text{greedy}}(0, T, r) \leq \log_2 |U(T,r,d)| + 1$. It is easy to check that $l(0, T, r) \leq 2$.

6 Problem of Recognition of Labels of Points in the Plain

In this section, we present a problem of recognition of colors of points in the plain (note that, we recognize labels attached to the points, and labels are named as colors), which illustrates the considered approach and the obtained bounds on precision of the greedy algorithm relative to the length of α-decision rules.

Let we have a finite set $S = \{(a_1, b_1), \dots, (a_n, b_n)\}$ of points in the plane and a mapping μ which corresponds to each point (a_p, b_p) a nonempty subset $\mu(a_p, b_p)$ of the set $\{green, yellow, red\}$. Colors are interpreted as decisions, and for each point from S we need to find a decision (color) from the set of decisions attached to this point. We denote this problem by (S, μ).

For the problem (S, μ) solving, we use attributes corresponding to straight lines which are given by equations of the kind $x = \beta$ or $y = \gamma$. These attributes are defined on the set S and take values from the set $\{0, 1\}$. Consider the line given by equation $x = \beta$. Then the value of corresponding attribute is equal to 0 on a point $(a, b) \in S$ if and only if $a < \beta$. Consider the line given by equation $y = \gamma$. Then the value of corresponding attribute is equal to 0 if and only if $b < \gamma$.

We now choose a finite set of straight lines which allow us to construct a decision rule with the minimum length for the problem (S, μ). It is possible that $a_i = a_j$ or $b_i = b_j$ for different i and j. Let $a_{i_1}, \ldots, a_{i_m}$ be all pairwise different numbers from the set $\{a_1, \ldots, a_n\}$ which are ordered such that $a_{i_1} < \ldots < a_{i_m}$. Let $b_{j_1}, \ldots, b_{j_t}$ be all pairwise different numbers from the set $\{b_1, \ldots, b_n\}$ which are ordered such that $b_{j_1} < \ldots < b_{j_t}$.

One can show that there exists a decision rule with minimum length which use only attributes corresponding to the straight lines defined by equations $x = a_{i_1} - 1$, $x = (a_{i_1} + a_{i_2})/2, \ldots$, $x = (a_{i_{m-1}} + a_{i_m})/2$, $x = a_{i_m} + 1$, $y = b_{j_1} - 1$, $y = (b_{j_1} + b_{j_2})/2, \ldots$, $y = (b_{j_{t-1}} + b_{j_t})/2$, $y = b_{j_t} + 1$.

Now, we describe a decision table $T(S, \mu)$ with $m+t+2$ columns and n rows. Columns of this table are labeled with attributes $f_1, \ldots, f_{m+t+2}$, corresponding to the considered $m + t + 2$ lines. Attributes $f_1, \ldots, f_{m+1}$ correspond to lines defined by equations $x = a_{i_1} - 1, x = (a_{i_1} + a_{i_2})/2, \ldots, x = (a_{i_{m-1}} + a_{i_m})/2, x = a_{i_m} + 1$ respectively. Attributes $f_{m+2}, \ldots, f_{m+t+2}$ correspond to lines defined by equations $y = b_{j_1} - 1$, $y = (b_{j_1} + b_{j_2})/2, \ldots, y = (b_{j_{t-1}} + b_{j_t})/2, y = b_{j_t} + 1$ respectively. Rows of the table $T(S, \mu)$ correspond to points $(a_1, b_1), \ldots, (a_n, b_n)$. At the intersection of the column f_l and row (a_p, b_p) the value $f_l(a_p, b_p)$ stays. For $p = 1, \ldots, n$, the row (a_p, b_p) is labeled with the set of decisions $\mu(a_p, b_p)$.

Example 2. A problem (S, μ) with four points and corresponding decision table $T(S, \mu)$ is depicted in Fig. 1. We write "*g*" instead of "*green*", "*r*" instead of "*red*", and "*y*" instead of "*yellow*".

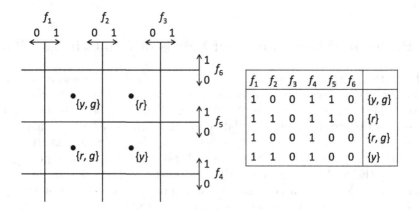

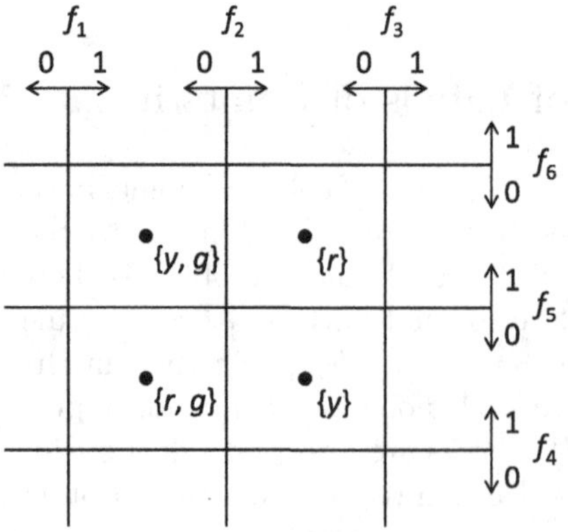

f_1	f_2	f_3	f_4	f_5	f_6	
1	0	0	1	1	0	{y, g}
1	1	0	1	1	0	{r}
1	0	0	1	0	0	{r, g}
1	1	0	1	0	0	{y}

Fig. 1. Problem (S, μ) and corresponding decision table $T(S, \mu)$

Let us evaluate the parameter $M(T(S,\mu))$.

Proposition 2. $M(T(S,\mu)) \leq 4$.

Proof. We denote $T = T(S,\mu)$. Let $\bar{\delta} = (\delta_1, \ldots, \delta_{m+t+2}) \in \{0,1\}^{m+t+2}$. If $\delta_1 = 0$, or $\delta_{m+1} = 1$, or $\delta_{m+2} = 0$, or $\delta_{m+t+2} = 1$, then $T(f_1,\delta_1)$, or $T(f_{m+1},\delta_{m+1})$, or $T(f_{m+2},\delta_{m+2})$, or $T(f_{m+t+2},\delta_{m+t+2})$ is empty table and $M(T,\bar{\delta}) \leq 1$. Let $\delta_1 = 1$, $\delta_{m+1} = 0$, $\delta_{m+2} = 1$ and $\delta_{m+t+2} = 0$. One can show that in this case there exist $i \in \{1,\ldots,m\}$ and $j \in \{m+2,\ldots,m+t+1\}$ such that $\delta_i = 1$, $\delta_{i+1} = 0$, $\delta_j = 1$, and $\delta_{j+1} = 0$. It is clear that the table $T(f_i,\delta_i)(f_{i+1},\delta_{i+1})(f_j,\delta_j)(f_{j+1},\delta_{j+1})$ contains exactly one row. So $M(T,\bar{\delta}) \leq 4$ and $M(T) \leq 4$. □

From Lemma 1, Theorem 2 and Proposition 2 the next statement follows:

Corollary 1. *For any real* α, $0 < \alpha < 1$, *and any row* r *of the table* $T(S,\mu)$,

$$L_{\text{greedy}}(\alpha, T(S,\mu), r) < 4\ln(1/\alpha) + 1.$$

Note that $4\ln(1/0.01)+1 < 19.43$, $4\ln(1/0.1)+1 < 10.22$, $4\ln(1/0.2)+1 < 7.44$, and $4\ln(1/0.5)+1 < 3.78$.

7 Results of Experiments

This section consists of three parts:

- experimental results for the many-valued decisions approach (Sect. 7.1),
- experimental results for the generalized decision approach (Sect. 7.2),
- comparative study (Sect. 7.3).

We consider a number of decision tables from UCI Machine Learning Repository [12]. In some tables there were missing values. Each such value was replaced with the most common value of the corresponding attribute. Some decision tables contain conditional attributes that take unique value for each row. Such attributes were removed. In some tables there were equal rows with, possibly, different decisions.

In this case each group of identical rows was replaced with a single row from the group which is labeled with the set of decisions attached to rows from the group. To obtain rows which are labeled with sets containing more than one decision we removed from decision tables more conditional attributes. The information about such decision tables can be found in Table 2. This table contains name of initial table, number of rows (column "Rows"), number of attributes (column "Attr"), spectrum of this table (column "Spectrum"), and list of names of removed attributes (column "Removed attributes"). Spectrum of a decision table with many-valued decisions is a sequence #1, #2,..., where #i, $i = 1,2,\ldots$, is the number of rows labeled with sets of decision with cardinality equals to i. All experiments are performed using DAGGER software tool [1] in C++.

Table 2. Characteristics of decision tables with many-valued decisions

Decision table	Rows	Attr	Spectrum						Removed attributes
			#1	#2	#3	#4	#5	#6	
balance-scale-1	125	3	45	50	30				left-weight
breast-cancer-1	193	8	169	24					tumor-size
breast-cancer-5	98	4	58	40					inv-nodes,node-caps,deg-malig,breast-quad,irradiat
cars-1	432	5	258	161	13				buying
flags-5	171	21	159	12					zone,language,religion,circles, sunstars
hayes-roth-data-1	39	3	22	13	4				marital_status
kr-vs-kp-5	1987	31	1564	423					katri,mulch,rimmx,skrxp,wknck
kr-vs-kp-4	2061	32	1652	409					katri,mulch,rimmx,wknck
lymphography-5	122	13	113	9					lymphatics,changes_in_node, changes_in_stru, special_forms,no_of_nodes_in
mushroom-5	4078	17	4048	30					odor,gill-size,stalk-root,stalk-surface-below-ring,habitat
nursery-4	240	4	97	96	47				parents,housing,finance,social
nursery-1	4320	7	2858	1460	2				parents
poker-hand-train-5	3324	5	156	1832	1140	188	7	1	S1,C1,C2,C4,C5
poker-hand-train-5a	3323	5	130	1850	1137	199	6	1	C1,S2,C2,C3,C4
poker-hand-train-5b	1024	5	0	246	444	286	44	4	C1,C2,C3,C4,C5
spect-test-1	164	21	161	3					F3
teeth-1	22	7	12	10					top incisors
teeth-5	14	3	6	3	0	5	0	2	bottom incisors,top canines,bottom canines,top premolars,bottom molars
tic-tac-toe-4	231	5	102	129					top-right-square,middle-middle-square,bottom-left-square,bottom-right-square
tic-tac-toe-3	449	6	300	149					middle-middle-square,bottom-left-square,bottom-right-square
zoo-data-5	42	11	36	6					feathers,backbone,breathes, legs,tail

7.1 Proposed Approach

We made four groups of experiments which are connected with:

- length of constructed α-decision rules,
- number of different α-decision rules,
- lower and upper bounds on the minimum length of α-decision rules,
- 0.5-hypothesis.

The first group of experiments is the following. For decision tables described in Table 2 and $\alpha \in \{0.0, 0.001, 0.01, 0.1, 0.2, 0.3\}$, we apply to each row of table the greedy algorithm. After that, among the constructed rules we find minimum (column "min"), average (column "avg") and maximum (column "max") length of such rules. Results can be found in Tables 3 and 4.

One can see that the length of constructed α-decision rules is decreasing when the value of α is increasing, and the greedy algorithm constructs relatively short α-decision rules.

Table 3. Length of α-decision rules for $\alpha \in \{0.0, 0.001, 0.01\}$ constructed by greedy algorithm

Decision table	$\alpha = 0.0$			$\alpha = 0.001$			$\alpha = 0.01$		
	min	avg	max	min	avg	max	min	avg	max
balance-scale-1	2	2.00	2	2	2.00	2	2	2.00	2
breast-cancer-1	1	2.94	5	1	2.94	5	1	2.81	5
breast-cancer-5	1	1.72	3	1	1.72	3	1	1.72	3
cars-1	1	1.38	4	1	1.38	4	1	1.36	3
flags-5	1	2.43	5	1	2.43	5	1	2.10	4
hayes-roth-data-1	1	1.59	2	1	1.59	2	1	1.59	2
kr-vs-kp-5	1	4.11	11	1	4.11	11	1	2.97	8
kr-vs-kp-4	1	4.14	11	1	4.14	11	1	2.98	8
lymphography-5	1	2.68	5	1	2.68	5	1	2.66	5
mushroom-5	1	1.52	5	1	1.52	5	1	1.46	4
nursery-4	1	1.33	2	1	1.33	2	1	1.33	2
nursery-1	1	2.05	5	1	2.00	5	1	1.80	3
poker-hand-train-5	2	2.35	5	2	2.35	5	2	2.27	4
poker-hand-train-5a	2	2.24	5	2	2.24	4	2	2.09	4
poker-hand-train-5b	2	2.01	5	2	2.01	5	2	2.01	4
spect-test-1	1	1.32	5	1	1.32	5	1	1.29	4
teeth-1	1	2.27	3	1	2.27	3	1	2.27	3
teeth-5	1	1.93	3	1	1.93	3	1	1.93	3
tic-tac-toe-4	2	2.24	4	2	2.24	4	2	2.24	4
tic-tac-toe-3	3	3.29	6	3	3.29	6	3	3.14	5
zoo-data-5	1	2.19	5	1	2.19	5	1	2.19	5

Table 4. Length of α-decision rules for $\alpha \in \{0.1, 0.2, 0.3\}$ constructed by greedy algorithm

Decision table	$\alpha = 0.1$			$\alpha = 0.2$			$\alpha = 0.3$		
	min	avg	max	min	avg	max	min	avg	max
balance-scale-1	1	1.06	2	1	1.00	1	1	1.00	1
breast-cancer-1	1	1.75	3	1	1.28	2	1	1.01	2
breast-cancer-5	1	1.15	2	1	1.00	1	1	1.00	1
cars-1	1	1.22	2	1	1.17	2	1	1.00	1
flags-5	1	1.20	2	1	1.02	2	1	1.00	1
hayes-roth-data-1	1	1.00	1	1	1.00	1	1	1.00	1
kr-vs-kp-5	1	1.77	4	1	1.17	2	1	1.04	2
kr-vs-kp-4	1	1.73	4	1	1.16	3	1	1.03	2
lymphography-5	1	1.80	3	1	1.37	2	1	1.01	2
mushroom-5	1	1.12	2	1	1.01	2	1	1.00	1
nursery-4	1	1.01	2	1	1.00	1	1	1.00	1
nursery-1	1	1.38	2	1	1.00	2	1	1.00	1
poker-hand-train-5	1	1.06	2	1	1.00	1	1	1.00	1
poker-hand-train-5a	1	1.01	2	1	1.00	1	1	1.00	1
poker-hand-train-5b	2	2.00	2	2	2.00	2	1	1.00	1
spect-test-1	1	1.25	3	1	1.08	2	1	1.07	2
teeth-1	1	1.50	2	1	1.14	2	1	1.05	2
teeth-5	1	1.71	3	1	1.29	2	1	1.07	2
tic-tac-toe-4	1	1.47	3	1	1.04	2	1	1.01	2
tic-tac-toe-3	2	2.00	3	1	1.50	2	1	1.01	2
zoo-data-5	1	1.69	4	1	1.26	3	1	1.12	2

Table 5. Number of different rules constructed by greedy algorithm

Decision table	values of α					
	0.0	0.001	0.01	0.1	0.2	0.3
balance-scale-1	51	51	51	45	27	23
breast-cancer-1	164	164	164	112	49	35
breast-cancer-5	61	61	61	38	25	25
cars-1	23	23	19	12	11	10
flags-5	159	159	155	111	101	98
hayes-roth-data-1	14	14	14	7	6	6
kr-vs-kp-5	856	856	577	188	93	65
kr-vs-kp-4	914	914	621	210	105	72
lymphography-5	59	59	58	44	32	23
mushroom-5	493	**508**	448	165	73	58
nursery-4	7	7	7	**13**	5	5
nursery-1	89	74	33	23	10	10
poker-hand-train-5	392	392	**402**	136	36	36
poker-hand-train-5a	333	333	255	93	34	34
poker-hand-train-5b	52	52	52	52	52	8
spect-test-1	29	29	29	28	22	22
teeth-1	22	22	22	22	22	22
teeth-5	14	14	14	14	14	14
tic-tac-toe-4	52	52	52	30	15	14
tic-tac-toe-3	157	157	137	63	57	16
zoo-data-5	16	16	16	16	15	15

Table 5 presents the number of different rules constructed by the greedy algorithm for $\alpha \in \{0.0, 0.001, 0.01, 0.1, 0.2, 0.3\}$. In the worst case, the number of different rules can be equal to the number of rows in decision table T. One can see that with the exception of three tables, the number of different rules is non-increasing when the value of α is increasing.

Next group of experimental results is connected with lower and upper bounds on the minimum length of α-decision rules. Figures 2 and 3 present average values of bounds $l(\alpha, T, r)$ and $u(\alpha, T, r)$ among all rows r of T for α, $0 \leq \alpha < 1$, with the step 0.01.

The last group of experiments is connected with 0.5-hypothesis. Table 6 contains, for $i = 1, 2, \ldots$, the average percentage of rows separated at i-th step of the greedy algorithm (average among all rows r and decisions $d \in D(r)$).

For decision tables described in Table 2 we find the number of rows for which 0.5-hypothesis is true. Table 7 contains name of decision table, number of rows and number of rows for which 0.5-hypothesis is true.

Results in Table 6 show that average percentage of rows separated at i-th step of the greedy algorithm during the exact decision rule construction is more than or equal to 50 % (7-th step of the greedy algorithm for "spect-test-1"). We say that 0.5-hypothesis is true for T if it is true for each row of T. Based on results in Table 7 we can see that 0.5-hypothesis is true for 12 decision tables

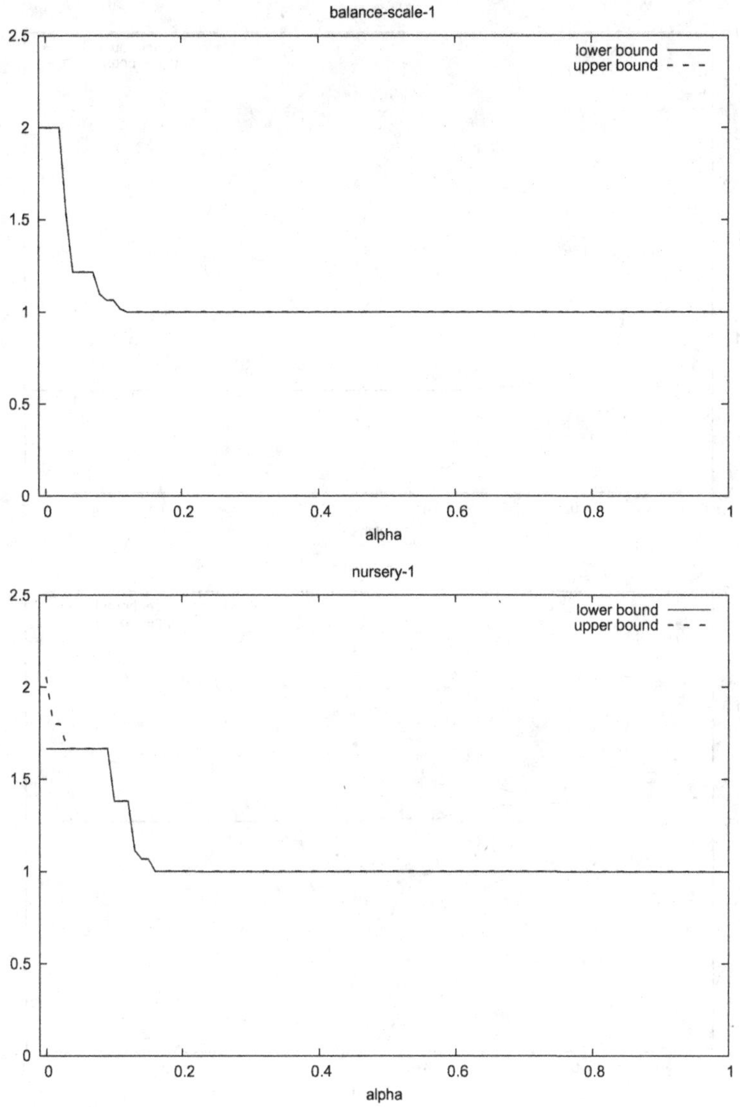

Fig. 2. Lower and upper bounds on $L_{\min}(\alpha, T, r)$ (“balance-scale-1” and “nursery-1”)

and is not true for 9 decision tables: “breast-cancer-1”, “kr-vs-kp-5”, “kr-vs-kp-4”, “lymphography-5”, “poker-hand-train-5”, “poker-hand-train-5a”, “spect-test-1”, “tic-tac-toe-3”, and “zoo-data-5”.

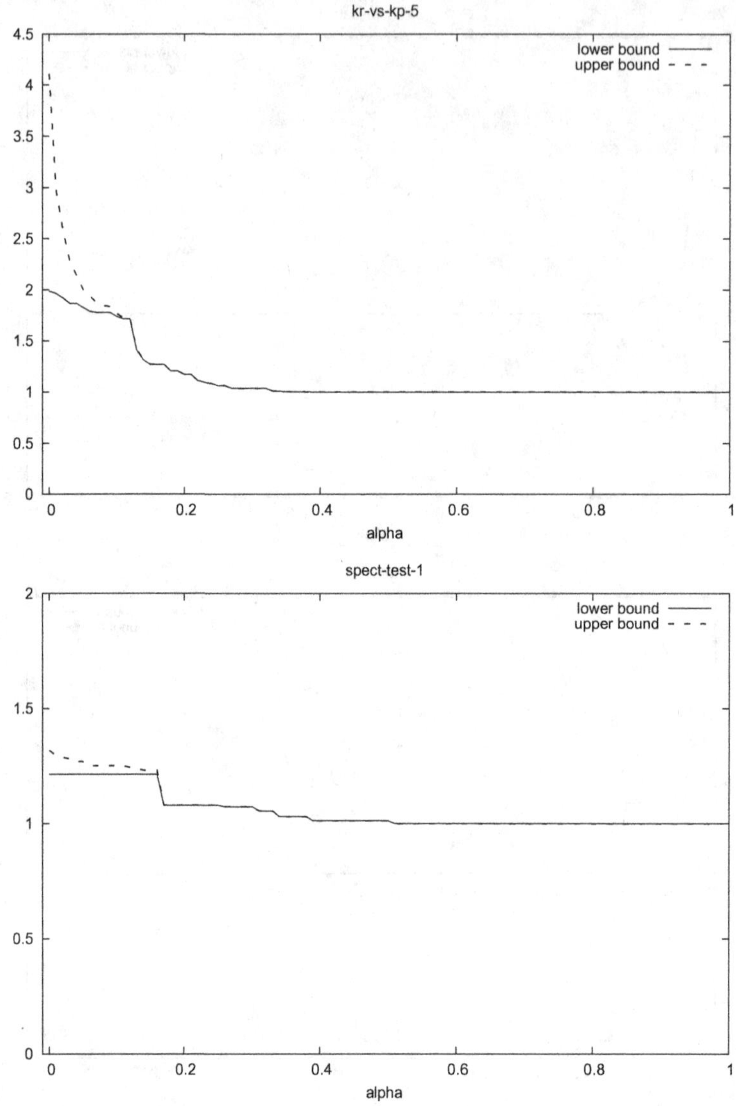

Fig. 3. Average values of lower and upper bounds on $L_{\min}(\alpha, T, r)$ ("kr-vs-kp-5" and "spect-test-1")

7.2 Generalized Decision Approach

In this section, we present experimental results for α-decision rules relative to:

- length of constructed α-decision rules,
- number of different α-decision rules,
- 0.5-hypothesis.

Table 6. Average percentage of rows separated at i-th step of the greedy algorithm work

Decision table	Number of step i									
	1	2	3	4	5	6	7	8	9	10
balance-scale-1	91.5	97.2	100							
breast-cancer-1	84.5	82.8	87.6	92.9	84.8	100				
breast-cancer-5	91.8	94.1	97.6	100						
cars-1	90.2	86.9	89.1	85.7	100					
flags-5	92.9	92.4	91.9	92.5	100					
hayes-roth-data-1	89.7	96.4	100							
kr-vs-kp-5	85.7	79.9	78.2	80.2	81.6	81.4	85.7	80.2	80.8	78.3
kr-vs-kp-4	85.8	79.5	78.4	79.8	81.9	82.1	85.3	82.1	82.3	80.0
lymphography-5	83.9	86.2	92.1	94.2	100					
mushroom-5	97.0	97.3	91.2	92.6	96.8	100				
nursery-4	94.8	99.8	90.9	100						
nursery-1	90.2	92.3	93.3	93.6	92.5	92.7	100			
poker-hand-train-5	92.3	87.2	91.0	86.3	100					
poker-hand-train-5a	92.3	88.7	91.1	84.9	100					
poker-hand-train-5b	75.5	92.2	81.5	87.5	100					
spect-test-1	94.3	89.4	80.5	85.6	77.4	75.0	50.0	100		
teeth-1	88.2	89.9	100							
teeth-5	87.0	92.0	100							
tic-tac-toe-4	86.7	91.6	90.6	94.9	100					
tic-tac-toe-3	78.5	83.6	88.8	88.1	91.2	100				
zoo-data-5	83.8	83.4	80.0	80.6	100					

Table 7. Number of rows in decision tables for which 0.5-hypothesis is true

Decision table	rows	rows with 0.5-hypothesis
balance-scale-1	125	125
breast-cancer-1	193	**191**
breast-cancer-5	98	98
cars-1	432	432
flags-5	171	171
hayes-roth-data-1	39	39
kr-vs-kp-5	1987	**1873**
kr-vs-kp-4	2061	**1949**
lymphography-5	122	**121**
mushroom-5	4078	4078
nursery-4	240	240
nursery-1	4320	4320
poker-hand-train-5	3324	**3321**
poker-hand-train-5a	3323	**3322**
poker-hand-train-5b	1024	1024
spect-test-1	164	**163**
teeth-1	22	22
teeth-5	14	14
tic-tac-toe-4	231	231
tic-tac-toe-3	449	**445**
zoo-data-5	42	**40**

Table 8. Transformation of the set of decisions for the generalized decision approach

$$T_0 = \begin{array}{|c|ccc|c|}\hline & f_1 & f_2 & f_3 & d \\ \hline r_1 & 1 & 1 & 1 & \{1\} \\ r_2 & 0 & 1 & 0 & \{1,3\} \\ r_3 & 1 & 1 & 0 & \{2\} \\ r_4 & 0 & 0 & 1 & \{2,3\} \\ r_5 & 1 & 0 & 0 & \{1,2\} \\ \hline \end{array} \Rightarrow \begin{array}{|c|}\hline d \\ \hline 1 \\ 2 \\ 3 \\ 4 \\ 5 \\ \hline \end{array}$$

In the generalized decision approach [23–25], the greedy algorithm constructs for each row one α-decision rule which has on the right-hand side the generalized decision (a number encoding the set of decisions attached to a given row) see Table 8.

For decision tables described in Table 2 and $\alpha \in \{0.0,0.001,0.01,0.1,0.2,0.3\}$ we apply to each row of table the greedy algorithm. After that, among the constructed rules we find minimum (column "min"), average (column "avg") and maximum (column "max") length of such rules. Results can be found in Tables 9 and 10.

We can say that for this approach the greedy algorithm constructs relatively short α-decision rules.

Table 9. Length of α-decision rules for $\alpha \in \{0.0, 0.001, 0.01\}$–generalized decision approach

Decision table	$\alpha = 0.0$			$\alpha = 0.001$			$\alpha = 0.01$		
	min	avg	max	min	avg	max	min	avg	max
balance-scale-1	3	3.00	3	3	3.00	3	2	2.92	3
breast-cancer-1	1	3.29	6	1	3.29	6	1	3.08	6
breast-cancer-5	1	2.38	4	1	2.38	4	1	2.38	4
cars-1	1	2.51	5	1	2.51	5	1	2.16	4
flags-5	1	2.49	5	1	2.49	5	1	2.15	4
hayes-roth-data-1	1	2.13	3	1	2.13	3	1	2.13	3
kr-vs-kp-5	1	4.81	12	1	4.42	11	1	3.14	8
kr-vs-kp-4	1	4.81	12	1	4.42	11	1	3.11	8
lymphography-5	1	2.94	6	1	2.94	6	1	2.89	6
mushroom-5	1	1.54	8	1	1.53	6	1	1.47	4
nursery-4	1	2.02	4	1	2.02	4	1	1.69	4
nursery-1	1	3.32	7	1	2.98	6	1	2.35	4
poker-hand-train-5	4	4.78	5	4	4.09	5	3	3.00	3
poker-hand-train-5a	4	4.78	5	4	4.08	5	3	3.00	3
poker-hand-train-5b	4	4.96	5	4	4.96	5	3	3.86	4
spect-test-1	1	1.72	10	1	1.72	10	1	1.68	9
teeth-1	1	2.27	3	1	2.27	3	1	2.27	3
teeth-5	1	1.93	3	1	1.93	3	1	1.93	3
tic-tac-toe-4	3	3.79	5	3	3.79	5	3	3.41	5
tic-tac-toe-3	3	4.55	6	3	4.55	6	3	3.61	5
zoo-data-5	1	2.55	6	1	2.55	6	1	2.55	6

Table 10. Length of α-decision rules for $\alpha \in \{0.1, 0.2, 0.3\}$–generalized decision approach

Decision table	$\alpha = 0.1$			$\alpha = 0.2$			$\alpha = 0.3$		
	min	avg	max	min	avg	max	min	avg	max
balance-scale-1	2	2.00	2	1	1.02	2	1	1.00	1
breast-cancer-1	1	1.70	3	1	1.30	2	1	1.08	2
breast-cancer-5	1	1.30	2	1	1.00	1	1	1.00	1
cars-1	1	1.44	2	1	1.44	2	1	1.00	1
flags-5	1	1.22	2	1	1.01	2	1	1.00	1
hayes-roth-data-1	1	1.59	2	1	1.26	2	1	1.00	1
kr-vs-kp-5	1	1.71	4	1	1.24	3	1	1.04	2
kr-vs-kp-4	1	1.68	4	1	1.23	3	1	1.04	2
lymphography-5	1	1.82	3	1	1.33	2	1	1.07	2
mushroom-5	1	1.12	2	1	1.01	2	1	1.00	1
nursery-4	1	1.67	2	1	1.07	2	1	1.00	1
nursery-1	1	1.67	2	1	1.15	2	1	1.00	1
poker-hand-train-5	1	1.00	1	1	1.00	1	1	1.00	1
poker-hand-train-5a	1	1.00	1	1	1.00	1	1	1.00	1
poker-hand-train-5b	2	2.00	2	2	2.00	2	1	1.00	1
spect-test-1	1	1.59	3	1	1.14	2	1	1.05	2
teeth-1	1	1.50	2	1	1.14	2	1	1.05	2
teeth-5	1	1.71	3	1	1.29	2	1	1.07	2
tic-tac-toe-4	2	2.03	3	2	2.00	2	1	1.15	2
tic-tac-toe-3	2	2.03	3	2	2.00	2	1	1.27	2
zoo-data-5	1	1.81	4	1	1.33	3	1	1.21	2

Table 11. Number of different rules–generalized decision approach

Decision table	values of α					
	0.0	0.001	0.01	0.1	0.2	0.3
balance-scale-1	125	125	118	83	27	27
breast-cancer-1	169	169	164	117	65	48
breast-cancer-5	82	82	82	44	33	33
cars-1	141	141	94	40	40	15
flags-5	159	159	154	115	106	106
hayes-roth-data-1	25	25	25	22	19	18
kr-vs-kp-5	1133	1073	783	256	140	87
kr-vs-kp-4	1205	1141	817	265	142	90
lymphography-5	73	73	72	58	43	33
mushroom-5	543	**556**	481	214	76	64
nursery-4	41	41	16	16	12	12
nursery-1	572	406	202	79	38	16
poker-hand-train-5	2883	2388	1459	213	213	213
poker-hand-train-5a	2904	2351	1439	205	205	205
poker-hand-train-5b	998	998	743	280	280	94
spect-test-1	42	42	42	41	29	23
teeth-1	22	22	22	22	22	22
teeth-5	14	14	14	14	14	14
tic-tac-toe-4	131	131	117	53	53	28
tic-tac-toe-3	347	347	262	96	95	49
zoo-data-5	24	24	24	24	21	20

Table 12. Average percentage of rows separated at i-th step of the greedy algorithm work–generalized decision approach

Decision table	Number of step i									
	1	2	3	4	5	6	7	8	9	10
balance-scale-1	83.9	86.0	100							
breast-cancer-1	85.5	82.6	83.8	88.0	87.0	100				
breast-cancer-5	93.3	90.5	92.1	100						
cars-1	90.1	77.4	82.8	83.8	100					
flags-5	92.9	92.4	93.6	91.2	100					
hayes-roth-data-1	87.7	85.3	100							
kr-vs-kp-5	86.1	80.5	77.1	77.9	75.1	73.4	76.9	76.6	74.4	87.1
kr-vs-kp-4	86.4	80.4	77.4	78.2	75.2	72.6	75.9	76.2	73.7	81.1
lymphography-5	83.5	85.5	88.8	86.4	88.9	100				
mushroom-5	96.9	97.4	91.2	94.4	90.5	48.9	77.8	100		
nursery-4	90.3	97.3	98.6	100						
nursery-1	89.4	83.5	86.7	89.3	91.3	93.0	100			
poker-hand-train-5	92.4	77.5	79.2	84.3	100					
poker-hand-train-5a	92.3	77.2	79.4	84.6	100					
poker-hand-train-5b	75.7	77.0	78.9	82.5	100					
spect-test-1	91.9	94.9	81.0	84.9	75.0	66.7	25.0	33.3	50.0	100
teeth-1	87.4	89.8	100							
teeth-5	86.3	92.4	100							
tic-tac-toe-4	73.7	79.8	84.7	89.5	100					
tic-tac-toe-3	73.4	76.2	79.2	82.1	85.8	100				
zoo-data-5	83.9	80.7	72.6	81.1	90.0	100				

Table 13. Number of rows in decision tables for which 0.5-hypothesis is true–generalized decision approach

Decision table	rows	rows with 0.5-hypothesis true
balance-scale-1	125	125
breast-cancer-1	193	**192**
breast-cancer-5	98	98
cars-1	432	432
flags-5	171	171
hayes-roth-data-1	39	39
kr-vs-kp-5	1987	**1799**
kr-vs-kp-4	2061	**1871**
lymphography-5	122	**120**
mushroom-5	4078	**4072**
nursery-4	240	240
nursery-1	4320	4320
poker-hand-train-5	3324	3324
poker-hand-train-5a	3323	3323
poker-hand-train-5b	1024	1024
spect-test-1	164	**163**
teeth-1	22	22
teeth-5	14	14
tic-tac-toe-4	231	231
tic-tac-toe-3	449	**446**
zoo-data-5	42	**39**

We computed the number of different rules constructed by the greedy algorithm for $\alpha \in \{0.0, 0.001, 0.01, 0.1, 0.2, 0.3\}$. Results can be found in Table 11. For the generalized decision approach, in the worst case, the number of different rules can be equal to the number of rows in decision table T. With the exception of one table, the number of different rules is nonincreasing with the growth of α.

The last group of experiments is connected with 0.5-hypothesis. Table 12 contains, for $i = 1, 2, \ldots$, the average percentage of rows separated at i-th step of the greedy algorithm work (average among all rows r). For two decision tables, the average percentage of separated rows is less than 50 %: for "spect-test-1" – at the 7-th and the 8-th step of the greedy algorithm work, for "mushroom-5"– at the 6-th step of the greedy algorithm work.

We say that 0.5-hypothesis is true for T, if is true for each row of T. Table 13 contains, for decision tables described in Table 2, number of rows for which 0.5-hypothesis is true. From 21 decision tables, the 0.5-hypothesis is not true for 8 of them: "breast-cancer-1", "kr-vs-kp-5", "kr-vs-kp-4", "lymphography-5", "mushroom-5", "spect-test-1", "tic-tac-toe-3" and "zoo-data-5".

7.3 Comparative Study

In this section, we make comparative study of α-decision rules for the proposed approach and the generalized decision approach, relative to:

- length of constructed α-decision rules,
- number of different α-decision rules,
- 0.5-hypothesis.

Table 14, based on results from Tables 3 and 9 presents, for $\alpha \in \{0.0, 0.001, 0.01\}$, a comparison of minimum (column "min"), average (column "avg") and maximum (column "max") length of α-decision rules for both approaches. Each input of Table 14 is equal to the (min, avg, max) length of α-decision rules for the generalized decision approach divided by the (min, avg, max) length of α-decision rules for proposed approach.

We can find decision tables for which minimum, average and maximum length of α-decision rules constructed using the proposed approach is two or more times shorter than minimum, average and maximum length of α-decision rules constructed using generalized decision approach. However, for the maximum values of length of 0.01-decision rules for decision tables "poker-hand-train-5" and "poker-hand-train-5a" we have an opposite situation.

Table 15, based on results from Tables 4 and 10 presents, for $\alpha \in \{0.1, 0.2, 0.3\}$, comparison of minimum (column "min"), average (column "avg") and maximum (column "max") length of α-decision rules for both approaches. Each input of Table 15 is equal to the corresponding input of Table 10 divided by the input of Table 4.

Results are similar to the results from Table 14.

Table 16, based on results from Tables 5 and 11 presents, for $\alpha \in \{0.0, 0.001, 0.01, 0.1, 0.2, 0.3\}$, a comparison of the number of different α-decision rules

Table 14. Comparison of length of α-decision rules for $\alpha \in \{0.0, 0.001, 0.01\}$

Decision table	$\alpha = 0.0$			$\alpha = 0.001$			$\alpha = 0.01$		
	min	avg	max	min	avg	max	min	avg	max
balance-scale-1	1.50	1.50	1.50	1.50	1.50	1.50	1.00	1.46	1.50
breast-cancer-1	1.00	1.12	1.20	1.00	1.12	1.20	1.00	1.10	1.20
breast-cancer-5	1.00	1.38	1.33	1.00	1.38	1.33	1.00	1.38	1.33
cars-1	1.00	1.82	1.25	1.00	1.82	1.25	1.00	1.59	1.33
flags-5	1.00	1.02	1.00	1.00	1.02	1.00	1.00	1.02	1.00
hayes-roth-data-1	1.00	1.34	1.50	1.00	1.34	1.50	1.00	1.34	1.50
kr-vs-kp-5	1.00	1.17	1.09	1.00	1.08	1.00	1.00	1.06	1.00
kr-vs-kp-4	1.00	1.16	1.09	1.00	1.07	1.00	1.00	1.04	1.00
lymphography-5	1.00	1.10	1.20	1.00	1.10	1.20	1.00	1.09	1.20
mushroom-5	1.00	1.01	1.60	1.00	1.01	1.20	1.00	1.01	1.00
nursery-4	1.00	1.52	**2.00**	1.00	1.52	**2.00**	1.00	1.27	**2.00**
nursery-1	1.00	1.62	1.40	1.00	1.49	1.20	1.00	1.31	1.33
poker-hand-train-5	**2.00**	**2.03**	1.00	**2.00**	1.74	1.00	1.50	1.32	**0.75**
poker-hand-train-5a	**2.00**	**2.13**	1.00	**2.00**	1.82	1.25	1.50	1.44	**0.75**
poker-hand-train-5b	**2.00**	**2.47**	1.00	**2.00**	**2.47**	1.00	1.50	1.92	1.00
spect-test-1	1.00	1.30	**2.00**	1.00	1.30	**2.00**	1.00	1.30	**2.25**
teeth-1	1.00	1.00	1.00	1.00	1.00	1.00	1.00	1.00	1.00
teeth-5	1.00	1.00	1.00	1.00	1.00	1.00	1.00	1.00	1.00
tic-tac-toe-4	1.50	1.69	1.25	1.50	1.69	1.25	1.50	1.52	1.25
tic-tac-toe-3	1.00	1.38	1.00	1.00	1.38	1.00	1.00	1.15	1.00
zoo-data-5	1.00	1.16	1.20	1.00	1.16	1.20	1.00	1.16	1.20

Table 15. Comparison of length of α-decision rules for $\alpha \in \{0.1, 0.2, 0.3\}$

Decision table	$\alpha = 0.1$			$\alpha = 0.2$			$\alpha = 0.3$		
	min	avg	max	min	avg	max	min	avg	max
balance-scale-1	**2.00**	1.89	1.00	1.00	1.02	**2.00**	1.00	1.00	1.00
breast-cancer-1	1.00	**0.97**	1.00	1.00	1.02	1.00	1.00	1.07	1.00
breast-cancer-5	1.00	1.13	1.00	1.00	1.00	1.00	1.00	1.00	1.00
cars-1	1.00	1.18	1.00	1.00	1.23	1.00	1.00	1.00	1.00
flags-5	1.00	1.02	1.00	1.00	**0.99**	1.00	1.00	1.00	1.00
hayes-roth-data-1	1.00	1.59	**2.00**	1.00	1.26	**2.00**	1.00	1.00	1.00
kr-vs-kp-5	1.00	**0.97**	1.00	1.00	1.06	1.50	1.00	1.00	1.00
kr-vs-kp-4	1.00	**0.97**	1.00	1.00	1.06	1.00	1.00	1.01	1.00
lymphography-5	1.00	1.01	1.00	1.00	**0.97**	1.00	1.00	1.06	1.00
mushroom-5	1.00	1.00	1.00	1.00	1.00	1.00	1.00	1.00	1.00
nursery-4	1.00	1.65	1.00	1.00	1.07	**2.00**	1.00	1.00	1.00
nursery-1	1.00	1.21	1.00	1.00	1.15	1.00	1.00	1.00	1.00
poker-hand-train-5	1.00	**0.94**	**0.50**	1.00	1.00	1.00	1.00	1.00	1.00
poker-hand-train-5a	1.00	**0.99**	**0.50**	1.00	1.00	1.00	1.00	1.00	1.00
poker-hand-train-5b	1.00	1.00	1.00	1.00	1.00	1.00	1.00	1.00	1.00
spect-test-1	1.00	1.27	1.00	1.00	1.06	1.00	1.00	**0.98**	1.00
teeth-1	1.00	1.00	1.00	1.00	1.00	1.00	1.00	1.00	1.00
teeth-5	1.00	1.00	1.00	1.00	1.00	1.00	1.00	1.00	1.00
tic-tac-toe-4	**2.00**	1.38	1.00	**2.00**	1.92	1.00	1.00	1.14	1.00
tic-tac-toe-3	1.00	1.02	1.00	**2.00**	1.33	1.00	1.00	1.26	1.00
zoo-data-5	1.00	1.07	1.00	1.00	1.06	1.00	1.00	1.08	1.00

Table 16. Comparison of number of different rules

Decision table	values of α					
	0.0	0.001	0.01	0.1	0.2	0.3
balance-scale-1	**2.45**	**2.45**	**2.31**	1.84	1.00	1.17
breast-cancer-1	1.03	1.03	1.00	1.04	1.33	1.37
breast-cancer-5	1.34	1.34	1.34	1.16	1.32	1.32
cars-1	**6.13**	**6.13**	**4.95**	**3.33**	**3.64**	1.50
flags-5	1.00	1.00	0.99	1.04	1.05	1.08
hayes-roth-data-1	1.79	1.79	1.79	**3.14**	**3.17**	**3.00**
kr-vs-kp-5	1.32	1.25	1.36	1.36	1.51	1.34
kr-vs-kp-4	1.32	1.25	1.32	1.26	1.35	1.25
lymphography-5	1.24	1.24	1.24	1.32	1.34	1.43
mushroom-5	1.10	1.09	1.07	1.30	1.04	1.10
nursery-4	**5.86**	**5.86**	**2.29**	1.23	**2.40**	**2.40**
nursery-1	**6.43**	**5.49**	**6.12**	**3.43**	**3.80**	1.60
poker-hand-train-5	**7.35**	**6.09**	**3.63**	1.57	**5.92**	**5.92**
poker-hand-train-5a	**8.72**	**7.06**	**5.64**	**2.20**	**6.03**	**6.03**
poker-hand-train-5b	**19.19**	**19.19**	**14.29**	**5.38**	**5.38**	**11.75**
spect-test-1	1.45	1.45	1.45	1.46	1.32	1.05
teeth-1	1.00	1.00	1.00	1.00	1.00	1.00
teeth-5	1.00	1.00	1.00	1.00	1.00	1.00
tic-tac-toe-4	**2.52**	**2.52**	**2.25**	1.77	**3.53**	**2.00**
tic-tac-toe-3	**2.21**	**2.21**	1.91	1.52	1.67	**3.06**
zoo-data-5	1.50	1.50	1.50	1.50	1.40	1.33

for both approaches. Each input of Table 16 is equal to the number of different α-decision rules for the generalized decision approach divided by the number of different α-decision rules for the proposed approach. We can see that often the number of different α-decision rules for the generalized decision approach is two or more times greater than the number of different rules for the proposed approach.

The last group of results is connected with 0.5-hypothesis. Based on results from Tables 7 and 13 we can see that, for the proposed approach, the 0.5-hypothesis is not true for 9 decision tables, for generalized decision approach, the 0.5-hypothesis is not true for 8 decision tables. So, the difference is not significant.

8 Conclusions

We studied the greedy algorithm for construction of approximate decision rules. This algorithm has polynomial time complexity for the whole set of decision tables with many-valued decisions. We obtained a bound on precision of this algorithm relative to the length of rules, and considered lower and upper bounds on the minimum length of α-decision rules. We studied binary decision tables with many-valued decisions but the considered approach can be used also for decision tables with more than two values of attributes, as presented in Sect. 7.

Experimental results are connected with the construction of exact and approximate decision rules. Based on them, we can see, that the greedy algorithm constructs relatively short α-decision rules. We also presented results relative to length, number of different α-decision rules and 0.5-hypothesis for the approach based on generalized decision.

Based on results connected with comparison of two approaches we can see that the length and number of different rules constructed in the framework of our approach (one decision from the set of decisions attached to row) are usually smaller than the length and number of different rules constructed in the framework of the generalized decision approach (all decisions from the set of decisions attached to row).

Future investigations will be connected with the study of other greedy algorithms and construction of classifiers for decision tables with many-valued decisions.

Acknowledgements. Research reported in this publication was supported by the King Abdullah University of Science and Technology (KAUST).

The authors wish to express their gratitude to anonymous reviewers for useful comments.

References

1. Alkhalid, A., Amin, T., Chikalov, I., Hussain, S., Moshkov, M., Zielosko, B.: Dagger: a tool for analysis and optimization of decision trees andrules. Comput. Inf. Soc. Factors New Inf. Technol. Hypermedia Perspect. Avant-Garde Experiences Eraof Communicability Expansion, 29–39 (2011)
2. Azad, M., Chikalov, I., Moshkov, M., Zielosko, B.: Greedy algorithms for construction of approximate tests for decision tables with many-valued decisions. Fundamenta Informaticae **120**(3–4), 231–242 (2012)
3. Blockeel, H., Schietgat, L., Struyf, J., Džeroski, S., Clare, A.: Decision trees for hierarchical multilabel classification: a case study in functional genomics. In: Fürnkranz, J., Scheffer, T., Spiliopoulou, M. (eds.) PKDD 2006. LNCS (LNAI), vol. 4213, pp. 18–29. Springer, Heidelberg (2006). doi:10.1007/11871637_7
4. Boutell, M.R., Luo, J., Shen, X., Brown, C.M.: Learning multi-label scene classification. Pattern Recogn. **37**(9), 1757–1771 (2004)
5. Cheriyan, J., Ravi, R.: Lecture notes on approximation algorithms for network problems (1998). http://www.math.uwaterloo.ca/~jcheriya/lecnotes.html
6. Chikalov, I., Zielosko, B.: Decision rules for decision tables with many-valued decisions. In: Yao, J.T., Ramanna, S., Wang, G., Suraj, Z. (eds.) RSKT 2011. LNCS (LNAI), vol. 6954, pp. 763–768. Springer, Heidelberg (2011). doi:10.1007/978-3-642-24425-4_95
7. Clare, A., King, R.D.: Knowledge discovery in multi-label phenotype data. In: Raedt, L., Siebes, A. (eds.) PKDD 2001. LNCS (LNAI), vol. 2168, pp. 42–53. Springer, Heidelberg (2001). doi:10.1007/3-540-44794-6_4
8. Comité, F., Gilleron, R., Tommasi, M.: Learning multi-label alternating decision trees from texts and data. In: Perner, P., Rosenfeld, A. (eds.) MLDM 2003. LNCS, vol. 2734, pp. 35–49. Springer, Heidelberg (2003). doi:10.1007/3-540-45065-3_4

9. Feige, U.: A threshold of ln n for approximating set cover. J. ACM (JACM) **45**(4), 634–652 (1998)
10. Greco, S., Matarazzo, B., Słowiński, R.: Rough sets theory for multicriteria decision analysis. Eur. J. Oper. Res. **129**(1), 1–47 (2001)
11. Kryszkiewicz, M.: Rules in incomplete information systems. Inf. Sci. **113**(34), 271–292 (1999)
12. Lichman, M.: UCI Machine Learning Repository (2013)
13. Lipski, W.: On databases with incomplete information. J. ACM (JACM) **28**(1), 41–70 (1981)
14. Lipski Jr., W.: On semantic issues connected with incomplete information databases. ACM Trans. Database Syst. **4**(3), 262–296 (1979)
15. Mencia, E.L., Furnkranz, J.: Pairwise learning of multilabel classifications with perceptrons. In: IEEE International Joint Conference on Neural Networks, 2008, IJCNN 2008 (IEEE World Congress on Computational Intelligence), pp. 2899–2906 (2008)
16. Moshkov, M.J., Piliszczuk, M., Zielosko, B.: Partial Covers, Reducts and Decision Rules in Rough Sets–Theory and Applications. SCI, vol. 145. Springer, Heidelberg (2008)
17. Moshkov, M., Zielosko, B.: Combinatorial Machine Learning–A Rough Set Approach. SCI, vol. 360. Springer, Heidelberg (2011)
18. Moshkov, M., Zielosko, B.: Construction of α-decision trees for tables with many-valued decisions. In: Yao, J.T., Ramanna, S., Wang, G., Suraj, Z. (eds.) RSKT 2011. LNCS (LNAI), vol. 6954, pp. 486–494. Springer, Heidelberg (2011). doi:10.1007/978-3-642-24425-4_63
19. Moshkov, M.J.: Greedy algorithm for decision tree construction in context of knowledge discovery problems. In: Tsumoto, S., Słowiński, R., Komorowski, J., Grzymała-Busse, J.W. (eds.) RSCTC 2004. LNCS (LNAI), vol. 3066, pp. 192–197. Springer, Heidelberg (2004). doi:10.1007/978-3-540-25929-9_22
20. Nguyen, H.S., Slezak, D.: Approximate reducts and association rules - correspondence and complexity results. In: Proceedings of the 7th International Workshop on New Directions in Rough Sets, Data Mining, and Granular-Soft Computing. RSFDGrC 1999, pp. 137–145. Springer, London (1999)
21. Orowska, E., Pawlak, Z.: Representation of nondeterministic information. Theoret. Comput. Sci. **29**(12), 27–39 (1984)
22. Pawlak, Z.: Rough Sets-Theoretical Aspects of Reasoning about Data. Kluwer Academic Publishers, Dordrecht (1991)
23. Pawlak, Z., Skowron, A.: Rough sets and boolean reasoning. Inf. Sci. **177**(1), 41–73 (2007)
24. Pawlak, Z., Skowron, A.: Rough sets: some extensions. Inf. Sci. **177**(1), 28–40 (2007)
25. Pawlak, Z., Skowron, A.: Rudiments of rough sets. Inf. Sci. **177**(1), 3–27 (2007)
26. Rissanen, J.: Modeling by shortest data description. Automatica **14**(5), 465–471 (1978)
27. Sakai, H., Ishibashi, R., Koba, K., Nakata, M.: Rules and apriori algorithm in non-deterministic information systems. In: Peters, J.F., Skowron, A., Rybiński, H. (eds.) Transactions on Rough Sets IX. LNCS, vol. 5390, pp. 328–350. Springer, Heidelberg (2008). doi:10.1007/978-3-540-89876-4_18
28. Sakai, H., Nakata, M., Ślęzak, D.: Rule generation in lipski's incomplete information databases. In: Szczuka, M., Kryszkiewicz, M., Ramanna, S., Jensen, R., Hu, Q. (eds.) RSCTC 2010. LNCS (LNAI), vol. 6086, pp. 376–385. Springer, Heidelberg (2010). doi:10.1007/978-3-642-13529-3_40

29. Sakai, H., Nakata, M., Ślęzak, D.: A prototype system for rule generation in lipski's incomplete information databases. In: Kuznetsov, S.O., Ślęzak, D., Hepting, D.H., Mirkin, B.G. (eds.) RSFDGrC 2011. LNCS (LNAI), vol. 6743, pp. 175–182. Springer, Heidelberg (2011). doi:10.1007/978-3-642-21881-1_29
30. Skowron, A., Rauszer, C.: The discernibility matrices and functions in information systems. In: Intelligent Decision Support. Handbook of Applications and Advances of the Rough Set Theory, pp. 331–362. Kluwer Academic Publishers (1992)
31. Ślęzak, D.: Normalized decision functions and measures for inconsistent decision tables analysis. Fundamenta Informaticae **44**(3), 291–319 (2000)
32. Ślęzak, D.: Approximate entropy reducts. Fundamenta Informaticae **53**(3–4), 365–390 (2002)
33. Tsoumakas, G., Katakis, I.: Multi-label classification: an overview. Int. J. Data Warehouse. Min. **3**(3), 1–13 (2007)
34. Tsoumakas, G., Katakis, I., Vlahavas, I.: Mining multi-label data. In: Data Mining and Knowledge Discovery Handbook, pp. 667–685. Springer, US (2010)
35. Wieczorkowska, A., Synak, P., Lewis, R., Raś, Z.W.: Extracting emotions from music data. In: Hacid, M.-S., Murray, N.V., Raś, Z.W., Tsumoto, S. (eds.) ISMIS 2005. LNCS (LNAI), vol. 3488, pp. 456–465. Springer, Heidelberg (2005). doi:10.1007/11425274_47
36. Zhou, Z.H., Jiang, K., Li, M.: Multi-instance learning based web mining. Appl. Intell. **22**(2), 135–147 (2005)
37. Zhou, Z.H., Zhang, M.L., Huang, S.J., Li, Y.F.: Multi-instance multi-label learning. Artif. Intell. **176**(1), 2291–2320 (2012)

Algebraic Semantics of Proto-Transitive Rough Sets

A. Mani(✉)

Department of Pure Mathematics, University of Calcutta,
9/1B, Jatin Bagchi Road, Kolkata 700029, India
a.mani.cms@gmail.com
http://www.logicamani.in

Abstract. Rough Sets over generalized transitive relations like proto-transitive ones have been initiated recently by the present author. In a recent paper, approximation of proto-transitive relations by other relations was investigated and the relation with rough approximations was developed towards constructing semantics that can handle *fragments of structure*. It was also proved that difference of approximations induced by some approximate relations need not induce rough structures. In this research, the structure of rough objects is characterized and a theory of dependence for general rough sets is developed and used to internalize the Nelson-algebra based approximate semantics developed earlier by the present author. This is part of the different semantics of PRAX developed in this paper by her. The theory of rough dependence initiated in earlier papers is extended in the process. This paper is reasonably self-contained and includes proofs and extensions of representation of objects that have not been published earlier.

Keywords: Proto-transitive relations · Generalized transitivity · Rough dependence · Rough objects · Granulation · Algebraic semantics · Approximate relations · Approximate semantics · Nelson algebras · Axiomatic theory of granules · Contamination problem · Knowledge

1 Introduction

Proto-transitivity is one of the infinite number of possible generalizations of transitivity. These types of generalized relations happen often in application contexts. Failure to recognize them causes mathematical models to be inadequate or underspecified and tends to unduly complicate algorithms and approximate methods. From among the many possible alternatives that fall under *generalized transitivity*, the present author's preference for *proto-transitivity* is because of application contexts, its simple set theoretic definition, connections with factor relations and consequent generative value among such relations. It has a special role in modelling knowledge as well.

Proto-transitive approximation spaces PRAX have been introduced by the present author in [26] and the structure of definite objects has been characterized in it to a degree. It is relatively a harder structure from a semantic

J.F. Peters and A. Skowron (Eds.): TRS XX, LNCS 10020, pp. 51–108, 2016.
DOI: 10.1007/978-3-662-53611-7_3

perspective as the representation of rough objects is involved [26]. Aspects of knowledge interpretation in PRAX contexts have been considered in [26] and in [28] the relation of approximations resulting from approximation of relations to the approximations from the original relation are studied in the context of PRAX. These are used for defining an approximate semantics for PRAX and their limitations are explored by the present author in the same paper. All of these are expanded upon in this paper.

Rough objects as explained in [22,24] are collections of objects in a classical domain (Meta-C) that appear to be indistinguishable among themselves in another rough semantic domain (Meta-R). But their representation in most RSTs in purely order theoretic terms is not known. For PRAX, this is solved in [26]. Rough objects in a PRAX need not correspond to intervals of the form $]a, b[$ with the definite object b covering (in the ordered set of definite objects) the definite object a.

If R is a relation on a set S, then R can be approximated by a wide variety of partial/quasi-order relations in both classical and rough set perspective [11]. Though the methods are essentially equivalent for binary relations, the latter method is more general. When the relation R satisfies proto-transitivity, then many new properties emerge. This aspect is developed further in the present paper and most of [28] is included.

When R is a quasi-order relation, then a semantics for the set of ordered pairs of lower and upper approximations $\{(A^l, A^u);\ A \subseteq S\}$ has recently been developed in [13,14]. Though such a set of ordered pairs of lower and upper approximations are not rough objects in the PRAX context, they can be used for an additional semantic approach. In this paper, it is shown that differences of consequent lower and upper approximations suggest partial structures for *measuring* structured deviation. The developed method should also be useful for studying correspondences between the different semantics [23,25]. Because of this some space is devoted to the nature of transformation of granules by the relational approximation process.

In this research paper, the nature of possible concepts of *rough dependence* are also investigated in a comparative way. Though the concept of independence is well studied in probability theory, the concept of dependence is rarely explored in any useful way. It has been shown to be very powerful in classical probability theory [7] - the formalism is valid over probability spaces, but its axiomatic potential is left unexplored. Connections between rough sets and probability theory have been explored from rough measure and information entropy viewpoint in a number of papers [1,9,36,39,43]. The nature of rough independence is also explored in [29] by the present author and there is some overlap with the paper. Apart from problems relating to contamination, it is shown that the comparison by way of corresponding concepts of dependence fails in a very essential way. Further, using the introduced concepts of rough dependence correspondences are done away with in the approximate semantics. This allows for richer variants of the earlier semantics of rough objects. During the time that this research paper had been under review, the axiomatic approach to probabilist dependence has been extended in [31,32] by the present author. The highlights of her work include a new deviant probability theory and a less intrusive application to three way decision making.

This paper is reasonably self-contained and is organized as follows: In the rest of this section the basics of proto-transitivity are introduced and relevant information of Nelson algebras, granules and granulations are recapitulated. In the following section, relevant approximations are defined in PRAX and their basic properties are studied (including those of definite elements). In the third section, an abstract and three other extended examples are proposed to justify this study. In the following section, the algebraic structures that can be associated with the semantic properties of definite objects in a PRAX are described. The representation of rough objects is done from an interesting perspective in the fifth section. In the sixth section, new derived operators in a PRAX are defined and their connection with non monotonic reasoning is investigated. These are of relevance in representation again. In the following section, atoms in the partially ordered set of rough object are described. This is followed by an algebraic semantics that relies on multiple types of aggregation and commonality operations. In the ninth section, a partial semantics similar to the increasing Nelson algebraic semantics is formulated. This semantics is completed in three different ways in the fourteenth section after internalization of dependency. In the tenth and eleventh sections approximate relations and approximate semantics are considered - the material in these sections includes expansions of the results in [28]. In the following two sections, concepts of rough dependence are defined, compared with those of probabilistic dependence and their stark differences are demonstrated - the material in these sections are expansions of [29]. The knowledge interpretation of PRAX is revisited in the fifteenth section.

1.1 Basic Concepts, Terminology

Definition 1. *A binary relation R on a set S is said to be* weakly-transitive, transitive or proto-transitive *respectively on S if and only if S satisfies*

- *If whenever Rxy, Ryz and $x \neq y \neq z$ holds, then Rxz. (i.e. $(R \circ R) \setminus \Delta_S \subseteq R$ (where $\circ$ is relation composition), or*
- *Whenever Rxy & Ryz holds then Rxz (i.e. $(R \circ R) \subseteq R$), or*
- *Whenever Rxy, Ryz, Ryx, Rzy and $x \neq y \neq z$ holds, then Rxz follows, respectively. Proto-transitivity of R is equivalent to $R \cap R^{-1} = \tau(R)$ being weakly transitive.*

The simpler example below will be used to illustrate many of the concepts and situations in the paper. For detailed motivations the reader is advised to see Sect. 3 on motivation and examples.

Persistent Example 1. *A simple real-life example of a proto-transitive, non transitive relation would be the relation* $\mathbb{P}$, *defined by*

$$\mathbb{P}xy \text{ if and only if } x \text{ thinks that } y \text{ thinks that color of object } O \text{ is a maroon.}$$

The following simple example from databases will also be used as a persistent one (especially in the sections on approximation of relations) to illustrate a number of mathematical concepts. It possesses other attributes, distinct from those of the previous one, for illustrating more involved aspects.

Let $\mathcal{I}$ *be survey data in table form with column names being for sex, gender, sexual orientations, other personal data and opinions on sexist contexts with each row corresponding to a person. Let*

$$Rab \text{ if and only if person } a \text{ agrees with } b's \text{ opinions.}$$

The predicate agrees with *can be constructed empirically or from the data by a suitable heuristic. Often* R *is a proto-transitive, reflexive relation and this condition can be imposed to complete partial data as well (as a rationality condition). If* a *agrees with the opinions of* b, *then it will be said that* a *is an* ally *of* b *- if* b *is also an ally of* a, *then they are* comrades. *Finding optimal subsets of allies can be an interesting problem in many contexts especially given the fact that responses may have some vagueness in them.*

Definition 2. *A binary relation* R *on a set* S *is said to be* semi-transitive *on* S *if and only if* S *satisfies*

- *Whenever* $\tau(R)ab$ & Rbc *holds then* Rac *follows and*
- *Whenever* $\tau(R)ab$ & Rca *holds then* Rcb *follows.*

Henceforth Rxy will be used for $(x, y) \in R$ uniformly. $Ref(S), Sym(S), Tol(S), r\tau(S)$, $w\tau(S)$, $p\tau(S)$, $s\tau(S)$, $EQ(S)$ will respectively denote the set of reflexive, symmetric, tolerance, transitive, weakly transitive, pseudo transitive, semi-transitive and equivalence relations on the set S respectively.

The following proposition has steep ontological commitments.

Proposition 1. *For a relation* R *on a set* S, *the following are satisfied:*

- R *is* weakly transitive *if and only if* $(R \cap R^{-1}) \setminus \Delta_S \subseteq R$.
- R *is* transitive *if and only if* $(R \cap R^{-1}) \subseteq R$.

Relations that are antisymmetric and reflexive are called *pseudo order* relations. A *quasi-order* is a reflexive, transitive relation, while a partial order a reflexive, antisymmetric and transitive relation.

Let $\alpha \subseteq \rho$ be two binary relations on S, then $\rho|\alpha$ will be the relation on $S|\rho$ defined via $(x, y) \in \rho|\alpha$ if and only if $(\exists b \in x, c \in y)(b, c) \in \rho$. The relation $Q|\tau(Q)$ for a relation Q will be denoted by $\sigma(Q)$.

The following are known:

Proposition 2. *If* Q *is a quasi-order on* S, *then* $Q|\tau(Q)$ *is a partial order on* $S|\tau(Q)$.

Proposition 3. *If* $R \in Ref(S)$, *then* $R \in p\tau(S)$ *if and only if* $\tau(R) \in EQ(S)$.

Proposition 4. *In general,*

$$w\tau(S) \subseteq s\tau(S) \subseteq p\tau(S).$$

Proposition 5. *If* $R \in p\tau(S) \cap Ref(S)$, *then the following are equivalent:*

A1 $([a],[b]) \in R|\tau(R)$ *if and only if* $(a,b) \in R$.
A2 R *is semi-transitive.*

In [4], it is proved that

Theorem 1. *If* $R \in Ref(S)$, *then the following are equivalent:*

A3 $R|\tau(R)$ *is a pseudo order on* $S|\tau(R)$ *and A1 holds.*
A2 R *is semi-transitive.*

Note that *Weak transitivity* of [4] is *proto-transitivity* here. $Ref(S)$, $r\tau(S)$, $w\tau(S)$, $p\tau(S)$, $EQ(S)$ will respectively denote the set of reflexive, transitive, weakly transitive, proto transitive, and equivalence relations on the set S respectively. Clearly, $w\tau(S) \subseteq p\tau(S)$.

Proposition 6. $\forall R \in Ref(S)(R \in p\tau(S) \leftrightarrow \tau(R) \in EQ(S))$.

Definition 3. *A* Partial Algebra P *is a tuple of the form*

$$\langle \underline{P}, f_1, f_2, \ldots, f_n, (r_1, \ldots, r_n) \rangle$$

with $\underline{P}$ *being a set,* f_i*'s being partial function symbols of arity* r_i. *The interpretation of* f_i *on the set* $\underline{P}$ *should be denoted by* $f_i^{\underline{P}}$, *but the superscript will be dropped in this paper as the application contexts are simple enough. If predicate symbols enter into the signature, then* P *is termed a* Partial Algebraic System. *(see [2, 17] for the basic theory)*

In a partial algebra, for term functions p, q,

$$p \stackrel{\omega}{=} q \text{ iff } (\forall x \in dom(p) \cap dom(q))p(x) = q(x).$$

The *weak strong equality* is defined via,

$$p \stackrel{\omega^*}{=} q \text{ iff } (\forall x \in dom(p) = dom(q))p(x) = q(x).$$

For two terms s, t, $s \stackrel{\omega}{=} t$ shall mean, if both sides are defined then the two terms are equal (the quantification is implicit). $s \stackrel{\omega^*}{=} t$ shall mean if either side is defined, then the other is and the two sides are equal (the quantification is implicit).

1.2 Nelson Algebras

A *De Morgan lattice* ΔML is an algebra of the form $L = \langle \underline{L}, \vee, \wedge, c, 0, 1 \rangle$ with $\vee, \wedge$ being distributive lattice operations and c satisfying

- $x^{cc} = x$; $(x \vee y)^c = x^c \wedge y^c$;
- $(x \leq y \leftrightarrow y^c \leq x^c)$; $(x \wedge y)^c = x^c \vee y^c$;

It is possible to define a partial unary operation $\star$, via $x^\star = \bigwedge\{x : x \leq x^c\}$ on any ΔML. If it is total, then the ΔML is said to be *complete*. In a complete ΔML L, all of the following hold:

- $x^\star \nleq x^c$; $x^{\star\star} = x$;
- $(x \leq y \longrightarrow y^\star \leq x^\star)$.
- $x^c = \bigvee\{y : x^\star \nleq y\}$.

A ΔML is said to be a *Kleene algebra* if it satisfies $x \wedge x^c \leq y \vee y^c$. If $L^+ = \{x \vee x^c : x \in L\}$ and $L^- = \{x \wedge x^c : x \in L\}$, then in a Kleene algebra we have

- $(L^-)^c = L^+$ is a filter and $(L^+)^c = L^-$ is an ideal.
- $(\forall a, b \in L^-)\, a \leq b^c$; $(\forall a, b \in L^+)\, a^c \leq b$.
- $x \in L^-$ if and only if $x \leq x^c$.

A *Heyting algebra* K, is a relatively pseudo-complemented lattice, that is $(\forall a, b)\, a \Rightarrow b = \bigvee\{x; a \wedge x \leq b\} \in K$.

A *Quasi-Nelson* algebra Q is a Kleene algebra that satisfies $(\forall a, b)\, a \Rightarrow (a^c \vee b) \in Q$. $a \Rightarrow (a^c \vee b)$ is abbreviated by $a \rightarrow b$ below. Such an algebra satisfies all of the sentences N1–N4:

$$x \rightarrow x = 1 \tag{N1}$$

$$(x^c \vee y) \wedge (x \rightarrow y) = x^c \vee y \tag{N2}$$

$$x \wedge (x \rightarrow y) = x \wedge (x^c \vee y) \tag{N3}$$

$$x \rightarrow (y \wedge z) = (x \rightarrow y) \wedge (x \rightarrow z) \tag{N4}$$

$$(x \wedge y) \rightarrow z = x \rightarrow (y \rightarrow z). \tag{N5}$$

A Nelson algebra is a quasi-Nelson algebra satisfying N5. A Nelson algebra can also be defined directly as an algebra of the form $\langle A, \vee, \wedge, \rightarrow, c, 0, 1 \rangle$ with $\langle A, \vee, \wedge, c, 0, 1 \rangle$ being a Kleene algebra with the binary operation $\rightarrow$ satisfying N1–N5.

1.3 Granules and Granular Computing Paradigms

The idea of granular computing is as old as human evolution. Even in the available information on earliest human habitations and dwellings, it is possible to identify a primitive granular computing process (PGCP) at work. This can for example be seen from the stone houses, dating to 3500 BCE, used in what is present-day Scotland. The main features of this and other primitive versions of the paradigm may be seen to be

- Problem requirements are not rigid.
- Concept of granules may be vague.
- Little effort on formalization right up to approximately the middle of the previous century.
- Scope of abstraction is very limited.
- Concept of granules may be concrete or abstract (relative all materialist viewpoints).

The precision based granular computing paradigm, traceable to Moore and Shannon's paper [33], will be referred to as the *classical granular computing paradigm* CGCP is usually understood as the granular computing paradigm (The reader may note that the idea is vaguely present in [38]). The distinct terminology would be useful to keep track of the differences with other paradigms. CGCP has since been adapted to fuzzy and rough set theories in different ways.

Granules may be assumed to subsume the concept of information granules – information at some level of precision. In granular approaches to both rough and fuzzy sets, we are usually concerned with such types of granules. Some of the fragments involved in applying CGCP may be:

- Paradigm Fragment-1: Granules can exist at different levels of precision.
- Paradigm Fragment-2: Among the many precision levels, choose a precision level at which the problem at hand is solved.
- Paradigm Fragment-3: Granulations (granules at specific levels or processes) form a hierarchy (later development).
- Paradigm Fragment-4: It is possible to easily switch between precision levels.
- Paradigm Fragment-5: The problem under investigation may be represented by the hierarchy of multiple levels of granulations.

The not so independent stages of development of the different granular computing paradigms is stated below:

- Classical Primitive Paradigm till middle of previous century.
- CGCP: Since Shannon's information theory
- CGCP in fuzzy set theory. It is natural for most real-valued types of fuzzy sets, but even in such domains unsatisfactory results are normal. Type-2 fuzzy sets have an advantage over type-1 fuzzy sets in handling data relating to emotion words, for example, but still far from satisfactory. For one thing linguistic hedges have little to do with numbers. A useful reference would be [45].
- For a long period (up to 2008 or so), the adaptation of CGCP for RST has been based solely on precision and related philosophical aspects. The adaptation is described for example in [42]. In the same paper the hierarchical structure of granulations is also stressed. This and many later papers on CGCP (like [16]) in rough sets speak of structure of granulations.
- Some Papers with explicit reference to multiple types of granules from a semantic viewpoint include [20–22,40,41].

- The axiomatic approach to granularity initiated in [22] has been developed by the present author in the direction of contamination reduction in [24]. From the order-theoretic/algebraic point of view, the deviation is in a very new direction relative the precision-based paradigm. The paradigm shift includes a new approach to measures.

There are other adaptations of CGCP to soft computing like [15] that we will not consider.

Unless the underlying language is restricted, granulations can bear upon the theory with unlimited diversity. Thus for example in classical RST, we can take any of the following as granulations: collection of equivalence classes, complements of equivalence classes, other partitions on the universal set S, other partition in S, set of finite subsets of $\mathcal{S}$ and set of finite subsets of $\mathcal{S}$ of cardinality greater than 2. This is also among the many motivations for the axiomatic approach.

A formal simplified version of the axiomatic approach to granules is in [23]. The axiomatic theory is capable of handling most contexts and is intended to permit relaxation of set-theoretic axioms at a later stage. The axioms are considered in the framework of Rough Y-Systems (RYS) that maybe seen as a generalized form of *abstract approximation spaces* [3] and approximation framework [6]. It includes relation-based RST, cover-based RST and more. These structures are provided with enough structure so that a classical semantic domain (Meta-C) and at least one rough semantic domain (called Meta-R) of roughly equivalent objects along with admissible operations and predicates are associable. But the exact way of association is not something absolute as there is no real end to recursive approximation processes of objects.

In the present paper we will stick to successor, predecessor and related granules generated by elements and will avoid the precision based paradigm.

2 Approximations and Definite Elements in PRAX

Definition 4. *By a* Proto Approximation Space S *(PRAS for short), we will mean a pair of the form* $\langle \underline{S}, R\rangle$ *with* $\underline{S}$ *being a set and* R *being a proto-transitive relation on it. If* R *is also reflexive, then it will be called a* Reflexive Proto Approximation Space *(PRAX) for short).* $\underline{S}$ *may be infinite.*

If S is a PRAX or a PRAS, then we will respectively denote *successor neighborhoods, inverted successor or predecessor neighborhoods* and *symmetrized successor neighborhoods* generated by an element $x \in S$ as follows:

$$[x] = \{y;\ Ryx\}.$$

$$[x]_i = \{y;\ Rxy\}.$$

$$[x]_o = \{y;\ Ryx\ \&\ Rxy\}.$$

Taking these as granules, the associated granulations will be denoted by $\mathcal{G} = \{[x] : x \in S\}$, $\mathcal{G}_i$ and $\mathcal{G}_o$ respectively. In all that follows S will be a PRAX unless indicated otherwise.

Definition 5. *Definable approximations on S include ($A \subseteq S$):*

$$A^u = \bigcup_{[x] \cap A \neq \emptyset} [x]. \quad \text{(Upper Proto)}$$

$$A^l = \bigcup_{[x] \subseteq A} [x]. \quad \text{(Lower Proto)}$$

$$A^{u_o} = \bigcup_{[x]_o \cap A \neq \emptyset} [x]_o. \quad \text{(Symmetrized Upper Proto)}$$

$$A^{l_o} = \bigcup_{[x]_o \subseteq A} [x]_o. \quad \text{(Symmetrized Lower Proto)}$$

$$A^{u+} = \{x : [x] \cap A \neq \emptyset\}. \quad \text{(Point-wise Upper)}$$

$$A^{l+} = \{x : [x] \subseteq A\}. \quad \text{(Point-wise Lower)}$$

Persistent Example 2. *In the context of our example 1,* $[x]$ *is the* set of allies x*, while* $[x]_o$ *is the set of comrades of* x*.* A^l *is the* union of the set of all allies of at least one of the members of A *if they are all in* A*.* A^u *is the union of the set of all allies of persons having at least one ally in* A*.* A^{l+} *is the set of all those persons in* A *all of whose allies are within* A*.* A^{u+} *is the set of all those persons having allies in* A*.*

Definition 6. *If* $A \subseteq S$ *is an arbitrary subset of a PRAX or a PRAS* S*, then*

$$A^{ux} = \bigcup_{[x]_o \cap A \neq \emptyset} [x]. \tag{1}$$

$$A^{lx} = \bigcup_{[x]_o \subseteq A} [x]. \tag{2}$$

$$A^{u*} = \bigcup \{[x] : [x] \cap A \neq \emptyset \;\&\; (\exists y)([x],[y]) \in \sigma(R),\, (x,y) \in R,\, x \neq y,\, [y] \subseteq A\}. \tag{3}$$

$$A^{l*} = \bigcup \{[x] : [x] \subseteq A \;\&\; (\exists y)(([x],[y]) \in \sigma(R),\, x \neq y,\, [y] \subseteq A)\}. \tag{4}$$

The following inverted approximations are also of relevance as they provide Galois connections in case of point-wise approximations (see [12]) under particular assumptions. Our main approximations of interest will be l, u, l_o, u_o.

Definition 7. *In the context of the above definition, the following will be referred to as inverted approximations:*

$$A^{ui} = \bigcup_{[x]_i \cap A \neq \emptyset} [x]_i$$

$$A^{li} = \bigcup_{[x]_i \subseteq A} [x]_i$$

$$A^{\Delta} = \{x : [x]_i \cap A \neq \emptyset\}$$

$$A^{\nabla} = \{x : [x]_i \subseteq A\}$$

Proposition 7. *In a PRAX S and for a subset $A \subseteq S$, all of the following hold:*

- $(\forall x)\, [x]_o \subseteq [x]$
- *It is possible that $A^l \neq A^{l+}$ and in general, $A^l \parallel A^{lo}$.*

Proof. The proof of the first two parts are easy. For the third, we chase the argument up to a trivial counter example (see the following section).

$$\bigcup_{[x]\subseteq A} [x] \subseteq \bigcup_{[x]_o\subseteq A} [x] \supseteq \bigcup_{[x]_o\subseteq A} [x]_o$$

$$\bigcup_{[x]_o\subseteq A} [x]_o \supseteq \bigcup_{[x]\subseteq A} [x]_o \subseteq \bigcup_{[x]\subseteq A} [x].$$

Proposition 8. *For any subset A of S,*

$$A^{u_o} \subseteq A^u.$$

Proof. Since $[x]_o \cap A \neq \emptyset$, therefore

$$A^{u_o} = \bigcup_{[x]_o\cap A\neq\emptyset} [x]_o \subseteq \bigcup_{[x]\cap A\neq\emptyset} [x]_o \subseteq A^{u_o} = \bigcup_{[x]\cap A\neq\emptyset} [x] = A^u.$$

Definition 8. *If X is an approximation operator, then by a X*-definite element, *we will mean a subset A satisfying $A^X = A$. The set of all X-definite elements will be denoted by $\delta_X(S)$, while the set of X and Y-definite elements (Y being another approximation operator) will be denoted by $\delta_{XY}(S)$. In particular,* lower proto-definite, upper proto definite *and* proto-definite *elements (those that are both lower and upper proto-definite) will be spoken of.*

Theorem 2. *In a PRAX S, the following hold:*

- *$\delta_u(S) \subseteq \delta_{u_o}(S)$, but $\delta_{l_o}(S) = \delta_{u_o}(S)$ and $\delta_u(S)$ is a complete sub-lattice of $\wp(S)$ with respect to inclusion.*
- *$\delta_l(S) \parallel \delta_{l_o}(S)$ in general. ($\parallel$ means* is not comparable.*)*
- *It is possible that $\delta_u \nsubseteq \delta_{u_o}$.*

Proof.

- As R is reflexive, if A, B are upper proto definite, then $A \cup B$ and $A \cap B$ are both upper proto definite. So $\delta_u(S)$ is a complete sub-lattice of $\wp(S)$.
- If $A \in \delta_u$, then $(\forall x \in A)[x] \subseteq A$ and $(\forall x \in A^c)[x] \cap A = \emptyset$.
- So $(\forall x \in A^c)\, [x]_o \cap A = \emptyset$. But as $A \subseteq A^{u_o}$ is necessary, so $A \in \delta_{u_o}$ follows. □

A^{u+}, A^{l+} have relatively been more commonly used in the literature and have also been the only kind of approximation studied in [12] for example (the inverse relation is also considered from the same perspective).

Definition 9. *A subset $B \subseteq A^{l+}$ will be said to be* skeleton *of A if and only if*

$$\bigcup_{x \in B} [x] = A^l,$$

and the set skeletons of A will be denoted by $\mathbf{sk}(A)$.

The skeleton of a set A is important because it relates all three classes of approximations.

Theorem 3. *In the context of the above definition, we have*

- $\mathbf{sk}(A)$ *is partially ordered by inclusion with greatest element* A^{l+}.
- $\mathbf{sk}(A)$ *has a set of minimal elements* $\mathbf{sk}_m(S)$.
- $\mathbf{sk}(A) = \mathbf{sk}(A^l)$
- $\mathbf{sk}(A) = \mathbf{sk}(B) \leftrightarrow A^l = B^l \;\&\; A^{l+} = B^{l+}$.
- *If* $B \in \mathbf{sk}(A)$, *then* $A^l \subseteq B^u$.
- *If* $\cap\mathbf{sk}(A) = B$, *then* $A^{l_o} \cap \bigcup_{x\in B}[x] = \emptyset$.

Proof. Much of the proof is implicit in other results proved earlier in this section.

- If $x \in A^l \setminus A^{l+}$, then $[x] \nsubseteq A^l$ and many subsets B of A^{l+} are in $\mathbf{sk}(A)$. If $B \subset K \subset A^{l+}$ and $B \in \mathbf{sk}(A)$, then $K \in \mathbf{sk}(A)$. Further minimal elements exist in the inclusion order (even if A is infinite) by the induced properties of inclusion in $\wp(S)$.
- has been proved above.
- More generally, if $A^l \subseteq B \subseteq A$, then $B^l = A^l$. So $\mathbf{sk}(A) = \mathbf{sk}(A^l)$.
- Follows from definition.
- If $B \in \mathbf{sk}(A)$, then $A^l = B^l \subseteq B^u$. □

Theorem 4. *All of the following hold in PRAX:*

- $(\forall A)\, A^{cl+} = A^{u+c},\ A^{cu+} = A^{l+c}$ *- that is $l+$ and $u+$ are mutually dual*
- $u+$ *($l+$ resp.) is a monotone $\vee$- (complete $\wedge$- resp.) morphism.*
- $\partial(A) = \partial(A^c)$, *where ∂ stands for the boundary operator.*
- $\Im(u+)$ *(the image of $u+$) is an interior system while $\Im(l+)$ is a closure system.*
- $\Im(u+)$ *and $\Im(l+)$ are dually isomorphic lattices.*

Theorem 5. *In a PRAX,* $(\forall A \in \wp(S))\, A^{l+} \subseteq A^l,\ A^{u+} \subseteq A^u$ *and all of the following hold.*

$$(\forall A \in \wp(S))\, A^{ll} = A^l \;\&\; A^u \subseteq A^{uu}. \tag{Bi}$$

$$(\forall A, B \in \wp(S))\, A^l \cup B^l \subseteq (A \cup B)^l. \tag{l-Cup}$$

$$(\forall A, B \in \wp(S))\, (A \cap B)^l \subseteq A^l \cap B^l. \tag{l-Cap}$$

$$(\forall A, B \in \wp(S))\, (A \cup B)^u = A^u \cup B^u. \tag{u-Cup}$$

$$(\forall A, B \in \wp(S))\, (A \cap B)^u \subseteq A^u \cap B^u. \tag{u-Cap}$$

$$(\forall A \in \wp(S))\, A^{lc} \subseteq A^{cu}. \tag{Dual}$$

Proof.

l-Cup For any $A, B \in \wp S$, $x \in (A \cup B)^l$
$\Leftrightarrow (\exists y \in (A \cup B))\, x \in [y] \subseteq A \cup B$.
$\Leftrightarrow (\exists y \in A)\, x \in [y] \subseteq A \cup B$ or $(\exists y \in B)\, x \in [y] \subseteq A \cup B$.
$\Leftrightarrow (\exists y \in A)\, x \in [y] \subseteq A$ or $(\exists y \in A)\, x \in [y] \subseteq B$ or $(\exists y \in B)\, x \in [y] \subseteq A$ or $(\exists y \in B)\, x \in [y] \subseteq B$ - this is implied by $x \in A^l \cup B^l$.

l-Cap For any $A, B \in \wp S$, $x \in (A \cap B)^l$
$\Leftrightarrow x \in A \cap B$
$\Leftrightarrow (\exists y \in A \cap B)\, x \in [y] \subseteq A \cap B$ and $x \in A$, $x \in B$
$\Leftrightarrow (\exists y \in A)\, x \in [y] \subseteq A$ and $(\exists y \in B)\, x \in [y] \subseteq B$ - Clearly this statement implies $x \in A^l$ & $x \in B^l$, but the converse is not true in general.

u-Cup $x \in (A \cup B)^u$
$\Leftrightarrow x \in \bigcup_{[y] \cap (A \cup B) \neq \emptyset} [y]$
$\Leftrightarrow x \in \bigcup_{([y] \cap A) \cup ([y] \cap B) \neq \emptyset}$
$\Leftrightarrow x \in \bigcup_{[y] \cap A \neq \emptyset} [y]$ or $x \in \bigcup_{[y] \cap B \neq \emptyset} [y]$
$\Leftrightarrow x \in A^u \cup B^u$.

u-Cap By monotonicity, $(A \cap B) \subseteq A^u$ and $(A \cap B) \subseteq B^u$, so $(A \cap B)^u \subseteq A^u \cap B^u$.

Dual If $z \in A^{lc}$, then $z \in [x]^c$ for all $[x] \subseteq A$ and either, $z \in A \setminus A^l$ or $z \in A^c$. If $z \in A^c$ then $z \in A^{cu}$. If $z \in A \setminus A^l$ and $z \neq A^{cu \setminus A^c}$ then $[z] \cap A^c = \emptyset$. But this contradicts $z \notin A^{cu} \setminus A^c$. So $(\forall A \in \wp(S))\, A^{lc} \subseteq A^{cu}$. □

Theorem 6. *In a PRAX S, all of the following hold:*

$$(\forall A, B \in \wp(S))\, (A \cap B)^{l+} = A^{l+} \cap B^{l+}. \tag{5}$$

$$(\forall A, B \in \wp(S))\, A^{l+} \cup B^{l+} \subseteq (A \cup B)^{l+}. \tag{6}$$

$$(\forall A \in \wp(S))\, (A^{l+})^c = (A^c)^{u+} \;\&\; A^{l+} \subseteq A^{l_o}. \tag{7}$$

Proof.

1. $x \in (A \cap B)^{l+}$
$\Leftrightarrow [x] \subseteq A \cap B$
$\Leftrightarrow [x] \subseteq A$ and $[x] \subseteq B$
$\Leftrightarrow x \in xA^{l+}$ and $x \in B^{l+}$.
2. $x \in A^{l+} \cup B^{l+}$
$\Leftrightarrow [x] \subseteq A^{l+}$ or $[x] \subseteq B^{l+}$
$\Leftrightarrow [x] \subseteq A$ or $[x] \subseteq B$
$\Rightarrow [x] \subseteq A \cup B \Leftrightarrow x \in (A \cup B)^{l+}$.
3. $z \in A^{l+c}$
$\Leftrightarrow z \notin A^{l+}$
$\Leftrightarrow [z] \nsubseteq A$
$\Leftrightarrow z \cap A^c \neq \emptyset$ □

Theorem 7. *If $u+, l+$ are treated as self maps on the power-set $\wp(S)$, S being a PRAX or a PRAS then all of the following hold:*

- $(\forall x)\, x^{cl+} = x^{u+c},\ x^{cu+} = x^{l+c}$ *- that is $l+$ and $u+$ are mutually dual*
- *$l+$, $u+$ are monotone.*
- *$l+$ is a complete $\wedge$-morphism, while $u+$ is a $\vee$-morphism.*
- $\partial(x) = \partial(x^c)$*, where partial stands for the boundary operator.*
- *$\Im(u+)$ is an interior system while $\Im(l+)$ is a closure system.*
- *$\Im(u+)$ and $\Im(l+)$ are dually isomorphic lattices.*

Theorem 8.

$$\text{In a PRAX } S, (\forall A \subseteq S)\, A^{l+} \subseteq A^{l},\ \ A^{u+} \subseteq A^{u}.$$

Proof.

- If $x \in A^{l+}$, then $[x] \subseteq A$ and so $[x] \subseteq A^l$, $x \in A^l$.
- If $x \in A^l$, then $(\exists y \in A)[y] \subseteq A$, Rxy. But it is possible that $[x] \nsubseteq A$, therefore it is possible that $x \notin A^{l+}$ and $A^l \nsubseteq A^{l+}$.
- If $x \in A^{u+}$, then $[x] \cap A \neq \emptyset$, so $x \in A^u$.
- So $A^{u+} \subseteq A^u$.
- Note that $x \in A^u$, if and only if $(\exists z \in S)\, x \in [z]$, $[z] \cap A \neq \emptyset$, but this does not imply $x \in A^{u+}$. □

Theorem 9. *In a PRAX S, all of the following hold:*

$$(\forall A \in \wp(S))\, A^{l+} \subseteq A^{l_o}. \tag{8}$$

$$(\forall A \in \wp(S))\, A^{u_o} \subseteq A^{u+}. \tag{9}$$

$$(\forall A \in \wp(S))\, A^{lc} \subseteq A^{cu}. \tag{10}$$

Proof.

1. • If $x \in A^{l+}$, then $[x] \subseteq A$.
   • But as $[x]_o \subseteq [x]$, $A^{l+} \subseteq A^{l_o}$.
2. This follows easily from definitions.
3. • If $z \in A^{lc}$, then $z \in [x]^c$ for all $[x] \subseteq A$ and either, $z \in A \setminus A^l$ or $z \in A^c$.
   • If $z \in A^c$ then $z \in A^{cu}$.
   • If $z \in A \setminus A^l$ and $z \neq A^{cu \setminus A^c}$ then $[z] \cap A^c = \emptyset$.
   • But this contradicts $z \notin A^{cu} \setminus A^c$.
   • So $(\forall A \in \wp(S))\, A^{lc} \subseteq A^{cu}$. □

From the above, the relation between approximations in general is as follows ($A^{u+} \longrightarrow A^u$ should be read as A^{u+} *is included in* A^u) (Fig. 1):

If a relation R is purely reflexive and not proto-transitive on a set S, then the relation $\tau(R) = R \cap R^{-1}$ will not be an equivalence and for a $A \subset S$, it is possible that $A^{u_o l} \subseteq A$ or $A^{u_o l} \parallel A$ or $A \subseteq A^{u_o l}$.

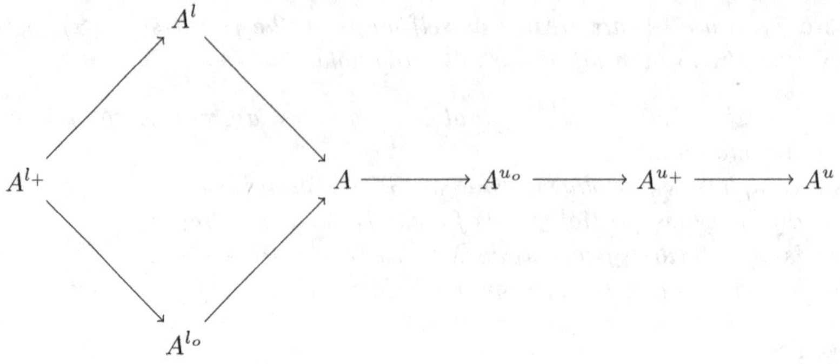

Fig. 1. Relationship between approximations

3 Motivation and Examples

Generalized transitive relations occur frequently in general information systems, but are often not recognized as such and there is hope for improved semantics and KI relative the situation for purely reflexive relation based RST. Not all of the definable approximations have been investigated in even closely related structures of general RST. Contamination-free semantics [24] for the contexts are also not known. Finally these relate to RYS and variants. A proper characterization of roughly equal (requal) objects is also motivated by [24].

3.1 Abstract Example

Let $\S = \{a, b, c, e, f, g, h, l, n\}$ and let R be a binary relation on it defined via

$$R = \{(a,\ a),\ (l,\ l),\ (n,\ n),\ (n,\ h),\ (h,\ n),\ (l,\ n),\ (g,\ c),\ (c,\ g)\\ (g,\ l),\ (b,\ g),\ (g,\ b),\ (h,\ g),\ (a,\ b),\ (b,\ c),\ (h,\ a),\ (a,\ c)\}.$$

Then $\langle S,\ R\rangle$ is a PRAS.

If P is the reflexive closure of R (that is $P = R \cup \Delta_S$), then $\langle S,\ P\rangle$ is a PRAX. The successor neighborhoods associated with different elements of S are as follows (**E** is a variable taking values in S) (Table 1):

Table 1. Successor neighborhoods

E	a	b	c	g	e	f	h	l	n
$[E]$	$\{a, h\}$	$\{b, c, g\}$	$\{b, c, g\}$	$\{b, c, g, h\}$	$\{e\}$	$\{f\}$	$\{h, n\}$	$\{l, g\}$	$\{n, l, g, h\}$
$[E]_o$	$\{a\}$	$\{b, c, g\}$	$\{b, c, g\}$	$\{b, c, g\}$	$\{e\}$	$\{f\}$	$\{h, n\}$	$\{l\}$	$\{n, h\}$

$$\text{If } A = \{a, h, f\},$$
$$\text{then } A^l = \{a, h, f\},$$
$$A^{l_o} = \{a, f\} \text{ and } A^{l_o} \subset A^l.$$
$$\text{If } F = \{l\},$$
$$\text{then } F^l = \emptyset,\ F^{l_o} = F$$
$$\text{and } F^l \subset F^{l_o}.$$

Now let $Z = N \cup S \cup X$, where N is the set of naturals, X is the set of elements of the infinite sequences $\{x_i\}$, $\{y_j\}$. Let Q be a relation on Z such that

$$Q \cap S^2 = P, \tag{11}$$
$$Q \cap N^2 \text{ is some equivalence,} \tag{12}$$
$$(\forall i \in N)(i, x_{3i+1}), (x_{2i}, i), (x_i, x_{i+1}), (y_i, y_{i+1}) \in Q. \tag{13}$$

Q is then a proto-transitive relation. For any $i \in N$, let $P_i = \{y_k : k \neq 2j \ \&\ k < i\} \cup \{x_{2j} : 2j < i\}$ - this will be used in later sections. The extension of the example to involve nets and densely ordered subsets is standard.

3.2 Caste Hierarchies and Interaction

The caste system and religion are among the deep-seated evils of Indian society that often cut across socio-economic classes and level of education. For the formulation of strategies aimed at large groups of people towards the elimination of such evils it would be fruitful to study interaction of people belonging to different castes and religions on different social fronts.

Most of these castes would have multiple sub caste hierarchies in addition. Social interactions are necessarily constrained by their type and untouchability perception. If x, y are two castes, then with respect to a possible social interaction α, people belonging to x will either regard people belonging to y as untouchable or otherwise. As the universality is so total, it is possible to write $\mathbb{U}_\alpha xy$ to mean that y is untouchable for x for the interaction α. Usually this is a asymmetric relation and y would be perceived as a *lower caste* by members of x and many others.

Other predicates will of course be involved in deciding on the possibility of the social interaction, but if $\mathbb{U}_\alpha xy$ then the interaction is forbidden relative x. If α is "context of possible marriage", then the complementary relation ($\mathbb{C}_\alpha$ say) is a reflexive proto-transitive relation. For various other modes of interaction similar relations may be found.

In devising remedial educational programmes targeted at mixed groups, it would be important to understand approximate perceptions of the group and the semantics of PRAX would be very relevant.

3.3 Compatibility Prediction Models

For predicting the compatibility among individuals or objects, then the following model can be used. Specific examples include situations involving data from dating sites like OK-Cupid.

Let one woman be defined by a sequence of sets of features $a_1, \ldots, a_n$ at different temporal instants and another woman by $b_1, \ldots, b_n$. Let $\omega(a_i, b_i)$ be the set of features that are desired by a_i, but missing in b_i. Let ρ be an equivalence relation on a subset K of S – the set of all features, that determines the classical rough approximations l_ρ, u_ρ on $\wp(K)$.

Let $(a, b) \in R$ if and only if $(\omega(a_n, b_n))^{l_\rho}$ is *small* (for example, that can mean being an atom of $\wp(K)$). The predicate R is intended to convey *may like to be related.* In dating sites, this is understood in terms of profile matches: if a woman's profile matches another woman's and conversely and similarly with another woman's, then the other two woman are assumed to be mutually compatible.

Proposition 9. *R is a proto-transitive relation and $\langle \underline{S}, R\rangle$ is a PRAS.*

Proof. Obviously R need not be reflexive or symmetric in general.

If $(a, b), (b, c), (b, a), (c, b) \in R$, then $(a, c), (c, a) \in R$ is a reasonable rule. □

This is a concrete example of a PRAS that is suggestive of many more practical contexts.

3.4 Indeterminate Information System Perspective

It is easy to derive PRAX from population census, medical, gender studies and other databases and these correspond to information systems. These connections are made clearer through this example.

If the problem is to classify a specific population O, for a purpose based on scientific data on sex, gender continuum, sexual orientation and other factors, then it is very likely that the data base is an indeterminate information system of the form $\mathcal{I} = \langle O, At, \{V_a : a \in At\}, \{\varphi_a : a \in At\}\rangle$, where At is a set of attributes, V_a a set of possible values corresponding to the attribute a and $\varphi_a : O \longmapsto \wp(V_a)$ the valuation function. Sex is determined by many attributes corresponding to hormones, brain structure, karyotypes, brain configuration, anatomy, clinical sex etc. Hormonal data can be in the form of free/bound values of over six hormones, the values of which vary widely over populations. The focus can be on a subset of attributes for which the inclusion/ordering of values (corresponding to any one of the attributes in the subset) of an object in another is relevant. For example, interest in patterns in sexual compatibility/relationships may be corresponded to such subsets. This relation is proto-transitive. Formally for a $B \subseteq At$, if we let $(x, y) \in \rho_B$ if and only if $(\exists a \in B)\varphi_a x \subseteq \varphi_a y$, then ρ_B is often proto-transitive via another predicate on B.

4 Algebras of Rough Definite Elements

In this section key results are proved on the fine structure of definite elements.

Theorem 10. *On the set of proto definite elements $\delta_{lu}(S)$ of a PRAX S, all of the following can be defined:*

$$x \wedge y \triangleq x \cap y. \tag{14}$$

$$x \vee y \triangleq x \cup y. \tag{15}$$

$$0 \triangleq \emptyset. \tag{16}$$

$$1 \triangleq S. \tag{17}$$

$$x^c \triangleq S \setminus x. \tag{18}$$

Proof. It is required to show that the operations are well defined. Suppose x, y are proto-definite elements, then

1.
$$(x \cap y)^u \subseteq x^u \cap y^u = x \cap y.$$
$$(x \cap y)^l = (x^u \cap y^u)^l = (x \cap y)^{ul} = (x \cap y)^u = x \cap y.$$
Since $a^{ul} = a^u$ for any a.
2.
$$(x \cup y)^u = x \cup y = x^l \cup y^l \subseteq (x \cup y)^l.$$
3. $0 \triangleq \emptyset$ is obviously well defined.
4. Obvious.
5. Suppose $A \in \delta_{lu}(S)$, then $(\forall z \in A^c)\, [z] \cap A = \emptyset$ is essential, else $[z]$ would be in A^u. This means $[z] \subseteq A^c$ and so $A^c = A^{cl}$. If there exists a $a \in A$ such that $[a] \cap A^c \neq \emptyset$, then $[a] \subseteq A^u = A$. So $A^c \in \delta_{lu}(S)$. □

Theorem 11. *The algebra $\delta_{proto}(S) = \langle \delta_{lu}(S), \vee, \wedge, c, 0, 1\rangle$ is a Boolean lattice.*

Proof. Follows from the previous theorem. The lattice order can be defined via, $x \leq y$ if and only if $x \cup y = y$ and $x \cap y = x$.

5 The Representation of Roughly Equal Elements

The representation of roughly equal elements in terms of definite elements are well known in case of classical rough set theory. In case of more general spaces including tolerance spaces [24], most authors have been concerned with describing the interaction of rough approximations of different types and not of the interaction of roughly equal objects. Higher order approaches, developed by the present author as in [22] for bitten approximation spaces, permit constructs over sets of roughly equal objects. In the light of the contamination problem [22,24], it would be an improvement to describe without higher order constructs. In this section a new method of representing roughly equal elements based on expanding concepts of definite elements is developed.

Definition 10. *On $\wp(S)$, the following relations can be defined:*

$$A \preceq B \text{ if and only if } A^l \subseteq B^l \;\&\; A^u \subseteq B^u. \qquad \text{(Rough Inclusion)}$$
$$A \approx B \text{ if and only if } A \preceq B \;\&\; B \preceq A. \qquad \text{(Rough Equality)}$$

Proposition 10. *The relation $\preceq$ defined on $\wp(S)$ is a bounded partial order and $\approx$ is an equivalence. The quotient $\wp(S)| \approx$ will be said to be the set of* roughly equivalent objects.

Definition 11. *A subset A of $\wp(S)$ will be said to a set of* roughly equal *elements if and only if*

$$(\forall x, y \in A)\, x^l = y^l \;\&\; x^u = y^u.$$

It will be said to be full *if no other subset properly including A has the property.*

Relative the situation for a general RYS, the following result has already been proved.

Theorem 12 (Meta-Theorem). *In a PRAX S, full set of roughly equal elements is necessarily a union of intervals in $\wp(S)$.*

Definition 12. *A non-empty set of non singleton subsets $\alpha = \{x : x \subseteq \wp(S)\}$ will be said to be a* upper broom *if and only if all of the following hold:*

$$(\forall x, y \in \alpha)\, x^u = y^u.$$
$$(\forall x, y \in \alpha)\, x \parallel y.$$

If $\alpha \subset \beta$, then β fails to satisfy at least one of the above two conditions.

The set of upper brooms of S will be denoted by $\pitchfork(S)$.

Definition 13. *A non-empty set of non singleton subsets $\alpha = \{x : x \subseteq \wp(S)\}$ will be said to be a* lower broom *if and only if all of the following hold:*

$$(\forall x, y \in \alpha)\, x^l = y^l \neq x. \tag{19}$$
$$(\forall x, y \in \alpha)\, x \parallel y. \tag{20}$$
$$\text{If } \beta \subset \alpha \;\&\; Card(\beta) \geq 2, \text{ then } \beta \text{ fails to satisfy condition (1) or (2).} \tag{21}$$

The set of lower brooms of S will be denoted by $\psi(S)$.

Proposition 11. *If $x \in \delta_{lu}(S)$ then $\{x\} \notin \pitchfork(S)$ and $\{x\} \notin \psi(S)$.*

In the next definition, the concept of union of intervals in a partially ordered set is modified in a way for use with specific types of objects.

Definition 14. *By a* bruinval, *will be meant a subset of $\wp(S)$ of one of the following forms:*

- *Bruinval-0: Intervals of the form* (x, y), $[x, y)$, $[x, x]$, $(x, y]$ *for* $x, y \in \wp(S)$.
- *Open Bruinvals: Sets of the form* $[x, \alpha) = \{z : x \leq z < b \;\&\; b \in \alpha\}$, $(x, \alpha] = \{z : x < z \leq b \;\&\; b \in \alpha\}$ *and* $(x, \alpha) = \{z : x < z < b, b \in \alpha\}$ *for* $\alpha \in \wp(\wp(S))$.
- *Closed Bruinvals: Sets of the form* $[x, \alpha] = \{z : x \leq z \leq b \;\&\; b \in \alpha\}$ *for* $\alpha \in \wp(\wp(S))$.
- *Closed Set Bruinvals: Sets of the form* $[\alpha, \beta] = \{z : x \leq z \leq y \;\&\; x \in \alpha \;\&\; y \in \beta\}$ *for* $\alpha, \beta \in \wp(\wp(S))$
- *Open Set Bruinvals: Sets of the form* $(\alpha, \beta) = \{z : x < z < y, x \in \alpha \;\&\; y \in \beta\}$ *for* $\alpha, \beta \in \wp(\wp(S))$.
- *Semi-Closed Set Bruinvals: Sets of the form* $[[\alpha, \beta]]$ *defined as follows:* $\alpha = \alpha_1 \cup \alpha_2$, $\beta = \beta_1 \cup \beta_2$ *and* $[[\alpha, \beta]] = (\alpha_1, \beta_1) \cup [\alpha_2, \beta_2] \cup (\alpha_1, \beta_2] \cup [\alpha_2, \beta_1)$ *for* $\alpha, \beta \in \wp(\wp(S))$.

In the example of the second section, the representation of the rough object (P_i^l, P_i^u) requires set bruinvals.

Proposition 12. *If* S *is a PRAX, then a set of the form* $[x, y]$ *with* $x, y \in \delta_{lu}(S)$ *will be a set of roughly equal subsets of* S *if and only if* $x = y$.

Proposition 13. *A bruinval-0 of the form* (x, y) *is a full set of roughly equal elements if*

- $x, y \in \delta_{lu}(S)$,
- x *is covered by* y *in the order on* $\delta_{lu}(S)$.

Proposition 14. *If* $x, y \in \delta_{lu}(S)$ *then sets of the form* $[x, y)$, $(x, y]$ *cannot be a non-empty set of roughly equal elements, while those of the form* $[x, y]$ *can be if and only if* $x = y$.

Proposition 15. *A bruinval-0 of the form* $[x, y)$ *is a full set of roughly equal elements if*

- $x^l, y^u \in \delta_{lu}(S)$, $x^l = y^l$ *and* $x^u = y^u$,
- x^l *is covered by* y^u *in* $\delta_{lu}(S)$ *and*
- $x \setminus (x^l)$ *and* $y^u \setminus y$ *are singletons.*

Remark 1. In the above proposition the condition $x^l, y^u \in \delta_{lu}(S)$, is not necessary.

Theorem 13. *If a bruinval-0 of the form* $[x, y]$ *satisfies*

$$x^l = y^l = x \;\&\; x^u = y^u.$$
$$Card(y^u \setminus y) = 1.$$

then $[x, y]$ *is a full set of roughly equal objects.*

Proof. Under the conditions, if $[x, y]$ is not a full set of roughly equal objects, then there must exist at least one set h such that $h^l = x$ and $h^u = y^u$ and $h \notin [x, y]$. But this contradicts the order constraint $x^l \leq h\, y^u$. Note that $y^u \notin [x, y]$ under the conditions. □

Theorem 14. *If a bruinval-0 of the form $(x, y]$ satisfies*

$$x^l = y^l = x \;\&\; (\forall z \in (x, y])\, z^u = y^u,$$
$$Card(y^u \setminus y) = 1.$$

then $(x, y]$ is a full set of roughly equal objects, that does not intersect the full set $[x, x^u]$.

Proof. By monotonicity it follows that $(x, y]$ is a full set of roughly equal objects. then there must exist at least one set h such that $h^l = x$ and $h^u = y^u$ and $h \notin [x, y]$. But this contradicts the order constraint $x^l \leq h\, y^u$. Note that $y^u \notin [x, y]$ under the conditions. □

Theorem 15. *A bruinval-0 of the form (x^l, x^u) is not always a set of roughly equal elements, but will be so when $x^{uu} = x^u$. In the latter situation it will be full if $[x^l, x^u)$ is not full.*

The above theorems essentially show that the description of rough objects depends on too many types of sets and the order as well. Most of the considerations extend to other types of bruinvals as is shown below and remain amenable.

Theorem 16. *An open bruinval of the form (x, α) is a full set of roughly equal elements if and only if*

$$\alpha \in \pitchfork (S).$$
$$(\forall y \in \alpha)\, x^l = y^l, x^u = y^u$$
$$(\forall z)(x^l \subseteq z \subset x \longrightarrow z^u \subset x^u).$$

Proof. It is clear that for any $y \in \alpha$, (x, y) is a convex interval and all elements in it have same upper and lower approximations. The third condition ensures that $[z, \alpha)$ is not a full set for any $z \in [x^l, x)$. □

Definition 15. *An element $x \in \wp(S)$ will be said to be a* weak upper critical element relative $z \subset x$ *if and only if $(\forall y \in \wp(S))\, (z = y^l \;\&\; x \subset y \longrightarrow x^u \subset y^u)$.*

An element $x \in \wp(S)$ will be said to be an upper critical element relative $z \subset x$ *if and only if $(\forall v, y \in \wp(S))\, (z = y^l = v^l \;\&\; v \subset x \subset y \longrightarrow v^u = x^u \subset y^u)$. Note that the inclusion is strict.*

An element a will be said to be bi-critical relative b *if and only if $(\forall x, y \in \wp(S))(a \subset x \subseteq y \subset b \longrightarrow x^u = y^u \;\&\; x^l = y^l \;\&\; x^u \subset b^u \;\&\; a^l \subset x^l)$.*

If x is an upper critical point relative z, then $[z, x)$ or (z, x) is a set of roughly equivalent elements.

Definition 16. *An element $x \in \wp(S)$ will be said to be an* weak lower critical element relative $z \supset x$ *if and only if $(\forall y \in \wp(S))\, (z = y^u \;\&\; y \subset x \longrightarrow y^l \subset x^l)$.*

An element $x \in \wp(S)$ will be said to be an lower critical element relative $z \supset x$ *if and only if $(\forall y, v \in \wp(S))\, (z = y^u = v^u \;\&\; y \subset x \subset v \longrightarrow y^l \subset x^l = v^l)$.*

An element $x \in \wp(S)$ will be said to be an lower critical element *if and only if* $(\forall y \in \wp(S))\,(y \subset x \longrightarrow y^l \subset x^l)$ *An element that is both lower and upper critical will be said to be* critical. *The set of upper critical, lower critical and critical elements respectively will be denoted by $UC(S)$, $LC(S)$ and $CR(S)$.*

Proposition 16. *In a PRAX, every upper definite subset is also upper critical, but the converse need not hold.*

The most important thing about the different lower and upper critical points is that they help in determining full sets of roughly equal elements by determining the boundaries of intervals in bruinvals of different types.

5.1 Types of Associated Sets

Because of reflexivity it might appear that lower approximations in PRAX and classical RST are too similar at least in the perspective of lower definite objects. It is necessary to classify subsets of a PRAX S, to see the differences relative the behavior of lower approximations in classical RST. All this will be used in some of the semantics as well.

Definition 17. *For each element $x \in \wp(S)$ the following sets can be associated:*

$$\begin{aligned}
F_0(x) &= \{y : (\exists a \in x^c)\, Rya \;\&\; y \in x\} && \text{(Forward Looking)}\\
F_1(x) &= \{y : (\exists a \in x^c)\, Rya \;\&\; Rzy \;\&\; z \in x\} && \text{(1-Forward Looking)}\\
\pi_0(x) &= \{y : y \in x \;\&\; (\exists a \in x^c)\, Ray\} && \text{(Progressive)}\\
St(x) &= \{y : [y] \subseteq x \;\&\; \neg(y \in F_0(x))\} && \text{(Stable)}\\
Sym(x) &= \{y : y \in x \;\&\; (\forall z \in x)(Ryz \leftrightarrow Rzy)\} && \text{(Relsym)}
\end{aligned}$$

Forward looking set associated with a set x includes those elements not in x whose successor neighborhoods intersect x. Elements of the set may be said to be relatively forward looking. *Progressive* set of x includes those elements of x whose successor neighborhoods are not included in x. It is obvious that progressive elements are all elements of $x \setminus x^l$. Stable elements are those that are strongly within x and are not directly reachable in any sense from outside. $Sym(x)$ includes those elements in x which are symmetrically related to all other elements within x.

Even though all these are important these cannot be easily represented in the rough domain. Their approximations have the following properties:

Proposition 17. *In the above context, all of the following hold:*

$$\begin{gathered}
(\pi_0(x))^l = \emptyset \;\&\; (\pi_0(x))^u \subseteq x^u \setminus x^l\\
(F_0(x))^u \subseteq x^u\\
St(x)^l \subseteq x^l \;\&\; F_0(x) = \emptyset \longrightarrow St(x) = x^{l+}\\
Sym(x)^u \subseteq x^u \;\&\; (Sym(x))^l \subseteq x^l.
\end{gathered}$$

Proof. Proof is fairly direct. □

6 More Results on Representation of Rough Objects

It has already been shown in the previous section that the representation of rough objects by definite objects is not possible in a PRAX. So it is important to look at possibilities based on other types of derived approximations. This problem is solved right up to representation theorems for the derived operators in this section.

Definition 18. *If $x \in \wp(S)$, then*

- *Let $\Pi^o_{\heartsuit}(x) = \{y\,;\, x \subseteq y \;\&\; x^l = y^l \;\&\; y^u \subseteq x^{uu}\}$.*
- *Form the set of maximal elements $\Pi_{\heartsuit}(x)$ of $\Pi^o_{\heartsuit}(x)$ with respect to the inclusion order.*
- *Select a unique element $\chi(\Pi_{\heartsuit}(x))$ through a fixed choice function χ.*
- *Form $(\chi(\Pi_{\heartsuit}(x)))^u$.*
- *$x^{\heartsuit\chi} = (\chi(\Pi_{\heartsuit}(x)))^u$ will be said to be the* almost upper approximation *of x relative χ.*
- *$x^{\heartsuit\chi}$ will be abbreviated by $x^{\heartsuit}$ when χ is fixed.*

The choice function will be said to be regular *if and only if $(\forall x, y)\,(x \subseteq y \;\&\; x^l = y^l \longrightarrow \chi(\Pi_{\heartsuit}(x)) = \chi(\Pi_{\heartsuit}(y)))$. Regularity will be assumed unless specified otherwise in what follows.*

Definition 19. *If $x \in \wp(S)$, then*

- *Let $\Pi^o_{\diamondsuit}(x) = \{y\,;\, x \subseteq y \;\&\; x^l = y^l\}$.*
- *Form the set of maximal elements $\Pi_{\diamondsuit}(x)$ of $\Pi^o_{\diamondsuit}(x)$ with respect to the inclusion order.*
- *Select a unique element $\chi(\Pi_{\diamondsuit}(x))$ through a fixed choice function χ.*
- *$x^{\diamondsuit\chi} = \chi(\Pi_{\diamondsuit}(x))$ will be said to be the* lower limiter *of x relative χ.*
- *$x^{\diamondsuit\chi}$ will be abbreviated by $x^{\diamondsuit}$ when χ is fixed.*

Definition 20. *If $x \in \wp(S)$, then*

- *Let $\Pi^o_{\flat}(x) = \{y\,;\, y \subseteq x \;\&\; x^u = y^u\}$.*
- *Form the set of maximal elements $\Pi_{\flat}(x)$ of $\Pi^o_{\flat}(x)$ with respect to the inclusion order.*
- *Select a unique element $\xi(\Pi_{\flat}(x))$ through a fixed choice function ξ.*
- *$x^{\flat\xi} = \xi(\Pi_{\flat}(x))$ will be said to be the* upper limiter *of x relative χ.*
- *$x^{\flat\xi}$ will be abbreviated by $x^{\flat}$ when ξ is fixed.*

Proposition 18. *In the context of the above definition, the almost upper approximation satisfies all of the following:*

$$(\forall x)\, x \subseteq x^{\heartsuit} \tag{Inclusion}$$

$$(\forall x)\, x^{\heartsuit} \subseteq x^{\heartsuit\heartsuit} \tag{Non-Idempotence}$$

$$(\forall x\, y)\,(x \subseteq y \subseteq x^{\heartsuit} \longrightarrow x^{\heartsuit} \subseteq y^{\heartsuit}) \tag{Cautious Monotony}$$

$$(\forall x)\, x^u \subseteq x^{\heartsuit} \tag{Supra Pseudo Classicality}$$

$$S^{\heartsuit} = S \tag{Top.}$$

Proof.

- Inclusion: Follows from the construction. If one element granules or successor neighborhoods are included in x, then these must be in the lower approximation. If a granule y is not included in x, but intersects it in f, then it is possible to include f in each of $\Pi_{\heartsuit}(x)$. So inclusion follows.
- Non-Idempotence: The reverse inclusion does not happen as $x^u \subseteq x^{uu}$.
- Cautious monotony: It is clear that monotony can fail in general because of the choice aspect, but if $x \subseteq y \subseteq x^{\heartsuit}$ holds, then $x^l \subseteq y^l$ and $y^{\heartsuit}$ has to be equal to $x^{\heartsuit}$ or include more granules because of regularity of the choice function.
- Supra Pseudo Classicality: The adjective *pseudo* is used because u is not a classical consequence operator. In the construction of $x^{\heartsuit}$, the selection is from super-sets of x^l that can generate maximal upper approximations. Upper approximation of the selected sets are done next. So that includes x^u in general. □

The conditions have been named in relation to the standard terminology used in non-monotonic reasoning. The upper approximation operator u is similar to classical consequence operator, but lacks idempotence. So the fourth property has been termed as *supra pseudo classicality* as opposed to *supra classicality.* This means the present domain of reasoning is more general than that of [18].

Theorem 17. *In the context of 18, the following additional properties hold:*

$$(\forall x)\, x^{\heartsuit} \subseteq x^{u\heartsuit} \quad \text{(Sub Left Absorption)}$$

$$(\forall x)\, x^{\heartsuit} \subseteq x^{\heartsuit u} \quad \text{(Sub Right Absorption)}$$

$$\boxdot(\forall x, y)\, (x^u = y^u \nrightarrow x^{\heartsuit} = y^{\heartsuit}) \quad \text{(No Left Logical Equivalence)}$$

$$\boxdot(\forall x, y)\, (x^{\heartsuit} = y^{\heartsuit} \nrightarrow x^l = y^l) \quad \text{(No Jump Equivalence)}$$

$$\boxdot(\forall x, y, z)\, (x \subseteq y^{\heartsuit} \;\&\; z \subseteq x^u \nrightarrow z \subseteq y^{\heartsuit}) \quad \text{(No Weakening)}$$

$$\boxdot(\forall x, y)\, (x \subseteq y \subseteq x^u \nrightarrow x^{\heartsuit} = y^{\heartsuit}) \quad \text{(No subclassical cumulativity)}$$

$$(\forall x, y)\, x^{\heartsuit} \cap y^{\heartsuit} \subseteq (x^u \cap y^u)^{\heartsuit} \quad \text{(Distributivity)}$$

$$(\forall x, y, z)\, (x \cup z)^{\heartsuit} \cap (y \cup z)^{\heartsuit} \subseteq (z \cup (x^u \cap y^u))^{\heartsuit} \quad \text{(Weak Distributivity)}$$

$$(\forall x, y, z)\, (x \cup y)^{\heartsuit} \cap (x \cup z)^{\heartsuit} \subseteq (x \cup (y \oplus z))^{\heartsuit}$$

(Disjunction in Antecedent)

$$(\forall x, y)\, (x \cup y)^{\heartsuit} \cap (x \cup y^c)^{\heartsuit} \subseteq x^{\heartsuit} \quad \text{(Proof by Cases)}$$

$$\text{If } y \subseteq (x \cup z)^{\heartsuit}, \text{ then } x \implies y \subseteq z^{\heartsuit} \quad \text{(Conditionalization.)}$$

Proof. **Sub Left Absorption.** For any x, $x^{\heartsuit}$ is the upper approximation of a maximal subset y containing x such that $x^l = y^l$ and $x^{u\heartsuit}$ is the upper approximation of a maximal subset z containing x^u such that $x^{ul} = x^u = z^l$. Since, $x^l \subseteq x^{ul}$ and $x \subseteq x^u$, so $x^{\heartsuit} \subseteq x^{u\heartsuit}$ follows.

Sub Right Absorption. Follows from the properties of u.

No Left Logical Equivalence. Two subsets x, y can have unequal lower approximations and equal upper approximations and so the implication does not hold in general. $\boxdot$ should be treated as an abbreviation for *in general.*

No Jump Equivalence. The reason is similar to that of the previous negative result.

No Weakening. In general if $x \subseteq y^{\heartsuit}$ & $z \subseteq x^u$, then it is possible that $x^u \subseteq y^{\heartsuit}$ or $y^{\heartsuit} \subseteq x^u$. So one cannot be sure about $z \subseteq y^{\heartsuit}$.

No Sub Classical Cumulativity. If $x \subseteq y \subseteq x^u$, then $x^l \subseteq y^l$ in general and so elements of $\Pi_{\heartsuit}(x)$ may be included in $\Pi_{\heartsuit}(y)$, the two may be unequal and it may not be possible to use a uniform choice function on them. So it need not happen that $x^{\heartsuit} = y^{\heartsuit}$.

Distributivity. If $z \in x^{\heartsuit} \cap y^{\heartsuit}$, then $z \in (\chi(\Pi_{\heartsuit}(x)))^u$ and $z \in (\chi(\Pi_{\heartsuit}(y)))^u$. So if $z \in x^l$ and $z \in y^l$, then $z \in (x^u \cap y^u)^{\heartsuit}$. Since in general, $(a \cap b)^u \subseteq a^u \cap b^u$ and $(a^u \cap b^u)^l = (a^u \cap b^u)$, the required inclusion follows.

$$(\forall x, y, z)\,(x \cup z)^{\heartsuit} \cap (y \cup z)^{\heartsuit} \subseteq (z \cup (x^u \cap y^u))^{\heartsuit} \quad \text{(Weak Distributivity)}$$

$$(\forall x, y, z)\,(x \cup y)^{\heartsuit} \cap (x \cup z)^{\heartsuit} \subseteq (x \cup (y \oplus z))^{\heartsuit}$$

(Disjunction in Antecedent)

$$(\forall x, y)\,(x \cup y)^{\heartsuit} \cap (x \cup y^c)^{\heartsuit} \subseteq x^{\heartsuit} \quad \text{(Proof by Cases)}$$

$$\text{If } y \subseteq (x \cup z)^{\heartsuit}, \text{ then } x \implies y \subseteq z^{\heartsuit} \quad \text{(Conditionalization.)}$$

$\square$

Proposition 19.

$$(\forall x, y)(x^{\diamond} = y^{\diamond} \longrightarrow x^l = y^{l};)$$
$$(\forall x, y)(x^{\flat} = y^{\flat} \longrightarrow x^u = y^u.)$$

Discussion: In non monotonic reasoning, if C is any consequence operator: $\wp(S) \longmapsto \wp(S)$, then the following named properties of crucial importance in semantics (in whatever sense, [18,19]):

$$A \subseteq B \subseteq C(A) \longrightarrow C(B) \subseteq C(A) \quad \text{(Cut)}$$
$$A \subseteq B \subseteq C(A) \longrightarrow C(B) = C(A) \quad \text{(Cumulativity)}$$
$$x \subseteq y \subseteq x^u \longrightarrow x^{\heartsuit} = y^{\heartsuit} \quad \text{(subclassical subcumulativity)}$$

Proposition 20. *In the context of the above definition, the lower limiter satisfies all of the following:*

$$(\forall x)\, x \subseteq x^{\diamondsuit} \quad \text{(Inclusion)}$$
$$(\forall x)\, x^{\diamondsuit\diamondsuit} = x^{\diamondsuit} \quad \text{(Idempotence)}$$
$$(\forall x\, y)\, (x \subseteq y \subseteq x^{\diamondsuit} \longrightarrow x^{\diamondsuit} = y^{\diamondsuit}) \quad \text{(Cumulativity)}$$
$$(\forall x)\, x^{u} \subseteq x^{\diamondsuit} \quad \text{(Upper Inclusion)}$$
$$S^{\diamondsuit} = S \quad \text{(Top)}$$
$$(\forall x\, y)\, (x \subseteq y \subseteq x^{\diamondsuit} \longrightarrow x^{\diamondsuit} = y^{\diamondsuit}) \quad \text{(Cumulativity)}$$

The above proposition means that the upper limiter corresponds to ways of reasoning in a stable way in the sense that the aggregation of conclusions does not affect inferential power or cut-like amplification.

A limited concrete representation theorem for operators like $\heartsuit$ in special cases and $\diamondsuit$ is proved next. The representation theorem is valid for similar operators in non-monotonic reasoning. The representation theorem permits us to identify cover based formulations of PRAX.

Definition 21. *A collection of sets $\mathcal{S}$ will be said to be a* closure system *of a type as per the following conditions:*

$$(\forall \mathcal{H} \subseteq \mathcal{S})\, \cap \mathcal{H} \in \mathcal{S}. \quad \text{(Closure System)}$$
$$(\forall \mathcal{H} \subseteq \mathcal{S})\, (\cap \mathcal{H})^{u} \in \mathcal{S}. \quad \text{(U-Closure System)}$$
$$(\forall \mathcal{H} \subseteq \mathcal{S})\, (\cap \mathcal{H})^{l} \in \mathcal{S}. \quad \text{(L-Closure System)}$$
$$(\forall \mathcal{H} \subseteq \mathcal{S})\, (\cap \mathcal{H})^{l}, (\cap \mathcal{H})^{u} \in \mathcal{S}. \quad \text{(LU-Closure System)}$$
$$(\exists 0, \top \in \mathcal{S})(\forall X \in \mathcal{S})\, 0 \subseteq X \subseteq \top. \quad \text{(Bounded)}$$

Proposition 21. *In a PRAX S, the set $\mathcal{U}(S) = \{x^{u};\, x \in \wp(S)\}$ is not a bounded U-closure system.*

Proposition 22.

$$(\forall x)\, x^{\heartsuit u} \subseteq x^{u \heartsuit}.$$

Proof. Because $x^{l} \subseteq x^{u}$, an evaluation of possible granules involved in the construction of $x^{\heartsuit u}$ and $x^{u\heartsuit}$ proves the result. □

Theorem 18. *In a PRAX S, the set $\heartsuit(S) = \{x^{\heartsuit};\, x \in \wp(S)\}$ is a bounded LU-closure system if the choice operation is regular.*

Proof.

- $x^{\heartsuit}$ is the upper approximation of a specific y containing x that is maximal subject to $x^{l} = y^{l}$.

- $x^{\heartsuit u}$ is the upper approximation of the upper approximation of a specific y containing x that is maximal subject to $x^l = y^l$ and its upper approximation.
- Clearly,

$$(\chi(\Pi_{\heartsuit}(x)) \cap \chi(\Pi_{\heartsuit}(y)))^u \subseteq (\chi(\Pi_{\heartsuit}(x)))^u \cap (\chi(\Pi_{\heartsuit}(y)))^u.$$

- The expression on the right of the inclusion is obviously a union of granules in the PRAX.
- From a constructive bottom-up perspective, let $p_1, p_2, \ldots p_s$ be a collection of subsets of $x \setminus x^l$ such that

$$\cup p_i \subseteq x \setminus x^l$$
$$(\exists z)\, p_i^u = [z]$$
$$\cup_{i \neq j}(p_i \cap p_j) \text{ is minimal on all such collections.}$$

- Now add subsets $k(p_i)$ of $x^{uu} \setminus x^c$ to x to form the required maximal subset.
- For the lower approximation part, it suffices to use the preservation of l by $\cap$. □

Proposition 23. *For each $x \in \wp(S)$ let $x^{\curlyvee} = (x^{\heartsuit})^u$, then the following properties hold:*

$$(\forall x)\, x \subseteq x^{\curlyvee}.$$
$$(\forall x)\, x^{\curlyvee\curlyvee} = x^{\curlyvee}.$$

Proof. $x^{\heartsuit u \heartsuit u} = x^{\heartsuit u}$. Because if a part of a class that retains the equality of lower approximations could be added, then that should be adjoinable in the construction of $x^{\heartsuit}$ as well. □

The following limited representation theorem can be useful for connections with covers.

Definition 22. *Let X be a set and $C : \wp(X) \longmapsto \wp(X)$ a map satisfying all the following conditions:*

$$(\forall A \in \wp(S))\, A \subseteq C(A) \quad \text{(Inclusion)}$$
$$(\forall A \in \wp(S))\, C(C(A)) = C(A) \quad \text{(Idempotence)}$$
$$(\forall A, B \in \wp(S))\, (A \subseteq B \subseteq C(A) \longrightarrow C(A) \subseteq C(B)) \quad \text{(Cautious Monotony,)}$$

then C will said to be a cautious closure operator *(CCO) on X.*

Definition 23. *Let $H = \langle \underline{H}, \preceq \rangle$, be a partially ordered set over a set $\underline{H}$. A subset $\mathcal{K}$ of the set of order ideals $\mathcal{F}(H)$ of H will be said to be* relevant *for a subset $B \subseteq H$ (in symbols $\rho(\mathcal{K}, H)$) if and only if the following hold:*

$$(\exists G \in \mathcal{K})(\forall P \in \mathcal{K})\, P \subseteq G.$$
$$(\forall P \in \mathcal{K})\, P \subseteq B.$$
$$\text{For any } \mathcal{L} \subseteq \mathcal{F}(H), \text{ if } \mathcal{K} \subseteq \mathcal{L}, \text{ then}$$
$$(\exists \top \in \mathcal{L})(\forall Y \in \mathcal{L})\, Y \subseteq \top \neq H \;\&\; \cap\mathcal{L} = \cap\mathcal{K}.$$

Definition 24. *In the context of Definition 23, a map $\jmath : \wp(L) \longmapsto \wp(L)$ defined as below will be said to be* safe

$$\jmath(Z) = \begin{cases} \cap\mathcal{K}, \text{ if all relevant collections for Z have same intersection.} \\ \cap\{\alpha : Z \subseteq \alpha \in \mathcal{F}(H)\}, \text{ else.} \end{cases}$$

Proposition 24. *A safe map $\jmath$ is a cautious closure operator.*

Proof. The verification of idempotence and inclusion is direct.

- For $A, B \in \wp(L)$, if it is the case that $A \subseteq B \subseteq \jmath(A)$,
- then either $A \subseteq B \subseteq \jmath(B) \subseteq \jmath(A)$ or $A \subseteq B \subseteq \jmath(A) \subseteq \jmath(B)$ must be true.
- If the former inclusions hold, then it is necessary that $\jmath(A) = \jmath(B)$.
- If $\jmath(B)$ is defined as the the intersection of order ideals and $\jmath(A)$ as that of relevant subcollections, then it is necessary that $\jmath(A) \subseteq \jmath(B)$. So cautious monotony holds. It can also be checked that monotonicity fails in this kind of situation. □

Theorem 19. *On every Boolean ordered unary algebra of the form*

$$\mathcal{H} = \langle \wp(H), \subseteq, C \rangle,$$

there exists a partial order $\leq$ on K such that $\langle \wp(K), \subseteq, \jmath \rangle$ is isomorphic to $\mathcal{H}$.

7 On Atoms in the POSET of Roughly Equivalent Sets

Definition 25. *For any two elements $x, y \in \wp(S)| \approx$, let*

$$x \leq y \text{ if and only if } (\forall a \in x)(\forall b \in y) a^l \subseteq b^l \;\&\; a^u \subseteq b^u.$$

$\wp(S)| \approx$ will be denoted by H in what follows.

Proposition 25. *The relation $\leq$ defined on H is a bounded and directed partial order. The least element will be denoted by 0 $(0 = \{\emptyset\})$ and the greatest by 1 $(1 = \{S\})$.*

Definition 26. *For any $a, b \in H$, let $UB(a, b) = \{x : a \leq x \;\&\; b \leq x\}$ and $LB(a, b) = \{x : x \leq a \;\&\; x \leq b\}$. By a* s-ideal *(strong ideal) of H, will be meant a subset K that satisfies all of*

$$(\forall x \in H)(\forall a \in K)(x \leq a \longrightarrow x \in K),$$
$$(\forall a, b \in K)\, UB(a, b) \cap K \neq \emptyset.$$

An atom *of H is any element that covers 0. The set of atoms of H will be denoted by $At(H)$.*

Theorem 20. *Atoms of H will be of one of the following types:*

Type-0. *Elements of the form $(\emptyset, [x])$, that intersect no other set of roughly equivalent sets.*
Type-1. *Bruinvals of the form $(\emptyset, \alpha)$, that do not contain full sets of roughly equivalent sets.*
Type-2. *Bruinvals of the form (α, β), that do not contain full sets of roughly equivalent sets and are such that $(\forall x)x^l = \emptyset$.*

Proof. It is obvious that a bruinval of the form (α, β) can be an atom only if α is the $\emptyset$. If not, then each element x of the bruinval $(\emptyset, \alpha)$ will satisfy $x^l = \emptyset \subset x^u$, thereby contradicting the assumption that (α, β) is an atom.

If $[x]$ intersects no other successor neighborhood, then

$$(\forall y \in (\emptyset, [x]))y^l = \emptyset \;\; \& \;\; x^u = [x]$$

and it will be a minimal set of roughly equal elements containing 0.

The other part can be verified based on the representation of possible sets of roughly equivalent elements. □

Theorem 21. *The partially ordered set H is atomic.*

Proof. It is required to prove that any element x greater than 0 is either an atom or there exists an atom a such that $a \leq x$, that is

$$(\forall x)(\exists a \in At(H))(0 < x \longrightarrow a \leq x).$$

Suppose the bruinval (α, β) represents a non-atom, then it is necessary that

$$(\forall x \in \alpha)\, x^l \neq \emptyset \;\; \& \;\; x^u \subseteq S.$$

Suppose the neighborhoods included in x^u are $\{[y] \,:\, y \in B \subseteq S\}$. If all combinations of bruinvals of the form $(\emptyset, \gamma)$ formed from these neighborhoods are not atoms, then it is necessary that the upper approximation of every singleton subset of a set in γ properly contains another non-trivial upper approximation. This is impossible.

So H is atomic. □

8 Algebraic Semantics-1

An algebraic semantics is a complete description of reasoning about rough objects involved in the context of PRAX or PRAS or any particular instances thereof. In the present author's view the objects of interest should be roughly equal elements in some sense and the semantics should avoid objects of other kinds (from other semantic domains) thereby contaminating the semantics. But in any perspective, semantics relative any semantic domain is of interest. When it comes to the question of defining sensible operations over rough objects, given the ontological constraints, there is scope for much variation.

If $A, B \in \wp(S)$ and $A \approx B$ then $A^u \approx B^u$ and $A^l \approx B^l$, but $\neg(A \approx A^u)$ in general. It has already been seen that $\leq$ is a partial order relation on $\wp(S)| \approx$. In this section elements of $\wp(S)| \approx$ would still be denoted by lower case Greek alphabets.

Theorem 22. *The following operations can be defined on $\wp(S)| \approx$ ($A, B \in \wp(S)$ and $[A]$, $[B]$ are corresponding classes):*

$$L[A] \triangleq [A^l] \tag{22}$$

$$[A] \odot [B] \triangleq [\bigcup_{X\in[A],\, Y\in[B]} (X \cap Y)] \tag{23}$$

$$[A] \oplus [B] \triangleq [\bigcup_{X\in[A],\, Y\in[B]} (X \cup Y)] \tag{24}$$

$$U[A] \triangleq [A^u] \tag{25}$$

$$[A] \cdot [B] \triangleq \lambda(LB([A], [B])) \tag{26}$$

$$[A] \circledast [B] \triangleq \lambda(UB([A], [B])) \tag{27}$$

$$[A] + [B] \triangleq \{X : X^l = (A^l \cap B^l)^l \ \&\ X^u = A^u \cup B^u\} \tag{28}$$

$$[A] \times [B] \triangleq \{X : X^l = A^l \cup B^l \ \&\ X^u = A^l \cup B^l \cup (A^u \cap B^u) \tag{29}$$

$$[A] \otimes [B] \triangleq \{X : X^l = A^l \cup B^l \ \&\ X^u = A^u \cup B^u\}. \tag{30}$$

Proof. If $A \approx B$ then $A^u \approx B^u$ and $A^l \approx B^l$, but $\neg(A \approx A^u)$ in general.

1. If $B \in [A]$, then $B^l = A^l$, $B^u = A^u$ and $L[A] = L[B] = [A^l]$.
2. $[A] \odot [B] \triangleq [\bigcup_{X\in[A],\, Y\in[B]}(X \cap Y)]$ is obviously well defined as sets of the form $[A]$ are elements of partitions.
3. Similar to the above.
4. If $B \in [A]$, then $B^u = A^u$ and so $[B^u] = [A^u]$.
5. $[A] \cdot [B] \triangleq \lambda(LB([A], [B]))$.
6. $[A] \circledast [B] \triangleq \lambda(UB([A], [B]))$.
7. $[A] + [B] \triangleq \{X : X^l = A^l \cap B^l \ \&\ X^u = A^u \cup B^u\}$. As the definitions is in terms of A^l, B^l, A^u, B^u, so there is no issue.
8. Similar to above.
9. Similar to above.

$+$, $\times$ and $\otimes$ will be referred to as *pragmatic* aggregation, commonality and commonality operations as they are less ontologically committed to the classical domain and more dependent on the main rough domain of interest. $+$ and the other pragmatic operations cannot be compared by the $\leq$ relation and so do not confirm to intuitive understanding of the concepts of aggregation and commonality.

The following theorems summarize the essential properties of the defined operations:

Theorem 23.

$$LL(\alpha) = L(\alpha). \tag{L1}$$
$$(\alpha \leq \beta \longrightarrow L(\alpha) \leq L(\beta)). \tag{L2}$$
$$(L(\alpha) = [\alpha] \longrightarrow \alpha = \{\alpha^l\}). \tag{L3}$$
$$(U(\alpha) \cap UU(\alpha) \neq \emptyset \longrightarrow U(\alpha) = UU(\alpha)). \tag{U1}$$
$$(UU(\alpha) = \emptyset \nrightarrow U(\alpha) = \emptyset). \tag{U2}$$
$$(\alpha \leq \beta \longrightarrow U(\alpha) \leq U(\beta). \tag{U3}$$
$$(U(\alpha) = \alpha \longrightarrow \alpha = \alpha^l = \alpha^u). \tag{U4}$$
$$UL(\alpha) \leq U(\alpha). \tag{U5}$$
$$LU(\alpha) = U(\alpha). \tag{U6}$$

Proof. Let $\alpha \in \wp(S)| \approx$, then the pair of lower and upper approximations associated with it will be denoted by α_l and α_u respectively. By α^u and α^l is meant the result of global operations respectively on the set α (seen as an element of $\wp(S)$). These take singleton values and so there is no real need of the approximations α_l and α_u and the former will be used.

Proof of L1:

$$\alpha \in \wp(S)| \approx, \text{ so } \alpha = \{X\,;\, \alpha_l = X^l \;\&\; \alpha_u = X^u, \;\&\; X \in \wp(S)\}.$$
$$\alpha^l = \{X^l;\, X \in \alpha\} = \{\alpha_l\}$$
$$\text{So } [\alpha^l] = \{Y\,;\, Y^l = \alpha^l \;\&\; Y^u = \alpha^{lu}\}.$$
$$(L(\alpha))^l = \{Y^l\,;\, Y^l = \alpha^l \;\&\; Y^u = \alpha^{lu}\} = \{\alpha^l\}.$$
$$\text{This yields } LL(\alpha) = L(\alpha). \tag{L1}$$

Proof of U1:

$$\alpha^u = \{X^u\,;\, \alpha^l = X^l \;\&\; \alpha^u = X^u\} = \{\alpha^u\}.$$
$$U(\alpha) = [\alpha^u] = \{Y\,;\, Y^l = \alpha^u \;\&\; Y^u = \alpha^{uu}\}.$$
$$\text{So } U(\alpha)^u = \{\alpha^{uu}\}.$$
$$UU(\alpha) = [U(\alpha)^u] = [\alpha^{uu}] = \{Y\,;\, Y^l = \alpha^{uu} \;\&\; Y^u = \alpha^{uuu}\}.$$
$$\text{Since } \alpha \subseteq \alpha^u \subseteq \alpha^{uu} \subseteq \alpha^{uuu},$$
$$\text{therefore } (U(\alpha) \cap UU(\alpha) \neq \emptyset \longrightarrow U(\alpha) = UU(\alpha). \tag{U1}$$

The other parts can be proved from the above considerations. □

Theorem 24. *In the context of the above theorem, the following hold:*

$$\alpha \odot \beta = \beta \odot \alpha) \quad \text{(CO1)}$$
$$\alpha \leq \alpha \odot \alpha \quad \text{(CO2)}$$
$$\alpha \leq \alpha \odot \top \quad \text{(CO3)}$$
$$\alpha \odot \alpha = \alpha \odot (\alpha \odot \alpha) = \alpha \odot \top \quad \text{(CO4)}$$
$$\alpha \oplus \beta = \beta \oplus \alpha) \quad \text{(AO1)}$$
$$\alpha \leq \alpha \oplus \beta \quad \text{(AO2)}$$
$$\alpha \leq \alpha \oplus \bot \quad \text{(AO3)}$$
$$(\alpha \oplus \alpha) \oplus \alpha = \alpha \oplus \alpha \quad \text{(AO4)}$$
$$\textit{In general, } \alpha \oplus (\alpha \odot \beta) \neq \alpha. \quad \text{(AC)}$$

Proof. **CO1** The definition of $\odot$ does not depend on the order in which the arguments occur as set theoretic intersection and union are commutative. To be precise $\bigcup_{X \in [A], Y \in [B]}(X \cap Y) = \bigcup_{X \in [A], Y \in [B]}(Y \cap X)$.

CO2 $\bigcup_{X \in [A], Y \in [A]}(X \cap Y) = \bigcup_{X \in [A]} X$. But because $X^l \cup Y^l \subseteq (X \cup Y)^l$ in general, so equality fails.

CO3 Follows from the last inequality.

CO4 In $[\alpha \odot (\alpha \odot \alpha)]$, any new elements that are not in $[\alpha \odot \alpha]$ cannot be introduced as the inequality in [CO2] is due to the lower approximation and all possible subsets have already been included.

AO1 The definition of $\oplus$ does not depend on the order in which the arguments occur as set theoretic union is commutative.

AO2 Even when $\beta = \alpha$, inequality can happen for reasons mentioned earlier.

Proof of [AO3, AO4, AC] are analogous or direct. □

The above result means that $\odot$ is an imperfect commonality relation. It is a proper commonality among a certain subset of elements of H.

Theorem 25. *In the context of the above theorem, the following properties of* $+, \times, \otimes$ *are provable:*

$$\alpha + \alpha = \alpha, \quad \text{(+I)}$$
$$\alpha + \beta = \beta + \alpha, \quad \text{(+C)}$$
$$\alpha \times \alpha = \alpha, \quad \text{(cI)}$$
$$\alpha \times \beta = \beta \times \alpha, \quad \text{(cC)}$$
$$\alpha \leq \beta \longrightarrow \alpha + \gamma \leq \beta + \gamma, \quad \text{(+Is)}$$
$$\alpha \leq \beta \longrightarrow \alpha \times \gamma \leq \beta \times \gamma, \quad \text{(cIs)}$$
$$\alpha \leq \beta \longrightarrow \alpha \leq \alpha \times \beta \leq \beta, \quad \text{(+In)}$$
$$\alpha + \beta \leq \alpha \oplus \beta, \quad \text{(R1)}$$
$$\alpha \times \beta \leq (\alpha \times \beta) \oplus \alpha. \quad \text{(Mix1)}$$

Proof. Most of the proof is in Sect. 9, so they are not repeated. □

Definition 27. *By a* Concrete Pre-PRAX Algebraic System *(CPPRAXA), will be meant a system of the form*

$$\mathfrak{H} = \langle H, \leq, L, U, \oplus, \odot, +, \times, \otimes, \bot, \top \rangle,$$

with all of the operations being as defined in this section.

Apparently the algebraic properties of the rough objects of l_o, u_o need to be involved for a representation theorem. The operations can be improved by related operations of the following section. Results concerning this will appear separately. Definable filters in general have reasonable properties.

Definition 28. *Let K be an arbitrary subset of a CPPRAXA $\mathfrak{H}$. Consider the following statements:*

$$(\forall x \in K)(\forall y \in \mathfrak{H})(x \leq y \Rightarrow y \in K). \tag{F1}$$

$$(\forall x, y \in K)\, x \oplus y, Lx \in K. \tag{F2}$$

$$(\forall a, b \in \mathfrak{H})(1 \neq a \oplus b \in K \;\Rightarrow\; a \in K \text{ or } b \in K). \tag{F3}$$

$$(\forall a, b \in \mathfrak{H})(1 \neq UB(a, b) \in K \;\Rightarrow\; a \in K \text{ or } b \in K). \tag{F4}$$

$$(\forall a, b \in K)\, LB(a, b) \cap K \neq \emptyset. \tag{F5}$$

- *If K satisfies **F1** then it will be said to be an* order filter. *The set of such filters on $\mathfrak{H}$ will be denoted by $\mathfrak{O}_F(\mathfrak{H})$.*
- *If K satisfies **F1, F2** then it will be said to be a* filter. *The set of such filters on $\mathfrak{H}$ will be denoted by $\mathcal{F}(\mathfrak{H})$.*
- *If K satisfies **F1, F2, F3** then it will be said to be a* prime filter. *The set of such filters on $\mathfrak{H}$ will be denoted by $\mathcal{F}_P(\mathfrak{H})$.*
- *If K satisfies **F1, F4** then it will be said to be a* prime order filter. *The set of such filters on $\mathfrak{H}$ will be denoted by $\mathfrak{O}_{PF}(\mathfrak{H})$.*
- *If K satisfies **F1, F5** then it will be said to be an* strong order filter. *The set of such filters on $\mathfrak{H}$ will be denoted by $\mathfrak{O}_{SF}(\mathfrak{H})$.*

Dual concepts of ideals of different kinds can be defined.

Proposition 26. *Filters of different kinds have the following properties:*

- *Every set of filters of a kind is ordered by inclusion.*
- *Every filter of a kind is contained in a maximal filter of the same kind.*
- *$\mathfrak{O}_{SF}(\mathfrak{H})$ is an algebraic lattice, with its compact elements being the finitely generated strong order filters in it.*

Definition 29. *For $F, P \in \mathcal{F}(\mathfrak{H})$, the following operations can be defined:*

$$F \wedge P \triangleq F \cap P$$

$$F \vee P \triangleq \langle F \cup P \rangle,$$

where $\langle F \cup P \rangle$ denotes the smallest filter containing $F \cup P$.

Theorem 26. *$\langle \mathcal{F}(\mathfrak{H}), \vee, \wedge, \bot, \top \rangle$ is an atomistic bounded lattice.*

9 Algebraic Semantics-2

It has already been seen that ordered pairs of the form (A^l, A^u) do correspond to rough objects by definition. If the representation and finer aspects of possible reasonable aggregation and commonality operations is ignored, then an interesting order structure based fragment of semantic processes is the result. It is very useful in the approximation based semantics of following sections.

Definition 30. *In a PRAX S, let*

$$\mathcal{R}(S) = \{(A^l, A^u)\,;\, A \in \wp(S)\}.$$

Then all of the following operations on $\mathcal{R}(S)$ can be defined:

$$(A^l, A^u) \vee (B^l, B^u) \triangleq (A^l \cup B^l, A^u \cup B^u). \qquad \text{(Aggregation)}$$

$$\text{If } (A^l \cap B^l, (A^u \cap B^u)) \in \mathcal{R}(S) \text{ then}$$

$$(A^l, A^u) \wedge (B^l, B^u) \triangleq (A^l \cap B^l, (A^u \cap B^u)). \qquad \text{(Commonality)}$$

$$\text{If } (A^{uc}, A^{lc}) \in \mathcal{R}(S) \text{ then}$$

$$\sim (A^l, A^u) \triangleq (A^{uc}, A^{lc}). \qquad \text{(Weak Complementation)}$$

$$\bot \triangleq (\emptyset, \emptyset).\ \top \triangleq (S, S). \qquad \text{(Bottom, Top)}$$

$$(A^l, A^u) \barwedge (B^l, B^u) \triangleq ((A^l \cap B^l)^l, (A^u \cap B^u)^l). \qquad \text{(Proper Commonality)}$$

Definition 31. *In the context of the above definition, a partial algebra of the form $\mathfrak{R}(S) = \langle \mathcal{R}(S), \vee, \wedge, c, \bot, \top \rangle$ will be termed a* proto-vague algebra *and $\mathfrak{R}_f(S) = \langle \mathcal{R}(S), \vee, \wedge, \barwedge. c, \bot, \top \rangle$ will be termed a* full proto-vague algebra.

*More generally, if L, U are arbitrary rough lower and upper approximation operators over the PRAX, and if each occurrence of l is replaced by L and of u by U in the above defn then the resulting algebra of the above form will be called a LU-*proto-vague partial algebra. *Analogously, $l_o u_o$-proto-vague algebras and similar algebras can be defined.*

Theorem 27. *A full proto-vague partial algebra $\mathfrak{R}_f(S)$ satisfies all of the following:*

1. *$\vee, \bar{\wedge}$ are total operations.*
2. *$\vee$ is a semi-lattice operation satisfying idempotency, commutativity and associativity.*
3. *$\wedge$ is a weak semi-lattice operation satisfying idempotency, weak strong commutativity and weak associativity. With $\vee$ it forms a weak distributive lattice.*
4. *$\sim$ is a weak strong idempotent partial operation; $\sim\sim\sim \alpha \stackrel{\omega^*}{=} \sim \alpha$.*
5. *$\sim (\alpha \vee \beta) \stackrel{\omega}{=} \sim \alpha \wedge \sim \beta$ (Weak De Morgan condition) holds.*
6. *$\bar{\wedge}$ is an idempotent, commutative and associative operation that forms a lattice with $\vee$.*
7. *$\alpha \bar{\wedge} \bot = \alpha \wedge \bot = \bot$. $\alpha \vee \bot = \alpha$; $\alpha \bar{\wedge} \top = \alpha \wedge \top = \alpha$. $\alpha \vee \top = \top$.*
8. *$\sim (\alpha \wedge \beta) = (\sim \alpha \vee \sim \beta) \longrightarrow \sim (\alpha \bar{\wedge} \beta) = (\sim \alpha \vee \sim \beta)$.*
9. *$\alpha \vee (\beta \bar{\wedge} \gamma) \subseteq (\alpha \vee \beta) \bar{\wedge} (\alpha \vee \gamma)$, but distributivity fails.*

Proof. Let $\alpha = (X^l, X^u)$, $\beta = (Y^l, Y^u)$ and $\gamma = (Z^l, Z^u)$ for some $X, Y, Z \in \wp(S)$, then

1. $\alpha \vee \beta = (X^l \cup Y^l, X^u \cup Y^u)$ belongs to $\mathfrak{R}(S)$ because the components are unions of successor neighborhoods and $X^l \cup Y^l \subseteq X^u \cup Y^u$. The proof for $\wedge$ is similar.
2. $\alpha \vee (\beta \vee \gamma) = (X^l, X^u) \vee ((Y^l, Y^u) \vee (Z^l, Z^u)) = (X^l, X^u) \vee (Y^l \cup Z^l, Y^u \cup Z^u) = (X^l \cup Y^l \cup Z^l, X^u \cup Y^u \cup Z^u) = (\alpha \vee \beta) \vee \gamma$.
3. Weak absorptivity and weak distributivity are proved next. $(X^l \cap (X^l \cup Y^l)) = X^l$ and $(X^u \cap (X^u \cup Y^u)) = X^l$ hold in all situations. If $(X^l \cup (X^l \cap Y^l))$ is defined then it is equal to X^l and if $(X^u \cup (X^u \cup Y^u))$ is defined, then it is equal to X^u. So
$$\alpha \vee (\alpha \wedge \beta) \stackrel{\omega}{=} \alpha = \alpha \wedge (\alpha \vee \beta).$$
For distributivity $(\alpha \vee (\beta \wedge \gamma) \stackrel{\omega}{=} (\alpha \vee \beta) \wedge (\alpha \vee \gamma)$ and $\alpha \wedge (\beta \vee \gamma) \stackrel{\omega}{=} (\alpha \wedge \beta) \vee (\alpha \wedge \gamma))$ again it is a matter of definability working in coherence with set-theoretic distributivity.
4. If $\sim \alpha$ is defined then $\sim \alpha = (X^{uc}, X^{lc})$ and $\sim\sim \alpha = \sim (X^{uc}, X^{lc}) = (X^{lcc}, X^{ucc}) = (X^l, X^u)$, by definition. If $\sim\sim \alpha$ is defined, then $\sim \alpha$ is necessarily defined. So
$$\sim\sim\sim \alpha \stackrel{\omega^*}{=} \sim \alpha.$$
5. If $\sim (\alpha \vee \beta)$ and $\sim \alpha \wedge \sim \beta$ are defined then $\sim (\alpha \vee \beta) = \sim ((X^l \cup Y^l), (X^u \cup Y^u)) = ((X^{uc} \cap Y^{uc}), (X^{lc} \cap Y^{lc})) \stackrel{\omega^*}{=} (X^{uc}, X^{lc}) \wedge (Y^{uc}, Y^{lc}) = \sim \alpha \wedge \sim \beta$. So $\sim (\alpha \vee \beta) \stackrel{\omega^*}{=} \sim \alpha \wedge \sim \beta$.
6. $\alpha \bar{\wedge} \beta = \beta \bar{\wedge} \alpha$ & $\alpha \bar{\wedge} \alpha = \alpha$ are obvious.
$\alpha \bar{\wedge} (\beta \bar{\wedge} \gamma) = ((X^l \cap (Y^l \cap Z^l)^l)^l, (X^u \cap (Y^u \cap Z^u)^u)^u)$ The components are basically the unions of common granules among the three. No granule in the final evaluation is eliminated by choice of order of operations. So
$\alpha \bar{\wedge} (\beta \bar{\wedge} \gamma) = (\alpha \bar{\wedge} \beta) \bar{\wedge} \gamma$.
$\alpha \bar{\wedge} (\alpha \vee \beta) = ((X^l \cap (X^l \cup Y^l))^l, (X^u \cap (X^u \cup Y^u))^l) = \alpha$.
Further, $\alpha \vee (\alpha \bar{\wedge} \beta) = ((X^l \cup (X^l \cap Y^l)^l), (X^u \cup (X^u \cap Y^u)^l)) = \alpha$. So $\vee, \bar{\wedge}$ are lattice operations.

7. – Since $\bot = (\emptyset, \emptyset)$, $\alpha \bar{\wedge} \bot = \alpha \wedge \bot = \bot$ and $\alpha \vee \bot = \alpha$ follow directly.
   – Since $\top = (S, S)$, $\alpha \bar{\wedge} \top = \alpha \wedge \top = \alpha$ and $\alpha \vee \top = \top$ follow directly.
8. Follows from the previous proofs.
9. – $\alpha \vee (\beta \bar{\wedge} \gamma) = ((X^l \cup (Y^l \cap Z^l)^l), (X^u \cup (Y^u \cap Z^u)^l))$. If $a \in S$ and $[a] \subseteq X^l \cup (Y^l \cap Z^l)^l$, and $[a] \subseteq (Y^l \cap Z^l)^l$, then $[a] \subseteq Y^l$ and $[a] \subseteq Z^l$. So $[a] \subseteq X^l \cup Y^l$ and $[a] \subseteq X^l \cup Z^l$.
   – If $[a] \subseteq X^l \cup (Y^l \cap Z^l)^l$ and if $[a] = P \cup Q$, with $P \subseteq X^l$, $Q \subseteq (Y^l \cap Z^l)^l$ then $[a] \subseteq X^l \cup Y^l$ and $[a] \subseteq X^l \cup Z^l$. This proves $\alpha \vee (\beta \bar{\wedge} \gamma) \subseteq (\alpha \vee \beta) \bar{\wedge} (\alpha \vee \gamma)$.
   – If $[a] \subseteq ((X^l \cup Y^l) \cap (X^l \cup Y^l))^l$ then $[a] \subseteq X^l \cup Y^l$ and $[a] \subseteq X^l \cup Z^l$. This means $[a] = P \cup Q$, with $P \subseteq X^l$, $Q \subseteq Y^l$ and $Q \subseteq Z^l$ and Q is contained in union of some other granules. So $Q \subseteq Y^l \cap Z^l$, but it cannot be ensured that $Q \subseteq (Y^l \cap Z^l)^l$ (required counterexamples are easy to construct). It follows that $((X^l \cup Y^l) \cap (X^l \cup Y^l))^l \nsubseteq X^l \cup (Y^l \cap Z^l)^l$.

The following theorem provides a condition for ensuring that $\sim \alpha$ is defined.

Theorem 28. *If $X^{uu} = X^u$, then $\sim (X^l, X^u) = (X^{uc}, X^{lc})$ but the converse is not necessarily true.*

Proof.

- $\sim (X^l, X^u)$ is defined if and only if X^{uc} is a union of granules.
- If $X^{uu} = X^u$ then X^{uc} is a union of granules generated by *some* of the elements in X^{uc}, but the converse need not hold.
- So the result follows.

Let W be any quasi-order relation that approximates R, and let the granules $[x]_w$, $[x]_{wi}$ and l_w, u_w be lower and upper approximations defined by analogy with the definitions of l, u. If $R \subset W$, then $(\forall x \in S)\, [x] \subseteq [x]_w$ and $(A,\, B \in \wp(S))$. $A \parallel B$ in all that follows shall mean $A \nsubseteq B$ & $B \nsubseteq A$):

- If $A \subset B$ and $A^u = B^u$, then it is possible that $A^{u_w} \subset B^{u_w}$.
- If $A \subset B$ and $A^l = B^l$, then it is possible that $A^{l_w} \subset B^{l_w}$.
- If $A \subset B$ and $A^{u_w} = B^{u_w}$, then it is possible that $A^u \subset B^u$.
- If $A \subset B$ and $A^{l_w} = B^{l_w}$, then it is possible that $A^l \subset B^l$.
- If $A \parallel B$ and $A^l = B^l$, then it is possible that $A^{l_w} \parallel B^{l_w}$.
- If $A \parallel B$ and $A^{l_w} = B^{l_w}$, then it is possible that $A^l \parallel B^l$.
- If $A \parallel B$ and $A^u = B^u$, then it is possible that $A^{u_w} \parallel B^{u_w}$.
- If $A \parallel B$ and $A^{u_w} = B^{u_w}$, then it is possible that $A^u \parallel B^u$.
- If $A \subset B$, $A^l = B^l$ and $A^u = B^u$, then it is possible that $A^{u_w} \subset B^{u_w}$ & $A^{l_w} \subset B^{l_w}$.

The above properties mean that meaningful correspondences between vague partial algebras and Nelson algebras may be quite complex. Focusing on granular evolution alone, the following can be defined

$$(\forall x \in S)\, \varphi_o([x]) = \bigcup_{z\in[x]} [z]_w.$$

$$(\forall A \in \wp(S))\, \varphi(A^l) = \bigcup_{[x]\subseteq A^l} \varphi_o([x]).$$

$$(\forall A \in \wp(S))\, \varphi(A^u) = \bigcup_{[x]\subseteq A^u} \varphi_o([x]).$$

$\varphi(A^l \cup B^l) = \bigcup_{[x]\subseteq A^l \cup B^l}$.

If $[x] \subseteq A^l \cup B^l$

φ can be naturally extended by components to a map τ as per

$$\tau(A^l, A^u) = (\varphi(A^l), \varphi(A^u)).$$

Proposition 27. *If $R \subseteq R_w$ and R_w is transitive, then*

- *If $z \in [x]$ and $x \in [z]$, then $\varphi([z]) = \varphi([x])$.*
- *If $z \in [x]$, then $\varphi([z]) \subseteq \varphi([x])$.*
- $$(\forall A \in \wp(S))\, \varphi(A^l) = \bigcup_{[x]\subseteq A^l} \varphi([x]) = \bigcup_{[x]\subseteq A^l} [x]_w$$

Proof.

- $z \in [x]$ yields Rzx. So if Raz, then Rax and it is clear that $\varphi([z]) \subseteq \varphi([x])$. Rbx & Rzx & Rxz implies $R_w bz$.
- This is the first part of the above.
- Follows from the above.

Definition 32. *The following abbreviations will be used for handling different types of subsets of S:*

$$\Gamma_u(S) = \{A^u; A \in \wp(S)\}. \quad \text{(Uppers)}$$

$$\Gamma_{uw}(S) = \{A^{u_w}; A \in \wp(S)\}. \quad \text{(w-Uppers)}$$

$$\Gamma(S) = \{B; (\exists A \in \wp(S))\, B = A^l \text{ or } B = A^u\}. \quad \text{(lower definites)}$$

Note that $\delta_l(S)$ is the same as $\Gamma(S)$ and similarly for $\delta_{lw}(S)$.

τ has the following properties:

Proposition 28. *If $R \subseteq R_w$ and R_w is transitive, then*

$$\tau(\bot) = \bot_w.$$

$$\tau(\top) = \top_w.$$

$$(\forall \alpha, \beta \in \Re(S))\, \tau(\alpha \vee \beta) = \tau(\alpha) \vee \tau(\beta).$$

$$(\forall \alpha, \beta \in \Re(S))\, \tau(\alpha \wedge \beta) \overset{\omega}{=} \tau(\alpha) \wedge \tau(\beta).$$

Definition 33. *For each* $\alpha \in \mathfrak{R}_w(S)$*, the set of ordered pairs* $\tau^{\dashv}(\alpha)$ *will be termed as a* co-rough object *of* S*, where*

$$\tau^{\dashv}(\alpha) = \{\beta;\, \beta \in \mathfrak{R}(S)\ \&\ \tau(\beta) = \alpha\}.$$

The collection of all co-rough objects will be denoted by $\mathfrak{CR}(S)$.

This permits us to define a variety of closely related semantics of PRAX when $R \subseteq R_w$ and R_w is transitive. These include:

- The map $\tau : \mathfrak{R}_f(S) \longmapsto \mathfrak{R}_w(S)$. $\mathfrak{R}_w(S)$ being a Nelson algebra over an algebraic lattice.
- $\mathfrak{R}_f(S) \cup \mathfrak{CR}(S)$ along with induced operations yields another semantics of PRAX.
- $\mathfrak{R}(S) \cup \mathfrak{R}_w(S)$ enriched with algebraic and dependency operations described in Sect. 12.

10 Approximate Relations

If R is a binary relation on a set X, then we let $R^o \stackrel{\partial}{=} R \cup \Delta_X$. The weak transitive closure of R will be denoted by $R^\#$. If $R^{(i)}$ is the i-times composition $\underbrace{R \circ R \ldots \circ R}_{\text{i-times}}$, then $R^\# = \bigcup R^{(i)}$. R is *acyclic* if and only if $(\forall x)\, \neg R^\# xx$. The relation $R^{\cdot}$ is defined by $R^{\cdot}ab$ if and only if $Rab\ \&\ \neg(R^\# ab\ \&\ R^\# ba)$.

Definition 34. *If* R *is a relation on a set* S*, then the relations* $R^{\curlywedge}$, R^{cyc} *and* R^h *will be defined via*

$$R^{\curlywedge}ab \text{ if and only if } [b]_{R^o} \subset [a]_{R^o}\ \&\ [a]_{iR^o} \subset [b]_{iR^o} \tag{31}$$

$$R^{cyc}ab \text{ if and only if } R^\# ab\ \&\ R^\# ba \tag{32}$$

$$R^h ab \text{ if and only if } R^{\curlywedge}ab\ \&\ R^{\cdot}ab. \tag{33}$$

In case of PRAX, $R^o = R$*, so the definition of* $R^{\curlywedge}$ *would involve neighborhoods of the form* $[a]$ *and* $[a]_i$ *alone.* $R^{\curlywedge} \subset R$ *and* $R^{\curlywedge}$ *is a partial order.*

Persistent Example 3. *In the Example 1,* $R^\# ab$ *happens when* a *is an ally of an ally of* b. $R^{\curlywedge}ab$ *happens iff every ally of* b *is an ally of* a *and if* a *is ally of* c*, then* b *is an ally of* c *- this can happen, for example, when* b *is a Marxist feminist and* a *is a socialist feminist.* $R^{cyc}ab$ *happens when* a *is an ally of an ally of* b *and* b *is an ally of an ally of* a. $R^{\cdot}ab$ *happens whenever* a *is an ally of* b*, but* b *is not an ally of anybody who is an ally of* a.

Theorem 29. $R^h = \emptyset$.

Proof.

$$\begin{aligned} R^h ab &\Leftrightarrow R^{\curlywedge} ab \;\&\; R^{\cdot} ab \\ &\Leftrightarrow \tau(R)ab \;\&\; (R \setminus \tau(R))ab \\ &\text{But } \neg(\exists a)(R \setminus \tau(R))aa. \end{aligned}$$

So $R^h = \emptyset$. □

Proposition 29. *All of the following hold in a PRAX S:*

$$R^{\cdot} ab \leftrightarrow (R \setminus \tau(R))ab \tag{34}$$

$$(\forall a, b)\neg(R^{\cdot} ab \;\&\; R^{\cdot} ba) \tag{35}$$

$$(\forall a, b, c)(R^{\cdot} ab \;\&\; R^{\cdot} bc \longrightarrow \neg R^{\cdot} ac). \tag{36}$$

Proof.

- $R^{\cdot} ab \leftrightarrow Rab \;\&\; \neg(R^{\#} ab\, R^{\#} ba)$.
- But $\neg(R^{\#} ab\, R^{\#} ba)$ is possible only when both Rab and Rba hold.
- So $R^{\cdot} ab \leftrightarrow Rab \;\&\; \neg(\tau(R)ab) \leftrightarrow (R \setminus \tau(R))ab$. □

Theorem 30.

$$R^{\#\cdot} = R^{\#} \setminus \tau(R) \tag{37}$$

$$R^{\cdot\#} = (R \setminus \tau(R))^{\#} \tag{38}$$

$$(R \setminus \tau(R))^{\#} \subseteq R^{\#} \setminus \tau(R). \tag{39}$$

Proof.

1.

$$\begin{aligned} R^{\#\cdot} ab &\leftrightarrow R^{\#} ab \;\&\; \neg(R^{\#\#} ab \;\&\; R^{\#\#} ba) \\ &\leftrightarrow R^{\#} ab \;\&\; \neg(R^{\#} ab \;\&\; R^{\#} ba) \\ &\leftrightarrow R^{\#} ab \;\&\; \neg\tau(R)ab \\ &\leftrightarrow (R^{\#} \setminus \tau(R))ab. \end{aligned}$$

2.

$$\begin{aligned} R^{\cdot\#} ab &\leftrightarrow (R^{\cdot})^{\#} ab \\ &\leftrightarrow (R \setminus \tau(R))^{\#} ab. \end{aligned}$$

3. Can be checked by a contradiction or a direct argument. □

Possible properties that approximations of proto transitive relations may or should possess will be considered next. If $<$ is a strict partial order on S and R is a relation, then consider the conditions:

$$(\forall a, b)(a < b \longrightarrow R^{\#}ab). \tag{PO1}$$

$$(\forall a, b)(a < b \longrightarrow \neg R^{\#}ba). \tag{PO2}$$

$$(\forall a, b)(R^{\curlywedge}ab \,\&\, R^{\cdot}ab \longrightarrow a < b. \tag{PO3}$$

$$\text{If } a \equiv_R b, \text{ then } a \equiv_< b. \tag{PO4}$$

$$(\forall a, b)(a < b \longleftrightarrow Rab). \tag{PO5}$$

As per [11], $<$ is said to be a *partial order approximation* POA (resp. *weak partial order approximation* WPOA) of R if and only if **PO1, PO2, PO3, PO4** (resp. **PO1, PO3, PO4**) hold. A POA $<$ is *inner approximation* IPOA of R if and only if **PO5** holds. **PO4** has a role beyond that of approximation and depends on both successor and predecessor neighborhoods. R^h, $R^{\cdot\curlywedge}$ are IPOA, while $R^{\cdot\#}$, $R^{\#\cdot}$ are POAs.

By a *lean quasi order approximation* $<$ of R, will be meant a quasi order satisfying **PO1** and **PO2**. The corresponding sets of such approximations of R will be denoted by $POA(R)$, $WPOA(R)$, $IPOA(R), IWPOA(R)$ and $LQO(R)$

Theorem 31. *For any $A, B \in LQO(R)$, the operations $\&, \vee, \top$ can be defined via:*

$$(\forall x, y)(A \,\&\, B)xy \text{ if and only if } (\forall x, y)Axy \,\&\, Bxy.$$
$$(A \vee B) = (A \cup B)^{\#},$$
$$\top = R^{\#}.$$

Proof.

- If Aab then $R^{+}ab$ and if Bab then $R^{+}ab$.
- But if $(A \,\&\, B)ab$, then both Aab and Bab.
- So $R^{+}ab$.

Similarly it can be shown that $A \vee B \in LQO(R)$. It is always defined and contained within $R^{\#}$ as it is the transitive completion of $A \cup B$. $\top = R^{\#}$ as transitive closure is a closure operator. □

Theorem 32. *In a PRAX, $R^{\cdot\#} \,\&\, R^{\#\cdot}xy \leftrightarrow (R \setminus \tau(R))^{\#}xy$.*

10.1 Granules of Derived Relations

The behavior of approximations and rough objects corresponding to derived relations is investigated in this subsection.

Definition 35. *The relation* $R^{\#\cdot}$ *will be termed the* trans ortho-completion *of* R. *The following granules will be associated with each* $x \in S$:

$$[x]_{ot} = \{y;\, R^{\#\cdot}yx\} \tag{40}$$
$$[x]^{i}_{ot} = \{y;\, R^{\#\cdot}xy\} \tag{41}$$
$$[x]^{o}_{ot} = \{y;\, R^{\#\cdot}yx \;\&\; R^{\#\cdot}xy\}. \tag{42}$$

Let the corresponding approximations be l_{ot}, u_{ot} *and so on.*

Theorem 33. *In a PRAX* S, $(\forall x \in S)\, [x]^{o}_{ot} = \{x\}$.

Proof. $R^{\#\cdot}xy \;\&\; R^{\#\cdot}yx$ means that the pair (x, y) is in the transitive completion of R and not in $\tau(R)$. So $y \in [x]^{o}_{ot}$ if and only if

$$(\exists a, b)\, Rxa \;\&\; Ray \;\&\; (\neg Rax \vee \neg Rya) \;\&\; (Ryb \;\&\; Rbx) \;\&\; (\neg Rby \vee \neg Rxb).$$

If it is assumed that $x \neq y$, then each of the possibilities leads to a contradiction as is shown below. In the context of the above statement:

Case-1

- $Rxa \;\&\; Ray \;\&\; \neg Rax \;\&\; Rya \;\&\; Ryb \;\&\; Rbx \;\&\; \neg Rby \;\&\; Rxb$.
- This yields $R^{\#}xa \;\&\; R^{\#}bb \;\&\; R^{\#}ba \;\&\; R^{\#}ab$.
- So, $R^{\#}xb \;\&\; R^{\#}ya \;\&\; R^{\#}ax$ and this contradicts the original assumption.

Case-2

- $Rxa \;\&\; Ray \;\&\; Rax \;\&\; \neg Rya \;\&\; Ryb \;\&\; Rbx \;\&\; Rby \;\&\; \neg Rxb$.
- This yields the contradiction $R^{\#}ab$.

Case-3

- $Rxa \;\&\; Ray \;\&\; \neg Rax \;\&\; Rya \;\&\; Ryb \;\&\; Rbx \;\&\; Rby \;\&\; \neg Rxb$.
- This yields $R^{\#}ba \;\&\; R^{\#}ab \;\&\; R^{\#}aa \;\&\; R^{\#}bb$ and $R^{\#}yy \;\&\; R^{\#}xy \;\&\; R^{\#}yx \;\&\; Rya \;\&\; R^{\#}xa$.
- But such a $R^{\#}$ is not possible.

Somewhat similarly the other cases can be seen to lead to contradictions. □

By the *symmetric center* of a relation R, will be meant the set $K_R = \bigcup e_i(\tau(R) \setminus \Delta_S)$ - basically the union of elements in either component of $\tau(R)$ minus the diagonal relation on S.

Proposition 30. $(\forall x)\, [x] \triangle [x]_{ot} \neq \emptyset$ *as*

$$x \notin K_R \longrightarrow [x] \subset [x]_{ot}$$
$$x \in K_R \longrightarrow [x] \not\subseteq [x]_{ot} \;\&\; \{x\} \subset [x] \cap [x]_{ot}.$$

Proof.

$$\begin{aligned} z \in [x]_{ot} &\leftrightarrow R^{\#\cdot} zx \\ &\leftrightarrow R^{\#} zx \;\&\; \neg\tau(R)zx \\ &\leftrightarrow (Rzx \;\&\; \neg Rxz) \,\text{or}\, (\neg Rzx \;\&\; \neg Rxz \;\&\; (R^{\#} \setminus R)zx). \end{aligned}$$

□

K_R can be used to partially categorize subsets of S based on intersection.

Proposition 31. *$(R \setminus \tau(R))^{\#} \cup \tau(R)$ is not necessarily a quasi order.*

Proof. $(x, y) \in (R \setminus \tau(R))^{\#} \cup \tau(R)$ and $(x, y) \notin \tau(R)$ and $x \in K_R$ & $y \notin K_R$ and $\exists z \in K_R$ & $z \neq x$ & Rzx do not disallow Rzy. So $(R \setminus \tau(R))^{\#} \cup \tau(R)$ is not necessarily a quasi-order. The missing part is left for the reader to complete. □

Proposition 32. $((R \setminus \tau(R))^{\#} \cup \tau(R))^{\#} = R^{\#}$.

Proof. Clearly $R \subseteq ((R \setminus \tau(R))^{\#} \cup \tau(R))^{\#}$ and it can be directly checked that if $a \in ((R \setminus \tau(R))^{\#} \cup \tau(R))^{\#} \setminus R$ then $a \in R^{\#} \setminus R$ and conversely.

11 Transitive Completion and Approximate Semantics

The interaction of the rough approximations in a PRAX and the rough approximations in the transitive completion can be expected to follow some order. *The definite or rough objects most closely related to the difference of lower approximations and those related to the difference of upper approximations can be expected to be related in a nice way.* It is shown that this *nice way* is not really a *rough way*. But the results proved remain relevant for the formulation of semantics that involves that of the transitive completion as in [13,14]. A rough theoretical alternative is possible by simply starting from sets of the form $A^* = (A^l \setminus A^{l\#}) \cup (A^{u\#} \setminus A^u)$ and taking their lower ($l_{\#}$) and upper ($u_{\#}$) approximations - the resulting structure would be a partial algebra derived from a Nelson algebra over an algebraic lattice [28].

Proposition 33. *For an arbitrary proto-transitive reflexive relation R on a set S, ($\#$ subscripts will be used for neighborhoods, approximation operators and rough equalities of the weak transitive completion) all of the following hold:*

$$(\forall x \in S)\, [x]_R \subseteq [x]_{R\#} \tag{Nbd}$$

$$(\forall A \subseteq S)\, A^l \subseteq A^{l\#} \;\&\; A^u \subseteq A^{u\#} \tag{App}$$

$$(\forall A \subseteq S)(\forall B \in [A]_{\approx})(\forall C \in [A]_{\approx_{\#}})\, B^l \subseteq C^{l\#} \;\&\; B^u \subseteq C^{u\#} \tag{REq}$$

The reverse inclusions are false in general in the second assertion in a specific way. Note that the last condition induces a more general partial order $\preceq$ over $\wp(\wp(S))$ via $A \preceq B$ if and only if $(\forall C \in A)(\forall E \in B)\, C^l \subseteq E^{l\#}$ & $C^u \subseteq E^{u\#}$.

Proof. The first of these is direct. For simplicity, the successor neighborhoods of x will be denoted by $[x]$ and $[x]_\#$ respectively. The possibility of tracking of the second assertion in the first part is also considered.

- If $z \in A^{l\#}$ then $z \in A^l$ as $[x]_\# \subseteq A$ implies $[x] \subseteq A$.
- If $z \in A^l$ then $(\exists x)\, z \in [x] \subseteq A^l$.
- For this x, $z \in [x]_\#$, but it is possible that $[x]_\# \subseteq A$ or $[x]_\# \nsubseteq A$.
- If $[x]_\# \nsubseteq A$, and $(\exists b \notin A)\, R_\# ax \;\&\; Rab \;\&\; Rbx$ then a contradiction happens as Rbx means $b \in [x]$.
- If $[x]_\# \nsubseteq A$, and $(\exists b \in A)\, R_\# ax \;\&\; Rab \;\&\; Rbx$ all that is required is a $c \notin A \;\&\; Rcb$ that is compatible with $R_\# cx$ and $A^l \nsubseteq A^{l\#}$. □

Definition 36. *By the* l-scedastic approximation $\hat{l}$ *and the* u-scedastic approximation $\hat{u}$ *of a subset* $A \subseteq S$ *will be meant the following approximations:*

$$A^{\hat{l}} = (A^l \setminus A^{l\#})^l, \quad A^{\hat{u}} = (A^{u\#} \setminus A^u)^{u\#}.$$

The above cross difference approximation is the best possible from closeness to properties of rough approximations (Fig. 2).

Theorem 34. *For an arbitrary subset* $A \subseteq S$ *of a PRAX* S*, the following statements and diagram of inclusion* ($\rightarrow$) *hold:*

- $A^{l\#l} = A^{l\#} = A^{ll\#} = A^{l\#l\#}$
- *If* $A^u \subset A^{u\#}$ *then* $A^{uu\#} \subseteq A^{u\#u\#}$.

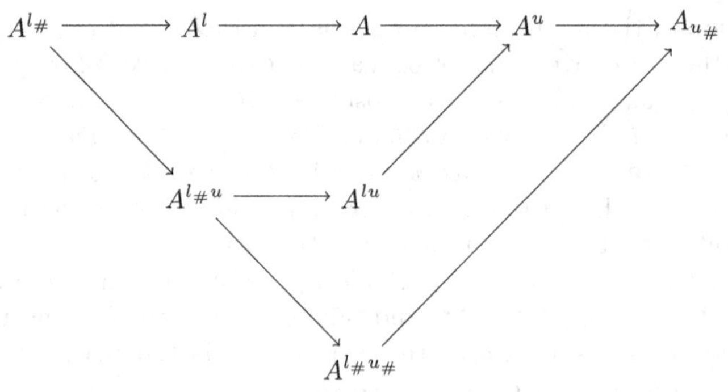

Fig. 2. Relation between approximate approximations

Proof. It is clear that $A^l \subseteq A^u \subseteq A^{u\#}$. So $A^l \nsubseteq A^{u\#} \setminus A^u$.

$$\begin{aligned} x \in (A^l \setminus A^{l\#})^l &\Rightarrow (\exists y)\, [y]_\# \nsubseteq A \;\&\; x \in [y] \subset A \;\&\; x \in [y]_\# \\ &\Rightarrow x \in A^{u\#} \;\&\; x \in A^u \\ &\Rightarrow x \notin A^{u\#} \setminus A^u. \end{aligned}$$

But $[y]_\# \subset A^{u\#}$ $(\exists z)\, z \in A^{u\#} \;\&\; z \notin A^u \;\&\; z \in [y]_\#$.

So $[y]_\# \subset (A^{u\#} \setminus A^u)^{u\#}$ and it is possible that $[y]_\# \nsubseteq (A^{u\#} \setminus A^u)^u$.

□

Theorem 35. *For an arbitrary subset $A \subseteq S$ of a PRAX S,*

$$(A^l \setminus A^{l\#})^l \nsubseteq (A^{u\#} \setminus A^u)^{u\#} \longrightarrow A^{u\#} = A^u.$$
$$A^{u\#} \neq A^u \longrightarrow A^l \setminus A^{l\#})^l \subseteq (A^{u\#} \setminus A^u)^{u\#}.$$

Proof.

- Let $S = \{a, b, c, e, f\}$ and
- R be the transitive completion satisfying Rab, Rbc, Ref.
- If $B = \{a, b\}$, $B^{\hat{l}} = B$, but $B^{u\#} = \{a, b, c\} = B^u$.
- So $B^{\hat{u}} = \emptyset$.
- The second part follows from the proof of the above proposition under the restriction in the premise. □

Theorem 36. *Key properties of the scedastic approximations follow:*

1. $(\forall B \in \wp(S))(B^{\hat{l}} = B \nrightarrow B^{\hat{u}} = B)$.
2. $(\forall B \in \wp(S))(B^{\hat{u}} = B \rightarrow B^{\hat{l}} = B)$.
3. $(\forall B \in \wp(S))\, B^{\hat{l}\hat{l}} = B^{\hat{l}}$.
4. $(\forall B \in \wp(S))\, B^{\hat{u}\hat{u}} \neq B^{\hat{u}}$.
5. *It is possible that* $(\exists B \in \wp(S)\, B^{\hat{u}\hat{u}} \subset B^{\hat{u}})$.

Proof.

1. The counter example in the proof of the above theorem works for this statement.
2. $x \in B \leftrightarrow x \in (B^{u\#} \setminus B^u)^{u\#} \leftrightarrow (\exists y \in B^{u\#})(\exists z \in B^{u\#} \setminus B^u)\, x,\, z \in [y]_\# \,\&\, z \in B^{u\#} \,\&\, z \notin B^u$. But this situation requires that elements of the form z be related to x and so it is essential that $B^{u\#} = B^u$.
3. $B^{\hat{l}\hat{l}} = (B^{\hat{l}l} \setminus B^{\hat{l}l\#})^l = ((B^l \setminus B^{l\#})^l \setminus \emptyset)^l = B^{\hat{l}}$. The missing step is of proving $(B^l \setminus B^{l\#})^{ll\#} = \emptyset$.

4-5 The last two assertions shall be proven together by way of a counterexample and an essential pattern of deviation.
Let $S = \{a, b, c, e, f\}$ and R be a reflexive relation s.t. Rab, Rbc, Ref.
If $A = \{a, e\}$, then $A^{u\#} = \{a, b, c, e\}$ and $A^u = \{a, b, e\}$.
Therefore $A^{\hat{u}} = \{c\} \,\&\, A^{\hat{u}\hat{u}} = \emptyset \,\&\, A^{\hat{u}\hat{u}} \subset A^{\hat{u}}$.
In general if B is some subset, then $x \in B^{\hat{u}} = (A^{u\#} \setminus A^u)^{u\#} \Rightarrow (\exists y \in A^{u\#})(\exists z)\, y \in [z]_\# \,\&\, y \notin A^u \,\&\, y \notin A \,\&\, z \in A \,\&\, y \notin [z] \,\&\, y \in [x]_\#$. □

An interesting problem can be given A for which $A^{u\#} \neq A^u$, when does there exist a B such that

$$B^l = (A^l \setminus A^{l\#})^l = A^{\hat{l}} \,\&\, B^u = (A^{u\#} \setminus A^u)^{u\#} = A^{\hat{u}}?$$

12 Rough Dependence

The concept of *rough dependence* was introduced in general rough set theory by the present author in [29]. By the term *rough dependence*, the present author seeks to capture the relation between two objects (crisp or rough) that have some representable rough objects in common. There is no process for similarity with the concept *mutual exclusivity* of probability theory and in rough evolution there temporality is not usually assumed. The present author would like to eventually analyze the extent to which ontology of not-necessarily-rough origin could be integrated in a seamless way and parts of this have been developed in [31,32]. In this paper, basic concepts will be introduced, compared with probabilistic concepts and the semantic value of introduced functions and predicates will be considered.

Overall the following problems are basic and relevant for use in semantics:

- Which concepts of rough dependence provide for an adequate semantics of rough objects in the PRAX context?
- More generally how does this relation vary over other RSTs?
- Characterize the connection between granularity and rough dependence?

As mentioned earlier, *relation based RST* refers to rough theories originating from generalized approximation spaces of the form $U = \langle \underline{U}, R \rangle$, with $\underline{U}$ being a set and R being any binary relation on $\underline{U}$.

Definition 37. *The* $\tau\nu$-infimal degree of dependence $\beta_{i\tau\nu}$ *of* A *on* B *will be defined as*

$$\beta_{i\tau\nu}(A, B) = \inf_{\nu(S)} \oplus\{C : C \in \tau(S) \,\&\, \mathbf{P}CA \,\&\, \mathbf{P}CB\}.$$

Here the infimum means the largest $\nu(S)$ *element contained in the aggregation.*

The $\tau\nu$-supremal degree of dependence $\beta_{s\tau\nu}$ *of* A *on* B *will be defined as*

$$\beta_{s\tau\nu}(A, B) = \sup_{\nu(S)} \oplus\{C : C \in \tau(S) \,\&\, \mathbf{P}CA \,\&\, \mathbf{P}CB\}.$$

Here the supremum means the least $\nu(S)$ *element containing the sets.*

The definition extends to RYS *[24] in a natural way.*

Note that all of the definitions do not use real-valued rough measures and the cardinality of sets in accord with one of the principles of avoiding contamination. The ideas of dependence are more closely related to certain semantic operations in classical RST. But these were never seen to be of much interest. The connections with probability theories has been part of a number of papers including [35–37,39,44], however neither dependence nor independence have received sufficient attention. This is the case with other papers on entropy. It should be noted that the idea of independence in statistics is seen in relation to probabilistic approaches, but dependence has largely not been given much importance in applications.

The positive region of a set X is X^l, while the negative region is X^{uc} – this region is independent from x in the sense of attributes being distinct, but not in the sense of derivability or inference by way of rules. In considerations of dependence or independence of a set relative another, a basic question would also be about possible balance between the two meta principles of independence in the rough theory and relation to the granular concepts of independence.

Definition 38. *Two elements x, y in a RBRST or CBRST S will be said to be* PN-independent $I_{PN}(xy)$ *if and only if*

$$x^l \subseteq y^{uc} \;\&\; y^l \subseteq x^{uc}.$$

Two elements x, y in a RBRST or CBRST S will be said to be PN-dependent $\varsigma_{PN}(xy)$ *if and only if*

$$x^l \nsubseteq y^{uc} \;\&\; y^l \nsubseteq x^{uc}.$$

Theorem 37. *Over the RYS corresponding to classical RST, the following properties of dependence degrees hold when $\tau(S) = \mathcal{G}(S)$ - the granulation of S and $\nu(S) = \delta_l(S)$ - the set of lower definite elements. The subscripts $\tau\nu$ and braces in $\beta_{i\tau\nu}(x, y)$ are omitted in the following:*

1. $\beta_i xy = x^l \cap y^l = \beta_s xy$ *(subscripts i, s on β can therefore be omitted).*
2. $\beta xx = x^l$.
3. $\beta xy = \beta yx$.
4. $\beta(\beta xy)x = \beta xy$.
5. $\mathbf{P}(\beta xy)(\beta x(y \oplus z))$.
6. $(\mathbf{P}y^l z \longrightarrow \mathbf{P}(\beta xy)(\beta xz))$.
7. $\beta xy = \beta x^l y^l = \beta xy^l$.
8. $\beta 0x = 0;\; \beta x1 = x^l$.
9. $(\mathbf{P}xy \longrightarrow \beta xy = x^l)$.

This is proved in the next subsection.

Theorem 38. *For classical RST, a semantics over the classical semantic domain can be formulated with no reference to lower and upper approximation operators using the operations $\cap$, c, β on the power-set of S, S being an approximation space.*

Proof. It has already been shown that l is representable in terms of β. So the result follows.

12.1 Dependence in PRAX

When $\nu(S) = \delta_l(S)$ and $\tau(S) = \mathcal{G}(S)$ - the successor neighborhood granulation, then the situation in PRAX contexts is similar, but it would not be possible to define u from l and complementation. However when $\nu(S) = \delta_u(S)$, then the situation is very different.

Theorem 39. *Over the RYS corresponding to PRAX with* $\mathbf{P} = \subseteq$, $\oplus = \cup$ *and* $\odot = \cap$, *the following properties of dependence degrees hold when* $\tau(S) = \mathcal{S}$ *- the granulation of* S *and* $\nu(S) = \delta_l(S)$ *- the set of lower definite elements. In fact this holds in any reflexive RBRST. The subscripts* $\tau\nu$ *and braces in* $\beta_{i\tau\nu}(x, y)$ *are omitted in the following:*

1. $\beta_i xy = x^l \cap y^l = \beta_s xy$ *(subscripts* i, s *on* β *can therefore be omitted).*
2. $\beta xx = x^l$; $\beta xy = \beta yx$.
3. $(x \odot y = 0 \longrightarrow \beta_i xy = 0)$, *but the converse is false.*
4. $\beta(\beta xy)x = \beta xy$.
5. $\mathbf{P}(\beta xy)(\beta x(y \oplus z))$.
6. $(\mathbf{P}y^l z \longrightarrow \mathbf{P}(\beta xy)(\beta xz))$.
7. $\beta xy = \beta x^l y^l = \beta xy^l$.
8. $\beta 0x = 0$; $\beta x1 = x^l$.
9. $(\mathbf{P}xy \longrightarrow \beta xy = x^l)$.

Proof.

1. $\beta_i xy$ is the union of the collection of successor neighborhoods generated by elements x and y that are included in both of them. So $\beta_i xy = x^l \cap y^l = \beta_s xy$.
2. $\beta xx = x^l$; $\beta xy = \beta yx$ is obvious.
3. If $(x \odot y = 0$, then x and y have no elements in common and cannot have common successor neighborhoods. If $\beta_i xy = 0$, then x, y have no common successor neighborhoods, but can still have common elements. So the statement follows.
4. $\beta xy \subseteq x^l \subseteq x$ by the first statement. So $\beta(\beta xy)x = \beta xy$.
5. $\mathbf{P}(\beta xy)(\beta x(y \oplus z))$ follows by monotonicity.
6. If $\mathbf{P}y^l z$ is the same thing as $y^l \subseteq z$. $\beta xy = x^l \cap y^l$ and $\beta xz = x^l \cap z^l$ by the first statement. So $(\mathbf{P}y^l z \longrightarrow \mathbf{P}(\beta xy)(\beta xz))$ holds.
7. $\beta xy = \beta x^l y^l = \beta xy^l$ holds because l is an idempotent operation in a PRAX.
8. Rest of the statements are obvious. □

Even though the properties are similar for reflexive RBRST when $\nu(S) = \delta_l(S)$ and $\tau(S) = \mathcal{G}(S)$, there are key differences that can be characterized in terms of special sets.

- $\beta xy = z$ if and only if $(\forall a \in z)(\exists b \in z)\, a \in [z] \subseteq x \cap y$.
- So a minimal $K_z \subseteq z$ satisfying $(\forall a \in z)(\exists b \in K_z)\, a \in [b] \subseteq x$ and $(\forall e \in K_z)\, [e] \subseteq x \cap y$ can be selected. Minimality being with respect to the inclusion order.
- Let $\mathcal{P}_z$ be the collection of all such K_z and let $\mathcal{B}_z$ be the subcollection of $\mathcal{P}_z$ satisfying the condition: if $K \in \mathcal{B}_z$ then $(\forall a \in K)(\forall b \in [a])(\exists J \in \mathcal{B}_z)\, b \in J$. $\mathcal{P}_z$ will be called the local basis and $\mathcal{B}_z$, the local super basis of z.

Proposition 34. *For classical RST* $(\forall z)\, \mathcal{B}_z = \mathcal{P}_z$ *and conversely.*

Theorem 40. *In the context of Theorem 39, if $\nu(S) = \delta_u(S)$ and $\tau(S)$ is as before, then all of the following hold (βxy is an abbreviation for $\beta_i xy$)*

1. $\mathbf{P}(\beta xy)(\beta_{i\delta_l(S)}xy)$,
2. $\mathbf{P}(\beta xx)(x^l)$; $\beta xy = \beta yx$.
3. $(x \odot y = 0 \longrightarrow \beta_i xy = 0)$, *but the converse is false.*
4. $\beta(\beta xy)x = \beta xy$.
5. $\mathbf{P}(\beta xy)(\beta x(y \oplus z))$.
6. $(\mathbf{P}y^l z \longrightarrow \mathbf{P}(\beta xy)(\beta xz))$.
7. $\beta xy = \beta x^l y^l$; $\mathbf{P}(\beta xy^l)(\beta x^u y^u)$.
8. $\beta 0x = 0$; $\mathbf{P}(\beta x1)(x^l)$.
9. $(\mathbf{P}xy \longrightarrow \mathbf{P}(\beta zx)(\beta zy))$
10. $(\beta xy)^l = \beta xy$.

Proof.

1. By definition $\beta_{i\tau\nu}(A, B) = \inf_{\nu(S)} \oplus\{C : C \in \tau(S) \ \&\ \mathbf{P}CA \ \&\ \mathbf{P}CB\}$, so βxy is the greatest upper definite set contained in the union of common successor neighborhoods included in x and y. So it is necessarily a subset of $x^l \cap y^l$. In a PRAX, u is not idempotent and in general $x^u \subseteq x^{uu}$. So $\mathbf{P}(\beta xy)(\beta_{i\delta_l(S)}xy)$.
2. The statements $\mathbf{P}(\beta xx)(x^l)$ and $\beta xy = \beta yx$ follow from the above.
3. The proof is similar to that of third statement of Theorem 39.
4. In constructing $\beta(\beta xy)x$ from βxy, no effort is made to look for upper definite subsets strictly contained in the latter. So the property follows.
5. $\mathbf{P}(\beta xy)(\beta x(y \oplus z))$ follows by monotonicity.
6. Obvious from previous statements.
7. Note that $\beta x^u y^u$ is a subset of $x^u \cap y^u$ and in general contains βxy.
8. Is a special case of the first statement. 0 is the empty set and 1 is the top.
9. Follows by monotonicity.
10. Upper definite subsets are necessarily lower definite, so $(\beta xy)^l = \beta xy$. □

The main properties of PN-dependence is as below:

Theorem 41. *In the context of Theorem 39, all of the following hold (the subscript 'PN' in ς_{PN} in the following):*

1. ςxx.
2. $(\varsigma xy \leftrightarrow \varsigma yx)$.
3. *In general, ςxy & ςzy does not imply ςxz. But $\neg\varsigma xz$ is more likely if a bit of frequentism is assumed.*
4. *In general, $\varsigma xy \nrightarrow \varsigma x^u y^u$ and $\varsigma x^u y^u \nrightarrow \varsigma xy$.*
5. $(x \cdot y = 0 \longrightarrow \neg\varsigma xy)$.
6. $(\mathbf{P}xy \longrightarrow \varsigma xy)$.

Theorem 42. *In the context of Theorem 39, if $\beta xy \neq 0$ then ςxy, but the converse need not hold. In classical RST, the converse holds as well.*

Proof. If $\beta xy \neq 0$, then it follows that $x^l \cap y^l \neq \emptyset$ under the assumptions. If it is assumed that $x^l \subseteq y^{uc} \vee y^l \subseteq x^{uc}$, then in each of the three cases a contradiction happens. So the first part of the result follows.

In the classical case, if $x^l \subseteq y^{uc}$ is not empty, then it should be a union of successor neighborhoods and similarly for $y^l \subseteq x^{uc}$. These two parts should necessarily be common to x^l and y^l. So the converse holds for classical RST. The proof does not work for PRAX and the reasons for failure have been made clear. □

13 Comparison with Dependence in Probabilistic Theories

Probability measures may not exist in the first place over any given collection of sets, so even CBRST is necessarily more general and the idea of mutual exclusivity is not the correct concept corresponding to rough dependence. The basic idea of probabilistic dependence is oriented because occurrence of an event can be favorable or unfavorable for another event. In standard versions of rough set theory this has no corresponding concept. The concept of dependence in probability is rarely considered in the literature. The version in [7] uses a not-so intuitive valuation but is nevertheless useful. The subjective aspect of the valuation is abstracted for comparison.

Among the different understandings of probabilistic causation, frequentism [10] and the tendency to omit necessary conditions are particularly problematic in various soft computing situations. A commitment to avoid the excesses of frequentism in rough sets is implicit in the present author's approach towards real-valued rough measures.

If $(X, \mathcal{S}, p)$ is a probability space with X being a set, $\mathcal{S}$ being a σ-algebra over X and p being a probability function (collections of probability functions can be used to handle more complex notions of dependence in 'probability structures', but these add little to the comparison), then the most natural dependence function $\delta : \mathcal{S}^2 \longmapsto \Re$ is defined by

$$\partial(x, y) = p(x \cap y) - p(x) \cdot p(y)$$

This function satisfies a number of properties that can be used to characterize dependence. In the subjective probability domain where p takes value in a bounded partially ordered partial semi-ring or your favorite partially ordered algebra, it will be required to replace δ with a pair of predicates. So orientation of dependence seems to be fundamental in general forms of probability theory as well.

Two events $x, y \in X$ are *mutually exclusive* if and only if $x \cap y \neq \emptyset$. This concept can be extended to countable sets of events in a natural way. Also it is worthwhile to modify the concept of mutual exclusivity as in following definition:

Definition 39. *Two events x, y will be said to be* weakly mutually exclusive *(WME) if and only if*

$$x \cap y \neq z \;\&\; p(z) = 0.$$

Most results of probability theory involving mutual exclusivity continue to hold with the weaker assumption of WME and importantly is a better (though artificial) concept for comparison with the situation for rough sets.

Definition 40. *In the above context, let*

- πxy *if and only if* $p(x) \cdot p(y) < p(x \cap y)$
- σxy *if and only if* $p(x \cap y) < p(x) \cdot p(y)$

Proposition 35. *All of the following hold in a probability space:*

- $\pi xy^c \leftrightarrow \sigma yx$
- $\pi xy \leftrightarrow \pi yx$
- $(x \cap y \neq \emptyset \longrightarrow (\pi xa \;\&\; \pi ya \longrightarrow \pi(x \cup y)a))$
- $(x \cap y \neq \emptyset \longrightarrow (\sigma xa \;\&\; \sigma ya \longrightarrow \sigma(x \cup y)a))$
- $(\emptyset \neq x \subseteq y \longrightarrow \pi xy)$
- $(x \cap y = \emptyset \longrightarrow \sigma xy)$

Instead of using the function $\partial(x, y)$, the relations π, σ can be used, as the former lacks a comparable contamination-free counterpart in rough set theory and also has peculiar properties like $\partial(x, x) \in [0, 1/4]$.

Proposition 36. *In the probability space above* $0 \leq \partial(x, x) \leq 0.25$, $-0.25 \leq \partial(x, x^c) \leq 0$ *and x, y are independent implies $\partial(x, y) = 0$, but not conversely.*

Proof. The proof of the inequalities follow by a simple application of real analysis. □

So it follows that the interpretation of the function $\partial(x, y)$ as in [7] is actually incomplete. It combines certainty of the event with dependence.

Even though it is possible to speak of positive, negative and neutral regions corresponding to an arbitrary subset A of a RBRST or CBRST S, natural ideas of dependence do not correspond to the scenario in probability space. In fact,

Theorem 43. *Predicates having properties identical with those of π and σ cannot be defined in the context of Theorem 39.*

Proof of this and more general results will appear separately.

14 Dependency Semantics of PRAX

Dependency based semantics are developed in at least two ways in this section. The *internalization based semantics* is essentially about adjoining predicates to the Nelson algebra corresponding to $\mathfrak{R}_w(S)$. The *cumulation based semantics* is essentially about cumulating both the semantics of $\mathfrak{R}(S)$, adjusting operations and adjoining predicates. Broader dependency based predicates are used in this case, but the value of the method is in fusion of the methodologies.

The central blocks of development of the cumulation based dependency semantics are the following:

- Take $\mathfrak{R}(S) \cup \mathfrak{R}_w(S)$ as the universal set of the intended partial/total algebraic system.
- Use a one point completion of τ to distinguish between elements of $\mathfrak{R}_w(S) \setminus \mathfrak{R}(S)$ and those in $\mathfrak{R}(S)$.
- Extend the idea of operational dependency to pairs of sets.
- Extend operations of aggregation, commonality and dual suitably.
- Interpret semantic dependence?

The first step is obvious, but involves elimination of other potential sets arising from the properties of the map τ.

One Point Completion

Since $R \subseteq R_w$ and R_w is transitive, so

Proposition 37.

$$\alpha \in \mathfrak{R}(S) \cap \mathfrak{R}_w(S) \text{ if and only if } \tau(\alpha) = \alpha.$$

Adjoin an element 0 to $\mathfrak{R}(S) \cup \mathfrak{R}_w(S)$ to form $\mathfrak{R}^*(S)$ and extend τ (interpreted as a partial operation) to $\overline{\tau}$ as follows:

$$\overline{\tau}(\alpha) = \begin{cases} \tau(\alpha) & \text{if } \alpha \in \mathfrak{R}(S), \\ 0 & \text{if } \alpha \notin \mathfrak{R}(S). \end{cases}$$

Note that this operation suffices to distinguish between elements common to $\mathfrak{R}(S)$ and $\mathfrak{R}_w(S)$, and those exclusively in $\mathfrak{R}(S)$ and not in $\mathfrak{R}_w(S)$.

Dependency on Pairs

It is possible to consider all dependencies relative to the Nelson algebra or $\mathfrak{R}(S)$. In the proposed approach the former is considered first towards avoiding references to the latter.

Definition 41. *By the* paired infimal degree of dependence $\beta^{+}_{i\tau_1\tau_2\nu_1\nu_2}$ *of α on β will be defined as*

$$(\beta_{i\tau_1\nu_1}(e_1\alpha, e_1\beta), \beta_{i\tau_2\nu_2}(e_2\alpha, e_2\beta)).$$

Here the infimums involved are the largest $\nu_1(S)$ and $\nu_2(S)$ elements contained in the aggregation and the $e_j\alpha$ is the j-th component of α.

The following well defined specialization with $\tau_1(S) = \tau_2(S) = \mathcal{G}_w(S)$, $\nu_1 = \delta_{lw}(S)$ and $\nu_2 = \Gamma_{uw}(S)$ will also be of interest. For specializing the dependencies between a element in $\mathfrak{R}(S)$ and its image in $\mathfrak{R}_w(S)$, it suffices to define:

Definition 42. *Under the above assumptions, by the* relative semantic dependence $\varrho(\alpha)$ *of* $\alpha \in \mathfrak{R}(S)$, *will be meant*

$$\varrho(\alpha) = \beta_i^+(\alpha, \tau(\alpha)).$$

The idea of relative semantic dependence refers to elements in $\mathfrak{R}(S)$ and it can be reinterpreted as a relation on $\mathfrak{R}_w(S)$.

Internalization Based Semantics

Definition 43. *By the* ϱ/σ-semantic dependences $\varrho(\alpha)$, $\sigma(\alpha)$ *of* $\alpha \in \mathfrak{R}(S)$, *will be meant* $\varrho(\alpha) = \beta_i^+(\alpha, \tau(\alpha))$ *and* $\sigma(\alpha) = \beta_i^+(\alpha, ((\varphi(e_1\alpha) \setminus e_1\alpha)^l, (\varphi(e_2\alpha) \setminus e_1\alpha)^u))$ *respectively. Such relations are optional in the internalization process.*

A relation Υ *on* $\mathfrak{R}_w(S)$ *will be said to be a* relsem-relation *if and only if*

$$\Upsilon\tau(\alpha)\gamma \leftrightarrow (\exists\beta \in \tau^{\dashv}\tau(\alpha))\,\gamma = \varrho(\beta).$$

Note that, $\tau(\alpha) = \tau(\beta)$ *by definition of* $\tau^{\dashv}$.

Through the above definitions the following internalized approximate definition has been arrived at:

Definition 44. *By an* Approximate Proto Vague Semantics *of a* PRAX S *will be meant an algebraic system of the form*

$$\mathfrak{P}(S) = \langle \mathfrak{R}_w(S), \Upsilon\vee, \wedge, c, \bot, \top \rangle,$$

with $\langle \mathfrak{R}_w(S), \vee_w, \wedge_w, c, \bot, \top \rangle$ *being a Nelson algebra over an algebraic lattice and* Υ *being as above.*

Theorem 44. Υ *has the following properties:*

$$\alpha = \tau(\alpha) \longrightarrow \Upsilon\alpha\alpha.$$
$$\Upsilon\alpha\gamma \longrightarrow \gamma \wedge_w \alpha = \gamma.$$
$$\Upsilon\alpha\gamma \;\&\; \Upsilon\gamma\alpha \longrightarrow \alpha = \gamma.$$
$$\Upsilon\bot\bot \;\&\; \Upsilon\top\top.$$
$$\Upsilon\alpha\gamma \;\&\; \Upsilon\beta\gamma \longrightarrow \Upsilon(\alpha \vee_w \beta)\gamma.$$

Proof.

- If $\alpha = \tau(\alpha)$, then $\alpha = \varrho(\alpha) = \beta_i^+(\alpha, \tau(\alpha))$. So $\Upsilon\alpha\alpha$.
- If $\Upsilon\alpha\gamma$, then it follows from the definition of β_i^+, that the components of gamma are respectively included in those of α. So $\gamma \wedge \alpha = \gamma$.
- Follows from the previous.
- Proof is easy.
- From the premise we have $(\exists\mu \in \tau^{\dashv}\tau(\alpha))\,\gamma = \varrho(\mu)$ and $(\exists\nu \in \tau^{\dashv}\tau(\beta))\,\gamma = \varrho(\nu)$. This yields $(\exists\lambda \in \tau^{\dashv}\tau(\alpha \vee_w \beta))\,\gamma = \varrho(\lambda)$ as can be checked from the components.

□

$\Upsilon_{\tau(\alpha)} = \{\gamma;\, \Upsilon\tau(\alpha)\gamma\}$ is the approximate reflection of the set of τ-equivalent elements in $\mathfrak{R}(S)$ identified by their dependence degree. In the approximate semantics aggregation and commonality are not lost track of as the above theorem shows. For a falls-down semantics, the natural candidates include the ones corresponding to largest equivalence or the largest semi-transitive contained in R. The latter will appear in a separate paper. For the former, the general technique (using $\sigma(\alpha)$) extends to PRAX as follows:

Definition 45.

- *Define a map from set of neighborhoods to l-definite elements* $\int([x]_o) = \cup_{y\in[x]_o}[y]$ *and extend it to images of* l_o, u_o *via,*

$$\oint(A^{l_o}) = \cup_{[y]_o\subseteq A^{l_o}} \int([y]_o).$$

- *Extend this to a map* $\ltimes : \mathfrak{R}_o(S) \longmapsto \mathfrak{R}(S)$ *via* $\ltimes(\alpha) = (\oint(e_1\alpha), \oint(e_2\alpha))$.
- *Define* $\Pi\alpha\nu$ *on* $\mathfrak{R}_o(S)$ *iff* $(\exists\gamma \in \ltimes^{\dashv}\ltimes(\alpha))\, \beta_i^+(\alpha,\gamma) = \nu$. *Let* $\Pi_\alpha = \{\nu\,;\Pi\alpha\nu\}$.
- *By a* Direct Falls Down *semantics of PRAX, will be meant an algebraic system of the form*

$$\mathfrak{I}(S) = \langle\mathfrak{R}_o(S), \Pi, \vee, \wedge, c, \bot, \top\rangle,$$

with $\langle\mathfrak{R}_o(S), \vee_o, \wedge_o, \rightarrow, c, \bot, \top\rangle$ *being a semi-simple Nelson algebra [34].*
- *The falls down semantics determines a cover* $\mathfrak{I}^*(S) = \{\Pi_\alpha\,;\, \alpha \in \mathfrak{R}_o(S)\}$

Theorem 45. *In the above context, all of the following hold:*

- $\Pi\alpha\alpha$.
- $(\Pi\alpha\mu \;\&\; \Pi\mu\alpha \longrightarrow \alpha = \mu)$.
- $(\Pi\alpha\gamma \longrightarrow \gamma \subseteq \alpha)$. *The converse is false.*
- $\alpha \neq \bot \;\&\; \Pi\alpha\gamma \;\&\; \Pi\alpha\mu \longrightarrow \beta_i^+(\gamma,\mu) \neq \bot$.
- $\mu \in \Pi_\alpha \;\&\; \mu \subseteq \nu \subseteq \alpha \longrightarrow \nu \in \Pi_\alpha$.

The theorems mean that a purely order theoretic representation theorem is not possible for the falls down semantics, but other possibilities remain open.

Cumulation Based Semantics

The idea of cumulation is correctly a way of enhancing our original semantics based on proto-vagueness algebras with the Nelson algebraic semantics and the operational dependence. This is defined for a central problem relating to the underlying semantic domains.

Definition 46. *By a* cumulative proto-vague algebra *will be meant a partial algebra of the form*

$$\mathfrak{C}(S) = \left\langle\mathfrak{R}^*(S), \overline{tau}, \oplus, \odot, \otimes, \dagger, \bot, \top\right\rangle.$$

Problem:

When can the cumulation based semantics be deduced from (that is the extra operations can be defined from the original ones) within a full proto-vagueness algebra?

15 Geometry of Granular Knowledge Interpretation

In this section aspects of knowledge interpretation in PRAX contexts will be considered in the light of the results on representation of rough objects. For details of the knowledge interpretation of rough sets, the reader is referred to [5,24,27,35]. Connections of rough sets with concept lattices provide for deducing concepts and related rules from data tables. This is obviously related at a level. In [8], applications corresponding to quasi-ordered and tolerances indiscernibility are considered with considerable intrusion into data. By an extension of all of these considerations any proto-transitive relation corresponds to knowledge. The most natural semantic domain for the present purpose is the one corresponding to the rough objects, but it is not the only one of interest and representation will be relative at least these two.

Any knowledge, however involved, may be seen as a collection of concepts with admissible operations of reasoning defined on them. Knowledges associated PRAX have various peculiarities corresponding to the semantic evolution of rough objects in it. To start with, the semantic domains of representation properly contain the semantic domains of interpretation. Not surprisingly, it is because the rough objects corresponding to l, u cannot be represented perfectly in terms of objects from $\delta_{lu}(S)$ alone. *In the nongranular perspective too, this representation aspect should matter - 'should'*, because it is matter of choice during generalization from the classical case in the non granular approach.

The natural rough semantic domains of l, u is Meta-R, while that of l_o, u_o is $\mathfrak{O}$ (say, corresponding rough objects of $\tau(R)$). These can be seen as separate domains or as parts of a minimal containing domain that permits enough expression. As demonstrated earlier, knowledge is correctly representable in terms of *atomic concepts of knowledge* at semantic domains placed between Meta-C and Meta-R and not at the latter. So the characterization of possible semantic domains and their mutual ordering - leading to their geometry is of interest.

The following will be assumed to be *part* of the interpretation:

- Two types of rough objects corresponding to Meta-R and $\mathfrak{O}$ and their natural correspondence correspond to concepts or weakenings thereof. A concept relative one semantic domain need not be one of the other.
- A granule of the rough semantic domain $\mathcal{O}$ is necessarily a concept of $\mathcal{O}$, but a granule of Meta-R may not be a concept of $\mathcal{O}$ or Meta-R.
- Critical points are not necessarily concepts of either semantic domain.
- Critical points and the representation of rough objects require the rough semantic domains to be extended.

The above obviously assumes that a PRAX S has at least two kinds of knowledge associated (in relation to the Pawlak-sense interpretation). To make the interpretations precise, these will be indicated by $\mathcal{I}_1(S)$ and $\mathcal{I}_o(S)$ respectively (corresponding to the approximations to l, u and l_o, u_o respectively). The pair $(\mathcal{I}_1(S), \mathcal{I}_o(S))$ will also be referred to as the *generalized* KI.

Definition 47. *Given two PRAX* $S = \langle \underline{S}, R\rangle$, $V = \langle \underline{S}, Q\rangle$, *S will be said to be* o-coarser *than V if and only if* $\mathcal{I}_o(S)$ *is coarser than* $\mathcal{I}_o(V)$ *in Pawlak-sense (that is* $\tau(R) \subseteq \tau(Q)$*). Conversely, V will be said to be a* o-refinement *of S.*

S will be said to be p-coarser *than V if and only if* $\mathcal{I}_1(S)$ *is coarser than* $\mathcal{I}_1(V)$ *in the sense* $R \subseteq Q$. *Conversely, V will be said to be a* p-refinement *of S.*

An extended concept of positive regions is defined next.

Definition 48. *If* $S_1 = \langle \underline{S}, Q\rangle$ *and* $S_2 = \langle \underline{S}, P\rangle$ *are two PRAX such that* $Q \subset R$, *then by the* granular positive region *of Q with respect to R is given by* $gPOS_R(Q) = \{[x]_Q^{l_R} : x \in \underline{S}\}$, *where* $[x]_Q^{l_R}$ *is the lower approximation (relative R) of the Q-related elements of x. Using this the* granular extent of dependence *of knowledge encoded by R on the knowledge encoded by Q can be defined by natural injections* : $gPOS_R(Q) \longmapsto \mathcal{G}_R$.

Lower critical points can be naturally interpreted as preconcepts that are definitely included in the discourse, while upper critical points are preconcepts that include most of the discourse. The problem with this interpretation is that it's representation requires a semantic domain at which critical points of different kinds can be found. A key requirement for such a domain would be the feasibility of rough counting procedures like IPC [24]. Semantic domains that have critical points of different types as basic objects will be referred to as a Meta-RC.

The following possible axioms of granular knowledge that also figure in [27] (due to the present author), get into difficulties with the present approach and even when attention is restricted to $\mathcal{I}_1(S)$:

1. Individual granules are atomic units of knowledge.
2. Maximal collections of granules subject to a concept of mutual independence are admissible concepts of knowledge.
3. Parts common to subcollections of maximal collections of granules are also knowledge.

The first axiom holds in weakened form as the granulation $\mathcal{G}$ for $\mathcal{I}_1(S)$ is only lower definite and affects the other. The possibility of other nice granulations being possible for the PRAX case appears to be possible at the cost of other nice properties. So it can be concluded that in proper KR happens at semantic domains like Meta-RC where critical points of different types are perceived. Further at Meta-R, rough objects may correspond to knowledge or conjectures - if the concept of proof is required to be an ontological concept or beliefs. The scenario can be made more complex with associations of $\mathfrak{O}$ knowledges.

From a non-granular perspective, in Meta-R rough objects must correspond to knowledge with some of them lacking a proper evolution - there is no problem here. When $\mathfrak{O}$ objects are permitted, then in the perspective two kinds of closely associated knowledges can be spoken of.

The connections with non-monotonic reasoning and the approximate Nelson algebra semantics developed in this paper suggest further enhancements to the above. These will be explored separately.

16 Further Directions and Remarks

In this research all of the following have been developed by the present author:

- The basic theory of rough sets over proto transitive relations,
- Characterization of the nature of rough objects and possible approximations,
- Two different algebraic semantics for the above. One of the semantics uses approximation strategies on relations to internalize the semantics within that of the semantics of quasi order relation based rough sets.
- Important connections between approximations and operators of generalized operators of non-monotonic reasoning.
- Knowledge interpretation in PRAX contexts have also been outlined.
- Various examples at the level of real-life applications are also outlined in the paper. Concepts have been illustrated through a persistent example.

In continuation of earlier work by the present author in [28], semantic consequences of the relation between proto transitivity and its approximations is developed in detail. The theory of rough dependence from the knowledge perspective is also specialized to PRAX and extended for the purposes of the semantics in this paper. Connections with probabilistic dependence is shown to be lacking any reasonable basis and questions are raised on unbridled frequentism in rough set theory.

The relation of the developed theory with entropy is strongly motivated by the knowledge interpretation [26,27] and will be part of future work. Connections of general rough sets with concept analysis as expounded in [8] and related papers is also motivated by the present work. Apart from rethinking the approach from a contamination avoidance perspective, extension to proto transitive relations in the context and correspondences of the semantic fragments (corresponding to knowledge) are of interest. These will also be taken up in the future.

The first algebraic semantics was seen to be inadequate in not being particularly elegant and requiring additional predicates for a reasonable abstract representation theorem. This was one reason for restricting derivations involving rough objects of $\tau(R)$. The internalization of a semantics of PRAX objects in Nelson algebras through ideas of rough dependence is shown to lead to a beautiful semantics. Further formulations of associated logics will be part of a future paper. The technique can be extended to define approximate semantics in various other rough set-theoretical contexts.

While this paper was in review, an antichain based semantics valid for the contexts of this paper and more general contexts have been developed by the present author in [30]. The logics that can be associated with the algebraic semantics is another research direction.

Acknowledgment. The present author would like to thank Prof Mihir Chakraborty for discussions on PRAX and the referee for drawing attention to [8] and related references in particular.

References

1. Bianucci, D., Cattaneo, G., Ciucci, D.: Entropies and co-entropies of coverings with application to incomplete information systems. Fundam. Inform. **75**, 77–105 (2007)
2. Burmeister, P.: A Model-Theoretic Oriented Approach to Partial Algebras. Akademie-Verlag, Berlin (1986, 2002)
3. Cattaneo, G., Ciucci, D.: Lattices with interior and closure operators and abstract approximation spaces. In: Peters, J.F., Skowron, A., Wolski, M., Chakraborty, M.K., Wu, W.-Z. (eds.) Transactions on Rough Sets X. LNCS, vol. 5656, pp. 67–116. Springer, Heidelberg (2009)
4. Chajda, I., Haviar, M.: Induced pseudo orders. Acta Univ. Palack. Olomou **30**(1), 9–16 (1991)
5. Chakraborty, M.K., Samanta, P.: Consistency-degree between knowledges. In: Kryszkiewicz, M., Peters, J.F., Rybiński, H., Skowron, A. (eds.) RSEISP 2007. LNCS (LNAI), vol. 4585, pp. 133–141. Springer, Heidelberg (2007)
6. Ciucci, D.: Approximation algebra and framework. Fundam. Inform. **94**, 147–161 (2009)
7. Dimitrov, B.: Some Obreshkov measures of dependence and their Use. Comptes Rendus Acad Bulg. Sci. **63**(1), 5–18 (2010)
8. Ganter, B., Meschke, C.: A formal concept analysis approach to rough data tables. In: Peters, J.F., Skowron, A., Sakai, H., Chakraborty, M.K., Slezak, D., Hassanien, A.E., Zhu, W. (eds.) Transactions on Rough Sets XIV. LNCS, vol. 6600, pp. 37–61. Springer, Heidelberg (2011)
9. Greco, S., Pawlak, Z., Slowinski, R.: Can Bayesian measures be useful for rough set decision making? Eng. Appl. AI **17**, 345–361 (2004)
10. Hajek, A.: Fifteen arguments against hypothetical frequentism. Erkenntnis **211**(70), 211–235 (2009)
11. Janicki, R.: Approximations of arbitrary binary relations by partial orders: classical and rough set models. In: Peters, J.F., Skowron, A., Chan, C.-C., Grzymala-Busse, J.W., Ziarko, W.P. (eds.) Transactions on Rough Sets XIII. LNCS, vol. 6499, pp. 17–38. Springer, Heidelberg (2011)
12. Järvinen, J.: Lattice theory for rough sets. In: Peters, J.F., Skowron, A., Düntsch, I., Grzymała-Busse, J.W., Orłowska, E., Polkowski, L. (eds.) Transactions on Rough Sets VI. LNCS, vol. 4374, pp. 400–498. Springer, Heidelberg (2007)
13. Jarvinen, J., Pagliani, P., Radeleczki, S.: Information completeness in Nelson algebras of rough sets induced by quasiorders. Stud. Logica. **101**, 1–20 (2012)
14. Jarvinen, J., Radeleczki, S.: Representation of Nelson algebras by rough sets determined by quasi-orders. Algebra Univers. **66**, 163–179 (2011)
15. Keet, C.M.: A formal theory of granules - Phd thesis. Ph.D. thesis, Faculty of Computer Science, Free University of Bozen (2008)
16. Lin, T.Y.: Granular computing -1: the concept of Granulation and its formal model. Int. J. Granular Comput. Rough Sets Int. Syst. **1**(1), 21–42 (2009)
17. Ljapin, E.S.: Partial Algebras and Their Applications. Kluwer Academic, Dordrecht (1996)
18. Makinson, D.: General Patterns in Nonmonotonic Reasoning, vol. 3, pp. 35–110. Oxford University Press, New York (1994)
19. Makinson, D.: Bridges between classical and nonmonotonic logic. Logic J. IGPL **11**, 69–96 (2003)

20. Mani, A.: Algebraic semantics of similarity-based bitten rough set theory. Fundam. Inform. **97**(1–2), 177–197 (2009)
21. Mani, A.: Meaning, choice and similarity based rough set theory. In: International Conference Logic and Application, January 2009, Chennai, pp. 1–12 (2009). http://arxiv.org/abs/0905.1352
22. Mani, A.: Choice inclusive general rough semantics. Inf. Sci. **181**(6), 1097–1115 (2011). http://dx.doi.org/10.1016/j.ins.2010.11.016
23. Mani, A.: Axiomatic granular approach to knowledge correspondences. In: Li, T., Nguyen, H.S., Wang, G., Grzymala-Busse, J., Janicki, R., Hassanien, A.E., Yu, H. (eds.) RSKT 2012. LNCS, vol. 7414, pp. 482–487. Springer, Heidelberg (2012)
24. Mani, A.: Dialectics of counting and the mathematics of vagueness. In: Peters, J.F., Skowron, A. (eds.) Transactions on Rough Sets XV. LNCS, vol. 7255, pp. 122–180. Springer, Heidelberg (2012)
25. Mani, A.: Contamination-free measures and Algebraic operations. In: Pal, N., et al. (ed.) Fuzzy Systems (FUZZ), 2013 IEEE International Conference on Fuzzy Systems, vol. F-1438, Hyderabad, India, pp. 1–8 (2013)
26. Mani, A.: Dialectics of knowledge representation in a granular rough set theory. In: Refereed Conference Paper: ICLA 2013, Institute of Mathematical Sciences, Chennai, pp. 1–12 (2013). http://arxiv.org/abs/1212.6519
27. Mani, A.: Towards logics of some rough perspectives of knowledge. In: Skowron, A., Suraj, Z. (eds.) Rough Sets and Intelligent Systems - Professor Zdzisław Pawlak in Memoriam. ISRL, vol. 43, pp. 419–444. Springer, Heidelberg (2013)
28. Mani, A.: Approximation dialectics of proto-transitive rough sets. In: Chakraborty, M.K., Skowron, A., Maiti, M., Kar, A. (eds.) Facets of Uncertainties and Applications. Springer Proceedings in Mathematics and Statistics, vol. 125. Springer, New Delhi (2015)
29. Mani, A.: Ontology, rough Y-systems and dependence. Int. J. Comput. Sci. Appl. **11**(2), 114–136 (2014). Special Issue of IJCSA on Comput. Intell
30. Mani, A.: Antichain based semantics for rough sets. In: Ciucci, D., et al. (eds.) RSKT 2015. LNCS, vol. 9436, pp. 335–346. Springer, Heidelberg (2015). doi:10.1007/978-3-319-25754-9_30
31. Mani, A.: Types of Probabilities Associated with Rough Membership Functions. In: Bhattacharyya, S., et al. (ed.) Proceedings of ICRCICN 2015: IEEE Xplore, pp. 175–180. IEEE Computer Society (2015). http://dx.doi.org/10.1109/ICRCICN.2015.7434231
32. Mani, A.: Probabilities, dependence and rough membership functions. Int. J. Comput. Appl. (Special Issue on Comput. Intell.), 1–27 (2016, accepted)
33. Moore, E.F., Shannon, C.E.: Reliable circuits using less reliable relays-I, II. Bell Syst. Tech. J. 191–208, 281–297 (1956)
34. Pagliani, P., Chakraborty, M.: A Geometry of Approximation: Rough Set Theory: Logic, Algebra and Topology of Conceptual Patterns. Springer, Berlin (2008)
35. Pawlak, Z.: Rough Sets: Theoretical Aspects of Reasoning About Data. Kluwer Academic Publishers, Dodrecht (1991)
36. Pawlak, Z.: Decision tables and decision spaces. In: Proceedings of the 6th International Conference on Soft Computing and Distributed Processing (SCDP 2002), 24–25 June 2002
37. Pawlak, Z.: Some issues on rough sets. In: Peters, J.F., Skowron, A., Grzymała-Busse, J.W., Kostek, B., Świniarski, R.W., Szczuka, M.S., et al. (eds.) Transactions on Rough Sets I. LNCS, vol. 3100, pp. 1–58. Springer, Heidelberg (2004)
38. Shannon, C.E.: A mathematical theory of communication. Bell Syst. Tech. J. **27**(379–423), 623–656 (1948)

39. Ślęzak, D.: Rough sets and bayes factor. In: Peters, J.F., Skowron, A. (eds.) Transactions on Rough Sets III. LNCS, vol. 3400, pp. 202–229. Springer, Heidelberg (2005)
40. Ślęzak, D., Wasilewski, P.: Granular sets – foundations and case study of tolerance spaces. In: An, A., Stefanowski, J., Ramanna, S., Butz, C.J., Pedrycz, W., Wang, G. (eds.) RSFDGrC 2007. LNCS (LNAI), vol. 4482, pp. 435–442. Springer, Heidelberg (2007)
41. Wasilewski, P., Slezak, D.: Foundations of rough sets from vagueness perspective. In: Hassanien, A., et al. (eds.) Rough Computing: Theories, Technologies and Applications, pp. 1–37. Information Science Reference, IGI, Global, Hershey (2008)
42. Yao, Y.: Information granulation and rough set approximation. Int. J. Intell. Syst. **16**, 87–104 (2001)
43. Yao, Y.: Probabilistic approach to rough sets. Expert Syst. **20**(5), 287–297 (2003)
44. Yao, Y.: Probabilistic rough set approximations. Int. J. Approximate Reasoning **49**, 255–271 (2008)
45. Zadeh, L.A.: Fuzzy sets and information granularity. In: Gupta, N., et al. (eds.) Advances in Fuzzy Set Theory and Applications, pp. 3–18. North Holland, Amsterdam (1979)

Covering Rough Sets and Formal Topology – A Uniform Approach Through Intensional and Extensional Constructors

Piero Pagliani(✉)

Rome, Italy
pier.pagliani@gmail.com

Abstract. Approximation operations induced by coverings are reinterpreted through a set of four "constructors" defined by simple logical formulas. The very logical definitions of the constructors make it possible to readily understand the properties of such operators and their meanings.

1 Introduction

Covering-based rough sets are receiving more and more interest by the data mining community, mathematicians and logicians. Indeed, a covering on a set U reflects many real-world situations that cannot be coped by partitions or by approaches based on particular relations. If partitions are connected to *equivalence relations*, coverings are connected to *tolerance (similarity) relations*, but only under some particular circumstances, so that the analysis of the properties of covering-based approximation operators requires specific investigations by means of different mathematical tools. The present paper is an attempt to provide a uniform framework for studying covering-based approximation operators. It is based on the intuition that, from a data-analysis point of view, a covering can be interpreted as the result of an *observation* (experiment, manifestation, and the like): if U is a set of *objects* (entities, points, *noumena*) and M a set of *properties* (characteristic features, and the like), and if R connects any object to the properties it fulfills, then the set $\{R^{\smile}(m) : m \in M\}$ is a covering $\mathbf{C}$ of U, where $R^{\smile}$ is the inverse relation of R. In other words, $\mathbf{C}$ is the set of the *extensions* of the properties in M.

This simple observation makes it possible to explore the properties of a number of approximation operators introduced so far in the literature on covering-based rough sets, by means of just two pairs of extensional and intensional operators. These operators have been introduced in Pointless (or Formal) Topology. They will be called *constructors* and are defined by simple logical formulas based on the relation R. Their nice mathematical properties make it possible to discover those of a number of covering-based approximation operators, uniformly and without complicated proofs. Moreover, the logical definitions of the constructors clarify the meaning of the approximation operations themselves.

J.F. Peters and A. Skowron (Eds.): TRS XX, LNCS 10020, pp. 109–145, 2016.
DOI: 10.1007/978-3-662-53611-7_4

Particularly, the second part of the paper is devoted to the algebraic status of the systems induced by the approximation operators in dependence of the properties of the given coverings.

The paper is self-contained. In Sect. 2 the notion of a "property system" (i.e. a Boolean matrix) is introduced together with the basic constructors definable over it. By means of these basic constructors, pre-topological operators can be defined. Moreover, by means of property systems, one can define *neighborhood systems* and their pre-topological operators. In Sect. 3, three pairs of neighborhood systems are introduced, along with the properties of their pre-topological operators. In Sect. 4, coverings are interpreted as property systems and neighborhood systems. In this way we shall be able to prove the properties of the main covering-based approximation operators studied in rough set literature by exploiting the properties of the pre-topological operators and the basic constructors provided by pointless topology. In Sect. 5, we use the same approach to study well-known and new algebraic properties of some kinds of covering-based approximation operators.

2 The Basic Constructors

In this section we define the basic constructors. They have been introduced in different fields: in formal (or pointless) topology by Sambin ([28], see also [29]), and after that they were used in the context of neighborhood systems by the author [20]. Independently, Düntsch and Gegida used them to define "property oriented concepts" in [6], while in [32] they have been used by Yao and Chen to define "object oriented concepts". Eventually, they were fully used in approximation theory by Pagliani and Chakraborty ([23], see also [24,25]). Moreover, these constructors are strictly connected to modal operators. The starting point is the idea that relations connect objects with their properties in what we call a *property system* or an *observation system*.

Definition 1. *A* property system *is a triple* $\mathbf{P} = \langle U, M, R\rangle$*, where* U *is a universe of objects,* M *a set of properties, and* $R \subseteq U \times M$ *is a "manifestation relation" so that* $\langle g, m\rangle \in R$ *means that object* g *fulfills property* m*, or is manifested through* m*.*[1]

Definition 2. *For any* $R \subseteq X \times Y, A \subseteq X, B \subseteq Y$*:*
$R^{\smile} = \{\langle m, g\rangle : \langle g, m\rangle \in R\}$*, is called the* inverse relation *of* R*.*
$R(A) = \{y \in Y : \exists x(\langle x, y\rangle \in R \wedge x \in A)\}$ *is called the* R-neighborhood of A*. If* $A = \{x\}$ *we shall write* $R(x)$*. Clearly,* $y \in R(x)$ *means* $\langle x, y\rangle \in R$*. If* R *is an order of some kind, then* $R(X)$ *will be denoted also with* $\uparrow X$*, the* order filter generated by X*, and if* $X = \{x\}$*, with* $\uparrow x$*.*[2]

[1] The members of U will be usually denoted by g after the German term *Gegenstand* which means an object before interpretation, while M is after *Merkmal*, which means "property". In Formal Topology, M is thought of as a set of *abstract neighborhoods*. This interpretation will be used later on in the paper.

[2] $R(A)$ and $R^{\smile}(B)$ are also called the *the left Peirce product of* R *and* A, and, respectively, the *right Peirce product of* R (left Peirce product of $R^{\smile}$) *and* B.

The relation R assembles *concepts*, by associating together in some way those properties which are observed through a given set of objects. Vice-versa, its inverse relation $R^{\smile}$ assembles *objects*, by grouping together in some way those which fulfill a given set of properties. In the former case an *intension* is derived from an *extension*. In the latter an opposite derivation occurs. This is why we shall decorate with an "i" a constructor which transforms sets of extensions into sets of intensions, and vice-versa, with an "e" the constructors which operates in the opposite direction. The constructors induce derived operators between sets of objects or between sets of properties, in the following way:

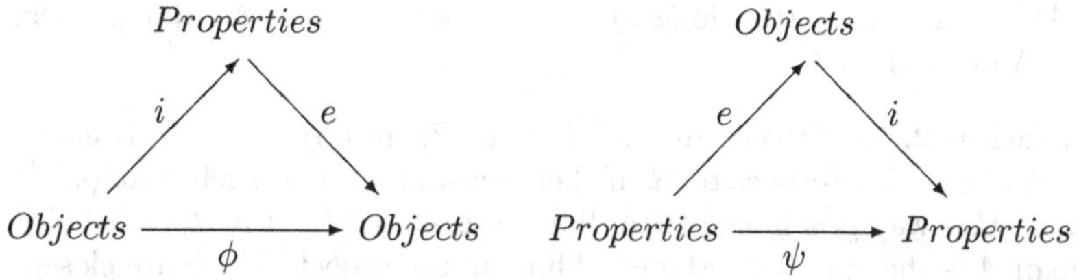

In turn, these operators induce relations $R \subseteq U \times U$ between objects, or relations $R \subseteq M \times M$ between properties. We shall call a system $\langle X, X, R\rangle$, where $R \subseteq X \times X$, a *square relational system*, SRS. On the basis of the above intuitions, the following constructors are defined:

Definition 3. *Let* $\mathbf{P} = \langle U, M, R\rangle$ *be a property system,* $A \subseteq U, B \subseteq M$*:*

$$\langle i\rangle : \wp(U) \mapsto \wp(M); \langle i\rangle(A) = \{m : \exists g(\langle g, m\rangle \in R \wedge g \in A)\} \quad (1)$$
$$\langle e\rangle : \wp(M) \mapsto \wp(G); \langle e\rangle(B) = \{g : \exists m(\langle g, m\rangle \in R \wedge m \in B)\} \quad (2)$$
$$[i] : \wp(U) \mapsto \wp(M); [i](A) = \{m : \forall g(\langle g, m\rangle \in R \Longrightarrow g \in A)\} \quad (3)$$
$$[e] : \wp(M) \mapsto \wp(G); [e](B) = \{g : \forall m(\langle g, m\rangle \in R \Longrightarrow m \in B)\} \quad (4)$$

Facts 1.

$$\langle i\rangle(A) = \{m : m \in R(A)\} = R(A) = \{m : R^{\smile}(m) \cap A \neq \emptyset\} \quad (5)$$
$$\langle e\rangle(B) = \{g : g \in R^{\smile}(B)\} = R^{\smile}(B) = \{g : R(g) \cap B \neq \emptyset\} \quad (6)$$
$$[i](A) = \{m : R^{\smile}(m) \subseteq A\} \quad (7)$$
$$[e](B) = \{g : R(g) \subseteq B\} \quad (8)$$

Example 1. $U = \{a, b, c, d, e, f, g\}$, $M = \{m_1, m_2, m_3, m_4, m_5, m_6\}$. R is:

R	m_1	m_2	m_3	m_4	m_5	m_6
a	1	0	1	0	0	0
b	1	1	1	1	0	0
c	1	1	1	1	0	0
d	1	1	0	1	1	1
e	0	0	0	1	1	0
f	0	0	0	0	1	0
g	0	0	0	0	0	1

$\langle i\rangle(\{f, g\}) = \{m_5, m_6\}$
$\langle e\rangle(\{m_2, m_6\}) = \{b, c, d, g\}$
$[i](\{b, c, d, f, g\}) = \{m_2, m_6\}$
$[e](\{m_1, m_5, m_6\}) = \{f, g\}$

Observation 1. A Kripke model for Modal Logic is an SRS where $U = M$. In such a model R is the accessibility relation between possible worlds in U and $\models$ is the forcing relation between possible worlds and formulas. If $B \subseteq U$ is interpreted as the domain of the validity of a formula α, that is, $B = \{w : w \models \alpha\}$, then the constructor $\langle e \rangle$ is the *possibility* modal operator $\Diamond$ and $[e]$ is the necessity operator $\Box$. A similar modal interpretation applies to our intensional and extensional constructors. Indeed, if $g \in A$ then it is *possible* that g fulfills properties in $\langle i \rangle(A)$, because if $m \in \langle i \rangle(A)$, then $R^{\smile}(m)$ has non void intersection with A, and this means that there are examples of objects in A which fulfill properties in $\langle i \rangle(A)$. Analogously, in order to fulfill the properties in $[i](A)$ it is necessary to belong to A (no objects outside A enjoy properties in $[i](A)$). And so on.

Observation 2. Intuitively, in a SRS $\langle U, U, R \rangle$ an object $g \in U$ is closed to a subset A of U, with respect to R, if there exists a g' in A such that $\langle g, g' \rangle \in R$, so that in this way g is linked to A. Thus g is closed to A if $R(g) \cap A \neq \emptyset$. The *closure* of A is the operation of embedding all the objects which are closed to A. On the contrary, the *interior* of A is the operation of collecting the elements of A which do not have links with elements outside A itself, that is, elements g such that $R(g) \subseteq A$. In view of Facts 1, the possibility constructors $\langle \cdot \rangle$ are likely to be closure operators while the necessity constructors $[\cdot]$ are likely to be interior operators.

Definition 4. *An operator ϕ on a preorder* $\mathbf{P}$ *is said to be a* closure *(resp.* interior*) operator if for any $x, y \in \mathbf{P}$ it is (i) increasing: $x \leq \phi(x)$ (resp. decreasing: $\phi(x) \leq x$), (ii) monotone: $x \leq y$ implies $\phi(x) \leq \phi(y)$, and (iii) idempotent: $\phi(\phi(x)) = \phi(x)$. Moreover, if* $\mathbf{P}$ *is a lattice, an interior (closure) operator ϕ is* topological *if it is (iv) additive: $\phi(x \vee y) = \phi(x) \vee \phi(y)$ (resp. multiplicative: $\phi(x \wedge y) = \phi(x) \wedge \phi(y)$) and (v) normal: $\phi(0) = 0$ (resp. conormal: $\phi(1) = 1$).*

Actually, $\langle \cdot \rangle$ and $[\cdot]$ in general are not closure and, respectively, interior operators, but for particular properties of R that will be discussed later on. However, they contribute to define genuine interior and closure operators. To understand that, we need a fundamental notion:

Definition 5. *Let* $\mathbf{O}$ *and* $\mathbf{O'}$ *be two preordered sets and $\sigma : \mathbf{O} \longmapsto \mathbf{O}'$ and $\iota : \mathbf{O}' \longmapsto \mathbf{O}$ be two maps such that for all $p \in \mathbf{O}$ and $p' \in \mathbf{O}'$*

$$\iota(p') \leq p \ \textit{ iff } \ p' \leq' \sigma(p) \tag{9}$$

then σ is called the upper adjoint *of ι and ι the* lower adjoint *of σ. This fact is denoted by*

$$\mathbf{O}' \dashv^{\iota,\sigma} \mathbf{O} \tag{10}$$

and we shall say that the two maps form an adjunction *between* $\mathbf{O}$ *and* $\mathbf{O}'$.

It is well known that in Classical Logic (which is our meta-logic) $\wedge$ is the lower adjoint of $\Longrightarrow$, and that $\exists$ and $\forall$ are the lower, respectively, upper adjoints to the pre-image f^{-1} of any function f. Recalling that the $\langle\cdot\rangle$ operators are defined by means of the logical combination $(\exists, \wedge)$, while the $[\cdot]$ operators by the combination $(\forall, \Longrightarrow)$, and, finally, considering the opposite directions of the intensional operators with respect to the extensional ones, it is not surprise that in any *P-system* $\mathbf{P} = \langle U, M, R\rangle$, for $B \subseteq M, A \subseteq U$:

$$\langle e\rangle(A) \subseteq B \text{ iff } A \subseteq [i](B); \quad \langle i\rangle(B) \subseteq A \text{ iff } B \subseteq [e](A) \tag{11}$$

From this we immediately obtain that $\langle\cdot\rangle$ operators are additive, since they are lower adjoints, while $[\cdot]$ operators are multiplicative, since they are upper adjoints[3]. Finally, one can prove by means of the equivalences $\neg\exists \equiv \forall\neg$ and $\neg(A \wedge \neg B) \equiv A \Longrightarrow B$:

$$\forall A \subseteq U, [e](A) = -\langle e\rangle(-A); \ \forall B \subseteq M, [i](B) = -\langle i\rangle(-B) \tag{12}$$

Using (5), (6) and (12), the elementary procedures for computing the constructors come straightforwardly (see [22]). From Observations 2 and (6), $\langle e\rangle(A)$ is the set of elements close to A. Dually, x is internal to A if all its neighbors belong to A, that is, if $R(x) \subseteq A$. Hence, from (8), $[e](A)$ is the interior of A, that is, the set of elements which are internal to A. Therefore, we are approaching the topological notions of a *closure* and an *interior* and the soft computing concepts of an *upper* and, respectively, a *lower approximation*. However, some problems arise.

First of all, in a generic SRS $\langle\cdot\rangle$ may fail to be increasing (so they cannot be used as veritable *upper approximations*) or may fail to be idempotent, although additivity gives monotonicity. Analogously, $[\cdot]$ might be neither decreasing (so that they cannot be used as veritable *lower approximations*) nor idempotent, although multiplicativity guarantees monotonicity. Eventually, $\langle\cdot\rangle$ ($[\cdot]$) fails to be co-normal (normal) if there is a g (an m) such that $R(g) = \emptyset$ ($R^{\smile}(m) = \emptyset$).

NOTE. In what follows, given two monadic operators ϕ, ψ, $\phi\psi(\alpha)$ means $\phi(\psi(\alpha))$.

Example 2. $U = \{a, b, c, d, e\}$. R is given by the following Boolean matrix.

R	a	b	c	d
a	1	0	0	1
b	0	1	1	1
c	0	1	0	0
d	0	1	0	1

$$\begin{array}{l} \langle i\rangle(\{a,c\}) = \{a,b,d\} \not\supseteq \{a,c\} \\ \langle e\rangle\langle e\rangle(\{a,c\}) = U \neq \{a,b\} = \langle e\rangle(\{a,c\}) \\ [e](\{a,b\}) = \{c\} \not\subseteq \{a,b\} \\ [i][i](\{a,b\}) = \{c\} \neq \{a,c\} = [i](\{a,b\}) \end{array}$$

The second problem concerns the fact that one would like to be able to obtain interiors and closures from a generic property system. Otherwise stated, we

[3] Often a lower adjoint is called "left adjoint" and an upper adjoint "right adjoint". We avoid the terms "right" and "left" because they could make confusion with the position of the arguments of the operations of binary relations. For the general notion of adjoint functors see for instance [3]. For Galois connections induced by binary relations a classic reference is [19]. For the present use in Rough Set Theory see [23] or [25].

would like to compute the interior or the closure of a set of objects by looking at the properties fulfilled by these objects or, vice-versa, find the interior or the closure of a set of properties by analysing their fulfillment relations.

To this end, let us set, for all $A \subseteq U, B \subseteq M$:

$$int : \wp(U) \mapsto \wp(U); int(A) = \langle e \rangle [i](A) = \bigcup \{R^{\smile}(m) : R^{\smile}(m) \subseteq A\} \tag{13}$$

$$cl : \wp(U) \mapsto \wp(U) : cl(A) = [e]\langle i \rangle (A) = \{g : R(g) \subseteq R(A)\} \tag{14}$$

$$\mathcal{C} : \wp(M) \mapsto \wp(M); \mathcal{C}(B) = \langle i \rangle [e](B) = \bigcup \{R(g) : R(g) \subseteq B)\} \tag{15}$$

$$\mathcal{A} : \wp(M) \mapsto \wp(M); \mathcal{A}(B) = [i]\langle e \rangle (B) = \{m : R^{\smile}(m) \subseteq R^{\smile}(B)\} \tag{16}$$

The rightmost equations are straightforward consequences of Facts 1.

Facts 2. From (12), one easily see that $\mathcal{C}$ and $\mathcal{A}$ are dual, that is, $\mathcal{C}(X) = -\mathcal{A}(-X)$. Similarly, *int* and *cl* are dual.

Example 3. The property system is the same as in Example 1.

$int(\{b, c, d, f, g\}) = \langle e \rangle (\{m_2, m_6\}) = \{b, c, d, g\}$
$\mathcal{C}(\{m_1, m_5, m_6\}) = \langle i \rangle (\{f, g\}) = \{m_5, m_6\}$
$cl(\{e, g\}) = [e](\{m_4, m_5, m_6\}) = \{e, f, g\}$
$\mathcal{A}(\{m_1\}) = [i](\{a, b, c, d\}) = \{m_1, m_2, m_3\}$

It is worth showing the easy proof of Facts 1.(5):

Lemma 1. *In a property systems $\langle U, M, R \rangle$, for any $A \subseteq U$, if $R^{\smile}(m) \cap A \neq \emptyset$ then $m \in R(A)$.*

Proof. Trivially, $R^{\smile}(m) \cap A \neq \emptyset$ iff $\exists g(g \in R^{\smile}(m) \wedge g \in A)$ iff $\exists g(m \in R(m) \wedge g \in A)$ iff $m \in R(A)$. □

Lemma 2. *Let $\mathbf{P} = \langle U, M, R \rangle$ be a property system, $\mathcal{Z} = \{R^{\smile}(m) : m \in M\}$ and $\mathcal{W} = \{R(g) : g \in U\}$. Them $\forall A \subseteq U, B \subseteq M$,*

(1) $int(A) = \bigcup \{X \in \mathcal{Z} : X \subseteq A\}$; $\mathcal{C}(B) \bigcup \{Y \in \mathcal{W} : Y \subseteq B\}$.
(2) $cl(A) = \bigcap \{-X \in \mathcal{Z} : X \cap A = \emptyset\}$; $\mathcal{A}(B) = \bigcap \{-Y \in \mathcal{W} : Y \cap B = \emptyset\}$.

Proof. (1) The equations are trivial translations of (13) and (15). (2) From duality, $cl(A) = -int(-A) = -\bigcup \{X \in \mathcal{Z} : X \subseteq -A\} = \bigcap \{-X \in \mathcal{Z} : X \subseteq -A\} = \bigcap \{-X \in \mathcal{Z} : X \cap A = \emptyset\}$. Similarly for $\mathcal{C}$ (for a direct proof see the Appendix). □

Observation 3. The operators $\mathcal{C}$, $\mathcal{A}$, *int and cl,* as combinations of adjoint functors, fulfill a number of properties (see [25]). We display them in a comparative table. $R \subseteq U \times M$ is assumed to be any binary relation. For any operator $\odot$ on sets, subadditivity means $\odot(X) \cup \odot(Y) \subseteq \odot(X \cup Y)$ and submultiplicativity means $\odot(X \cap Y) \subseteq \odot(X) \cap \odot(Y)$ (abbreviations: Unp(redictable), Mono(tonic), Add(itive), Mult(iplicative), Subad(ditive), Submul(tiplicative), Decr(easing). In turn, Norm(ality), Conor(mality), Incr(easing) and Idem(potence) apply to $\langle \cdot \rangle$ and $[\cdot]$ only in SRS).

Table 1. Properties of constructors and operators

Operator	Incr.	Decr.	Mono.	Idem.	Add.	Mult.	Subad.	Submul.	Norm.	Conor.
$\langle\cdot\rangle$	unp.	unp.	Yes	No	Yes	No	*a fortiori*	Yes	Yes	No
$[\cdot]$	unp.	unp.	Yes	No	No	Yes	Yes	*a fortiori*	No	Yes
$int, \mathcal{C}$	No	Yes	Yes	Yes	No	No	Yes	Yes	Yes	No
$cl, \mathcal{A}$	Yes	No	Yes	Yes	No	No	Yes	Yes	No	Yes

Observation 4. From the properties listed in Table 1, int and $\mathcal{C}$ are interior operators and cl *and* $\mathcal{A}$ are closure operators. This means that $int(A) \subseteq A \subseteq cl(A)$, any $A \subseteq U$, and $\mathcal{C}(B) \subseteq B \subseteq \mathcal{A}(B)$, any $B \subseteq M$, so that they are veritable approximations. However, they are not topological. Indeed, int and $\mathcal{C}$ are not multiplicative, because the external constructors $\langle\cdot\rangle$ are not, and are not additive, because the internal constructors are not. Symmetrically, cl and $\mathcal{A}$ are neither additive nor multiplicative. We call them *pretopological.*

A third problem gets us into the core of covering-based approximations, as we are going to see. How does all these machinery work if we consequently consider M as a set of *abstract neighborhoods*? To put it another way, what does it happen if $m \in M$ is considered as a "name" or a "proxy" for the set of objects which fulfill m, that is, if one substitutes $R^{\smile}(m)$ for m. By means of such a substitution, we enter in an intermediate realm between pointless (pre)topology and point (pre)topology. A first analysis of this intermediate situation was done in [20, 25]. In this paper we address a further aspect of the problem. Let us start with two subproblems:

- In a SRS, when $\langle e\rangle = \mathcal{C}$, $\langle i\rangle = cl$, $[e] = \mathcal{A}$ and $[i] = int$?
- Given a property system $\mathbf{P} = \langle U, M, R\rangle$ one obtains neighborhood systems by considering the *concrete* (filled with points) counterpart $R^{\smile}(m)$ of any abstract neighborhood $m \in M$. Over these concrete neighborhood systems one can define closure and interior operators. What are the relations between these "concrete" operators and the abstract ones defined on $\mathbf{P}$?

The first subproblem has been solved:

Facts 3 (see [22]). *In a SRS* $\langle U, U, R\rangle$, *the following are equivalent:* (i) R *is a preorder,* (ii) int *and* $\mathcal{C}$ *are topological interior operators,* cl *and* $\mathcal{A}$ *are topological closure operators.* (iii) $int = [i]$, $\mathcal{C} = [e]$, $\langle i\rangle = cl$, $\langle e\rangle = \mathcal{A}$.

Proofs about the equivalence of (i) and (ii) can be found in Modal or Intuitionistic Logic and Topology. In [25] they are obtained through a series of intermediate steps within pointless pre-topology. The equivalence between (i) and (iii) was proved in [25] and the entire picture have been proved in [22] by reasoning about specialisation preorders, which is a useful approach (for this notion see for instance [31]).

Definition 6. *Let* X *be a set and* $\odot$ *any monadic operator on* $\wp(X)$. *We set for all* $x, y \in X$: $x \preceq_{\odot} y$ *iff* $\forall A \subseteq X, x \in \odot(A) \Longrightarrow y \in \odot(A)$.

The relation $\preceq_{\odot}$ is a preorder: clearly it is reflexive because by substituting x for y we obtain a tautology, and it is transitive, because implication is transitive. Thus $\preceq_{\odot}$ is called the *specialisation preorder induced by* $\odot$. We recall that if $R \subseteq X \times X$ is a preorder, then the topology with bases the family $\{R(x) : x \in X\}$ is called the *Alexandrov topology induced by* R. The specialisation preorder induced by its interior operator coincides with R itself. We have now to solve the second subproblem. To this end the following results are useful.

Facts 4 (see [22]). *Let* $\langle U, U, R\rangle$ *be a SRS. Then the following are equivalent: (i)* R *is preorder. (ii)* $\preceq_{[i]}$ *and* $\preceq_{\mathcal{A}}$ *coincide with* $R^{\smile}$. *(iii) int and* $[i]$ *coincide with the interior operator of the Alexandrov topology induced by* $R^{\smile}$ *while* $\langle i\rangle$ *and cl are their dual closure operators. (iv)* $\preceq_{[e]}$ *and* $\preceq_{\mathcal{C}}$ *coincide with* R. *(v)* $[e]$ *and* $\mathcal{C}$ *coincide with the interior operator of the Alexandrov topology induced by* R *while* $\langle e\rangle$ *and* $\mathcal{A}$ *are their dual closure operators.*

The following are elementary albeit very useful results:

Lemma 3. *Let* $\mathbf{P} = \langle U, U, R\rangle$ *be a SRS such that* R *is a preorder: Then for any* $x, y \in X$, $A \subseteq X$:

(1) $y \in R(x) \Longrightarrow R(y) \subseteq R(x)$.
(2) $\{x : R(x) \subseteq A\} = \bigcup\{R(x) : R(x) \subseteq A\}$.

Proof. (1) The easy proof is in [22]. (2) The left set is included in the right one because, for reflexivity, $x \in R(x)$. Conversely, from (1) if $y \in R(x)$ then $R(y) \subseteq R(x)$. Since the inclusion relation is transitive, $R(y) \subseteq A$. Therefore, the right set is included in the left one. □

Observation 5. On the contrary, even if R is a partial order one cannot prove $\{x : A \subseteq R(x)\} = \bigcup\{R(x) : A \subseteq R(x)\}$. Indeed, if R is reflexive we can just prove that the left set is included in the right one. But the converse inclusion fails. For instance, in the poset on $\{1,2,3,4\}$ such that $\{1 \leq 3, 1 \leq 2 \leq 4\} \cup \{x \leq x\}$, if $A = \{2,4\}$ then $\bigcup\{R(x) : A \subseteq R(x)\} = \bigcup\{R(2)\} = \{2,4\}$. But $\{x : A \subseteq R(x)\} = \{2\}$.

Lemma 4. *Let* $\mathbf{P} = \langle U, U, R\rangle$ *be a SRS. Then for any* $X, A \subseteq U$, $\bigcup\{R(X) : R(X) \subseteq A\} = \bigcup\{R(x) : x \in X \wedge R(x) \subseteq A\}$

Proof. Independently of the properties of R, from elementary logic. In fact, since R is additive, $R(X) = \bigcup\{R(x)\}_{x \in X}$. Moreover, if $R(x) \subseteq A$ and $R(x') \subseteq A$, then $R(x) \cup R(x') \subseteq A$. □

Observation 6. The above proof is expressed in Natural Deduction by the $\vee$-elimination rule: $(\alpha \vdash \beta \;\&\; \gamma \vdash \beta) \longrightarrow (\alpha \vee \gamma \vdash \beta))$. On the contrary, $A \subseteq R(x) \cup R(x')$ does not implies $A \subseteq R(x)$ or $A \subseteq R(x')$. The logical reason is that $((\alpha \Longrightarrow \beta) \vee (\alpha \Longrightarrow \gamma)) \Longrightarrow (\alpha \Longrightarrow (\beta \vee \gamma))$ but the reverse implication does not hold, because $\Longrightarrow$ is not additive, but multiplicative *qua* upper adjoint[4]. This

[4] It is worth noticing that there are constructive logics between Intuitionistic and Classical logics such that the opposite of the above implication holds if the premise is a negated formula (see [14]).

is true independently of the properties of R (trivially, $\{a,b\} \subseteq \{a\} \cup \{b\}$ *but* $\{a,b\} \nsubseteq \{a\}$ and $\{a,b\} \nsubseteq \{b\}$).

A particular type of property system is given by relational structures $\mathbf{P} = \langle U, \mathcal{Z}, R\rangle$ such that $\mathcal{Z} \subseteq \wp(U)$. If $\langle g, A\rangle \in R$, then A can be considered a collection of neighbors of g. Hence we call such a property system $\mathbf{P}$ a *relational neighborhood structure*, RNSTR, and if $g' \in N \in R(g)$, we say that g' is a *neighbor* and N a *neighborhood* of g. We set $\mathcal{N}_g = R(g)$ and call it the *neighborhood family* of g. The family $\mathcal{N}(\mathbf{P}) = \{\mathcal{N}_g : g \in U\}$ is called a *neighborhood system*. Let us define on $\wp(U)$:

$$\text{(a) } G(X) = \{g : X \in \mathcal{N}_g\}; \text{(b) } F(X) = -G(-X) = \{g : -X \notin \mathcal{N}_g\}.$$

Consider the following conditions on $\mathcal{N}(\mathbf{P})$, for any $g \in U$, $A, N, N' \subseteq U$:

1: $U \in \mathcal{N}_g$; **0**: $\emptyset \notin \mathcal{N}_g$; **Id**: if $g \in G(A)$ then $G(A) \in \mathcal{N}_g$;
N1: $g \in N$, for all $N \in \mathcal{N}_g$; **N2**: if $N \in \mathcal{N}_g$ and $N \subseteq N'$, then $N' \in \mathcal{N}_g$;
N3: if $N, N' \in \mathcal{N}_g$, then $N \cap N' \in \mathcal{N}_g$. **N4**: $\exists N, \mathcal{N}_g = \uparrow N = \{N' : N \subseteq N'\}$.

They induce the following properties of the operators G and F (see [21] or [25])[5]:

Table 2. Properties of the operators G and F

Condition	Equivalent properties of G	Equivalent properties of F
1	$G(U) = U$	$F(\emptyset) = \emptyset$
0	$G(\emptyset) = \emptyset$	$F(U) = U$
Id	$G(X) \subseteq G(G(X))$	$F(F(X)) \subseteq F(X)$
N1	$G(X) \subseteq X$	$X \subseteq F(X)$
N2	$X \subseteq Y \Rightarrow G(X) \subseteq G(Y)$	$G(X \cap Y) \subseteq G(X) \cap G(Y)$
	$X \subseteq Y \Rightarrow F(X) \subseteq F(Y)$	$F(X \cup Y) \supseteq F(X) \cup F(Y)$
N3	$G(X \cap Y) \supseteq G(X) \cap G(Y)$	$F(X \cup Y) \subseteq F(X) \cup F(Y)$

Facts 5. *In the presence of* **N2**, **Id** *is equivalent to the following weaker condition:*

$$\textit{if } N \in \mathcal{N}_x, \textit{ then } \exists N' \in \mathcal{N}_x \textit{ such that for any } y \in N', N \in \mathcal{N}_y \qquad (\tau)$$

This is the familiar topological property usually explained by the sentence: "if X is a neighborhood of a point x, then it is also a neighborhood of all the points that are sufficiently close to x"[6].

We are now ready to face the second subproblem.

[5] In these works $\mathcal{N}(\mathbf{P})$ is denoted as $\mathcal{N}(U)$ and instead of $\mathcal{Z}$ the entire powerset $\wp(U)$ is considered. The present is a slight generalization.

[6] For the notions of a "neighborhood" and a "pretopology", see [5,16].

3 Towards Coverings: Neighborhood System

In this section we introduce two pairs of RNSTRs of the form $\langle U, \mathcal{Z}, \in\rangle$, and a pair of neighborhood systems, which we can derive by a property system $\langle U, M, R\rangle$. In each pair, in the first structure $\mathcal{Z}$ is a collections of subset of objects provided by the reverse relation $R^{\smile}$, while in the second one $\mathcal{Z}$ is a collections of subsets of abstract neighborhoods defined by means of R. Together with the basic constructors, these neighborhood systems will be used in the next section to define the covering-based approximation operators that have been introduced in rough set literature using different approaches, and to study their properties and relationships.

Definition 7. *Let* $\mathbf{P} = \langle U, M, R\rangle$ *be a property system. For any* $g \in U, m \in M$*:*

(a) *Let* $\mathcal{N}_g^{R^{\smile}} = \{R^{\smile}(m) : m \in R(g)\}$*. The family* $\mathcal{N}_{R^{\smile}}(\mathbf{P}) = \{\mathcal{N}_g^{R^{\smile}} : g \in U\}$ *will be called the* normal neighborhood system, NNS, *induced by* $\mathbf{P}$.
(a') *Let us set* $\mathbf{P}^{\mathbf{R}^{\smile}} = \langle U, \bigcup(\mathcal{N}_{R^{\smile}}(\mathbf{P})), \in\rangle$.
(b) *Let* $\mathcal{N}_m^{R} = \{R(g) : g \in R^{\smile}(m)\}$*. The family* $\mathcal{N}_R(\mathbf{P}) = \{\mathcal{N}_m^{R} : m \in M\}$ *will be called the* co-normal neighborhood system, CNNS, *induced by* $\mathbf{P}$.
(b') *Let us set* $\mathbf{P}^{\mathbf{R}} = \langle M, \bigcup(\mathcal{N}_R(\mathbf{P})), \in\rangle$.
(c) *Let* $\mathcal{N}_g^{\uparrow R^{\smile}} = \bigcup\{\uparrow R^{\smile}(m) : m \in R(g)\}$*. The family* $\mathcal{N}_{\uparrow R^{\smile}}(\mathbf{P}) = \{\mathcal{N}_g^{\uparrow R^{\smile}} : g \in U\}$ *will be called* principal neighborhood system, PNS, *induced by* $\mathbf{P}$.
(c') *Let us set* $\mathbf{P}^{\uparrow\mathbf{R}^{\smile}} = \langle U, \bigcup(\mathcal{N}_{\uparrow R^{\smile}}(\mathbf{P})), \in\rangle$.
(d) *Let* $\mathcal{N}_m^{\uparrow R} = \bigcup\{\uparrow R(g) : g \in R^{\smile}(m)\}$*. The family* $\mathcal{N}_{\uparrow R}(\mathbf{P}) = \{\mathcal{N}_m^{\uparrow R} : m \in M\}$ *will be called* co-principal neighborhood system, CPNS, *induced by* $\mathbf{P}$.
(d') *Let us set* $\mathbf{P}^{\uparrow\mathbf{R}} = \langle M, \bigcup(\mathcal{N}_{\uparrow R}(\mathbf{P})), \in\rangle$.
(e) *Let* $\mathbf{P}$ *be a SRS. Let us set* $\mathcal{N}_x^* = \uparrow R(x)$ *and* $\mathcal{N}_x^{*\smile} = \uparrow R^{\smile}(x)$ *The family* $\mathcal{N}_*(\mathbf{P}) = \{\mathcal{N}_x^* : x \in U\}$ *is called* R-neighborhood system, RNS, *induced by* $\mathbf{P}$ *and the family* $\mathcal{N}_*^{\smile}(\mathbf{P}) = \{\mathcal{N}_x^{*\smile} : x \in U\}$ *the* co-R-neighborhood system, CRNS, *induced by* $\mathbf{P}$.

Observation 7. The domain $\mathcal{N}_*(\mathbf{P})$ is justified by the fact that in a SRS if we consider an element g of U as an abstract neighborhood, then in moving towards a more concrete framework, it must represent the set of its neighbors.

Theorem 1. *Let* $\mathbf{P} = \langle U, U, R\rangle$ *be a SRS with* R *a preorder. Then* $\mathcal{N}_{\uparrow R}(\mathbf{P}) = \mathcal{N}_*(\mathbf{P})$ *and* $\mathcal{N}_{\uparrow R^{\smile}}(\mathbf{P}) = \mathcal{N}_*^{\smile}(\mathbf{P})$.

Proof. Let us prove that if R is a preorder then $\mathcal{N}_x^{\uparrow R} = \uparrow R(x)$, any $x \in U$. Because of reflexivity, $x \in R^{\smile}(x)$. Therefore, $\uparrow R(x) \in \{\uparrow R(z) : z \in R^{\smile}(x)\}$. Suppose $y \in R^{\smile}(x)$. By transitivity, $R^{\smile}(y) \subseteq R^{\smile}(x)$ so that $R(x) \subseteq R(y)$. It follows that $\uparrow R(y) \subseteq \uparrow R(x)$. We conclude that $\uparrow R(x) \cup \uparrow R(y) = \uparrow R(x)$. Analogously for $R^{\smile}$. □

NOTE: From now on, if an operator ϕ is defined on a structure $\mathbf{S}$, we shall write $\phi^{\mathbf{S}}$ if the structure must be distinguished.

Definition 8. *Let* $R \subseteq X \times Y$*.* R *is called* serial *if* $\forall x \in X, \exists y \in Y(\langle x, y\rangle \in R)$.

Lemma 5. *For any property system* $\mathbf{P} = \langle U, M, R\rangle$*: If* $R^{\smile}$ *is serial:*

(1) $\bigcup(\mathcal{N}_{R^{\smile}}(\mathbf{P})) = \{R^{\smile}(m) : m \in M\}$ *and* $\bigcup(\mathcal{N}_{\uparrow R^{\smile}}(\mathbf{P})) = \{\uparrow R^{\smile}(m) : m \in M\}$.
(2) $\bigcup(\mathcal{N}_{R}(\mathbf{P})) = \{R(g) : g \in U\}$ *and* $\bigcup(\mathcal{N}_{\uparrow R}(\mathbf{P})) = \{\uparrow R(g) : g \in U\} = \mathcal{N}_{*}$.

Proof. (1) By definition, $\mathcal{N}_{R^{\smile}}(\mathbf{P}) = \{\{R^{\smile}(m) : m \in R(g)\}\}_{g\in U}$. It follows that $\bigcup(\mathcal{N}_{R^{\smile}}(\mathbf{P})) = \{R^{\smile}(m) : m \in R(g)\}_{g\in U}$. Since $R^{\smile}$ is serial, $R(U) = M$. Hence, $\{R^{\smile}(m) : m \in R(g)\}_{g\in U} = \{R^{\smile}(m) : m \in M\}$. The second part follows similarly. (2) is just the reverse of (1) using the fact that if R is serial, then $R^{\smile}(M) = U$. □

Clearly, $\mathbf{P}^{\mathbf{R}^{\smile}}$, $\mathbf{P}^{\uparrow\mathbf{R}^{\smile}}$, $\mathbf{P}^{\uparrow\mathbf{R}}$ and $\mathbf{P}^{\mathbf{R}}$ are RNSTRs of the form $\langle U, \mathcal{Z}, \in\rangle$. In the first two structures, $\mathcal{Z} \subseteq \wp(U)$, while in the second two $\mathcal{Z} \subseteq \wp(M)$.

To avoid confusion, in a relational structures $\langle U, \mathcal{Z}, R\rangle$ with $\mathcal{Z} \subseteq \wp(U)$ and $R \subseteq U \times \mathcal{Z}$, instead of the short $R^{\smile}(X)$ we shall adopt the full notation $R^{\smile}(\{X\})$, to underline that X is an element, not a subset, of the codomain of R. On the contrary, we shall write $R(X)$ because in the domain of R, X is a subset not an element. The next result deals with the fact that in property systems like that a *subset* of the domain of R may appear as an *element* of the domain of $R^{\smile}$.

Theorem 2. *Let* $\mathbf{P} = \langle U, \mathcal{Z}, R\rangle$ *be such that* $\mathcal{Z} \subseteq \wp(U)$*,* $\mathcal{N}(\mathbf{P})$ *its induced neighborhood system, and G the operator defined on it. Then for any* $X \in \mathcal{Z}$*:*

(1) $R^{\smile}(\{X\}) = G(X)$*; (2) if* $\emptyset \notin \mathcal{Z}$*, then* **0** *holds in* $\mathcal{N}(\mathbf{P})$.
(3) *If* $R = \in$*, then:* (a) $R^{\smile}(\{X\}) = X$*;* (b) $G(X) = X$*;* (c) *for all* $g \in U$*, if* $g \in G(X)$*, then* $G(X) \in \mathcal{N}_g$*;* (d) *in* $\mathcal{N}(\mathbf{P})$*,* **0** *and* **N1** *hold. (see* [25]*).*

Proof. (1) (adapted from [22]): $G(X) = \{g : X \in \mathcal{N}_g\}$. But $X \in \mathcal{N}_g$ iff $\langle g, X\rangle \in R$ iff $g \in R^{\smile}(\{X\})$. Hence, $G(X) = \{g : g \in R^{\smile}(\{X\})\} = R^{\smile}(\{X\})$. (2) Obvious. (3): (a) Trivially, $g \in R^{\smile}(\{X\})$ iff $X \in R(g)$ iff $g \in X$. (b) From (1) and (a). (c) $g \in G(X)$ if $X \in \mathcal{N}_g$. But from (b), $G(X) = X$ so that $G(X) \in \mathcal{N}_g$. □

Observation 8. Point (3).(c) of the above Theorem 2 is a restricted form of **Id**. Notice that (1) and (2) do not require R to be serial. The fact that there can be a $g \in U$ such that $R(g) = \emptyset$ does not affect the results[7]. Since it is not required that $\bigcup \mathcal{Z} = U$, if $R = \in$ and $\bigcup \mathcal{Z} \neq U$, then surely a g exists such that g does not belong to any element of $\mathcal{Z}$. The results of (3) are not affected, though.

Theorem 3. *Let* $\mathbf{P} = \langle U, M, R\rangle$ *be a property system. Then,* (1) *In* $\mathcal{N}_{R^{\smile}}(\mathbf{P})$ *and* $\mathcal{N}_{R}(\mathbf{P})$*,* **0**, **N1** *and* (τ) *hold.* (2) *In* $\mathcal{N}_{\uparrow R^{\smile}}(\mathbf{P})$ *and* $\mathcal{N}_{\uparrow R}(\mathbf{P})$*,* **0**, **N1**, **N2** *and* **Id** *hold.*

---

[7] If R is not serial and $R(g) = \emptyset$, then $\emptyset$ does not belong to $\mathcal{N}_g$. Otherwise stated, $\emptyset$ is different from $\{\emptyset\}$. **0** does not hold if there exists $g \in G$ such that $\langle g, \emptyset\rangle \in R$.

Proof. (1) **N1**: if $m \in R(g)$ then $g \in R^{\smile}(m)$. Therefore, for all $N \in \mathcal{N}_g^{R^{\smile}}$, $g \in N$. **0**: it is an obvious consequence of **N1**. (τ): Let $R^{\smile}(m) \in \mathcal{N}_g^{R^{\smile}}$. By definition, for all $g' \in R^{\smile}(m), R^{\smile}(m) \in \mathcal{N}_{g'}^{\uparrow R^{\smile}}$. The proofs for $\mathcal{N}_R(\mathbf{P})$ are just the reverse. (2) **0**, **N1** and (τ) as before and the trivial fact that if $x \in X$ and $X \subseteq X'$, then $x \in X'$. **N2** is a direct consequence of the definition of a PNS and a CPNS. **1** is a consequence of **N2**. Since (τ) and **N2** hold, in view of Facts 5, we obtain **Id**. The proofs for $\mathcal{N}_{\uparrow R}(\mathbf{P})$ are just the reverse. □

Observation 9. Notice that **N2** and **N1** are not enough to prove **Id**. Indeed, the above theorem holds because of the particular assignment "$R^{\smile}(m) \in \mathcal{N}_g$ iff $m \in R(g)$". As a counterexample, consider $U = \{a, b, c\}$ *and set* $\mathcal{N}_a = \bigcup\{\uparrow \{a, b\}\}$, $\mathcal{N}_b = \bigcup\{\uparrow \{a, b\}\}$, $\mathcal{N}_c = \bigcup\{\uparrow \{a, c\}\}$. Then **N1** and **N2** hold, but if we take $N = \{a, c\}$, then $N \in \mathcal{N}_c$, a belongs to any element of $\mathcal{N}_c$, but $N \notin \mathcal{N}_a$. The problem here is the following: suppose $\uparrow X \subseteq \mathcal{N}_x$ if $X = R^{\smile}(m)$ for some $m \in M$, then $\{a, c\} = R^{\smile}(m)$ implies $m \in R(a)$. But $\{a, c\} \notin \mathcal{N}_a$. On the contrary, if $\uparrow R^{\smile}(m) \subseteq \mathcal{N}_x$ iff $m \in R(x)$, we would have $\mathcal{N}_a = \uparrow \{a, b\} \cup \uparrow \{a, c\}$. Notice, moreover, that because of **N2**, $F(A) = -\{x : \exists N(N \in \mathcal{N}_x \wedge N \subseteq -A\} = \{x : \forall N(N \in \mathcal{N}_x \Longrightarrow N \cap A \neq \emptyset)\}$.

In view of Theorem 3.(2), by looking at Table 2 one expects $G^{\uparrow R^{\smile}}$ and $G^{\uparrow R}$ to be interior operators. Indeed, we have the following result:

Theorem 4. *Let* $\mathbf{P} = \langle U, M, R\rangle$ *be a property system. Let* $int^{\mathbf{P}}$ *be the operator int defined on* $\mathbf{P}$ *and* $G^{\uparrow R^{\smile}}$ *the operator* G *defined on* $\mathcal{N}_{\uparrow R^{\smile}}(\mathbf{P})$. *Then, for any* $A \subseteq U, int^{\mathbf{P}}(A) = G^{\uparrow R^{\smile}}(A)$.

Proof. For all $g \in U$, $g \in G^{\uparrow R^{\smile}}(A)$ iff $A \in \mathcal{N}_g^{\uparrow R^{\smile}}$ iff $\exists m(m \in R(g) \wedge A \in \uparrow R^{\smile}(m))$ iff $\exists m(g \in R^{\smile}(m) \wedge R^{\smile}(m) \subseteq A)$ iff $x \in \bigcup\{R^{\smile}(m) : R^{\smile}(m) \subseteq A\} = int^{\mathbf{P}}(A)$. □

Anyway, apart from the operator G, the neighborhood system $\mathbf{P}^{\uparrow \mathbf{R}^{\smile}}$ is less meaningful for the other operators. For instance, for any $A \subseteq U$, if for some $g \in U$, $R(g) \subseteq A$, then $int^{\uparrow R^{\smile}}(A) = A$. In fact, in this case $A \in \mathcal{N}_g^{\uparrow R^{\smile}}$ so that $A \in \bigcup(\mathcal{N}_{\uparrow R^{\smile}}(U))$. From Lemma 2, it follows that $\in^{\smile}(\{A\}) = A$. Hence $\bigcup\{\in^{\smile}(\{X\}) : \in^{\smile}(\{X\}) \subseteq A\} = \in^{\smile}(\{A\}) = A$.

Definition 9. *Let* U *be a set and* $\mathcal{Z} \subseteq \wp(U)$. *Then by* $\Uparrow \mathcal{Z}$ *we denote the set* $\{\uparrow X : X \in \mathcal{Z}\}$.

Example 4. Consider the property system of Example 1.
$\mathcal{N}_a^{R^{\smile}} = \{R^{\smile}(m_1), R^{\smile}(m_3)\} = \{\{a, b, c, d\}, \{a, b, c\}\}$.
$\mathcal{N}_b^{R^{\smile}} = \mathcal{N}_b^{R^{\smile}} = \{\{a, b, c, d\}, \{b, c, d\}, \{a, b, c\}, \{b, c, d, e\}\}$.
$\mathcal{N}_d^{R^{\smile}} = \{\{a, b, c, d\}, \{b, c, d\}, \{b, c, d, e\}, \{d, e, f\}, \{d, g\}\}$.
One can easily verify, for instance, that $\{b, c, d\}$ is in the neighborhoods of all its elements (property (τ)). Since $R^{\smile}$ is serial, $\bigcup(\mathcal{N}_{R^{\smile}}(\mathbf{P})) = \{R^{\smile}(m) : m \in M\} = \{\{a, b, c, d\}, \{b, c, d\}, \{a, b, c\}, \{b, c, d, e\}, \{d, e, f\}, \{d, g\}\}$.
$\mathcal{N}_a^{\uparrow R^{\smile}} = \bigcup\{\uparrow \{a, b, c, d\}, \uparrow \{a, b, c\}\}$. $\mathcal{N}_e^{\uparrow R^{\smile}} = \bigcup\{\uparrow \{b, c, d, e\}, \uparrow \{d, e, f\}\}$.
In general, for $g \in U$, $\mathcal{N}_g^{\uparrow R^{\smile}} = \bigcup(\Uparrow \mathcal{N}_g^{R^{\smile}})$, and so on.

From Example 3 we know that $int^{\mathbf{P}}(\{b,c,d,f,g\}) = \{b,c,d,g\}$. Let us compute $G^{\uparrow R^{\smile}}(\{b,c,d,f,g\})$. Step 1: $\{b,c,d,f,g\} \in\uparrow \{b,c,d\}$ and $\{b,c,d\}$ belongs to $\mathcal{N}_b^{\uparrow R^{\smile}}$, $\mathcal{N}_c^{\uparrow R^{\smile}}$ and $\mathcal{N}_d^{\uparrow R^{\smile}}$. Moreover, $\{b,c,d,f,g\} \in\uparrow \{d,g\}$ and $\{d,g\}$ belongs to $\mathcal{N}_d^{\uparrow R^{\smile}}$ and $\mathcal{N}_g^{\uparrow R^{\smile}}$. On the contrary, $\{b,c,d,f,g\} \notin \mathcal{N}_f^{\uparrow R^{\smile}}$, because for all $N \in \mathcal{N}_f^{\uparrow R^{\smile}}, e \in N$. Clearly, $\{b,c,d,f,g\} \notin \mathcal{N}_a^{\uparrow R^{\smile}}$, because $a \notin \{b,c,d,f,g\}$. Step 2: we conclude $G^{\uparrow R^{\smile}}(\{b,c,d,f,g\}) = \{b,c,d,g\}$.

The operator G induced by an R-neighborhood system coincides with the constructor $[e]$ induced by R itself (see [22])[8]:

Facts 6. *Let $\mathbf{P} = \langle U,U,R\rangle$ be a SRS, $\mathcal{N}_*(\mathbf{P})$ its R-neighborhood system and G^* the operator G defined on it. Then for all $A \subseteq U, G^*(A) = [e]^{\mathbf{P}}(A)$.*

Facts 6 hold independently of the properties of R. But if, moreover, R is a preorder, then from Facts 3, $G^*(A) = [e]^{\mathbf{P}}(A) = \mathcal{C}^{\mathbf{P}}(A)$, all $A \subseteq U$.

Example 5. In the property system $\mathbf{P}$ of Example 2, $\{a,c,d\}$ belongs just to $\mathcal{N}_a^* = \uparrow \{a,d\}$. Hence, $G^*(\{a,c,d\}) = \{a\} = [e]^{\mathbf{P}}(\{a,c,d\})$. On the contrary, $\mathcal{C}^{\mathbf{P}}(\{a,c,d\}) = \langle i\rangle^{\mathbf{P}}(\{a\}) = \{a,d\}$. In fact, R is not a preorder (for instance, $\langle c,c\rangle \notin R$).

Finally, obviously any RNSTR is a property system. So it is possible to define *int* and *cl* on it. Under this respect one has (see [22,25]):

Facts 7. *Let a RNSTR $\langle U,\mathcal{Z},R\rangle$ induce a neighborhood system such that* **Id**, **N1** *and* **N2** *hold. Then for any $A \subseteq U, int(A) = G(A)$.*

4 Covering-Based Approximations

Now we have the logico-mathematical machinery to deal with covering-based approximation operators. We have mentioned in the Introduction that a covering can be supposed to be induced by some property system. Indeed, we have:

Theorem 5. *For any property system $\mathbf{P} = \langle U,M,R\rangle$, (1) $\bigcup(\mathcal{N}_{R^{\smile}}(\mathbf{P}))$ and $\bigcup(\mathcal{N}_{\uparrow R^{\smile}}(\mathbf{P}))$ are coverings of U, provided $R^{\smile}$ is serial. (2) $\bigcup(\mathcal{N}_R(\mathbf{P}))$ and $\bigcup(\mathcal{N}_{\uparrow R}(\mathbf{P}))$ are coverings of M, provided R is serial.*

Proof. (1) If R is serial $R^{\smile}(M) = U$. Thus $\bigcup\{R^{\smile}(m) : m \in M\} = R^{\smile}(M) = U$. From Lemma 5 we obtain the proof. The second part of (1) is similar and (2) is the reverse of (1). □

Given a property system $\mathbf{P}$ we shall denote the covering $\bigcup(\mathcal{N}_{R^{\smile}}(\mathbf{P}))$ as $\mathbf{C}(\mathbf{P})$ and $\bigcup(\mathcal{N}_{\uparrow R^{\smile}}(\mathbf{P}))$ as $\mathbf{C}(\uparrow \mathbf{P})$. $\mathbf{C}(\mathbf{P})$ will be called the *covering induced by* $\mathbf{P}$ and $\mathbf{C}(\uparrow \mathbf{P})$ the *principal covering induced by* $\mathbf{P}$. By K, K' and so on, we shall denote any element of a covering $\mathbf{C}$. Therefore, for any property system $\mathbf{P}$ and any $m \in M$, $R^{\smile}(m)$ is a block K of $\mathbf{C}(\mathbf{P})$ and any $N \in\uparrow R^{\smile}(m)$ is a block K of $\mathbf{C}(\uparrow \mathbf{P})$.

[8] In [22] $\mathcal{N}_*(U)$ is denoted as $\mathcal{N}_{F(R)}(U)$, and $\mathcal{N}_x^*$ as $\mathcal{N}_x^R$.

As we can see from Example 4, the covering $\mathbf{C}(\mathbf{P})$ induced by the property system of Example 1 is $\{\{a,b,c,d\},\{b,c,d\},\{a,b,c\},\{b,c,d,e\},\{d,e,f\},\{d,g\}\}$. Now, given a covering $\mathbf{C}$ on a set U we define the prototype of the property systems $\mathbf{P}$ such that $\mathbf{C} = \mathbf{C}(\mathbf{P})$.

Definition 10. *Let $\mathbf{C}$ be a covering of a set U, with both U and $\mathbf{C}$ at most countable. Let us set for all $x \in U, \langle x, K\rangle \in R$ iff $x \in K$. The resulting property system $\mathbf{P}(\mathbf{C}) = \langle U, \mathbf{C}, R\rangle$ will be called the* covering property system, *CPS, induced by $\mathbf{C}$.*

Clearly, $\mathbf{C} = \bigcup(\mathcal{N}_{R^{\smile}}(\mathbf{P}(\mathbf{C})))$ so that $\mathbf{C} = \mathbf{C}(\mathbf{P}(\mathbf{C}))$.

Example 6. Let $U = \{a,b,c,d,e\}$ and $\mathbf{C} = \{\{c,e\},\{d,e\},\{a,b,c\},\{b,d,e\}\}$.

$\mathbf{R}$	$\{c,e\}$	$\{d,e\}$	$\{a,b,c\}$	$\{b,d,e\}$
a	0	0	1	0
b	0	0	1	1
c	1	0	1	0
d	0	1	0	1
e	1	1	0	1

At this point, some considerations on the operators so far discussed are in order. First of all, notice that since int and $\mathcal{C}$ have the form $\langle\cdot\rangle[\cdot]$, their logical structure is $\exists\forall$. Indeed in any property system $\mathbf{P} = \langle U, G, R\rangle$, for all $A \subseteq U, B \subseteq M$:
$int(A) = \{g : \exists m(g \in R^{\smile}(m) \wedge R^{\smile}(m) \subseteq A)\}$
$= \{g : \exists m(g \in R^{\smile}(m) \wedge \forall g'(g' \in R^{\smile}(m) \Longrightarrow g' \in A))\}$.
Symmetrically, $\mathcal{C}(B) = \{m : \exists g(m \in R(g) \wedge \forall m'(m' \in R(g) \Longrightarrow m' \in B))\}$.

In turn, cl and $\mathcal{A}$ have the form $\forall\exists$:
$cl(A) = \{g : \forall m(m \in R(g) \Longrightarrow m \in R(A)\}$
$= \{g : \forall m(g \in R^{\smile}(m) \Longrightarrow \exists a(a \in A \wedge a \in R^{\smile}(m)))\}$
$= \{g : \forall m(g \in R^{\smile}(m) \Longrightarrow A \cap R^{\smile}(m) \neq \emptyset)\}$.
Symmetrically, $\mathcal{A}(B) = \{m : \forall g(m \in R(g) \Longrightarrow B \cap R(m) \neq \emptyset)\}$. Therefore, since $R^{\smile}(m)$ is a block of the induced covering $\mathbf{C}(\mathbf{P})$, $g \in cl(A)$ iff $K \cap A \neq \emptyset$ for all blocks K such that $g \in K$. This is different from operators such as $\phi(A) = \{g : \exists m(g \in R^{\smile}(m) \wedge R^{\smile}(m) \cap A \neq \emptyset)\}$ or, in a SRS, $\phi(A) = \{g : R(g) \cap A \neq \emptyset\}$.

In the next two definitions, some standard notions are introduced.

Definition 11. *For any covering $\mathbf{C}$ on a set U, let us set:*

$$R^{\mathbf{C}} = \{\langle x, y\rangle : \exists K \in \mathbf{C}(x \in K \wedge y \in K)\} \quad (17)$$

$$R_{\mathbf{C}} = \{\langle x, y\rangle : \forall K \in \mathbf{C}(x \in K \Longrightarrow y \in K)\} \quad (18)$$

Lemma 6. *For any covering $\mathbf{C}$ on a set U,*
(1) *$R^{\mathbf{C}}$ is a tolerance relation.* (2) *$R_{\mathbf{C}}$ is a preorder.*

Proof. (1) Because $\wedge$ is reflexive ($\alpha \wedge \alpha \equiv \alpha$) and commutative ($\alpha \wedge \beta \equiv \beta \wedge \alpha$). (2) Because $\Longrightarrow$ is reflexive ($\alpha \Longrightarrow \alpha$ is a tautology) and transitive. □

Definition 12. *Given a covering* $\mathbf{C}$ *on a set* U *and* $x \in U$ *let us set:*
(a) $i(x) = \bigcup\{K : x \in K\}$, (a') $i(\mathbf{C}) = \{i(x) : x \in U\}$.
(b) $n(x) = \bigcap\{K : x \in K\}$, (b') $n(\mathbf{C}) = \{n(x) : x \in U\}$.
(c) $md(x) = \{K : x \in K \wedge K$ *is minimal*$\}$, (c') $md(\mathbf{C}) = \{md(x) : x \in U\}$.
(d) $\mathbf{P}(R^{\mathbf{C}})) = \langle U, U, R^{\mathbf{C}}\rangle$. (e) $\mathbf{P}(R_{\mathbf{C}})) = \langle U, U, R_{\mathbf{C}}\rangle$.

The set $md(x)$ is called *set of minimal descriptions of* x.

Theorem 6. *For any covering* $\mathbf{C}$ *on a set* U *and any* $x \in U$,
(1) $\langle x, y\rangle \in R_{\mathbf{C}}$ *iff* $\langle i\rangle(x) \subseteq \langle i\rangle(y)$. (2) $\langle i\rangle(x) \subseteq \langle i\rangle(y)$ *iff* $md(x) \subseteq md(y)$. (3) $R^{\mathbf{C}}(x) = i(x)$. (4) $R_{\mathbf{C}}(x) = n(x) = \bigcap md(x)$.

Proof. The proofs come directly from the definitions: (1) Trivial: $\langle i\rangle(x) \subseteq \langle i\rangle(y)$ iff $\{K : x \in K\} \subseteq \{K : y \in K\}$ iff $\forall K(x \in K \Longrightarrow y \in K)$. (2) If $\langle i\rangle(x) \subseteq \langle i\rangle(y)$, trivially $md(x) \subseteq md(y)$. Conversely, suppose $md(x) \subseteq md(y)$. For any $K \in \mathbf{C}$ such that $x \in K$, there is a $K' \in md(x)$ such that $K' \subseteq K$. By assumption $x \in K'$, so that $y \in K$, too. (3) If $x \in K$, then $\forall y(y \in K \Longrightarrow y \in R^{\mathbf{C}})$. So, $\forall K(x \in K \Longrightarrow K \subseteq R^{\mathbf{C}}(x))$. Therefore, $i(x) \subseteq R^{\mathbf{C}}(x)$. Vice-versa, for all $y \in U$, if $y \in R^{\mathbf{C}}(x)$ then $\exists K(x \in K \wedge y \in K$. Hence $R^{\mathbf{C}}(x) \subseteq i(x)$. (4) If $y \in R_{\mathbf{C}}(x)$ then $y \in K$ for all K such that $x \in K$. Hence, $y \in n(x)$, so that $R_{\mathbf{C}}(x) \subseteq n(x)$. Vice-versa, if $y \in n(x)$ then $\forall K(x \in K \Longrightarrow y \in K)$. Thus, $n(x) \subseteq R_{\mathbf{C}}(x)$. The last equation derives directly from the minimality of the elements of $md(x)$. □

Obviously, if $md(x) = \{K\}$, then $K = n(x) = R_{\mathbf{C}}(x)$.

CONNECTIONS. For Theorem 6.(2) cf. also [15]. As to Theorem 6.(4) cf. [9].

Example 7. For the covering of Example 6 one has:

x	$n(x)$	$i(x)$	$md(x)$
a	$\{a,b,c\}$	$\{a,b,c\}$	$\{\{a,b,c\}\}$
b	$\{b\}$	U	$\{\{b,d,e\},\{a,b,c\}\}$
c	$\{c\}$	$\{a,b,c,e\}$	$\{\{a,b,c\},\{c,e\}\}$
d	$\{d,e\}$	$\{b,d,e\}$	$\{\{d,e\}\}$
e	$\{e\}$	$\{b,c,d,e\}$	$\{\{d,e\},\{c,e\}\}$

$\mathbf{R}^{\mathbf{C}}$	a	b	c	d	e
a	1	1	1	0	0
b	1	1	1	1	1
c	1	1	1	0	1
d	0	1	0	1	1
e	0	1	1	1	1

$\mathbf{R}_{\mathbf{C}}$	a	b	c	d	e
a	1	1	1	0	0
b	0	1	0	0	0
c	0	0	1	0	0
d	0	0	0	1	1
e	0	0	0	0	1

Definition 13. *Given* $R \subseteq X \times Y$, *let us set* $\mathbf{P}(R) = \langle X, Y, R\rangle$.

One has that $\mathbf{C} \neq \mathbf{C}(\mathbf{P}(R^{\mathbf{C}}))$ and $\mathbf{C} \neq \mathbf{C}(\mathbf{P}(R_{\mathbf{C}}))$. However:

Lemma 7. *Let* $\mathbf{C}$ *be a covering of* U. *Then for all* $X \subseteq U$,
(1) $\langle i\rangle^{\mathbf{P}(R^{\mathbf{C}})}(X) = \langle e\rangle^{\mathbf{P}(R^{\mathbf{C}})}(X) = \langle e\rangle^{\mathbf{P}(\mathbf{C})}\langle i\rangle^{\mathbf{P}(\mathbf{C})}(X)$.
(2) $[i]^{\mathbf{P}(R^{\mathbf{C}})}(X) = [e]^{\mathbf{P}(R^{\mathbf{C}})}(X) = [e]^{\mathbf{P}(\mathbf{C})}[i]^{\mathbf{P}(\mathbf{C})}(X)$.

Proof. In what follows, R is the relation of the property system $\mathbf{P}(\mathbf{C})$.

$$\begin{aligned}(1)\ \langle e\rangle^{\mathbf{P}(\mathbf{C})}\langle i\rangle^{\mathbf{P}(\mathbf{C})}(X) &= \{g : g \in R^{\smile}(\{K\}) \wedge K \in \langle i\rangle^{\mathbf{P}(\mathbf{C})}(X)\}\\ &= \{g : g \in K \wedge \exists g'(g' \in X \wedge K \in R(g'))\}\\ &= \{g : g \in K \wedge \exists g'(g' \in X \wedge g' \in K)\}\\ &= \{g : g \in K \wedge \exists g'(g' \in X \wedge g \in R^{\mathbf{C}}(g'))\}\\ &= R^{\mathbf{C}}(X) = \langle i\rangle^{\mathbf{P}(R^{\mathbf{C}})}(X) = \langle e\rangle^{\mathbf{P}(R^{\mathbf{C}})}(X)\end{aligned}$$

(because $R^{\mathbf{C}}$ is symmetric). (2) By duality (a direct proof is in the Appendix). □

The following operators induced by a covering $\mathbf{C}$ on a set U are taken from [9,15,26]. To make notation uniform, we superimpose our own symbols[9] (Table 3):

Table 3. Covering based approximation operators

[9]	[15]	[26]	New symbols and original definitions
	L		$(lC)_0(X) = \bigcup\{n(x) : n(x) \subseteq X\}$
	U		$(uC)_0(X) = \bigcup\{n(x) : x \in X\}$
CL	L_5	$\underline{C_1}$	$(lC)_1(X) = \bigcup\{K : K \subseteq X\}$
		$\overline{C_1}$	$(uC)_1(X) = \bigcap\{-K : K \cap X = \emptyset\}$
XL	L_1	$\underline{C_2}$	$(lC)_2(X) = \{x : n(x) \subseteq X\}$
XH	U_1	$\overline{C_2}$	$(uC)_2(X) = \{x : n(x) \cap X \neq \emptyset\}$
	L_2	$\underline{C_3}$	$(lC)_3(X) = \{x : \exists y(y \in n(x) \wedge n(y) \subseteq X)\}$
	U_2	$\overline{C_3}$	$(uC)_3(X) = \{x : \forall y(y \in n(x) \Longrightarrow n(y) \cap X \neq \emptyset)\}$
	L_3	$\underline{C_4}$	$(lC)_4(X) = \{x : \forall y(x \in n(y) \Longrightarrow n(y) \subseteq X)\}$
	U_3	$\overline{C_4}$	$(uC)_4(X) = \bigcup\{n(x) : n(x) \cap X \neq \emptyset)\}$
	L_4	$\underline{C_5}$	$(lC)_5(X) = \{x : \forall y(x \in n(y) \Longrightarrow y \in X)\}$
	U_4	$\overline{C_5}$	$(uC)_5(X) = \bigcup\{n(x) : x \in X\}$
SL			$(lC)_6(X) = \{x : \forall K(x \in K \Longrightarrow K \subseteq X)\} = \{x : i(x) \subseteq X\}$
SH			$(uC)_6(X) = \bigcup\{K : K \cap X \neq \emptyset\} = \bigcup\{i(x) : x \in X\}$
IH	U_5		$(uC)_7(X) = \bigcup\{n(x) : x \in X\}$
FH			$(uC)_8(X) = (lC)_1(X) \cup (\bigcup(\bigcup\{Md(x) : x \in X \cap -(lC)_1(X)\}))$

Now we shall prove that the above operators can be defined by means of the intensional and extensional constructors introduced in the previous section[10]. Let us start with some observations about the operator $n(x)$.

Observation 10. Notice that $\{g : \forall K(g \in K \Longrightarrow K \cap A \neq \emptyset)\}$ is not equal to $\{g : n(g) \cap A \neq \emptyset\}$. In fact the former says that $g \in K \wedge g \in K' \Longrightarrow (\exists a(a \in$

[9] A wider reference about covering-based approximation operators and the scientific literature about the topic can be found in Sect. 5 of [33].

[10] The original definition of $(uC)_7$ is $(lC)_1(X) \cup (\bigcup\{n(x) : x \in X \cap -(lC)_1(X)\})$.

$A \wedge a \in K) \wedge \exists a'(a' \in A \wedge a' \in K'))$, while the latter says that $g \in K \wedge g \in K' \Longrightarrow (\exists a(a \in A \wedge a \in K \wedge a \in K')$. But $\exists$ is additive, not multiplicative. Thus, $\exists a(a \in K) \wedge \exists a(a \in K') \not\equiv \exists a(a \in K \wedge a \in K')$.

Observation 11. In general topology it is proved that given a covering $\mathbf{C}$ on a set U, the family $\{n(x) : x \in U\}$ is a basis of the coarsest topology among the topologies in which the elements of $\mathbf{C}$ are open sets. Actually, under our hypotheses we can prove something more precise. Indeed, in view of the above results and Facts 4, it is immediate to see that $\{n(x) : x \in U\}$ is a basis for the Alexandrov topology induced by $R_{\mathbf{C}}$.

Let us start with the operators defined on CPSs. They are the operators in which $n(x)$ does not appear.

Theorem 7. *Let* $\mathbf{P}(\mathbf{C}) = \langle U, \mathbf{C}, R\rangle$ *be a CPS, int, cl,* $\langle\cdot\rangle$ *and* $[\cdot]$*, the operators and constructors defined on* $\mathbf{P}(\mathbf{C})$*. Then, for all* $A \subseteq U$*:*

(1) $(lC)_1(A) = int(A)$.
(2) $(lC)_1$ *is a pretopological interior operator.*
(3) *The dual operator of* $(lC)_1$ *is* $(uC)_1$*. Hence,* $(uC)_1(A) = cl(A)$.
(4) *For any* $K \in \mathbf{C}, (lC)_1(K) = int(K) = K$.
(5) $(lC)_6(A) = [e][i](A)$, $(uC)_6(A) = \langle e\rangle\langle i\rangle(A)$.
(6) $(lC)_1^{md(\mathbf{C})}(A) = (lC)_1(A)$, *where* $(lC)_1^{md(\mathbf{C})}(A)$ *is defined as* $\{K : K \in md(\mathbf{C}) \wedge K \subseteq A\}$.

Proof.(1) Since, from Theorem 2.(3), $R^{\smile}(\{K\}) = K$, the result is immediate from Lemma 2.(1).
(2) From (1) and Observation 3.
(3) Since $R^{\smile}(\{K\}) = K$, immediate from Lemma 2.(2).
(4) $\bigcup\{K' : K' \subseteq K\} = K$, trivially.
(5) From Theorem 6.(3), $(uC)_6(A) = \langle i\rangle^{\mathbf{P}(R^{\mathbf{C}})}(A)$ and $(lC)_6(A) = [i]^{\mathbf{P}(R^{\mathbf{C}})}(A)$. Therefore, from Lemma 7 one obtains the proof immediately (a direct proof is in the Appendix).
(6) Since $md(\mathbf{C}) \subseteq \mathbf{C}$, $(lC)_1^{md(\mathbf{C})}(A) \subseteq (lC)_1(A)$. Let $K \notin md(x)$ for $x \in K$. The $\exists K' \in md(x), K' \subsetneq K$. Hence, $K = \bigcup_{x\in K}\{K' : K' \in md(x)\}$. Therefore, $\bigcup\{K \in \mathbf{C} : K \subseteq A\} = \bigcup\{K' : K' \in md(\mathbf{C}) \wedge K' \subseteq A\}$. That is, $(lC)_1(A) \subseteq (lC)_1^{md(\mathbf{C})}(A)$. □

Observation 12. The last proof informs us that $\mathbf{C}$ and $md(\mathbf{C})$ as subbases induce the same topology. Moreover, $\forall K \in \mathbf{C}, \exists \mathcal{K} \subseteq md(\mathbf{C})$ such that $K = \bigcup \mathcal{K}$. Thus, we shall say that $md(\mathbf{C})$ is a finer subbasis than $\mathbf{C}$.

Notice that two non comparable subbases $\mathcal{B}_1, \mathcal{B}_2 \subseteq \wp(U)$ may induce the same topology on U. For instance, the subbases $\mathcal{B}_1 = \{\{2,4\}, \{3,4\}, \{3,5\}, \{4,5\}, \{1,2,3,4\}\}$ and $\mathcal{B}_2 = \{\{2,4\}, \{3\}, \{5\}, \{4,5\}, \{1,2,3,4\}\}$ induce the same base $\{\{2,4\}, \{4\}, \{3\}, \{5\}, \{1,2,3,4\}\}$, *but* $\{3,4\}$ is not union of elements of $\mathcal{B}_2$ *and* $\{3\}$ is not union of elements of $\mathcal{B}_1$.

Theorem 8. *Let* $\mathbf{P}(\mathbf{C}) = \langle U, \mathbf{C}, R\rangle$ *be a CPS, and* $\mathcal{N}_{\uparrow R^{\smile}}(\mathbf{P})$ *its PNS. Then for all* $A \subseteq U$, $(lC)_1(A) = G^{\uparrow R^{\smile}}(A)$.

Proof. Directly from Theorem 4 and Theorem 7.(1). □

Theorem 9. *Let* $\mathbf{P}(\mathbf{C}) = \langle U, \mathbf{C}, R\rangle$ *be a CPS and* $\preceq_{int}$ *be the specialisation preorder induced by int. Then* $\preceq_{int}$ *coincides with* $R_{\mathbf{C}}$.

Proof. Let $x \preceq_{int} y$. Thus, $x \in int(A) \Longrightarrow y \in int(A)$, all $A \subseteq U$. In particular A may be any $K \in \mathbf{C}$. From Theorem 7.(4), $int(K) = K$. Hence $x \in K \Longrightarrow y \in K$, for any $K \in \mathbf{C}$. Therefore $y \in n(x)$, so that from Theorem 6, $\langle x, y\rangle \in R_C$. Vice-versa, let $\langle x, y\rangle \in R_C$. Then, by definition, $y \in n(x)$ so that for all $K, x \in K \Longrightarrow y \in K$. If $x \in int(A)$, Then there is K such that $x \in K$ and $K \subseteq A$. But from hypothesis $y \in K$, too. We conclude $y \in int(A)$. □

Observation 13. The above theorem does not imply that $int^{\mathbf{P}(\mathbf{C})} = \mathcal{C}^{\mathbf{P}(R_{\mathbf{C}})}$. Indeed, this equation holds only if $int^{\mathbf{P}(\mathbf{C})}$ is a topological interior operator (cf. Facts 4).

Now we move to another set of approximation operators. As a matter of fact, all the approximation operators based on the operator $n(x)$ are actually induced not by CPSs but by SRSs of the form $\langle U, U, R_{\mathbf{C}}\rangle$. We have denoted such SRSs as $\mathbf{P}(R_{\mathbf{C}})$. In what follows, since $(uC)_0, (uC)_5$ and $(uC)_7$ have the same definition, we shall write collectively $(uC)_{0,5,7}$. The same convention will be adopted for operators proved to coincide.

Theorem 10. *Let* $\mathbf{P}(R_{\mathbf{C}}) = \langle U, U, R_{\mathbf{C}}\rangle$ *be an SRS induced by a CPS and int, cl,* $\mathcal{A}$, $\mathcal{C}$, $\langle\cdot\rangle$ *and* $[\cdot]$ *the operators and constructors defined on* $\mathbf{P}(R_{\mathbf{C}})$. *Then for all* $A \subseteq U$:

(1) $(lC)_0(A) = (lC)_2(A) = [e](A) = \mathcal{C}(A)$.
(2) $(uC)_2(A) = \langle e\rangle(A) = \mathcal{A}(A)$.
(3) *The dual operator of* $(lC)_{0,2}$ *is* $(uC)_2$.
(4) $(uC)_{0,5,7}(A) = \langle i\rangle(A) = cl(A)$
(5) $(lC)_0$ *is upper adjoint to* $(uC)_{0,5,7}$.
(6) $(lC)_5(A) = [i](A) = int(A)$.
(7) $(lC)_5$ *is the dual of* $(uC)_{0,5,7}$.
(8) $(lC)_3(A) = \langle e\rangle[e](A)$.
(9) $(uC)_3(A) = [e]\langle e\rangle(A)$.
(10) $(lC)_4(A) = [i][e](A)$.
(11) $(uC)_4(A) = \langle i\rangle\langle e\rangle(A)$.

Proof.(1) From Theorem 6.(4) for any $x, n(x) = R_{\mathbf{C}}(x)$. Thus:
$(lC)_0(A) = \bigcup\{R_{\mathbf{C}}(x) : R_C(x) \subseteq A\} = \mathcal{C}(A)$;
$(lC)_2(A) = \{x : R_{\mathbf{C}}(X) \subseteq A\} = [e](A)$.
Since R_C is a preorder, from Facts 3 we complete the proof.
(2) $(uC)_2(A) = \{x : R_{\mathbf{C}}(x) \cap A \neq \emptyset\} = \{x : x \in R_{\mathbf{C}}^{\smile}(A)\} = \langle e\rangle(A)$.
(3) $[e]$ is the dual of $\langle e\rangle$.

(4) Clearly $(uC)_0(A) = R(A) = \langle i\rangle(A)$.
(5) From (1), (4) and the adjoint relations (11) we obtain the result.
(6) $(uC)_5(A) = \{x : \forall y(x \in R_{\mathbf{C}}(y) \Longrightarrow y \in A)\} =$
$= \{x : \forall y(y \in R_{\mathbf{C}}^{\smile}(x) \Longrightarrow y \in A)\} = \{x : R_{\mathbf{C}}^{\smile}(x) \subseteq A\} = [i](A)$.
(7) From (4), (6) and the fact that $[i]$ is the dual of $\langle i\rangle$.

$$\begin{aligned}
(8)\ (lC)_3(A) &= \{x : \exists y(y \in R_{\mathbf{C}}(x) \wedge R_{\mathbf{C}}(y) \subseteq A)\} = \\
&= \{x : \exists y(x \in R_{\mathbf{C}}^{\smile}(y) \wedge y \in [e](A))\} = \\
&= \{x : \exists y(x \in \langle e\rangle(y) \wedge y \in [e](A))\} = \langle e\rangle[e](A) \\
(9)\ (uC)_3(A) &= \{x : \forall y(y \in R_{\mathbf{C}}(x) \Longrightarrow R_{\mathbf{C}}(y) \cap A \neq \emptyset)\} = \\
&= \{x : \forall y(y \in R_{\mathbf{C}}(x) \Longrightarrow y \in \langle e\rangle(A))\} = \\
&= \{x : R_C(x) \subseteq \langle e\rangle(A)\} = [e]\langle e\rangle(A) \\
(10)\ (lC)_4(A) &= \{x : \forall y(y \in R_{\mathbf{C}}^{\smile}(x) \Longrightarrow R_C(y) \subseteq A)\} = \\
&= \{x : \forall y(y \in R_{\mathbf{C}}^{\smile}(x) \Longrightarrow y \in [e](A))\} = \\
&= \{x : R_{\mathbf{C}}^{\smile}(x) \subseteq [e](A)\} = [i][e](A) \\
(11)\ (uC)_4(A) &= \bigcup\{R_{\mathbf{C}}(x) : R_{\mathbf{C}}(x) \cap A \neq \emptyset\} = \bigcup\{R_{\mathbf{C}}(x) : x \in R_C^{\smile}(A)\} = \\
&= \bigcup\{R_{\mathbf{C}}(x) : x \in \langle e\rangle(A)\} = \bigcup\{\langle i\rangle(x) : x \in \langle e\rangle(A)\} = \langle i\rangle\langle e\rangle(A).
\end{aligned}$$

□

The last operator to be interpreted is a tricky one, because it is computed on two different domains. One is the usual $\mathbf{P}(\mathbf{C})$, the other is $\mathbf{P}(\mathbf{C}^\star) = \langle U, \mathbf{C}^\star, R\rangle$, where $\mathbf{C}^\star = \{K : K \in \{Md(x) : x \in U\}\}$, that is, $\mathbf{P}(\mathbf{C}^\star)$ is $\mathbf{P}(\mathbf{C})$ restricted to the minimal elements of $\mathbf{C}$. In what follows, if an operator $\odot$ is computed on $\mathbf{P}(\mathbf{C}^\star)$, then it will be denoted with $\odot^\star$. The members of $\mathbf{C}^\star$ are denoted as $K^\star$.

Theorem 11. $(uC)_8(A) = int(A) \cup (uC)_6^\star(A \cap -int(A))$.

Proof. From Theorem 7.(1), $(uC)_8(A) = int(A) \cup (\bigcup(\bigcup\{Md(x) : x \in A \cap -int(A)\}))$. On $\mathbf{P}(\mathbf{C}^\star)$, $\bigcup\{Md(x) : x \in A \cap -int(A)\} = \bigcup\{K^\star : x \in K^\star \wedge x \in A \cap -int(A)\} = \{K^\star : K^\star \cap A \cap -int(A) \neq \emptyset\} = (uC)_6^\star(A \cap -int(A))$. □

Usually, $(uC)_6^\star(A \cap -int(A))$ is called the *boundary* of A.

Example 8. Given the covering $\mathbf{C}$ of Example 6:
$(lC)_1(\{a,c,e\}) = \{c,e\}$, $(lC)_1(\{a,b,e\}) = \emptyset$
$(uC)_1(\{a,c,e\}) = U$, $(uC)_1(\{b,c\}) = \{a,b,c\}$
$(lC)_6(\{a,c,e\}) = \emptyset$, $(lC)_6(\{b,d,e\}) = \{d\}$
$(uC)_6(\{a,c,e\}) = U$, $(uC)_6(\{a,b,e\}) = U$
$(lC)_{0,2}(\{a,c,e\}) = \{c,e\}$, $(lC)_{0,2}(\{a,b,e\}) = \{b,e\}$
$(uC)_2(\{a,c,e\}) = \{a,c,e,d\}$, $(uC)_2(\{b,c\}) = \{a,b,c\}$
$(lC)_3(\{a,c,e\}) = \{a,c,d,e\}$, $(lC)_3(\{a,b,e\}) = \{a,b,e,d\}$
$(uC)_3(\{a,c,e\}) = \{c,d,e)\}$, $(uC)_3(\{a,b,e\}) = \{b,d,e\}$
$(lC)_4(\{a,c,e\}) = \emptyset$, $(lC)_4(\{a,b,e\}) = \emptyset$, $(lC)_4(\{c,d,e\}) = \{d,e\}$
$(uC)_4(\{a,c,e\}) = U$, $(uC)_4(\{a,b,e\}) = U$, $(uC)_4(\{c\}) = \{a,b,c\}$
$(lC)_5(\{a,c,e\}) = \{a\}$, $(lC)_5(\{a,b,e\}) = \{a\}$, $(lC)_5(\{c,d,e\}) = \{d,e\}$
$(uC)_5(\{a,c,e\}) = \{a,b,c,e\}$, $(uC)_5(\{a,b,c,e\}) = U$, $(uC)_5(\{c\}) = \{c\}$

Let us display the above results in a synoptic table:

Table 4. Properties of covering-based approximation operators

Operator	System	Formula	Dual	Adj.	Incr.	Decr.	Mono.	Idem.	Mult.	Add.
$(lC)_1$	$\mathbf{P}(\mathbf{C})$	*int*	$(uC)_1$	Note 1	No	Yes	Yes	Yes	No	No
$(uC)_1$	$\mathbf{P}(\mathbf{C})$	*cl*	$(lC)_1$	Note 1	Yes	No	Yes	Yes	No	No
$(lC)_6$	$\mathbf{P}(\mathbf{C})$	$[e][i]$	$(uC)_6$	$(uC)_6$	No	Yes	Yes	No	Yes	No
$(uC)_6$	$\mathbf{P}(\mathbf{C})$	$\langle e\rangle\langle i\rangle$	$(lC)_6$	$(lC)_6$	Yes	No	Yes	No	No	Yes
$(lC)_6$	$\mathbf{P}(R^{\mathbf{C}})$	$[i]$, $[e]$	$(uC)_6$	$(uC)_6$	No	Yes	Yes	No	Yes	No
$(uC)_6$	$\mathbf{P}(R^{\mathbf{C}})$	$\langle e\rangle$, $\langle i\rangle$	$(lC)_6$	$(lC)_6$	Yes	No	Yes	No	No	Yes
$(lC)_1$	$\mathcal{N}_{\uparrow R^{\smile}}(\mathbf{P})$	G	$(uC)_1$		No	Yes	Yes	Yes	No	No
$(uC)_1$	$\mathcal{N}_{\uparrow R^{\smile}}(\mathbf{P})$	F	$(lC)_1$		Yes	No	Yes	Yes	No	No
$(lC)_{0,2}$	$\mathbf{P}(R_{\mathbf{C}})$	$[e], \mathcal{C}$	$(uC)_2$	$(uC)_{0,5,7}$	No	Yes	Yes	Yes	Yes	No
$(uC)_2$	$\mathbf{P}(R_{\mathbf{C}})$	$\langle e\rangle, \mathcal{A}$	$(lC)_{0,2}$	$(lC)_5$	Yes	No	Yes	Yes	No	Yes
$(lC)_5$	$\mathbf{P}(R_{\mathbf{C}})$	$[i], int$	$(uC)_{0,5,7}$	$(uC)_2$	No	Yes	Yes	Yes	Yes	No
$(uC)_{0,5,7}$	$\mathbf{P}(R_{\mathbf{C}})$	$\langle i\rangle, cl$	$(lC)_5$	$(lC)_{0,2}$	Yes	No	Yes	Yes	No	Yes
$(lC)_3$	$\mathbf{P}(R_{\mathbf{C}})$	$\langle e\rangle[e]$	$(uC)_3$	Note 2	Unp.	Unp.	Yes	Yes	No	Yes
$(uC)_3$	$\mathbf{P}(R_{\mathbf{C}})$	$[e]\langle e\rangle$	$(lC)_3$	Note 3	Unp.	Unp.	Yes	Yes	Yes	No
$(lC)_4$	$\mathbf{P}(R_{\mathbf{C}})$	$[i][e]$	$(uC)_4$	$(uC)_4$	No	Yes	Yes	Yes	Yes	No
$(uC)_4$	$\mathbf{P}(R_{\mathbf{C}})$	$\langle i\rangle\langle e\rangle$	$(lC)_4$	$(lC)_4$	Yes	No	Yes	Yes	No	Yes

Note 1: Both *int* and *cl* cannot have lower or upper adjoint because they are neither additive nor multiplicative, in general. Note 2: $(uC)_3$ is monotone and multiplicative, hence it has the following lower adjoint: $\phi(X) = \bigcap\{Y : X \subseteq (uC)_3(Y)\}$. Note 3: $(lC)_3$ is monotone and additive, hence it has the following upper adjoint: $\psi(X) = \bigcup\{Y : (lC)_3(Y) \subseteq X\}$.

Remarks.

(A) The fact that $(lC)_6$ is both upper adjoint and dual of $(lC)_6$, and vice-versa, is given by the duality between $[i]^{\mathbf{P}(R^{\mathbf{C}})}$ and $\langle i\rangle^{\mathbf{P}(R^{\mathbf{C}})}$, the fact that in any property system $[i]$ is upper adjoint of $\langle e\rangle$ and, finally, that $\langle e\rangle^{\mathbf{P}(R^{\mathbf{C}})} = \langle i\rangle^{\mathbf{P}(R^{\mathbf{C}})}$. In $\mathbf{P}(\mathbf{C})$ the proof of the duality runs as follows: $\langle e\rangle\langle i\rangle(X) \subseteq Y$ iff $\langle i\rangle(X) \subseteq [i](Y)$ iff $X \subseteq [e][i](Y)$. A similar argument applies to $(lC)_4$ and $(uC)_4$.
(B) It is not difficult to prove the properties not established or directly derivable by the preceding theorems. For instance, as to $(lC)_6$, suppose R is symmetric. Then $(lC)_6 = [i][i]$. But $[i]$ might fail to be idempotent. Therefore, $([i][i])([i][i](A)) \neq [i]([i][i](A)) \neq [i][i](A)$. In turn, $(uC)_4$ is increasing because $R_{\mathbf{C}}$ is reflexive, so is $R_{\mathbf{C}}^{\smile}$. Therefore, $A \subseteq R_{\mathbf{C}}^{\smile}(A)$, so that $A \subseteq R_{\mathbf{C}}^{\smile}(R_{\mathbf{C}}^{\smile}(A)) = \langle i\rangle\langle e\rangle(A) = (uC)_4$. Similarly, from transitivity one obtains that $(lC)_4$ is decreasing. Moreover, one can notice that $(lC)_4$ and $(uC)_4$ are topological interior and,

respectively, closure operators. On the contrary, the behaviour of $(uC)_3$ cannot be predicted. In fact, it amounts to $\mathcal{I}\mathcal{C}(A)$, where $\mathcal{I}$ and $\mathcal{C}$ are the interior, respectively, closure operators of the Alexandrov topology with bases $\{R_{\mathbf{C}}(x) : x \in U\}$. Thus from a logical point of view it is a double pseudo-complementation, $\neg\neg(A)$ where the pseudocomplementation (i.e. intuitionistic negation $\neg A$ of A is $\mathcal{I}(-A) = [e](-A)$. It is well known that in Intuitionistic logic $\alpha \leq \neg\neg(\alpha)$, all formula α. But a formula α is interpreted on open sets. Indeed, $A \subseteq (uC)_3(A)$ if A is such that $[e](A) = A$. So, in this case $(uC)_3$ is increasing. On the contrary, if A is a closed set, that is, $\langle e\rangle(A) = A$, then trivially $(uC)_3(A) \subseteq A$ because $[e]$ is decreasing on $\mathbf{P}(R_{\mathbf{C}})$. But, if A is neither open nor closed, the result is unpredictable, besides trivial cases (such as $(uC)_3(A) = \emptyset$). Dually for $(lC)_3$.
(C) There are other combinations. Obviously, since in $\mathbf{P}(R_{\mathbf{C}})$ the constructors are idempotent, all the sequences $\odot\odot, ..., \odot$, of the same constructor reduce to $\odot$. The combinations $[e][i], \langle e\rangle\langle i\rangle, \langle i\rangle[i]$ and $[i]\langle i\rangle$ are the inverse of $(lC)_4, (uC)_4, (lC)_3$ and $(uC)_3$, respectively. The last operator was briefly discussed in [25], p. 102.
(D) One problem is the interpretation of the above operators. Here are some hints:
$(lC)_3(A)$ takes all the objects linked only to the internal kernel (necessary part) of A. Otherwise stated, all the unnecessary links are discharged.
Since $(uC)_3(A)$ is the interior of the closure of A (see above (B)), it collects the necessary part inside the topological closure of A.
$(uC)_4(A)$ collects the objects linked to A in both directions.
$(lC)_4(A)$ collects the elements of A which are linked in both directions only to elements of A itself.

Connections. In Table 4 one can recognise, or easily derive, a number of properties about adjointness and duality between approximation operators like those established in [27]. Moreover, one easily derives comparative results between approximation operators, such as those presented in Sect. 4 of [26].

5 Some Algebra

The above results have a number of algebraic consequences. Some are new while others are well-established. Nonetheless in the present framework all of them are easy corollaries of a strict number of properties of the four constructors and the relation which define the given property system. That proves the potentiality of the present approach. Apart from some details, the key point is that $R_{\mathbf{C}}$ is a preorder.
From now on, let $\mathbf{C}$ be a covering on a countable set U.

Definition 14. *Let $\mathbf{P} = \langle U, M, R\rangle$ be a property system. If ϕ is a constructor or combination of constructors with domain X, for $X = U$ or $X = M$, we set:*
(i) $\mathbf{S}_\phi(\mathbf{P}) = \{Y \subseteq X : \phi(Y) = Y\}$*;* (ii) $\mathbf{L}_\phi(\mathbf{P}) = \{\phi(Y) : Y \subseteq X\}$.
If $\mathbf{P}$ is either $\mathbf{P}(\mathbf{C})$ or $\mathbf{P}(R_{\mathbf{C}})$ then we set:

(iii) $\mathbf{S^n(P)} = \{Y \subseteq U : (uC)_n(Y) = Y\}$; (iv) $\mathbf{L^n(P)} = \{(uC)_n(Y) : Y \subseteq U\}$; (v) $\mathbf{S_n(P)} = \{Y \subseteq U : (lC)_n(Y) = Y\}$; (vi) $\mathbf{L_n(P)} = \{(lC)_n(Y) : Y \subseteq U\}$.

Facts 8. *If ϕ is idempotent, then $\mathbf{L}_\phi(\mathbf{P}) = \mathbf{S}_\phi(\mathbf{P})$.*

Facts 9. *If $\mathbf{O'} \dashv^{\iota,\sigma} \mathbf{O}$ is an adjoint pair (see (9) and (10)), then:* (i) *$\iota\sigma$ is an interior operator (see Table 1);* (ii) *$\sigma\iota$ is a closure operator (see Table 1);* (iii) *$\iota\sigma\iota = \iota$;* (iv) *$\sigma\iota\sigma = \sigma$;* (v) *$\sigma : \mathbf{S}_{\iota\sigma}(\mathbf{O'}) \longmapsto \mathbf{S}_{\sigma\iota}(\mathbf{O})$ is an isomorphism;* (vi) *$\iota : \mathbf{S}_{\sigma\iota}(\mathbf{O}) \longmapsto \mathbf{S}_{\iota\sigma}(\mathbf{O'})$ is an isomorphism.*

Recalling that in any property system $\mathbf{P} = \langle U, M, R\rangle$, $\wp(M) \dashv^{\langle e\rangle,[i]} \wp(U)$ and $\wp(U) \dashv^{\langle i\rangle,[e]} \wp(M)$, one obtains:

Corollary 1. (1) *In any property system $\mathbf{P}$, $\mathbf{L}_{int}(\mathbf{P}) = \mathbf{S}_{int}(\mathbf{P})$, $\mathbf{L}_{cl}(\mathbf{P}) = \mathbf{S}_{cl}(\mathbf{P})$, $\mathbf{L}_{\mathcal{C}}(\mathbf{P}) = \mathbf{S}_{\mathcal{C}}(\mathbf{P})$, $\mathbf{L}_{\mathcal{A}}(\mathbf{P}) = \mathbf{S}_{\mathcal{A}}(\mathbf{P})$.*
(2) *In any SRS such that R is a preorder, $\mathbf{L}_{\langle i\rangle}(\mathbf{P}) = \mathbf{S}_{\langle i\rangle}(\mathbf{P})$, $\mathbf{L}_{\langle e\rangle}(\mathbf{P}) = \mathbf{S}_{\langle e\rangle}(\mathbf{P})$, $\mathbf{L}_{[i]}(\mathbf{P}) = \mathbf{S}_{[i]}(\mathbf{P})$.*

Proof. (1) From Facts 9.(i) and (ii), in any property system int, cl, $\mathcal{C}$ and $\mathcal{A}$ are idempotent. (2) In a SRS, if R is a preorder, then $\langle\cdot\rangle$ and $[\cdot]$ are idempotent. □

Lemma 8. *In any property system $\mathbf{P}$, the set-theoretic complement "$-$" is an antisomorphism between $\mathbf{L}_{\langle\cdot\rangle}(\mathbf{P})$ and $\mathbf{L}_{[\cdot]}(\mathbf{P})$, and between $\mathbf{S}_{\langle\cdot\rangle}(\mathbf{P})$ and $\mathbf{S}_{[\cdot]}(\mathbf{P})$, where "$\cdot$" are either both "$i$" or both "$e$".*

Proof. Provided "$\cdot$" is either "i" or "e" in both constructors, for any $X \subseteq U$, $\langle\cdot\rangle(X) \leq \langle\cdot\rangle(Y)$ iff $-\langle\cdot\rangle(Y) \leq -\langle\cdot\rangle(X)$ and by duality the latter is equivalent to $[\cdot](-Y) \leq [\cdot](-X)$. Moreover, if $X \in \mathbf{L}_{\langle\cdot\rangle}(\mathbf{P})$ then for some $Y \subseteq U$, $X = [\cdot](Y)$, so that $-X = -[\cdot](Y) = \langle\cdot\rangle(-Y)$ and vice-versa. We conclude that $\forall X \in \mathbf{L}_{\langle\cdot\rangle}(\mathbf{P})$, $\exists Y \in \mathbf{L}_{[\cdot]}(\mathbf{P})$ and vice-versa. Thus the set-theoretic complementation is an order-reversing isomorphism between $\mathbf{L}_{\langle\cdot\rangle}(\mathbf{P})$ and $\mathbf{L}_{[\cdot]}(\mathbf{P})$. The latter antisomorphism is a straightforward consequence of the former. □

Corollary 2. *In any property system $\langle U, M, R\rangle$, the set-theoretic complement "$-$" is an anti order-isomorphism between $\mathbf{S}_{cl}(\mathbf{P})$ and $\mathbf{S}_{int}(\mathbf{P})$, and between $\mathbf{S}_{\mathcal{C}}(\mathbf{P})$ and $\mathbf{S}_{\mathcal{A}}(\mathbf{P})$.*

Proof. From a double application of Lemma 8 and the fact that the operators are idempotent. □

Consider the covering $\mathbf{C}$ of Example 5. The following are the diagrams of the lattices induced by the approximation operators defined on $\mathbf{P}(\mathbf{C})$. Notice that they are not distributive lattices. However, since in $\mathbf{P}(\mathbf{C})$ the relation R is serial, they are bounded by U and $\emptyset$ (or by M and $\emptyset$ for the opposite side, since $R^{\smile}$ is serial, too):

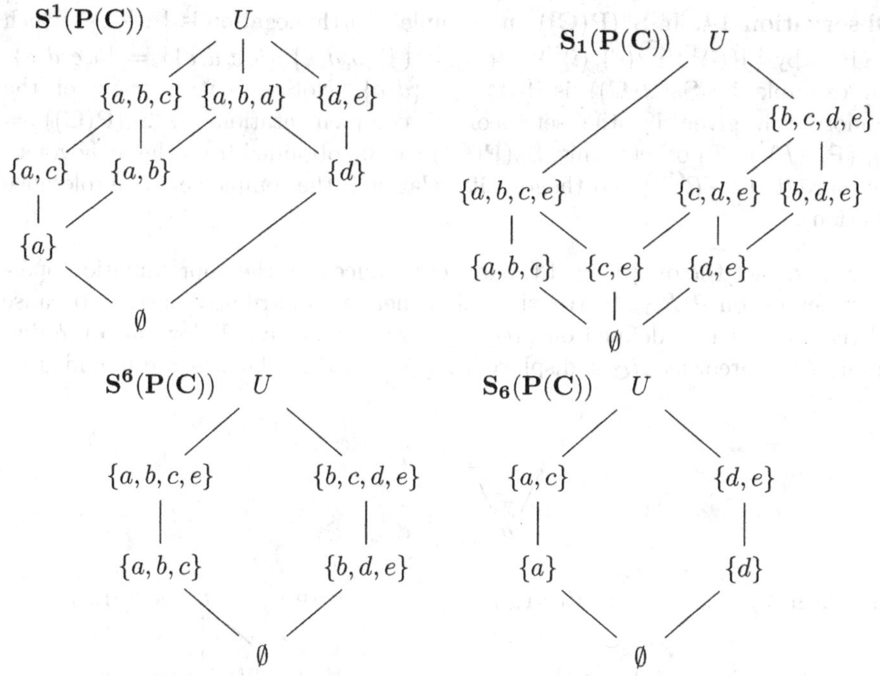

The above diagrams suggest some further investigations.

Lemma 9. [25]. *Let* $\mathbf{P}$ *be a property system. Then,*

(1) $\langle \mathbf{S}_{int}(\mathbf{P}), \cup, \wedge, \emptyset, U\rangle$, *where* $\bigwedge_{i\in I} X_i = int(\bigcap_{i\in I} X_i)$, *is a complete lattice.*

(2) $\langle \mathbf{S}_{cl}(\mathbf{P}), \vee, \cap, \emptyset, U\rangle$, *where* $\bigvee_{i\in I} X_i = cl(\bigcup_{i\in I} X_i)$, *is a complete lattice.*

Corollary 3. *For any covering* $\mathbf{C}$, $\mathbf{S^1}(\mathbf{P}(\mathbf{C}))$ *and* $\mathbf{S_1}(\mathbf{P}(\mathbf{C}))$, *equipped with the above operations are complete lattices.*

From [2], every complete ortholattice is isomorphic to some $\mathbf{S}_{\langle i\rangle}(\mathbf{P})$ with R a tolerance relation. Therefore:

Lemma 10. *Let* $\mathbf{P} = \langle U, U, R\rangle$ *be a SRS such that* R *is a tolerance relation. Then* $\mathbf{S}_{\langle i\rangle}(\mathbf{P})$ *equipped with the operations* $\cup$, $\wedge$ *and* *, *for* $\bigwedge\{X_i\}_{i\in I} = \bigcup\{\langle i\rangle(x) : \langle i\rangle(x) \subseteq \bigcup\{X_i\}_{i\in I}\}$ *and* $X^\star = \bigcup\{\langle i\rangle(x) : x \notin X\}$, *is a complete ortholattice.*

Corollary 4. *For any covering* $\mathbf{C}$, $\mathbf{S_6}(\mathbf{P}(\mathbf{C}))$ *and* $\mathbf{S^6}(\mathbf{P}(\mathbf{C}))$ *can be made into complete ortholattices.*

Proof. From Lemma 6.(1), the proof of Theorem 7.(5) and Lemma 10, $\mathbf{S^6}(\mathbf{P}(\mathbf{C}))$ can be made into a complete ortholattice. The same happens to the isomorphic $\mathbf{S_6}(\mathbf{P}(\mathbf{C}))$ in view of the same Lemma 10. □

Observation 14. In $\mathbf{S}^{6}(\mathbf{P}(\mathbf{C}))$ an example of orthonegation is $\{a,b,c\}^{\star}$ which is given by $\bigcup\{\langle i\rangle^{\mathbf{P}(R^{\mathbf{C}})}(d), \langle i\rangle^{\mathbf{P}(R^{\mathbf{C}})}(e)\} = \bigcup\{\{b,d,e\},\{b,c,d,e\}\} = \{b,c,d,e\}$. An example in $\mathbf{S}_{6}(\mathbf{P}(\mathbf{C}))$ is $\{a\}^{\star} = \{d,e\}$. Notice that in view of the isomorphism given by the set-theoretic complementation $-$, $\mathbf{S}_{6}(\mathbf{P}(\mathbf{C})) = \mathbf{S}_{\langle i\rangle}(\mathbf{P}(-R^{\mathbf{C}}))$. In other terms, $\mathbf{S}_{6}(\mathbf{P}(\mathbf{C}))$ can be obtained from the orthogonality space $\langle U, U, -R^{\mathbf{C}}\rangle$ (an orthogonality relation is the complement of a tolerance relation).

Now we depict the diagrams of the lattices induced by the approximation operators defined on $\mathbf{P}(R_{\mathbf{C}})$. Notice that all of them are distributive lattices because all the constructors defined on preorders are distributive. Below, also the diagram of the preorder $R_{\mathbf{C}}$ is displayed. Again, all these lattices are bounded by U and $\emptyset$:

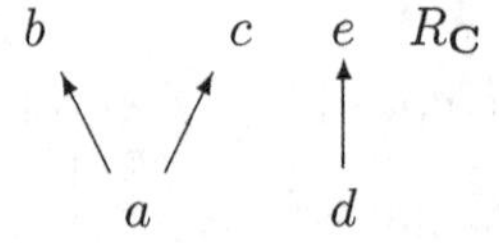

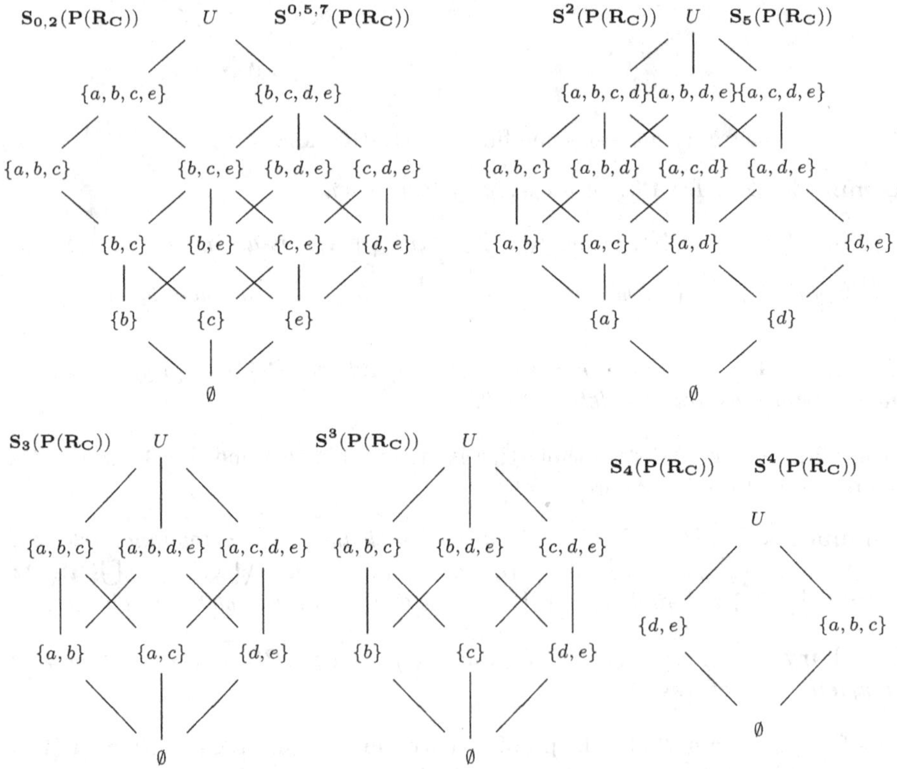

We have to prove the properties that are suggested by the above diagrams. A number of them are straightforwardly derivable from the adjunction properties of the constructors.

Lemma 11. *Let* $\mathbf{P} = \langle U, M, R\rangle$ *be a property system. Then:*
(1) $\mathbf{S}_{[i]\langle e\rangle}(\mathbf{P})$ *and* $\mathbf{S}_{\langle e\rangle[i]}(\mathbf{P})$ *are order isomorphic;* (2) $\mathbf{S}_{[e]\langle i\rangle}(\mathbf{P})$ *and* $\mathbf{S}_{\langle i\rangle[e]}(\mathbf{P})$ *are order isomorphic;* (3) $\mathbf{S}_{[e]}(\mathbf{P}) = \mathbf{S}_{[e]\langle i\rangle}(\mathbf{P})$; (4) $\mathbf{S}_{\langle e\rangle}(\mathbf{P}) = \mathbf{S}_{\langle e\rangle[i]}(\mathbf{P})$; (5) $\mathbf{S}_{[i]}(\mathbf{P}) = \mathbf{S}_{[i]\langle e\rangle}(\mathbf{P})$; (6) $\mathbf{S}_{\langle i\rangle}(\mathbf{P}) = \mathbf{S}_{\langle i\rangle[e]}(\mathbf{P})$.

The proofs of (1) and (2) are trivially obtained from Facts 9.(v) and (vi). The proofs of (3)–(6) derive from Facts 9.(iii) and (iv) according to the following diagrams:

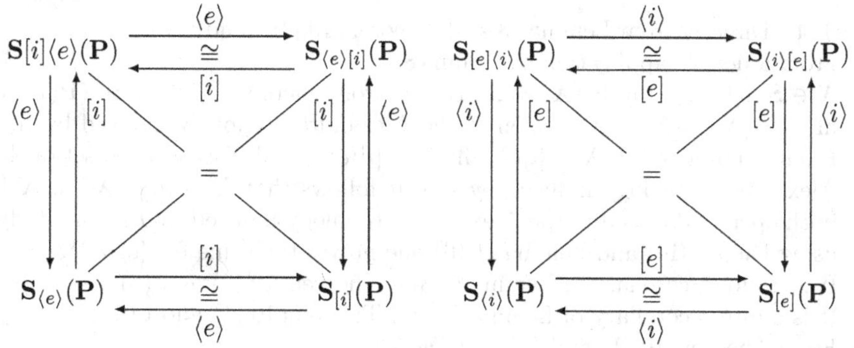

Lemma 12. *In any SRS, if R is reflexive then* $\mathbf{S}_{[e]}(\mathbf{P}) = \mathbf{S}_{\langle i\rangle}(\mathbf{P})$ *and* $\mathbf{S}_{[i]}(\mathbf{P}) = \mathbf{S}_{\langle e\rangle}(\mathbf{P})$.

Proof. If $\langle i\rangle(X) = X$ then $\langle i\rangle(X) \subseteq X$. It follows that $X \subseteq [e](X)$. But if R is reflexive $[e](X) \subseteq X$ so that $[e](X) = X$. The reverse implication is proved in the same way. The proof for the second equation is analogous (a direct proof can be found in the Appendix). □

Lemma 13. *In any SRS,* $\mathbf{P}$, $\mathbf{S}_{\langle i\rangle\langle e\rangle}(\mathbf{P}) = \mathbf{S}_{[i][e]}(\mathbf{P})$ *and* $\mathbf{S}_{\langle e\rangle\langle i\rangle}(\mathbf{P}) = \mathbf{S}_{[e][i]}(\mathbf{P})$, *provided R is reflexive.*

Proof. Suppose $\langle i\rangle\langle e\rangle(X) = X$. Then $\langle e\rangle(X) = X$. In fact, if $\langle e\rangle(X) \neq X$, then since R is reflexive, $X \subsetneq \langle e\rangle(X)$. But since $\langle i\rangle$ is increasing, too, $X \subsetneq \langle i\rangle\langle e\rangle(X)$, contradiction. Hence X belongs to $\mathbf{S}_{\langle e\rangle}(\mathbf{P})$ and for Lemma 12, $X \in \mathbf{S}_{[i]}(\mathbf{P})$. Moreover, $\langle i\rangle\langle e\rangle(X) = \langle i\rangle(X) = X$, so that $X \in \mathbf{S}_{\langle i\rangle}(\mathbf{P})$ and from Lemma 12, $X \in \mathbf{S}_{[e]}(\mathbf{P})$. It follows that $[i][e](X) = [i](X) = X$. Therefore, $X \in \mathbf{S}_{[i][e]}(\mathbf{P})$. The reverse implication is proved similarly, using the fact that if R is reflexive, then $[\cdot]$ is decreasing. The proof for the second equation is analogous. □

Theorem 12. *For any covering* $\mathbf{C}$,

(1) $\mathbf{S_4}(\mathbf{P}(R_\mathbf{C})) = \mathbf{S^4}(\mathbf{P}(R_\mathbf{C}))$;
(2) *The set-theoretic complementation is an anti order-isomorphism between* $\mathbf{S_4}(\mathbf{P}(R_\mathbf{C}))$ *and* $\mathbf{S^4}(\mathbf{P}(R_\mathbf{C}))$ *and between* $\mathbf{S_6}(\mathbf{P}(\mathbf{C}))$ *and* $\mathbf{S^6}(\mathbf{P}(\mathbf{C}))$.
(3) $\mathbf{S_4}(\mathbf{P}(R_\mathbf{C}))$ *and* $\mathbf{S^4}(\mathbf{P}(R_\mathbf{C}))$ *are the Boolean algebras of the clopen (closed and open) sets of the Alexandrov topology induced by* $R_\mathbf{C}$ *(or* $R_{\breve{\mathbf{C}}}$*), that is, with sub-basis* $\{R_\mathbf{C}(x) : x \in U\}$ *(or* $\{R_{\breve{\mathbf{C}}}(x) : x \in U\}$*).*
(4) $\mathbf{S^2}(\mathbf{P}(R_\mathbf{C})) = \mathbf{S_5}(\mathbf{P}(R_\mathbf{C}))$.

(5) $\mathbf{S_{0,2}}(\mathbf{P}(R_\mathbf{C})) = \mathbf{S^{0,5,7}}(\mathbf{P}(R_\mathbf{C}))$.
(6) $\mathbf{S_{0,2}}(\mathbf{P}(R_\mathbf{C}))$ *is the family of open sets of the Alexandrov topology induced by* $R_\mathbf{C}$ *and* $\mathbf{S^{0,5,7}}(\mathbf{P}(R_\mathbf{C}))$ *is the family of closed sets of the Alexandrov topology induced by* $R_\mathbf{C}^{\smile}$*;*
(7) $[e]$ *is an anti order-isomorphism from* $\mathbf{S_3}(\mathbf{P}(R_\mathbf{C}))$ *to* $\mathbf{S^3}(\mathbf{P}(R_\mathbf{C}))$*;* $\langle e\rangle$ *is an antisomorphism from* $\mathbf{S^3}(\mathbf{P}(R_\mathbf{C}))$ *to* $\mathbf{S_3}(\mathbf{P}(R_\mathbf{C}))$.
(8) $\mathbf{S^3}(\mathbf{P}(R_\mathbf{C}))$ *and* $\mathbf{S_3}(\mathbf{P}(R_\mathbf{C}))$ *are the Boolean algebras of the regular elements of the Alexandrov topologies induced by* $R_\mathbf{C}$ *and, respectively,* $R_\mathbf{C}^{\smile}$.

Proof.(1) Directly from Lemma 13 and Theorem 10.(10) and (11).
(2) From a double application of Lemma 8.
(3) $X \in \mathbf{S}_{[e]}(\mathbf{P}(R_\mathbf{C}))$ iff $[e](X) = X$. Hence, from Facts 3.(iii), $X \in \mathbf{S}_{[e]}(\mathbf{P}(R_\mathbf{C}))$ iff $\langle i\rangle[e](X) = X$ iff X is open in the Alexandrov topology induced by $R_\mathbf{C}$. From Lemma 11.(5), $X = [i](X)$ iff $X = [i]\langle e\rangle(X)$ iff X is a closed set of the Alexandrov topology induced by $R_\mathbf{C}$. It follows that $X = [i][e](X)$ iff X is both open and closed in the Alexandrov topology induced by $R_\mathbf{C}$. Similarly, using Facts 3.(iii) and Lemma 11.(6) one proves the same for $\langle e\rangle\langle i\rangle$[11].
(4) It is a direct corollary of Lemma 12 and Theorem 10.(2) and (6).
(5) It is a direct corollary of Lemma 12 and Theorem 10.(1) and (4).
(6) From Theorem 10.(1) and (4) and Facts 4.
(7) Suppose $X \in \mathbf{S_3}(\mathbf{P}(R_\mathbf{C}))$. Therefore, $X = \langle e\rangle[e](X)$. Since $\langle e\rangle$ is idempotent, $[e]\langle e\rangle[e](X) = [e]\langle e\rangle\langle e\rangle[e](X) = [e]\langle e\rangle(X)$. But the latter is an element of $\mathbf{S}_{[e]\langle e\rangle}(\mathbf{P}(R_\mathbf{C}))$. The proof for $\langle e\rangle$ is analogous (a direct proof is in the Appendix). Finally, let $X \in \mathbf{S^3}(\mathbf{P}(R_\mathbf{C}))$. Hence, $-X = -[e]\langle e\rangle(X) = \langle e\rangle[e](-X)$. Thus $-X \in \mathbf{S_3}(\mathbf{P}(R_\mathbf{C}))$. Similar argument for the reverse direction. From Lemma 8 we obtain the conclusion.
(8) In a topological space on a set U, interior operator $\mathcal{I}$ and closure operator $\mathcal{C}$, a subset X of U is called *regular* if $\mathcal{IC}(X) = X$. It is well-known that in any topological space the set of all the regular elements equipped with the operations $\neg(X) := \mathcal{I}(-X)$, $\cap$, and $X \vee Y := \mathcal{IC}(X \cup Y)$ is a Boolean algebra. But from Facts 4 and 3, $[e]$ is the interior operator of the Alexandrov topology induced by R_C, while $\langle e\rangle$ is its closure. Therefore, $\mathbf{S}_{[e]\langle e\rangle}(\mathbf{P}(R_\mathbf{C}))$ is the set of the regular elements of the topology. As to $\mathbf{S_3}(\mathbf{P}(R_\mathbf{C}))$, one has just to notice that the topology is given by the reverse relation $R_\mathbf{C}^{\smile}$. Therefore, *sup* coincides with $\cup$, while *inf* is given by $\bigwedge_{i\in I} X_i = \mathcal{IC}(\bigcap_{i\in I} X_i)$. □

Corollary 5. *For any covering* $\mathbf{C}$ *and any* $n \in \{0, 2, 3, 4, 5, 7\}$, $\mathbf{S_n}(\mathbf{P}(R_\mathbf{C}))$ *and* $\mathbf{S^n}(\mathbf{P}(R_\mathbf{C}))$ *are bounded distributive lattices*[12].

[11] Actually, from Facts 3.(iii), Corollary 1 and Lemma 11.(5) and (6) one trivially derives that in any SRS $\mathbf{P}$ with R a preorder: $\mathbf{S}_{\langle i\rangle}(\mathbf{P}) = \mathbf{S}_{[e]\langle i\rangle}(\mathbf{P}) = \mathbf{S}_{[e]}(\mathbf{P}) = \mathbf{S}_{\langle i\rangle[e]}(\mathbf{P})$; $\mathbf{S}_{\langle e\rangle}(\mathbf{P}) = \mathbf{S}_{[i]\langle e\rangle}(\mathbf{P}) = \mathbf{S}_{[i]}(\mathbf{P}) = \mathbf{S}_{\langle e\rangle[i]}(\mathbf{P})$.

[12] In general, from Facts 3, if R is a preorder then the set of fixpoints of the constructors $\langle\cdot\rangle$ and $[\cdot]$ coincides with the sets of fixpoint of their derived operators $\langle\cdot\rangle[\cdot]$ and $[\cdot]\langle\cdot\rangle$ (where the directions, intension or intension, alternate). Since the sets of fixpoints of the derived operators form distributive lattices, the same happens for the constructors.

Lemma 14. *Let* $\mathbf{P} = \langle U, U, R \rangle$ *be an SRS. Then the following statements are equivalent:* (1) $R^{\smile} = R$; (2) $\langle i \rangle = \langle e \rangle$; (3) $[i] = [e]$. *Moreover, if* R *is an equivalence relation, then* (4) $\mathbf{S}_{\langle i \rangle}(\mathbf{P}) = \mathbf{S}_{int}(\mathbf{P}) = \mathbf{S}_{[e]}(\mathbf{P})$; (5) $\mathbf{S}_{[e]}(\mathbf{P}) = \mathbf{S}_{\mathcal{A}}(\mathbf{P}) = \mathbf{S}_{[i]}(\mathbf{P})$. (6) $\mathbf{S}_{\langle i \rangle}(\mathbf{P}) = \mathbf{S}_{[i]}(\mathbf{P})$; (7) $\mathbf{S}_{\langle e \rangle}(\mathbf{P}) = \mathbf{S}_{[e]}(\mathbf{P})$.

Proof. From Facts 1.(5) and 1.(6), $R^{\smile} = R$ iff $\langle i \rangle = \langle e \rangle$ (or $[i] = [e]$). Therefore the other equalities are described by the following diagrams (the double lines mean equality):

$$\begin{array}{ccccccc}
\mathbf{S}_{[e]\langle i \rangle}(\mathbf{P}) & \overset{Facts\ 4}{=\!=\!=\!=\!=} & \mathbf{S}_{\langle i \rangle}(\mathbf{P}) & = & \mathbf{S}_{\langle e \rangle}(\mathbf{P}) & \overset{Facts\ 4}{=\!=\!=\!=\!=} & \mathbf{S}_{[i]\langle e \rangle}(\mathbf{P}) \\
\Vert\ Lemma\ 10 & & Lemma\ 10\ \Vert & & \Vert\ Lemma\ 10 & & Lemma\ 10\ \Vert \\
\mathbf{S}_{[e]}(\mathbf{P}) & \underset{Facts\ 4}{=\!=\!=\!=\!=} & \mathbf{S}_{\langle i \rangle [e]}(\mathbf{P}) & & \mathbf{S}_{\langle e \rangle [i]}(\mathbf{P}) & \underset{Facts\ 4}{=\!=\!=\!=\!=} & \mathbf{S}_{[i]}(\mathbf{P})
\end{array}$$

Corollary 6. *Given a covering* $\mathbf{C}$ *on a set* U*, the following statements are equivalent:* (1) $R_{\mathbf{C}}$ *is an equivalence relation;* (2) $(uC)_2 = (uC)_{0,5,7}$; (3) $(lC)_{0,2} = (lC)_5$; (3) $\mathbf{S}_5(\mathbf{P}(R_{\mathbf{C}})) = \mathbf{S}^{0,5,7}(\mathbf{P}(R_{\mathbf{C}}))$; (4) $\mathbf{S}_{0,2}(\mathbf{P}(R_{\mathbf{C}})) = \mathbf{S}^2(\mathbf{P}(R_{\mathbf{C}}))$.

Proof. Since $R_{\mathbf{C}}$ is always a preorder, it is an equivalence relation iff $R^{\smile} = R$. Thus the results come trivially from the previous Lemma. □

Observation 15. From Theorem 12 and Corollary 6 one easily derives a number of results established in Sect. 3 of [9] and in Sect. 4 of [26].

Definition 15. *Let* $\mathbf{C}$ *be a covering on* U*:*

(i) *An element* $x \in U$ *such that* $md(x) = \{K\}$ *for some* $K \in \mathbf{C}$ *is called* representative. *We also say that* K *is* represented *by* x.
(ii) *If* $\forall K \in \mathbf{C}, \exists x \in U$ *such that* $md(x) = \{K\}$*, then* $\mathbf{C}$ *is called* representative.
(iii) *If* $\forall x \in U, \exists K \in \mathbf{C}$ *such that* $md(x) = \{K\}$*, then* $\mathbf{C}$ *is called* unary.
(iv) *If* $\mathbf{C}$ *is representative and for any* $K \neq K'$*, neither* $K \subseteq K'$ *or* $K' \subseteq K$*, then* $\mathbf{C}$ *is called* reduced.
(v) *If* $\mathbf{C}$ *is representative and reduced, then it is called* irredundant.
(vi) *If* $\mathbf{C}$ *is both representative and unary, then it is said to be* specialised.
(vii) *If* $K \in \mathbf{C}$ *is represented, by* r_K *we denote an arbitrary representative of* K*. If* x *represents* K*, in some cases we shall denote* K *as* K_x.

CONNECTIONS. For the notion of a "unary covering", see for instance [9] or [26], for those of "representative" and "reduced" covering see [7]. However, in that paper the term "reduced" is applied to the covering here called "irredundant". In turn, this term was introduced in [13] to define a covering $\mathbf{C}$ of U such that for no $K \in \mathbf{C}$, $\mathbf{C} \cap -\{K\}$ is still a covering of U. The equivalence with our definition is proved in the Appendix. The notion of a "specialised covering" seems to be new.

The following examples show that all the above situations are independent.

Example 9. $\mathbf{C_1} = \{\{b,c,d\},\{a,b,d\},\{c,d\}\}$; $\mathbf{C_2} = \{\{a,b\},\{a,c\},\{a,c,d\}\}$; $\mathbf{C_3} = \{\{a\},\{b,c\},\{b,c,d\}\}$; $\mathbf{C_4} = \{\{a,c\},\{b,c\},\{d\}\}$; $\mathbf{C_5} = \{\{a\},\{b\},\{a,b\}\}$. $\mathbf{C_6} = \{\{a,b\},\{a,c\},\{b,c\}\}$.

$\mathbf{C_1}$ is not representative ($\{b,c,d\}$ is not represented), not unary ($md(b) = \{\{d,c\},\{a,b,d\}\}$), not reduced ($\{c,d\} \subsetneq \{b,c,d\}$). $\mathbf{C_2}$ is representative ($\{\{a,b\}\} = md(b)$, $\{\{a,c\}\} = md(c)$, $\{\{a,c,d\}\} = md(d)$), but it is neither unary ($md(a) = \{\{a,b\},\{a,c\}\}$) nor reduced. $\mathbf{C_3}$ is specialised but not reduced. $\mathbf{C_4}$ is representative, reduced, but not unary. $\mathbf{C_5}$ is unary but not representative. $\mathbf{C_6}$ is reduced but neither representative nor unary.

However, there are some relationships among combinations of features:

Lemma 15. *Let* $\mathbf{C}$ *be a covering. (1) If* $\mathbf{C}$ *is unary and reduced, then it is specialised. (2) If* $\mathbf{C}$ *is unary but not representative, then it is not reduced. (3) If* $\mathbf{C}$ *is reduced but not representative, it is not unary.*

Proof. (1) Assume K is not represented and $x \in K$. Therefore, since $\mathbf{C}$ is unary, $md(x) = \{K'\}$ for some $K' \neq K$. Since $\mathbf{C}$ is reduced, $K \nsubseteq K'$ and $K' \nsubseteq K$. Therefore, $md(x) \supseteq \{K, K'\}$. Contradiction. (2) and (3) are logical consequences of (1). If $\mathbf{C}$ is specialised then it is representative. Therefore, from (1), if it is unary and reduced, it is representative. It follows that if $\mathbf{C}$ is not representative, then it is not both unary and reduced. This is equivalent to the following facts: (a) if $\mathbf{C}$ is not representative, then if it is unary it is not reduced; (b) if $\mathbf{C}$ is not representative, then if it is reduced it is not unary: But (a) is equivalent to (2) and (b) to (3). □

One easily obtains the useful:

Lemma 16. *If* $\mathbf{C}$ *is unary, then* $\forall K \in \mathbf{C}$, $K = \bigcup\{R_{\mathbf{C}}(x) : x \in K\}$.

Proof. $\forall x \in K$, $K_x \subseteq K$, otherwise $md(x) \supseteq \{K_x, K\}$, which is a contradiction. Since $\forall x(x \in K \Longrightarrow x \in K_x)$, because $\mathbf{C}$ is unary, we obtain $K = \bigcup\{K_x\}_{x\in K}$. But from Theorem 6, $K_x = n(x) = R_{\mathbf{C}}(x)$, so that we obtain the result. □

Corollary 7. *Let* $\mathbf{C}$ *be a unary covering. Then* $int^{\mathbf{P}(\mathbf{C})} = \mathcal{C}^{\mathbf{P}(R_{\mathbf{C}})}$.

Proof. Indeed, $\forall X \subseteq U$, $int^{\mathbf{P}(\mathbf{C})}(X) = \bigcup\{R^{\smile}(\{K\}) : R^{\smile}(\{K\}) \subseteq X\}$. But from Lemma 16 this amounts to saying that $int^{\mathbf{P}(\mathbf{C})}(X) = \bigcup\{R_{\mathbf{C}}(x) : R_{\mathbf{C}}(x) \subseteq X\} = \mathcal{C}^{\mathbf{P}(R_{\mathbf{C}})}(X)$. □

Therefore, using Facts 3 one proves:

Theorem 13. *If* $\mathbf{C}$ *is a unary covering on a set* U*, then* $int^{\mathbf{P}(\mathbf{C})}$ *is a topological interior operator on* U *and* $\mathbf{S}_{int}(\mathbf{P}(\mathbf{C}))$ *is the distributive lattice of the open subsets of* U.

In the Appendix an alternative proof of Theorem 13 is provided.

Now we show that if a covering $\mathbf{C}$ on a set U is representative, then there is a specialised covering which is isomorphic to $\mathbf{S}_{int}(\mathbf{C})$.

Definition 16. *Let* $\mathbf{C}$ *be a representative covering on a set* U. *Let* $U^{\blacktriangleleft} = \{x \in U : |md(x)| = 1\}$. *Otherwise stated,* $U^{\blacktriangleleft}$ *is the set of all representative elements in* $\mathbf{C}$. *Let us set* $\mathbf{C}^{\blacktriangleleft} = \langle U^{\blacktriangleleft}, \mathbf{C}, R^{\blacktriangleleft}\rangle$, *where for any* $x \in U^{\blacktriangleleft}$, $K \in \mathbf{C}$, $\langle x, K\rangle \in R^{\blacktriangleleft}$ *iff* $\langle x, K\rangle \in R$.

Clearly, $\mathbf{C}^{\blacktriangleleft}$ is a specialised covering on U, hence unary.

Lemma 17. *Given a representative covering* $\mathbf{C}$ *on a set* U:

(1) *If* $x \in -U^{\blacktriangleleft}$ *then for all* $y \in U^{\blacktriangleleft}$, $y \notin R_{\mathbf{C}}(x)$.
(2) $\forall x \in -U^{\blacktriangleleft}$, $\exists y \in U^{\blacktriangleleft}$ *such that* $x \in R_{\mathbf{C}}(y)$.

Proof. (1) Trivially, if $x \notin U^{\blacktriangleleft}$, then $R_{\mathbf{C}}(x) = n(x) \notin \mathbf{C}$, while for all $x \in U^{\blacktriangleleft}$, $R_{\mathbf{C}}(x) = n(x) = K_x \in \mathbf{C}$. (2) Since $\mathbf{C}$ is representative, for any $K \in \mathbf{C}$, K_y for some $y \in U$. But from (1), necessarily $y \in U^{\blacktriangleleft}$. □

Definition 17. *Given a representative covering* $\mathbf{C}$ *on a set* U, *let us define the functions* $\phi^{\blacktriangleleft} : \wp(U) \longmapsto \wp(U^{\blacktriangleleft}); \phi^{\blacktriangleleft}(X) = X \cap U^{\blacktriangleleft}$ *and* $\phi^{\blacktriangleright} : \wp(U^{\blacktriangleleft}) \longmapsto \wp(U); \phi^{\blacktriangleright}(X) = R_{\mathbf{C}}(X)$.

Corollary 8. *For any representative covering* $\mathbf{C}$, $\mathbf{S}_{int}(\mathbf{P}(\mathbf{C}^{\blacktriangleleft}))$ *is isomorphic to* $\mathbf{S}_{int}(\mathbf{P}(\mathbf{C}))$.

Proof. From Lemma 17, $\forall x \in -U^{\blacktriangleleft}$, $\forall K \in \mathbf{C}$, $y \in \langle e\rangle(\{K\})$ iff $\exists y \in U^{\blacktriangleleft}$ such that $y \in \langle e\rangle(\{K\})$. Therefore, $\forall X \subseteq U$, $\forall x \in -U^{\blacktriangleleft}$, $x \in int(X)$ iff $\exists y(y \in U^{\blacktriangleleft} \wedge y \in K)$. We conclude that the function $\phi^{\blacktriangleleft}$ is an isomorphism between $\mathbf{S}_{int}(\mathbf{P}(\mathbf{C}))$ and $\mathbf{S}_{int}(\mathbf{P}(\mathbf{C}^{\blacktriangleleft}))$. □

Theorem 14. *If* $\mathbf{C}$ *is a representative covering:*
(1) $\mathbf{S}_{int}(\mathbf{P}(\mathbf{C}))$ *is a distributive lattice.* (2) $\mathbf{S}_{[i]}(\mathbf{P}(\mathbf{C}))$ *is a distributive lattice.*

Proof. (1) is obtained immediately from Theorem 13 and Corollary 8. (2) is a straightforward consequence of (1) and the adjointness isomorphisms. □

However, if $\mathbf{C}$ is just representative but not specialised, then in $\mathbf{P}(\mathbf{C})$ the operator *int* fails to be a topological interior operator on U, so that $\mathcal{C}$ fails to be a topological interior operator on $\mathbf{C}$. Remember that in $\mathbf{S}_{int}(\mathbf{P}(\mathbf{C}))$, *int* is defined as in Lemma 9. Indeed, $int^{\mathbf{P}(\mathbf{C})}$ does not distribute over $\cap$.

Example 10. Let $\mathbf{C}$ be the covering $\mathbf{C_2}$ of Example 9. As we know, it is representative, but neither reduced nor unary. $U^{\blacktriangleleft} = \{b, c, d\}$.

$\mathbf{P}(\mathbf{C})$	$\{a,b\}$	$\{a,c\}$	$\{a,b,d\}$
a	1	1	1
b	1	0	1
c	0	1	0
d	0	0	1

$R_{\mathbf{C}}$	a	b	c	d
a	1	0	0	0
b	1	1	0	0
c	1	0	1	0
d	1	1	0	1

$\mathbf{P}(\mathbf{C}^{\blacktriangleleft})$	$\{a,b\}$	$\{a,c\}$	$\{a,b,d\}$
b	1	0	1
c	0	1	0
d	0	0	1

$R_{\mathbf{C}^{\blacktriangleleft}}$	b	c	d
b	1	0	0
c	0	1	0
d	1	0	1

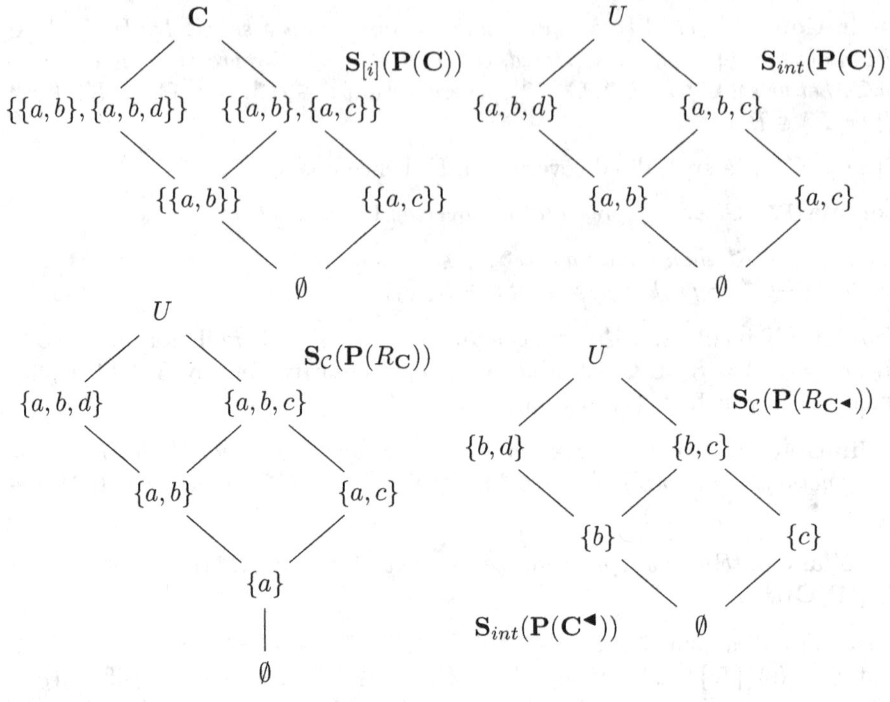

We have $U^{◀} = \{b, c, d\}$. Hence, for instance, $\phi^{◀}(\{a, b\}) = \{b\}$ and $\phi^{▶}(\{b, d\}) = \{a, b, d\}$. Moreover, $int^{\mathbf{P}(\mathbf{C})}(\{a, b\}) \cap int^{\mathbf{P}(\mathbf{C})}(\{a, c\}) = \{a\} \neq int^{\mathbf{P}(\mathbf{C})}(\{a, b\} \cap \{a, c\}) = int^{\mathbf{P}(\mathbf{C})}(\{a\}) = \emptyset$. Therefore, $int^{\mathbf{P}(\mathbf{C})}$ is not topological. On the contrary, $int^{\mathbf{P}(\mathbf{C}^{◀})}$ is topological.

Notice that the difference between $\mathbf{S}_{\mathcal{C}}(\mathbf{P}(R_{\mathbf{C}}))$ and $\mathbf{S}_{int}(\mathbf{P}(\mathbf{C}))$ is due to the non representative elements. In fact, if x is not representative, then $\bigcap md(x) \notin \mathbf{C}$, but $\bigcap md(x) = R_{\mathbf{C}}(x)$. In the Appendix we prove that if $\mathbf{C}$ is representative, then $\mathbf{C}$ is the set of coprime elements of $\mathbf{S}_{int}(\mathbf{P}(\mathbf{C}))$. Hence $R_{\mathbf{C}}(x)$ is not a coprime element of $\mathbf{S}_{int}(\mathbf{P}(\mathbf{C}))$, if x is not representative. On the contrary, the set of coprime elements of $\mathbf{S}_{\mathcal{C}}(\mathbf{P}(R_{\mathbf{C}}))$ is $\{R_{\mathbf{C}}(x) : x \in U\}$. That explains the importance for $\mathbf{C}$ to be unary in order to have $\mathbf{S}_{\mathcal{C}}(\mathbf{P}(R_{\mathbf{C}}))$ and $\mathbf{S}_{int}(\mathbf{P}(\mathbf{C}))$ isomorphic. In our example a is not representative, $R_{\mathbf{C}}(a) = \{a\}$, but $\{a\} \notin \mathbf{C}$.

Moreover, if $\mathcal{K} \subseteq \mathbf{C}$, then $\langle e \rangle(\mathcal{K}) = \{x \in R^{\smile}(\mathcal{K})\} = \{x : \exists K \in \mathcal{K} \wedge x \in R^{\smile}(K)\} = \{x : \exists K \in \mathcal{K} \wedge x \in K\} = \bigcup\{K : K \in \mathcal{K}\}$. This explains the isomorphism between $\mathbf{S}_{[i]}(\mathbf{P}(\mathbf{C}))$ and $\mathbf{S}_{int}(\mathbf{P}(\mathbf{C}))$.

In view of Theorem 4 the above results can be restated in terms of neighborhood systems. In what follow, notice that $\mathcal{N}_g^{\uparrow R^{\smile}} = \bigcup\{R^{\smile}(\{K\}) : K \in R(g)\} = \bigcup\{K : K \in R(g)\} = \bigcup\{K : g \in K\}$, so that $\mathcal{N}_{\uparrow R^{\smile}}(\mathbf{P}(\mathbf{C})) = \{\uparrow K : K \in \mathbf{C}\}$.

Corollary 9. *If* $\mathbf{C}$ *is a unary covering of a set* U*, then in* $\mathcal{N}_{\uparrow R^{\smile}}(\mathbf{P}(\mathbf{C}))$ *all the properties* **0**, **1**, **N1**, **N2**, **N3** *and* **N4** *hold.*

Since the proof is offered by Theorem 4, what it is interesting here are the connections between the algebraic properties and the properties fulfilled by the neighborhood system $\mathcal{N}_{\uparrow R^{\smile}}(\mathbf{P}(\mathbf{C}))$. In any such a system, **N1** and **1** derive from the properties of $R^{\smile}$, while **N2** holds by definition of an order filter. **0** holds because $\mathbf{C}$ is a covering of U; this means that in $\mathbf{P}(\mathbf{C})$ the relation $R^{\smile}$ is serial (moreover, R is serial, as well). If $\mathbf{C}$ is unary, then $x \in K$ iff $K \in \uparrow K_x$. Hence **N4** holds and **N3** as well, as a consequence. On the contrary, if $\mathbf{C}$ is not unary, **N3** does not hold because for some $x \in U$, $md(x) \subseteq \{K_1, K_2\}$ for $K_1 \not\subseteq K_2$ and $K_2 \not\subseteq K_1$. Thus, $K_1 \cap K_2 \notin \mathbf{C}$. Nonetheless, Theorem 14 together Theorem 8 informs us that if $\mathbf{C}$ is representative, then $\mathbf{S}_G(\mathcal{N}_{\uparrow R^{\smile}}(\mathbf{P}(\mathbf{C}))$ is a distributive lattice (which does not mean that G distributes over $\cap$ - cf. Lemma 9).

Observation 16. Results like those above and those of Theorem 12.(8) and Theorem 14 are connected to the properties of adjoint operators (see Proposition 1.4.9 of [25]) and the fact that there are topological formal systems which are not topological spaces and topological spaces which are not topological formal systems (see 15.14 of [25]).

Let $-$ be the set theoretic complementation with respect to U. As a topological space, on $\mathbf{S}_{int}(\mathbf{P}(\mathbf{C}^{\blacktriangleleft}))$ a pseudocomplementation $\neg$ can be defined as follows: $\neg(X) = int^{\mathbf{P}(\mathbf{C}^{\blacktriangleleft})}(U^{\blacktriangleleft} \cap -X)$. Thus $\phi^{\blacktriangleright}(\neg X) = int^{\mathbf{P}(\mathbf{C})}(-\phi^{\blacktriangleleft}(X)) = int^{\mathbf{P}(\mathbf{C})}(-(U^{\blacktriangleleft} \cap X))$. It follows that on $\mathbf{S}_{[i]}(\mathbf{P}(\mathbf{C}))$ the pseudocomplement of $X \subseteq U$ is $[i]^{\mathbf{P}(\mathbf{C})}(-(U^{\blacktriangleleft} \cap X))$.

Theorem 15. *If* $\mathbf{C}$ *is a specialised and reduced covering, then* $\mathbf{S}_{int}(\mathbf{P}(\mathbf{C}))$ *is a Boolean algebra.*

Proof. For all $x \in U$, $md(x) = R_{\mathbf{C}}(x)$. Assume $x \neq y$. If $md(x) = md(y)$, then $R_{\mathbf{C}}(x) = R_{\mathbf{C}}(y)$. Otherwise, if $md(y) \neq md(x)$, neither $x \in R_{\mathbf{C}}(y)$ nor $y \in R_{\mathbf{C}}(x)$ because $\mathbf{C}$ is unary and reduced, so that any $x \in U$ belongs just to a unique $K \in \mathbf{C}$. We conclude (*): if $R_{\mathbf{C}}(x) \neq R_{\mathbf{C}}(y)$, then $R_{\mathbf{C}}(x) \not\subseteq R_{\mathbf{C}}(y)$ and $R_{\mathbf{C}}(y) \not\subseteq R_{\mathbf{C}}(x)$. Let us now set $x \triangleq y$ iff $R_{\mathbf{C}}(x) = R_{\mathbf{C}}(y)$. Clearly, $\mathbf{S}_{int}(\mathbf{P}(\mathbf{C}))$ is isomorphic to $\mathbf{S}_{int}(\mathbf{P}(R_{\mathbf{C}_{\triangleq}}))$, where $\mathbf{P}(\mathbf{C})_{\triangleq} = \langle U/_{\triangleq}, \mathbf{C}_{\triangleq}, R_{\triangleq}\rangle$, for $\mathbf{C}_{\triangleq} = \{K/_{\triangleq} : K \in \mathbf{C}\}$ and $\langle [x]_{\triangleq}, K/_{\triangleq}\rangle \in R_{\triangleq}$ iff $x \in K$. In view of (*), $R_{\mathbf{C}_{\triangleq}}$ is the identity relation on $U/_{\triangleq}$. Therefore, $\mathbf{S}_{int}(\mathbf{P}(R_{\mathbf{C}_{\triangleq}}))$ is a Boolean algebra. As a consequence, $\mathbf{S}_{int}(\mathbf{P}(R_{\mathbf{C}}))$ is a Boolean algebra, and, thus, $\mathbf{S}_{int}(\mathbf{P}(\mathbf{C}))$ as well. Finally, as an immediate corollary, $\mathbf{S}_{[i]}(\mathbf{P}(\mathbf{C}))$ is a Boolean algebra. □

Corollary 10. *If* $\mathbf{C}$ *is irredundat, then* $\mathbf{S}_{int}(\mathbf{P}(\mathbf{C}))$ *is a Boolean algebra.*

The above Corollary is a direct consequence of Corollary 8 and Theorem 15. An alternative proof which exploits an interesting result of [13] is in the Appendix.

CONNECTIONS. Theorem 14.(2) is a proof of Theorem 4.1 of [7] by means of the constructor approach. In [7] given a covering $\mathbf{C}$ of a set U, the set $C_{in}(X) = \{K : K \subseteq X\}$ is defined. After that, an equivalence relation R_{in} is defined on $\wp(U)$ as $\langle X, Y\rangle \in R_{in}$ iff $C_{in}(X) = C_{in}(Y)$. Since $C_{in}(X) = [i]^{\mathbf{P}(\mathbf{C})}(X)$, one has that the lattice $\wp(U)/R_{in}$ is isomorphic to $\mathbf{S}_{[i]}(\mathbf{P}(\mathbf{C}))$. Theorem 15 corresponds to Theorem 4.4 of [7] (where $U^{\blacktriangleleft}$ is denoted by C_o). As to Corollary 7, it corresponds to Corollary 2 of [9] (see also Lemma 3.1 of [30]).

6 Conclusions

In our opinion, the operators acting properly on the basis of a covering are those induced by a CPS $\langle U, \mathbf{C}, R\rangle$ and not by an SRS $\langle U, U, R_{\mathbf{C}}\rangle$. Indeed, when one moves from operators based on the elements of a covering $\mathbf{C}$ to operators based on $R_{\mathbf{C}}$, one leaves the field of coverings proper and enters that of the operators defined by preorders. About rough sets based on preorders, the literature is abundant. See for instance [8,11,12,18,25] and others (not to mention the long-stated results in Modal Logic).

Coverings, instead, have strict, albeit not univocal, relationships with tolerance relations. Also in this case, the literature is large. We just mention [1,13]. By the way, in the latter paper the notion of an *injective* covering is introduced to study the relationships between coverings and tolerance relations. Using our terminology, a covering $\mathbf{C}$ of a set U is called injective if for all $x, y \in U$, $\langle i\rangle(x) = \langle i\rangle(y)$ implies $x = y$. This is a notion completely different from those discussed in the present paper. Indeed, $\{\{a\}, \{b, c\}\}$ is representative, unary and reduced, but not injective. On the other hand, $\{\{a, b\}, \{a, c\}, \{a, b, c\}\}$ is injective but neither representative, nor unary (and not reduced, which is obvious in view of Lemma 15.(1)).

In any case, it is worth underlining that the interpretation of covering-based approximation operators by means of the four basic constructors has a practical significance, because these constructors are easily computed (see [22] and the Appendix).

A Appendix

The following result is the basis of the calculus of constructors by means of Boolean matrices:

Duality between the constructors. In any SRS $\mathbf{P}$, $\langle e\rangle$ and $[e]$ are dual; $\langle i\rangle$ and $[i]$ are dual.

Proof. $-R^{\smile}(-A) = \{x : R(x) \subseteq A\}$ (aka: $-\langle e\rangle(-A) = [e](A)$).

$$\begin{aligned}-R^{\smile}(-A) &= -\{x : x \in R^{\smile}(-A)\} = -\{x : \exists y(y \notin A \wedge x \in R^{\smile}(y))\}\\ &= \{x :\neq \exists y(y \notin A \wedge x \in R^{\smile}(y))\}\\ &= \{x : \forall y \neg(y \notin A \wedge x \in R^{\smile}(y))\} = \{x : \forall y(x \in R^{\smile}(y) \Longrightarrow y \in A)\}\\ &= \{x : \forall y(y \in R(x) \Longrightarrow y \in A)\} = \{x : R(x) \subseteq A\}.\end{aligned}$$

Same for the intensional pair of constructors. □

Direct proof of Lemma 2.(2). The proof comes straightforwardly from (14). Indeed, $\bigcap\{-X \in \mathcal{Z} : X \cap A = \emptyset\} = \bigcap\{-R^{\smile}(m) : R^{\smile}(m) \subseteq -A\} = \{g : \forall m((R^{\smile}(m) \subseteq -A) \Longrightarrow g \notin R^{\smile}(m))\} = \{g : \forall m(g \in R^{\smile}(m) \Longrightarrow R^{\smile}(m) \nsubseteq -A)\} = \{g : \forall m(m \in R(g) \Longrightarrow R^{\smile}(m) \cap A \neq \emptyset)\}$. From Lemma 1, the latter set equals $\{g : \forall m(m \in R(g) \Longrightarrow m \in R(A))\} = \{g : R(g) \subseteq R(A)\}$.

Direct proof of Theorem 12.(7). Since R is a preorder, for any $X \subseteq U$, $[e](X) \subseteq X$ so that $[e]\langle e\rangle[e](X) \subseteq [e]\langle e\rangle(X)$. Suppose now $x \in [e]\langle e\rangle(X)$. Then

$x \in [e](R^{\smile}(X))$ so that $R(x) \subseteq R^{\smile}(X)$. If $x \notin [e]\langle e\rangle[e](X)$ then $R(x) \not\subseteq R^{\smile}(\{y : R(y) \subseteq X\})$. Therefore, $\exists z$ such that $z \in R(x)$ and $z \notin R^{\smile}(\{y : R(y) \subseteq X\})$. Therefore, $\forall y(R(y) \subseteq X \Longrightarrow y \notin R(z))$, so that $\forall y(y \in R(z) \Longrightarrow R(y) \not\subseteq X)$. But R is transitive. So $R(y) \subseteq R(z)$. It follows $R(z) \not\subseteq X$, which leads to a contradiction because $z \in R(x)$ so that $R(z) \subseteq R(x)$ which implies $R(x) \not\subseteq X$.

Direct proof of Lemma 7.(2):

$$\begin{aligned}[i]^{\mathbf{P}(R^{\mathbf{C}})}(X) &= \{g : R^{\mathbf{C}}(x) \subseteq X\} = \{g : \{g' : \exists K(g' \in K \wedge g \in K)\} \subseteq X\}\\ &= \{g : \forall K(g \in K \Longrightarrow K \subseteq X)\}\\ &= \{g : \forall K(K \in R(g) \Longrightarrow R^{\smile}(\{K\}) \subseteq X)\}\\ &= \{g : \forall K(K \in R(g) \Longrightarrow K \in [i](X))\}\\ &= \{g : R(g) \subseteq [i]^{\mathbf{P}(\mathbf{C})}(X)\} = [e]^{\mathbf{P}(\mathbf{C})}[i]^{\mathbf{P}(\mathbf{C})}(X).\end{aligned}$$

Direct proof of Lemma 7.(5):

$$\begin{aligned}(lC)_6(A) &= \{x : \bigcup\{K : x \in K\} \subseteq A\} = \{x : \forall K(x \in K \Longrightarrow K \subseteq A)\}\\ &= \{x : \forall K(K \in R(x) \Longrightarrow R^{\smile}(\{K\}) \subseteq A)\}\\ &= \{x : \forall K(K \in R(x) \Longrightarrow K \in [i](A))\} = [e][i](A).\end{aligned}$$

Moreover, $(uC)_6(A) = \bigcup\{K : K \cap A \neq \emptyset\}$. Thus $x \in (uC)_6(A)$ iff $\exists K(x \in K \wedge K \cap A \neq \emptyset)$ iff $\exists K(x \in R^{\smile}(K) \wedge R^{\smile}(\{K\}) \cap A \neq \emptyset)$ iff $\exists K(x \in \langle e\rangle(K) \wedge K \in \langle i\rangle(A))$ iff $x \in \langle e\rangle\langle i\rangle(A)$.

Direct proof of Lemma 12: $\langle i\rangle(X) = X$ iff $R(X) = X$ iff $\bigcup\{R(x) : x \in X\} = X$ which implies $\forall x \in X(R(x) \subseteq X)$. Let $y \notin X$. Since R is reflexive, $x \in R(x)$, so that $R(x) \not\subseteq X$. In sum, $x \in X \Longrightarrow R(x) \subseteq X$ and $x \notin X \Longrightarrow R(x) \not\subseteq X$. We conclude that $x \in X$ iff $R(x) \subseteq X$ which amounts to saying $R(X) = \{x : R(x) \subseteq X\} = [e](X)$. Notice that seriality, trivially, is not enough. Indeed seriality does not prevent from the existence of a $g \in X$ such that $R(g) \cap X = \emptyset$ - think of the relation $\{\langle a, b\rangle, \langle b, b\rangle\}$ on the set $\{a, b\}$ and put $X = \{a\}$.

Alternative proof of Theorem 13 (point (3) of Corollary 11 below):

Lemma 18. *Let* $\mathbf{C}$ *be unary and* $\mathbf{C}^{\blacktriangledown} = \{K \in \mathbf{C} : \exists x \wedge md(x) = \{K\}\}$. *Then* $R_{\mathbf{C}^{\blacktriangledown}} = R_{\mathbf{C}}$.

Proof. Since $\mathbf{C}$ is unary, $\mathbf{P}(\mathbf{C}^{\blacktriangledown}) = \langle U, C^{\blacktriangledown}, R^{\blacktriangledown}\rangle$, where $R^{\blacktriangledown}$ is R restricted to $U \times \mathbf{C}^{\blacktriangledown}$, is a specialised covering of U. So, let us assume $y \in R_{\mathbf{C}^{\blacktriangledown}}(x)$. Therefore, $\forall K \in \mathbf{C}^{\blacktriangledown}(x \in K \Longrightarrow y \in K)$. In particular this occurs for $\{K\} = md(x)$. Then for all $K \in \mathbf{C}(x \in K \Longrightarrow y \in K)$ so that $y \in R_{\mathbf{C}}(x)$. Thus $R_{\mathbf{C}^{\blacktriangledown}} \subseteq R_{\mathbf{C}}$. Vice-versa, since $\mathbf{C}^{\blacktriangledown} \subseteq \mathbf{C}$, $R_{\mathbf{C}} \subseteq R_{\mathbf{C}^{\blacktriangledown}}$. □

Corollary 11. *For any unary covering* $\mathbf{C}$*:*

(1) $\mathbf{S}_{int}(\mathbf{P}(\mathbf{C})) = \mathbf{S}_{int}(\mathbf{P}(\mathbf{C}^{\blacktriangledown}))$;
(2) $\mathbf{S}_{int}(\mathbf{P}(\mathbf{C})) = \mathbf{S}_{\mathcal{C}}(\mathbf{P}(R_{\mathbf{C}^{\blacktriangledown}}))$;
(3) $\mathbf{S}_{int}(\mathbf{P}(\mathbf{C}))$ *is a topological space.*

Proof. (1) From Theorem 9: $\preceq_{int^{\mathbf{P}(\mathbf{C})}} = R_{\mathbf{C}}$ and $\preceq_{int^{\mathbf{P}(\mathbf{C}^{\blacktriangledown})}} = R_{\mathbf{C}^{\blacktriangledown}}$. But from Lemma 18, $R_{\mathbf{C}} = R_{\mathbf{C}^{\blacktriangledown}}$. Therefore, $\preceq_{int^{\mathbf{P}(\mathbf{C})}} = \preceq_{int^{\mathbf{P}(\mathbf{C}^{\blacktriangledown})}}$ so that $\mathbf{S}_{int}(\mathbf{P}(\mathbf{C})) = \mathbf{S}_{int}(\mathbf{P}(\mathbf{C}^{\blacktriangledown}))$. (2) From Corollary 7, $int^{\mathbf{P}(\mathbf{C}^{\blacktriangledown})} = int^{\mathbf{P}(R_{\mathbf{C}^{\blacktriangledown}})}$, because $\mathbf{P}(\mathbf{C}^{\blacktriangledown})$

is a specialised covering whenever $\mathbf{C}$ is unary. In view of (1) one concludes $\mathbf{S}_{int}(\mathbf{P}(\mathbf{C})) = \mathbf{S}_{int}(\mathbf{P}(\mathbf{C}^{\blacktriangledown})) = \mathbf{S}_{\mathcal{C}}(\mathbf{P}(R_{\mathbf{C}^{\blacktriangledown}}))$. (3) is obtained from (1) and Theorem 13. □

Corollary 11.(3) is a proof, by means of the constructor approach, of Theorem 12 of [17].
The following Lemma helps clarifying the isomorphism stated above:

Lemma 19. *$\forall K \in \mathbf{C} \cap -\mathbf{C}^{\blacktriangledown}$, $\exists \mathcal{K} \subseteq \mathbf{C}^{\blacktriangledown}$ such that $K = \bigcup \mathcal{K}$.*

Proof. Let $K \in \mathbf{C} \cap -\mathbf{C}^{\blacktriangledown}$ and $x \in K$. Clearly, $K_x \subsetneq K$, for minimality of K_x. Therefore, $\bigcup\{K_x : x \in K\} \subseteq K$. But by definition, $x \in K_x$. Therefore, $\bigcup\{K_x : x \in K\} = K$. □

Example 11. Let $\mathbf{C} = \{\{a\}, \{b\}, \{a, b\}, \{a, c\}\}$. $\mathbf{C}^{\blacktriangledown} = \{\{a\}, \{b\}, \{a, c\}\}$.

$R_{\mathbf{C}} = R_{\mathbf{C}^{\blacktriangledown}}$	a	b	c
a	1	0	0
b	0	1	0
c	1	0	1

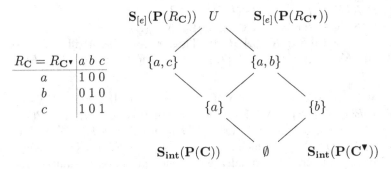

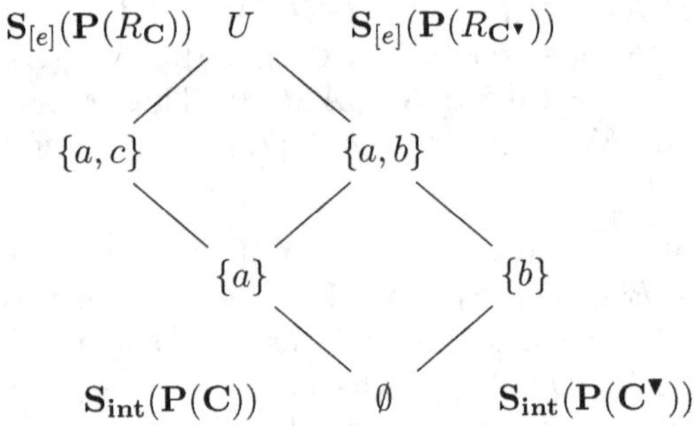

Theorem 16. *If $\mathbf{C}$ is a representative covering of a set U, then $\mathbf{C}$ is the set of coprime elements of $\mathbf{S}_{int}(\mathbf{P}(\mathbf{C}))$.*

Proof. Since $\mathbf{C}$ is representative, for all $K \in \mathbf{C}$, $K = K_x$ for some $x \in U$. Therefore, for all $K' \subsetneq K$, $x \notin K'$, by definition of K_x. It follows that if $\mathcal{K} \subseteq \mathbf{C}$ and $K \notin \mathcal{K}$, then $K \neq \bigcup \mathcal{K}$. □

Justification of the definition of an "irredundant covering". We recall that in [13] a covering $\mathbf{C}$ of a set U is called "irredundant" if for any $K \in \mathbf{C}$, $\mathbf{C} \cap -\{K\}$ is no longer a covering of U.

Theorem 17. *$\mathbf{C}$ is an irredundant covering if and only if it is representative and reduced.*

Proof. (A) If $\mathbf{C}$ is redundant, then either it is not reduced or it is not representative. Assume $K \in \mathbf{C}$ is redundant. Then for all $x \in K$, there is a $K' \in \mathbf{C}$ such that $x \in K'$. If $K \subseteq K'$ or $K' \subseteq K$, then $\mathbf{C}$ is not reduced. If it is reduced, then $K \cap K' \notin \mathbf{C}$, because $K \cap K' \subseteq K$ and $K \cap K' \subseteq K'$. It follows that for all $x \in K$, $K \neq K_x$. Hence K is not represented. (B) If $\mathbf{C}$ is either non-reduced or non-representative, then it is redundant. Trivially, if $\mathbf{C}$ is not reduced, and $K \subsetneq K'$, then K is redundant. Suppose $\mathbf{C}$ is reduced but there exists $K \in \mathbf{C}$ which is not represented. Then for all $x \in K$, $K \neq K_x$. It follows that for all $x \in K$, there is a $K' \neq K$ such that $x \in K'$. We conclude that K is redundant. □

The composition of two relations $R \subseteq X \times Y$ and $Z \subseteq Y \times W$, is defined as $R \otimes Z = \{\langle x, w\rangle : \exists y \in Y(\langle x, y\rangle \in R \wedge \langle y, w\rangle \in Z\}$.

In [13] it is stated that if R is a preorder, then $R^{\smile} \otimes R$, is a tolerance relation. We know that $R_{\mathbf{C}}$ is a preorder. Now we prove that if in addition $\mathbf{C}$ is reduced, then $R_{\mathbf{C}}^{\smile} \otimes R_{\mathbf{C}}(x) = R_{\mathbf{C}}(x)$ for all $x \in U^{\blacktriangleleft}$. Since $R_{\mathbf{C}}^{\smile} \otimes R_{\mathbf{C}}(x)$ is a tolerance relation, as a corollary we have that if $\mathbf{C}$ in addition is specialised then $R_{\mathbf{C}}$ is an equivalence relation. From that we can deduce that $\mathbf{S}_{int}(\mathbf{P}(\mathbf{C}))$ is a Boolean algebra (i.e. Theorem 15).

Lemma 20. *For any covering* $\mathbf{C}$ *of* U*, for all* $x, y \in U$*,* $\langle x, y\rangle \in R_{\mathbf{C}}^{\smile} \otimes R_{\mathbf{C}}$ *iff* $\exists z(\langle z, x\rangle \in R_{\mathbf{C}} \wedge \langle z, y\rangle \in R_{\mathbf{C}})$.

The proof is trivial.

Lemma 21. *For any covering* $\mathbf{C}$ *of* U*,* $R_{\mathbf{C}}^{\smile} \otimes R_{\mathbf{C}} \supseteq R_{\mathbf{C}}$.

Proof. Let $\langle x, y\rangle \in R_{\mathbf{C}}$. From reflexivity, $\langle x, x\rangle \in R_{\mathbf{C}}$, so that from Lemma 20, $\langle x, y\rangle \in R_{\mathbf{C}}^{\smile} \otimes R_{\mathbf{C}}$. □

Theorem 18. *For any reduced covering* $\mathbf{C}$ *of* U*,* $R_{\mathbf{C}}^{\smile} \otimes R_{\mathbf{C}}(x) = R_{\mathbf{C}}(x)$ *for all* $x \in U^{\blacktriangleleft}$.

Proof. Suppose $x \in U^{\blacktriangleleft}$ and $\langle x, y\rangle \in R_{\mathbf{C}}^{\smile} \otimes R_{\mathbf{C}}$. Then $\exists z$ such that $x, y \in R_{\mathbf{C}}(z) = n(z) = \bigcap md(z)$. Since x is representative there is a $K_x \in \mathbf{C}$ and since $x \in n(x)$, $K_x \subseteq K$ for all $K \in md(z)$. But $\mathbf{C}$ is reduced, so that it must be $K = K_x$ for all $K \in md(z)$. That is, $md(z) = \{K_x\}$, so that $R_{\mathbf{C}}(z) = R_{\mathbf{C}}(x)$. It follows that $y \in R_{\mathbf{C}}(x)$ and one concludes $R_{\mathbf{C}}^{\smile} \otimes R_{\mathbf{C}} \subseteq R_{\mathbf{C}}$. In view of Lemma 21 the proof is complete. □

References

1. Bartol, W., Miró, J., Pióro, K., Rosselló, F.: On the coverings by tolerance classes. Inf. Sci. **166**, 193–211 (2004)
2. Bell, J.L.: Orthologic, forcing, and the manifestation of attributes. In: Chong, C.T., Wicks, M.J. (eds.) Southeast Asian Conference on Logic, pp. 13–36. Elsevier Science Publisher, North-Holland (1983)
3. Blyth, T.S., Janowitz, M.F.: Residuation Theory. Pergamon Press, Oxford (1972)
4. Bonikowski, Z., Bryniarski, E., Wybraniec-Skardowska, U.: Extensions and intensions in the rough set theory. Inf. Sci. **107**, 149–167 (1998)
5. Čech, E.: Topological Spaces. Wiley, Hoboken (1966)
6. Düntsch, I., Gegida, G.: Modal-style operators in qualitative data analysis. In: Proceedings of the IEEE International Conference on Data Mining, pp. 155–162 (2002)
7. Ghanim, M.H., Mustafa, H.I., Abd El Aziz, S.: On lower and upper intension order relations by different cover concepts. Inf. Sci. **181**, 3723–3734 (2011)

8. Greco, S., Matarazzo, B., Słowinski, R.: Algebra and topology for dominance-based rough set approach. In: Ras, Z.W., Tsay, L.-S. (eds.) Advanced in Intelligent Systems. SCI, vol. 265, pp. 43–78. Springer, Heidelberg (2010)
9. Huang, A., Zhu, W.: Topological characterizations for three covering approximation operators. In: Ciucci, D., Inuiguchi, M., Yao, Y., Ślęzak, D., Wang, G. (eds.) RSFDGrC 2013. LNCS, vol. 8170, pp. 277–284. Springer, Heidelberg (2013)
10. Järvinen, J.: Lattice theory for rough sets. In: Peters, J.F., Skowron, A., Düntsch, I., Grzymała-Busse, J.W., Orłowska, E., Polkowski, L. (eds.) Transactions on Rough Sets VI. LNCS, vol. 4374, pp. 400–498. Springer, Heidelberg (2007)
11. Järvinen, J., Pagliani, P., Radeleczki, S.: Information completeness in nelson algebras of rough sets induced by quasiorders. Stud. Log. **101**(5), 1073–1092 (2013)
12. Järvinen, J., Radeleczki, S.: Representation of Nelson algebras by rough sets determined by quasiorders. Algebra Univers. **66**, 163–179 (2011)
13. Järvinen, J., Radeleczki, S.: Tolerances induced by irredundant coverings. Fundamenta Informaticae **137**, 341–353 (2015)
14. Kreisel, G., Putnam, H.: Eine Unableitbarkeitsbeweismethode für den intuitionistischen Aussegenkalkül. Archiv für mathematische Logik und Grundlagenferschung **3**, 74–78 (1957)
15. Kumar, A., Banerjee, M.: Definable and rough sets in covering-based approximation spaces. In: Li, T., Nguyen, H.S., Wang, G., Grzymala-Busse, J., Janicki, R., Hassanien, A.E., Yu, H. (eds.) RSKT 2012. LNCS, vol. 7414, pp. 488–495. Springer, Heidelberg (2012)
16. Langeron, C., Bonnevay, S.: Une approach pretopologique pour la structuration. Document de Recherche N° 1999 - 7. Centre de reserches economiques de l'Université de Saint Etienne
17. Li, Q., Zhu, W.: Lattice structures of fixed points of the lower approximations of two types of covering-based rough sets. http://arxiv.org/abs/1209.5569
18. Nagarajan, E.K.R., Umadevi, D.: A method of representing rough sets system determined by quasi orders. Order **30**(1), 313–337 (2013)
19. Ore, O.: Galois connexions. Trans. Am. Math. Soc. **55**, 493–513 (1944)
20. Pagliani, P.: Concrete neighbourhood systems and formal pretopological spaces (draft). In: Presented at the Calcutta Logical Circle Conference on Logic and Artificial Intelligence, Calcutta, 13–16 October 2003
21. Pagliani, P.: Pretopology and dynamic spaces. In: Wang, G., Liu, Q., Yao, Y., Skowron, A. (eds.) RSFDGrC 2003. LNCS, vol. 2639, pp. 146–155. Springer, Heidelberg (2003). Extended Version in Fundamenta **59**(2–3), 221–239 (2004)
22. Pagliani, P.: The relational construction of conceptual patterns - tools, implementation and theory. In: Kryszkiewicz, M., Cornelis, C., Ciucci, D., Medina-Moreno, J., Motoda, H., Raś, Z.W. (eds.) RSEISP 2014. LNCS, vol. 8537, pp. 14–27. Springer, Heidelberg (2014)
23. Pagliani, P., Chakraborty, M.K.: Information Quanta and approximation spaces. I: Non-classical approximation operators. In: Proceedings of the IEEE International Conference on Granular Computing, Beijing, China, 25–27 July 2005, vol. 2, pp. 605–610. IEEE, Los Alamitos, July 2005
24. Pagliani, P., Chakraborty, M.K.: Formal topology and information systems. In: Peters, J.F., Skowron, A., Düntsch, I., Grzymała-Busse, J.W., Orłowska, E., Polkowski, L. (eds.) Transactions on Rough Sets VI. LNCS, vol. 4374, pp. 253–297. Springer, Heidelberg (2007)
25. Pagliani, P., Chakraborty, M.K.: A Geometry of Approximation. Trends in Logic, 27th edn. Springer, Heidelberg (2008)

26. Qin, K., Gao, Y., Pei, Z.: On covering rough sets. In: Yao, J.T., Lingras, P., Wu, W.-Z., Szczuka, M.S., Cercone, N.J., Ślęzak, D. (eds.) RSKT 2007. LNCS (LNAI), vol. 4481, pp. 34–41. Springer, Heidelberg (2007)
27. Restrepo, M., Cornelis, C., Gmez, J.: Duality, conjugacy and adjointness of approximation operators in covering-based rough sets. Int. J. Approximate Reasoning **55**(1), 469–485 (2014)
28. Sambin, G.: Intuitionistic formal spaces - a first communication. In: Skordev, D. (ed.) Mathematical Logic and Its Applications, pp. 187–204. Plenum Press, New York (1987)
29. Sambin, G., Gebellato, S.: A preview of the basic picture: a new perspective on formal topology. In: Altenkirch, T., Reus, B., Naraschewski, W. (eds.) TYPES 1998. LNCS, vol. 1657, pp. 194–208. Springer, Heidelberg (1999). doi:10.1007/3-540-48167-2_14
30. Thuan, N.D.: Covering rough sets from a topological point of view. Int. J. Comput. Theor. Eng. **1**(5), 606–609 (2009)
31. Vickers, S.: Topology via Logic. Cambridge Tracts in Theoretical Computer Science, vol. 5. Cambridge University Press, Cambridge (1989)
32. Yao, Y.Y., Chen, Y.H.: Rough set approximations in formal concept analysis. In: Dick, S., Kurgan, L., Pedrycz, W., Reformat, M. (eds.) Proceedings of 2004 Annual Meeting of the North American Fuzzy Information Processing Society, (NAFIPS 2004). IEEE Catalog Number: 04TH8736, pp. 73–78 (2004)
33. Yao, Y., Yao, B.: Covering based rough set approximations. Inf. Sci. **200**, 91–107 (2012)
34. Zakowski, W.: Approximations in the space (U, Π). Demonstratio Mathematica **16**, 761–769 (1983)

Multiple-Source Approximation Systems, Evolving Information Systems and Corresponding Logics: A Study in Rough Set Theory

Md. Aquil Khan[1,2](✉)

[1] Discipline of Mathematics, Indian Institute of Technology Indore, Indore 453552, India
aquilk@iiti.ac.in

[2] Department of Mathematics and Statistics, Indian Institute of Technology Kanpur, Kanpur 208016, India

Abstract. Mathematical logic is used as a tool/language to reason about any kind of data. With the inception of rough set theory (RST), the question of a suitable logic for RST has attracted the attention of many researchers. One of the main contribution of the current article is the development of a logic that can describe aspects of information system such as attribute, attribute-values, as well as the induced concept approximations. Moreover, the current article relates RST to some important issues in artificial intelligence such as multiple-source (agent) knowledge-bases, temporal evolution of knowledge-bases, and information updates. For the multiple-source case, we explored counterparts of standard rough set-theoretic concepts such as concept approximations, definability of concepts, as well as corresponding logics that can express these notions. For the temporal situation, we proposed temporal logics for RST that bring temporal and approximation operators together, to enable reasoning about concept approximations relative to time. An update logic for RST is also introduced that can be used to study flow of information and its effect on concept approximations.

Keywords: Approximation spaces · Information systems · Rough sets · First-order logic · Modal logic · Temporal logic · Dynamic epistemic logic · Tableau-based proof procedure · Combination of modal logics · Boolean algebra with operators

1 Introduction

A lack of complete information results in a granularity of the domain of discourse – this is the premise of rough set theory [83, 84]. Due to the granularity,

Md.A. Khan—The article is based on the work carried out by the author for his Ph.D. degree at the Indian Institute of Technology Kanpur.

J.F. Peters and A. Skowron (Eds.): TRS XX, LNCS 10020, pp. 146–320, 2016.
DOI: 10.1007/978-3-662-53611-7_5

a concept on the domain may not be describable 'precisely', and the theory addresses 'approximations' of such concepts. Formally, the backdrop is constituted by an *approximation space*, taken as a tuple (U, R), R being an equivalence relation (called the *indiscernibility* relation) on the domain U[1] of objects. The partition induced by R reflects the granularity (mentioned above) of U that is a result of a lack of complete information about the objects in U. In other words, having 'complete information' about the domain U is identified with the case when R is the identity relation on U. The equivalence classes of R give the 'elementary' granules. A concept, represented extensionally by a subset X of U, may not, in general, be expressible in terms of the elementary granules using the set-theoretic operations of union, intersection and complementation. X is then approximated from within and outside: the union of all equivalence classes $[x]_R$ with $[x]_R \subseteq X$ gives the *lower approximation* of X, denoted as $\underline{X}_R$. On the other hand, the union of all equivalence classes having a non-empty intersection with X gives the *upper approximation* of X, denoted as $\overline{X}_R$. The elements belonging to $\underline{X}_R$, $\overline{X}_R$ and $(\overline{X}_R)^c$, where Y^c represents the set-theoretic complement of the set Y, are respectively called the *positive*, *possible* and *negative* elements of X. The set $B_R(X) := \overline{X}_R \setminus \underline{X}_R$ consists of the *boundary/undecidable* elements of X. A set X with non-empty boundary $B_R(X)$, would then be one that cannot be defined by equivalence classes as mentioned above. In that case, X is said to be *rough*. Otherwise, X is termed *definable*.

A practical source of approximation spaces is a *deterministic information system* [84]. Formally, a deterministic information system (DIS) $\mathcal{S} := (U, A, \bigcup_{a \in A} Val_a, f)$ comprises a non-empty set U of objects, A of attributes, Val_a of attribute values for each $a \in A$, and a map $f : U \times A \rightarrow \bigcup_{a \in A} Val_a$ such that $f(x, a) \in Val_a$. An equivalence (*indiscernibility*) relation $Ind_{\mathcal{S}}(B)$ on U is then induced for any subset B of A: objects x and y are related, if they cannot be distinguished using only the information provided by the attributes of the set B. In other words, $(x, y) \in Ind_{\mathcal{S}}(B)$, if and only if $f(x, a) = f(y, a)$ for all $a \in B$. Thus, given a deterministic information system $\mathcal{S}$ and a set $B \subseteq A$ of attributes, we obtain an approximation space $(U, Ind_{\mathcal{S}}(B))$. Moreover, it is established that given an approximation space, one can determine an information system with the same domain and a set of attributes such that the induced indiscernibility relation coincides with the relation of the approximation space.

The above notion of information system is termed *deterministic* as it provides precise information about each object regarding each attribute. Thus, the notion of *incomplete information systems* is considered to represent the situation when some of the information regarding an object's attribute is missing. Different interpretations are given to the absence of information, such as 'lost' or 'do not care'. In case of an incomplete information system, a *similarity* relation become relevant [60, 61]. It is assumed that, with more information, the attribute value gaps in the data table would be filled by values from some attribute value domain V that is given *a priori*. Different 'completions' of the information system may then be obtained by assigning attribute values from V in place of the

[1] In Pawlak's original definition, U is assumed to be finite.

missing values. The similarity relation is such that two objects are considered distinguishable, if and only if they are distinguished by the induced indiscernibility relation in *every* completion.

With time, Pawlak's simple rough set model has seen many generalizations due to demands from different practical situations. The variable precision rough set model [111], the rough set model based on covering [85], neighborhood system [68] and tolerance relation [95], the Bayesian rough set model [96], the fuzzy rough set model [21] are a few instances. In this article, we are interested in generalizations which arise in the following three kinds of situations.

(i) Information arrives from different agents.
(ii) There is inflow of information varying with time.
(iii) A combination of (i) and (ii).

Since, in Pawlak's rough set theory, the information about the domain is represented by a partition (equivalence relation), called the *knowledge base* (cf. [92]), these situations give rise to families of classifications of the domains of interest. For instance, in the multi-agent case, each agent has her own knowledge base and thus we obtain a family of partitions on the same domain representing the knowledge base of the group of agents under consideration. Similarly, when there is a flow of information with time, which will possibly be in terms of attributes and attribute values of the objects, we need to update the knowledge base (information system) to incorporate new information. So we obtain a family of partitions/information systems that are evolving with time, reflecting the information about the domain at different time points.
Thus, in either case, the basic structure of interest to us is a family of partitions of the same domain. Our interest is in studying the behavior of rough sets in such structures. We adopt a formal logical approach.

Families of classifications of the domains of interest, were considered by Pawlak himself at the very beginning of the development of rough set theory (cf. [84]). In fact, according to him, "knowledge consists of a family of various classification patterns of the domain of interest, which provide explicit facts about reality – together with the reasoning capacity able to deliver implicit facts derivable from explicit knowledge". In [80], Orłowska and Pawlak studied rough set theory in a multi-agent setting, although the term multi-agent is not mentioned explicitly. A structure with a collection of equivalence relations over the same domain was considered. These relations may be taken to represent the knowledge base of individual agents such that each agent perceives the same domain differently depending on what information she has about the domain. A multi-modal logic was proposed where one can express the approximations of the sets with respect to the knowledge base of the agents. Groups of agents also enter the work of Rasiowa and Marek [90]. Inspired by rough sets, Rasiowa and Marek investigated a first order logic with epistemic operators for groups of agents, including an expression for joint knowledge. A set consisting of intelligent agents, a 'perception' order (mathematically, a partial order on the set), along with a collection of indiscernibility (equivalence) relations on the domain observed by each agent are considered. The joint capability of all agents to discern objects of the domain is taken to be reflected by the intersection of the group's equivalence relations, and is incorporated into the syntax of the logic. One may note that this is similar to the notion of 'distributed knowledge' in epistemic logic [34]. Rauszer [92] also investigated the multi-agent situation in rough set theory in the lines of Rasiowa and Marek, calling the intersection of the equivalence relations, the 'strong distributed knowledge base' of the group of agents. A sound and complete propositional logic was proposed by Rauszer, which is able to express the approximation of the sets with respect to the strong distributed knowledge base of the group of agents. It is to be noted that the work of Orłowska and Pawlak [80], Rauszer [92] does not address the issue of counterparts of the standard rough set concepts such as approximations of sets, definability of sets in the multi-agent scenario.

Section 3 of the article presents our study of the multi-agent situation. Throughout the article, the term 'agent' and the more general term 'source' will be used synonymously. Our study of multi-agent situation is based on a *Multiple-source approximation system with distributed knowledge base* ($MSAS^D$), which also consists of a number of equivalence relations over the same domain. These relations represent the knowledge base and distributed knowledge base of individuals and groups of sources respectively. We begin our study by exploring the counterparts of the standard concepts of rough set theory mentioned above, viz. approximations of sets and definability of sets. Moreover, we have also given some expressions to dependency, which reflect how much the information provided by the MSAS^D depends on that of a source or a group of sources.

In Sect. 4, we focus on possible logics for MSAS^D. It is observed that the existing logical systems employed to study Pawlak approximation spaces including the epistemic logic $S5_n$ [22] and one given in [80, 92] are not strong enough to express the generalized notions of approximations and definability of sets. Thus, to facilitate formal reasoning with rough sets in MSAS^Ds, a quantified propositional modal logic LMSAS^D is proposed. It will be shown how the language of LMSAS^D may be used to express the properties of rough sets in the multiple-source situation. We will also explore some fragments of the logic and obtain the relationship of LMSAS^D with some known logics.

A deductive system for LMSAS^D is presented in Sect. 5 and the corresponding completeness theorem is established. The proof of the completeness theorem employs the technique of copying given by Vakarelov [101], and later used by Balbiani [1, 2] to obtain completeness of many modal logics with semantics based on relative accessibility relations. The issue of decidability is addressed in Sect. 6 and a decidable fragment of LMSAS^D is presented.

In Sect. 7, we give an algebraic semantics of LMSAS^D and the corresponding soundness and completeness theorems are proved with respect to a class of complete Boolean algebra with operators satisfying some additional properties. In order to obtain the completeness theorem, we have used the technique of completions of algebras. $Q - filters$ are used instead of ultra-filters, because the embedding given by Jónsson-Tarski Theorem may not preserve infinite joins and meets – which is what we require. The soundness and completeness theorems also establish a strong connection between the MSAS^{D}s and the algebraic counterpart of LMSAS^{D}.

Section 8 of the article deals with the scenario (ii), where the partition changes with time due to a flow of information. For this, we consider a structure, called *dynamic approximation space* (DAS), consisting of a finite sequence of equivalence relations over the same domain. The DAS represents the knowledge base, or the information that we have about the domain at different time points. Thus, we are dealing with a finite linear time line. In fact, in the line of generalizations of approximation space with relations other than equivalence (e.g. [58,65,95]), we consider the more general structure of a *dynamic space* defined by Pagliani [82], restricted to be finite in our study. So, a dynamic space for us is a finite sequence of Kripke frames (with any binary relation) over the same domain. The dynamic space consisting of Kripke I-frames, $I \in \{K, K4, T, B, S4, KTB, KB4, S5\}$, is called a *dynamic I space*, following standard modal logic nomenclature. Thus, the class of dynamic $S5$ spaces is just the class of DASs. As DASs are specially relevant to the classical rough sets, we begin our study by exploring their properties. It is found that different patterns of flow of information determine different types of DASs. The search for a suitable logic which can express approximations of sets relative to time leads us to the language $\mathcal{L}$, that has temporal operators and modal operators for 'necessity' and 'possibility'. Satisfiability of the wffs of $\mathcal{L}$ is defined in structures based on dynamic spaces. It is shown that the different dynamic I spaces as well as different types of DASs mentioned earlier can be characterized by $\mathcal{L}$-wffs. For each I, the class of dynamic I spaces determines a logic $L(T, I)$, and thus, in particular, we obtain the logic $L(T, S5)$ for DASs. The logics $L(T, I)$ have a close connection with multi-modal logics. Keeping in view the standard translation of the modal operator $\Box$ (for necessity), $L(T, I)$ can also be related to first order temporal logics (cf. e.g. [37–39]). Further, one can show that the $\mathcal{L}$-semantics can be determined through a kind of fibring over a combination of temporal and Kripke frames. In fact, this manner of combination of temporal and Kripke frames has also found application in modelling belief revision [10,11]. Moreover, it is distinct from known proposals of combination of modal logics, e.g. [8,20,27,29,35,62,63,106]. In Sect. 9, we shall discuss the proposed logics in the perspective of other closely related known systems.

In Sect. 10, we focus on the proof procedures of the logics $L(T, I)$. We present tableau-based proof procedures in a schematic manner, and corresponding soundness and completeness theorems are proved. A prefixed tableau is used, in the line of [28]. In the case of standard modal logic, well-formed formulae (wffs) are labeled with prefixes to name the world where each wff is supposed to hold. As the satisfiability of $\mathcal{L}$-wffs depends on time points as well as on the objects of the domain, the notion of prefix is modified so that it not only names an object, but also mentions the time point where the wff is supposed to hold. Prefixed tableau calculi for multi-modal logics have been studied, for example, in [3,32]. As there are connections between multi-modal logics and $L(T, I)$, one may wonder about possible syntactical links between the respective prefixed tableau calculi. However, on comparison, one can see that the notions of prefix used in these studies are all different from each other.

Section 11 discusses the decidability results relative to different classes of dynamic spaces. The general decidability result for $L(T, I)$ is still open, but the decidability of the validity problem of wffs in the class of dynamic I spaces with a given cardinality (number of Kripke frames in the dynamic I space) is shown. The same problem can also be decided for the class of all dynamic I spaces with domains of a given finite cardinality. Two decidable fragments of the logics $L(T, I)$ are also presented.

The language $\mathcal{L}$ can study the effects of flow of information on the partitions, but the flow of information itself does not appear in the picture. In other words, we cannot see how the information is provided to the system through $\mathcal{L}$. So, in the last part of the article, we are interested in a dynamic logic for information systems, where we can express the flow of information as well as its effect on the approximation of sets. The first step in this direction is to come up with a logic for information systems having the following desirable properties.

(I) The language should have attribute, attribute value constants.
(II) The semantics should be based on a structure having relative accessibility relations with the power set of the set A of attributes as the parameter set. These relations in the structure would be present syntactically as modalities.
(III) The relationship between the induced relations (indiscernibility, similarity etc.) with the attributes and attribute values should be reflected syntactically in the relationship between the modalities and the pairs of attribute, attribute value constants.
(IV) The logic should have a sound and complete deductive system.

In Sect. 12, we present logics $\mathrm{LIS_f}$ and $\mathrm{LDIS_f}$ for information systems (which could be deterministic or incomplete) and deterministic information systems respectively which have all these properties. In fact, $\mathrm{LIS_f}$ and $\mathrm{LDIS_f}$ have modal operators for similarity relations in addition to those for indiscernibility. It is shown how the presence of the features (I)–(III) enables $\mathrm{LDIS_f}$ to express various concepts related to dependencies in data and data reduction. Soundness, completeness and decidability for the logics are all established. Notions of information and information update for information systems are then defined. The logics $\mathrm{LIS_f}$ and $\mathrm{LDIS_f}$ are extended to define dynamic logics $\mathrm{DLIS_f}$ and $\mathrm{DLDIS_f}$ for information systems and deterministic information systems respectively, that accommodate these notions. The logics $\mathrm{DLIS_f}$ and $\mathrm{DLDIS_f}$ are related with $\mathrm{LIS_f}$ and $\mathrm{LDIS_f}$ in the same way as dynamic epistemic logic (DEL) is related with epistemic logic. A set of reduction axioms is provided which gives us the decidability and sound and complete deductive systems for $\mathrm{DLIS_f}$ and $\mathrm{DLDIS_f}$. Finally, a comparison with different dynamic epistemic logics is made.

In order to deal with situation (iii) mentioned in the beginning, where we have multi-agents as well as information flow with time, the logics $\mathrm{LIS_f}$, $\mathrm{LDIS_f}$, $\mathrm{DLIS_f}$, $\mathrm{DLDIS_f}$ are extended to a multi-agent scenario in Sect. 13. The extended logic not only can express the flow of information from outside, but can also express the situation where an agent borrows some information from another agent of the system. All the results of the single agent case can be carried over to this case.

Section 14 gives a summary of the article and discusses some issues coming out from the article which need further work. In the next section, we present the requisite preliminaries.

This article is based on the work done in [7,42,45–49,51–55].

2 Rough Set Theory: Preliminaries

The section is organized as follows. In Sect. 2.1, we give some basic concepts related with approximation spaces and deterministic information systems. Some generalizations of the deterministic information system and Pawlak's approximation space are also presented. In Sect. 2.2, we shall discuss some of the logics proposed for rough set theory. We only present the logics and generalizations which are relevant to this article. Finally, Sect. 2.3 gives some basic results related to Boolean algebras with operators.

2.1 Approximation Spaces and Information Systems

The notion of an approximation space plays a crucial role in the rough set theory.

Definition 1. *An* approximation space *is a pair* (U, R)*, where* U *is a non-empty set and* R *an equivalence relation on it.*

A subset X of U may not, in general, be exactly describable using (the partition induced by) R. It is 'approximated' by the lower and upper approximations ($\underline{X}_R, \overline{X}_R$ respectively) defined as follows.

Definition 2.

$$\underline{X}_R := \{x \in U : [x]_R \subseteq X\},$$
$$\overline{X}_R := \{x \in U : [x]_R \cap X \neq \emptyset\}.$$

So, given an approximation space (U, R) and a set $X \subseteq U$, the domain is divided into three disjoint sets, viz. $\underline{X}_R, \overline{X}_R \setminus \underline{X}_R$ and $(\overline{X}_R)^c$.

Definition 3. $x \in U$ *is said to be a*

positive element *of* X *if* $x \in \underline{X}_R$,
negative element *of* X *if* $x \in (\overline{X}_R)^c$,
boundary/undecidable *element of* X *if* $\overline{X}_R \setminus \underline{X}_R$.

The set X is called *definable* in (U, R) if there is no boundary element of X. X is *rough*, if it is not definable.

A realization of an approximation space is obtained through any deterministic information system.

Definition 4. *A tuple* $\mathcal{S} := (U, \mathcal{A}, \{Val_a\}_{a\in\mathcal{A}}, f)$ *is called a* deterministic information system *(DIS), where*

- U *is a non-empty set of* objects*;*
- $\mathcal{A}$ *is a non-empty set of* attributes*;*
- Val_a *for each attribute* a *is a non-empty set, the elements of which are called* values *of attribute* a*;*
- $f : U \times \mathcal{A} \to \bigcup \{Val_a : \ a \in \mathcal{A}\}$ *is such that* $f(x, a) \in Val_a$*, for any* $x \in U$*,* $a \in \mathcal{A}$.

Given a deterministic information system $\mathcal{S} := (U, \mathcal{A}, \{Val_a\}_{a\in\mathcal{A}}, f)$ and a set $B \subseteq \mathcal{A}$, the *indiscernibility relation* $Ind_{\mathcal{S}}(B)$ is an equivalence relation on U defined by:

$$(x, y) \in Ind_{\mathcal{S}}(B), \text{ if and only if } f(x, a) = f(y, a) \text{ for all } a \in B.$$

Let $\mathcal{C}$ be a map from the set of all deterministic information systems to the set of all approximation spaces defined by $\mathcal{C}(\mathcal{S}) := (U, Ind_{\mathcal{S}}(\mathcal{A}))$, with $\mathcal{S}$ as above. It is not difficult to show that $\mathcal{C}$ is onto.

Data Reduction and Dependency. Let $\mathcal{S} := (U, \mathcal{A}, \{Val_a\}_{a\in\mathcal{A}}, f)$ be a deterministic information system. Recall that the notion of indiscernibility relation is relative to a set of attributes. Thus, one may be interested in removing all 'superfluous' attributes, that is, those which do not affect the partition of the domain, and consequently, set approximations. This is the main idea of *reduction of knowledge*. Formally, we have the definitions as below.

Definition 5. *Let* $P, Q \subseteq \mathcal{A}$.

1. $POS_P(Q) := \bigcup_{X\in U/Ind_{\mathcal{S}}(Q)} \underline{X}_{Ind_{\mathcal{S}}(P)}$ *is the* P-positive region *of* Q*, where* $U/Ind_{\mathcal{S}}(Q)$ *denotes the quotient set for the equivalence relation* $Ind_{\mathcal{S}}(Q)$.
2. $b \in P$ *is said to be* $Q-$ dispensable *in* P *if* $POS_P(Q) = POS_{(P\setminus\{b\})}(Q)$*; otherwise* b *is* $Q-$ indispensable *in* P.
3. *If every* $b \in P$ *is* Q*-indispensable,* P *is* Q-independent*; otherwise* P *is* Q-dependent.
4. $S \subseteq P$ *will be called a* $Q-$reduct *of* P *if* S *is* $Q-$*independent and* $POS_S(Q) = POS_P(Q)$.

We note that $P \subseteq \mathcal{A}$ may have multiple $Q-$reducts. Moreover, if P is infinite, then it may not have any $Q-$reduct at all. In the special case that $P = Q$, we drop the prefix '$Q-$' in the above. In this case, observe that $POS_P(Q) = U$, and the condition under which b is dispensable in P, reduces to $Ind_{\mathcal{S}}(P) = Ind_{\mathcal{S}}(P\setminus\{b\})$.

The notion of dependency of sets of attributes is given as follows.

Definition 6. *Let* $P, Q \subseteq \mathcal{A}$.

1. Q *is said to* depend *on* P *(denoted* $P \Rightarrow Q$*), if* $Ind_{\mathcal{S}}(P) \subseteq Ind_{\mathcal{S}}(Q)$.

2. *P and Q are called* equivalent *(denoted $P \equiv Q$), if $Ind_{\mathcal{S}}(P) = Ind_{\mathcal{S}}(Q)$.*
3. *P and Q are* independent *($P \not\equiv Q$), if neither $P \Rightarrow Q$ nor $Q \Rightarrow P$ hold.*

Table 1. Deterministic information system $\mathcal{S}$

	T	H	N
x_1	Very high	No	Yes
x_2	High	Yes	No
x_3	No	No	Yes
x_4	No	No	Yes
x_5	Very high	No	Yes
x_6	Very high	Yes	Yes
x_7	High	Yes	Yes
x_8	High	Yes	No

Example 1. This example will serve as an illustration of ideas considered above. Let us consider the deterministic information system $\mathcal{S}$ of Table 1 below, which provides information about eight patients $x_1 - x_8$ regarding the attributes 'Temperature (T)', 'Headache (H)' and 'Nausea (N)'. This table is a modified form of the one given in [33]. The indiscernibility relations corresponding to attribute sets $\{a\}$, $\{b\}$, $\{c\}$, and $P := \{a, b, c\}$ are obtained as follows:

$$
\begin{aligned}
Ind_{\mathcal{S}}(\{a\}) &:= \{\{x_1, x_5, x_6\}, \{x_2, x_7, x_8\}, \{x_3, x_4\}\};\\
Ind_{\mathcal{S}}(\{b\}) &:= \{\{x_2, x_6, x_7, x_8\}, \{x_1, x_3, x_4, x_5\}\};\\
Ind_{\mathcal{S}}(\{c\}) &:= \{\{x_1, x_3, x_4, x_5, x_6, x_7\}, \{x_2, x_8\}\};\\
Ind_{\mathcal{S}}(\{a, b\}) &:= \{\{x_1, x_5\}, \{x_2, x_7, x_8\}, \{x_3, x_4\}, \{x_6\}\};\\
Ind_{\mathcal{S}}(\{a, c\}) &:= \{\{x_1, x_5, x_6\}, \{x_2, x_8\}, \{x_3, x_4\}, \{x_7\}\};\\
Ind_{\mathcal{S}}(\{b, c\}) &:= \{\{x_1, x_3, x_4, x_5\}, \{x_2, x_8\}, \{x_6, x_7\}\};\\
Ind_{\mathcal{S}}(P) &:= \{\{x_1, x_5\}, \{x_3, x_4\}, \{x_2, x_8\}, \{x_6\}, \{x_7\}\}.
\end{aligned}
$$

Let Q be an attribute set such that the corresponding indiscernibility relation $Ind_{\mathcal{S}}(Q)$ is given with the equivalence classes

$$Ind_{\mathcal{S}}(Q) := \{\{x_1, x_5, x_6\}, \{x_3, x_4\}, \{x_2, x_7\}, \{x_8\}\}.$$

The positive region of Q relative to different sets of attributes are obtained as follows:

$$
\begin{aligned}
POS_P(Q) &:= \{x_1, x_3, x_4, x_5, x_6, x_7\};\\
POS_{\{a\}}(Q) &:= \{x_1, x_3, x_4, x_5, x_6\} = POS_{\{a,c\}}(Q);\\
POS_{\{b\}}(Q) &:= \emptyset = POS_{\{c\}}(Q) = POS_{\{b,c\}}(Q);\\
POS_{\{a,b\}}(Q) &:= \{x_1, x_3, x_4, x_5\}.
\end{aligned}
$$

Since $POS_{\{a,c\}}(Q) = POS_{\{a\}}(Q)$, it follows that c is $Q-$dispensable in $\{a,c\}$. But, a is $Q-$indispensable in $\{a,c\}$ as $POS_{\{a,c\}}(Q) \neq POS_{\{a\}}(Q)$. Also note that a, b, c are all $Q-$indispensable in P, and hence P is $Q-$independent.

The notion of deterministic information system has been generalized in many ways to consider different practical situations. Some of these generalizations are presented in the next section.

Extensions of the Notion of Information System. A temporal dimension is added to the study of information system by Orłowska. In [75], the notion of an information system is extended by adding the concept of time. A set T of 'time points' and a linear order $<$ on T is included to give

Definition 7. *A tuple* $\mathcal{DS} := (U, \mathcal{A}, \{Val_a\}_{a\in\mathcal{A}}, T, <, f)$ *is called a* dynamic information system, *where*

- U, A, Val_a *are as in Definition 4;*
- T *is a non-empty set of* time points*;*
- $<$ *is a linear order on* T*;*
- $f : U \times T \times \mathcal{A} \to \cup \{Val_a : a \in \mathcal{A}\}$ *is such that* $f(x,t,a) \in Val_a$*, for any* $x \in U$, $t \in T$, $a \in \mathcal{A}$.

Thus in case of a dynamic information system, the value that f assigns to an object x for any attribute a, becomes dependent on the chosen time point t. Let us note that the attribute set $\mathcal{A}$ does not vary with time. In [36], a dynamic information system based on *time sequence* (finite set of time points) is defined, of which the dynamic information system defined above with finite time line is obtained as a special case.

From the definition of deterministic information system, it is clear that for each object of the domain, we have precise information about the attribute value of every attribute in $\mathcal{A}$. But in reality, this may not be the case. For instance, it could be the case that (i) we know some possible attribute values that an object may take for an attribute, but do not know exactly which one, or worse, (ii) we may not have any information about some objects regarding some of the attributes. In order to represent the situation (i), the notion of deterministic information system is extended to define the following.

Definition 8. *A tuple* $\mathcal{K} := (U, \mathcal{A}, \{Val_a\}_{a\in\mathcal{A}}, f)$ *is called a* non-deterministic information system *(NIS) where* $U, \mathcal{A}$ *and* Val_a *are same as in Definition 4 and* $f : U \times \mathcal{A} \to 2^V$, $V := \cup\{Val_a : a \in \mathcal{A}\}$*, satisfying* $f(x,a) \subseteq Val_a$*, for* $x \in U$, $a \in \mathcal{A}$.

We note that a deterministic information system is a special case of a non-deterministic information system, where the set $f(x,a)$ is a singleton, for any object x and attribute a. In the case of a non-deterministic information system, one can find induced relations other than indiscernibility in literature [79].

For instance, we have the similarity '$Sim_{\mathcal{K}}$' and inclusion '$\subseteq_{\mathcal{K}}$' relations defined as follows [81].

$Sim_{\mathcal{K}}$: x $Sim_{\mathcal{K}}$ y, if and only if $f(x,a) \cap f(y,a) \neq \emptyset$, for all $a \in \mathcal{A}$.
$\subseteq_{\mathcal{K}}$: $x \subseteq_{\mathcal{K}} y$, if and only if $f(x,a) \subseteq f(y,a)$, for all $a \in \mathcal{A}$.

Let us move to the situation (ii) where some of attribute values for an object are missing. To indicate such a situation, a distinguished value $*$ is taken as an attribute value for each attribute such that $f(x,a) = *$ signifies that we do not have information about the object regarding attribute a. Thus we have the following definitions.

Definition 9. *A tuple* $\mathcal{S} := (U, \mathcal{A}, \{Val_a\}_{a \in \mathcal{A}} \cup \{*\}, f)$ *is called an* information system *(IS), where*

- $W, \mathcal{A}, Val_a$ *are as in Definition 4 and* $* \notin \bigcup_{a \in \mathcal{A}} Val_a$*;*
- $f : U \times \mathcal{A} \rightarrow \bigcup \{Val_a : a \in \mathcal{A}\} \cup \{*\}$ *such that* $f(x,a) \in Val_a \cup \{*\}$.

An information system which satisfies $f(x,a) = *$ *for some* $x \in U$ *and* $a \in \mathcal{A}$ *will be called an* incomplete information system *(IIS).*

We would like to mention that a deterministic information system can be identified with the information system $\mathcal{S} := (U, \mathcal{A}, \{Val_a\}_{a \in \mathcal{A}} \cup \{*\}, f)$, where $f(x,a) \neq *$ for all $x \in U$ and $a \in \mathcal{A}$.

Let us consider the indiscernibility relation $Ind_{\mathcal{S}}(B)$ corresponding to an incomplete information system. If $(x,y) \notin Ind_{\mathcal{S}}(B)$, then x and y are distinguishable using the *information provided by the system* regarding the attributes of the set B. But it can be the case that in reality x and y take the same value for the attributes of B and so in reality they are indistinguishable with respect to attribute set B. For instance, suppose we have information that a patient x has symptom C but we do not have any information about the another patient y. Then these two patients are distinguishable with respect to information what we have at the moment, but in reality, it could be the case that y also has the symptom C.

One can give different interpretations for the absence of the information. For instance, [97] considered a situation where objects may be described 'incompletely' not only because of our imperfect knowledge, but also because definitely impossible to describe them on all the attributes. On the other hand, in [33,60,61], a situation is taken into account where each object has complete description but we may not have information about some objects due to our imperfect/partial knowledge of the objects. For instance, we may not have the test report of a patient at the moment but we know that the test report is going to be either positive or negative for the disease.

In [60,61], similarity relation defined below is considered as the distinguishability relation instead of indiscernibility relation in the case of incomplete information system. The assumption here is that the real value of missing attributes is one from the attribute domain.
$(x,y) \in Sim_{\mathcal{S}}(B)$ if and only if , $f(x,a) = f(y,a)$ or $f(x,a) = *$, or $f(y,a) = *$, for all $a \in B$.

We have used the same notation for the similarity relations defined on non-deterministic and incomplete information systems. It will not create any confusion as in the rest of the article, we will only work with the similarity relations defined on incomplete information systems.

Note that similarity is a tolerance relation. Moreover,

$$Sim_{\mathcal{S}}(B) = \bigcap_{b \in B} Sim_{\mathcal{S}}(\{b\}).$$

Definition 10. *Let $\mathcal{S} := (U, \mathcal{A}, \bigcup_{a \in \mathcal{A}} Val_a \cup \{*\}, f)$ be an IS. A deterministic information system $\mathcal{S}' := (U, \mathcal{A}, \bigcup_{a \in \mathcal{A}} Val_a \cup \{*\}, f')$ is said to be a* completion *of $\mathcal{S}$ if $f(x, a) \neq *$ implies $f'(x, a) = f(x, a)$ for all $a \in \mathcal{A}$ and $x \in U$.*

Proposition 1. *$(x, y) \notin Sim_{\mathcal{S}}(B)$ if and only if $(x, y) \notin Ind_{\mathcal{S}'}(B)$ for all completions $\mathcal{S}'$ of $\mathcal{S}$.*

Thus two objects are distinguishable with respect to the similarity relation if and only if these are distinguishable with respect to the indiscernibility relations in all the deterministic information systems obtained by assigning whatever attribute value from the domain we wish, to the missing attributes. In other words, if two objects are related by the similarity, then there is a possibility that they are indistinguishable. So, in the case of deterministic information systems, indiscernibility and similarity relations coincide.

The notion of lower and upper approximation in an approximation space is generalized to obtain the following. Let $\mathcal{S} := (U, \mathcal{A}, \bigcup_{a \in \mathcal{A}} Val_a \cup \{*\}, f)$ be an IS and $B \subseteq \mathcal{A}$, $X \subseteq U$. Let $Sim_{\mathcal{S}}(B)(x) := \{y \in U : (x, y) \in Sim_{\mathcal{S}}(B)\}$.

Definition 11. *The lower and upper approximations of X with respect to similarity relation $Sim_{\mathcal{S}}(B)$, denoted by $\underline{X}_{Sim_{\mathcal{S}}(B)}$ and $\overline{X}_{Sim_{\mathcal{S}}(B)}$, are defined as follows:*

$$\underline{X}_{Sim_{\mathcal{S}}(B)} := \{x \in U : Sim_{\mathcal{S}}(B)(x) \subseteq X\},$$
$$\overline{X}_{Sim_{\mathcal{S}}(B)} := \{x \in U : Sim_{\mathcal{S}}(B)(x) \cap X \neq \emptyset\}.$$

Using Proposition 1, it is not difficult to obtain the following.

Proposition 2.

(i) $x \in \underline{X}_{Sim_{\mathcal{S}}(B)}$ if and only if $x \in \underline{X}_{Ind_{\mathcal{S}'}(P)}$ for all completions $\mathcal{S}'$ of $\mathcal{S}$.
(ii) $x \in \overline{X}_{Sim_{\mathcal{S}}(B)}$ if and only if $x \in \overline{X}_{Ind_{\mathcal{S}'}(P)}$ for some completion $\mathcal{S}'$ of $\mathcal{S}$.

Generalizations of Approximation Space. In this section, we present generalizations of the notion of approximation space based on a relation or a collection of relations.

Approximation Space with Relation other than Equivalence. The most useful natural generalization is the one where the relation R is not necessarily an equivalence. For instance, in [58,95], a tolerance approximation space is considered which is a tuple (U, R) with a tolerance, that is, reflexive and symmetric relation R. Similarity relations obtained from non-deterministic and incomplete information systems are realizations of such approximation spaces. The notion of lower and upper approximations in these generalized approximation spaces is defined in a standard way as follows. Let $R(x) := \{y \in U : (x, y) \in R\}$. Then for $X \subseteq U$,

Definition 12.

- $\underline{X}_R := \{x \in U : R(x) \subseteq X\}$ *and*
- $\overline{X}_R := \{x \in U : R(x) \cap X \neq \emptyset\}$.

Multiple Relation Approximation Space. Another natural generalization of approximation space arises when we have a number of relations instead of just one. In [84], a relational system $K := (U, \mathbf{R})$ is studied, where $\mathbf{R}$ is a family of equivalence relations. Moreover, for any $P \subseteq \mathbf{R}$ and $P \neq \emptyset$, the equivalence relation $IND(P)$, which is the intersection of all equivalence relations belonging to P, is considered. One can view $\mathbf{R}$ as the collection of indiscernibility relations corresponding to individual attributes and so for $P \subseteq \mathbf{R}$, $IND(P)$ is the indiscernibility relation corresponding to the set consisting of precisely those attributes, the indiscernibility relation of which are in P. This intuition is more explicitly represented by the *information structure* proposed by Orłowska [78].

Definition 13. *A tuple $(U, \{R_B\}_{B \subseteq \mathcal{A}})$ is called* information structure, *where $\mathcal{A}$ is a non-empty set of parameters or attributes and for each $B \subseteq \mathcal{A}$, R_B is an equivalence relation on U satisfying,*

(I1) $R_\emptyset = U \times U$ *and*
(I2) $R_{B \cup C} = R_B \cap R_C$.

Condition (I1) signifies that we can distinguish objects only using the information about the objects regarding the attributes. We note that given an information system $\mathcal{S} := (W, \mathcal{A}, \bigcup_{a \in \mathcal{A}} Val_a, f)$, the structure $(W, \{Ind_{\mathcal{S}}(B)\}_{B \subseteq \mathcal{A}})$ is an information structure. For every information structure $(W, \{R_B\}_{B \subseteq \mathcal{A}})$, can we determine an information system $\mathcal{S} := (W, \mathcal{A}, \bigcup_{a \in \mathcal{A}} Val_a, f)$ such that $Ind_{\mathcal{S}}(B) = R_B$ for all $B \subseteq \mathcal{A}$? The answer is yes, *provided $\mathcal{A}$ is finite.* This is due to the fact that an information structure may not have the property $R_B = \bigcap_{b \in B} R_{\{b\}}$ for infinite $B \subseteq \mathcal{A}$, as shown in Example 1 below, but we always have $Ind_{\mathcal{S}}(B) = \bigcap_{b \in B} Ind_{\mathcal{S}}(\{b\})$.

Example 2. Consider $\mathfrak{F} := (U := \{x, y\}, \{R_B\}_{B \subseteq \mathcal{A}})$ where $R_B := U \times U$ for any finite subset B of $\mathcal{A}$, while for infinite B, R_B is the identity relation on U. Note that for any infinite B, we have $R_B \neq \bigcap_{b \in B} R_{\{b\}}$.

Given the collection of indiscernibility relations corresponding to different sets of attributes, one can define different operations on this collection to obtain new properties. For instance, in [26], the intersection and transitive closure of union of indiscernibility relations is considered. A structure of the form $(U, \{R_a\}_{a \in REL})$ is defined, where REL is a set of relational expressions built inductively using a set $\mathcal{R}$ of relational variables and two binary operations $\cap$ and $\uplus$ such that

- for each $a \in REL$, R_a is an equivalence relation on U,
- for $a, b \in REL$, $R_{a \cap b} = R_a \cap R_b$ and $R_{a \uplus b} = (R_a \cup R_b)^*$ (transitive closure of $R_a \cup R_b$).

Let us call the above structure, a DAL-*structure*. A variant of DAL-structure is considered in [30], where $R_{a \uplus b}$ is taken as $R_a \cup R_b$, keeping all the other conditions same as in DAL-structure. Let us call it a DALLA-*structure*. So every DALLA-structure is a DAL-structure, but not conversely.

A collection of equivalence relations over the same domain also appears in [92], but the motivation is different from the structures discussed above. It represents a multiple agent scenario where each agent has its own knowledge base represented by an equivalence relation. Thus in [92], a structure of the form $(U, \{R_t\}_{t \in \mathcal{T}})$ is considered, where $\mathcal{T}$ is a set of terms built using a set T of individual agents and two binary operations $\wedge$ and $\vee$ such that

- for each $a \in \mathcal{T}$, R_a is an equivalence relation on U,
- for $a, b \in \mathcal{T}$,
  $U/R_{a \vee b} := \{[x]_{R_a} \cap [y]_{R_b} : [x]_{R_a} \cap [y]_{R_b} \neq \emptyset\}$ and
  $U/R_{a \wedge b} := \{[x]_{R_a} \cup [y]_{R_b} : [x]_{R_a} \cap [y]_{R_b} \neq \emptyset\}$.

If R_a and R_b represent the knowledge base of the agents a and b respectively, then $R_{a \vee b}$ and $R_{a \wedge b}$ are respectively called the *strong distributed knowledge base* and *weak distributed knowledge base* of the group $\{a, b\}$ of agents. Note that $R_{a \vee b} = R_a \cap R_b$ and hence notion of strong distributed knowledge base can be identified with the notion of distributed knowledge in epistemic logic [22].

All the structures discussed so far in this section, are based on a collection of equivalence relations over the same domain. This can be further generalized by considering a collection of relations other than equivalence, or even a collection of different types of relations over the same domain. For instance, in [82], a structure of the form $(U, \{R_i\}_{i \in I})$, called *dynamic space*, is considered, where $\{R_i\}_{i \in I}$ is a family of binary relations. On the other hand, in [100], a *NIL-structure* is defined to be of the form (U, S, R), where R and S are binary relations on U satisfying the following.

1. $(x, x) \in R \cap S$.
2. If $(x, y) \in R$ and $(y, z) \in R$ then $(x, z) \in R$.
3. If $(x, y) \in S$, then $(y, x) \in S$.
4. If $(x, y) \in S$, $(x, u) \in R$, $(y, v) \in R$, then $(u, v) \in S$.

Note that given a non-deterministic information system $\mathcal{K} := (U, \mathcal{A}, \{Val_a\}_{a\in\mathcal{A}}, f)$, $(U, Sim_{\mathcal{K}}, \subseteq_{\mathcal{K}})$ is a NIL-structure, called *standard* NIL*-structure.* Moreover, in [100], the following proposition is proved.

Proposition 3. *Every* NIL*-structure is standard, that is, given a* NIL*-structure* (U, S, R)*, there exists a non-deterministic information system* $\mathcal{K} := (U, \mathcal{A}, \{Val_a\}_{a\in\mathcal{A}}, f)$ *such that* $Sim_{\mathcal{K}} = S$ *and* $\subseteq_{\mathcal{K}} = R$.

2.2 Some Logics from Rough Set Theory

In this section, we will survey some of the logics relevant to this article. The main reference is [6]. For a detailed study on the logics inspired by rough set theory, we refer to [18].

Normal Modal Systems. The modal nature of the lower and upper approximations of rough sets was evident from the start. Hence, it is no surprise that normal modal systems were focussed upon, during investigations on logics for rough sets. In particular, in case of Pawlak rough sets, the two approximations considered as operators clearly obey all the $S5$ laws. The formal connection between the syntax of $S5$ and its semantics in terms of rough sets is given as follows [5].

According to the Kripke semantics for $S5$, a well-formed formula (wff) α is interpreted by a function v as a subset in a non-empty domain U, the subset representing the extension of the formula – i.e. the collection of objects/worlds where the wff holds. Moreover, in a $S5$-model $\mathcal{M} := (U, R, v)$ (say), the accessibility relation R is an equivalence on U. Further, if $\Box$, $\Diamond$ denote the necessity and possibility operators respectively then for any wff α, $v(\Box\alpha) = \underline{v(\alpha)}_R$ and $v(\Diamond\alpha) = \overline{v(\alpha)}_R$.

A wff α is *true* in $\mathcal{M}$, if $v(\alpha) = U$. Now it can easily be seen that all the $S5$ theorems involving $\Box$ and $\Diamond$ translate into valid properties of lower and upper approximations.

Taking a cue from this connection, a multi-modal logic is defined in [80]. The language of the logic contains a set $CONREL$ of constants representing indiscernibility relations. Using the standard Boolean connectives, the set of all wffs is defined following the scheme:

$$p \in PV \mid \neg\alpha \mid \alpha \wedge \beta \mid [R]\alpha,$$

where PV is the set of propositional variables and $R \in CONREL$.

The semantics is based on a structure of the form $\mathfrak{F} := (U, \{R_i\}_{i\in I})$, where $\{R_i\}_{i\in I}$ is a family of equivalence relations over U. The satisfiability relation is defined using the meaning functions $m : CONREL \to \{R_i\}_{i\in I}$ and $v : PV \to 2^U$ in a standard way. For instance, for $\mathfrak{M} := (\mathfrak{F}, m, v)$ and $w \in U$,

$$\mathfrak{M}, w \models [R]\alpha \text{ if and only if for all } w' \text{ such that } (w, w') \in m(R),\ \mathfrak{M}, w' \models \alpha.$$

The collection $\{R_i\}_{i\in I}$ of equivalence relations, as mentioned in [80], is intended to represent the family of indiscernibility relations corresponding to a family of information systems over the same domain.

A sound and complete deductive system for the logic, consisting of the following axioms, is also presented.

1. $[R](\alpha \to \beta) \to ([R]\alpha \to [R]\beta)$.
2. $[R]\alpha \to \alpha$.
3. $\alpha \to [R]\langle R\rangle\alpha$, where $\langle R\rangle\alpha := \neg[R]\neg\alpha$.
4. $[R]\alpha \to [R][R]\alpha$.

Observe that this axiomatic system is the same as the axiomatic system of the n-agent epistemic logic $S5_n$ without the common knowledge and distributed knowledge operator [22]. However, there is a difference in the language of the two logics. $S5_n$ has n modal operators representing the epistemic state of n agents. On the other hand, nothing is said about the cardinality of $CONREL$. On the side of semantics, in case of $S5_n$, we have structures with exactly n equivalence relations. But in the case of the logic defined in [80], we do not have any such restriction.

Logic DAL and its Variant. As mentioned in Sect. 2.1, one can obtain different properties by defining operations on the set of indiscernibility relations, and we obtained a structure such as DAL-structure (cf. Sect. 2.1). A logic DAL for DAL-structure is proposed in [26]. The language of DAL, as in the case of the logic defined in [80], contains a set $\mathcal{R}$ of relation variables representing indiscernibility relations. Moreover, there are binary operations $\cap, \uplus$, and a collection REL of *relational expressions* is built inductively out of the members of $\mathcal{R}$ with these operations. Thus the set of all DAL wffs is defined following the scheme:

$$p \in PV \mid \neg\alpha \mid \alpha \wedge \beta \mid [a]\alpha,$$

where $a \in REL$. For a DAL-structure $(U, \{R_a\}_{a\in REL})$, the satisfiability relation is defined in a similar way as in case of the logic in [80], using a meaning function $m : REL \to \{R_a\}_{a\in REL}$ such that $m(a) = R_a$.

As mentioned in [18], DAL is the paradigm logic for reasoning about indiscernibility relation, but unfortunately, very few results have been obtained for DAL. Decidability as well as a Hilbert-style axiomatization of DAL are still open. In [30], a variant of DAL, called DALLA is proposed. The language of DALLA is the same as that of DAL, but the semantics is based on DALLA-structures (cf. Sect. 2.1). A sound and complete deductive system consisting of the following two axioms in addition to $S5$ axioms for the operators is given.

1. $[a \uplus b]\alpha \leftrightarrow [a]\alpha \wedge [b]\alpha$.
2. $[a \wedge b]\alpha \leftrightarrow [a]\alpha \vee [b]\alpha$.

Information Structures and the Logic by Balbiani. As pointed out by Orłowska in [78], when we say two objects are indistinguishable in an information system, we actually mean that these are indistinguishable not absolutely, but with respect to certain properties/attributes. Thus in the study of indiscernibility relations, it seems important to bring the attribute set also into the picture. But the logics discussed so far lack this feature. In order to achieve this, Orłowska proposed the notion of an information structure (cf. Definition 13), but cited the axiomatization of a logic with semantics based on information structures as an open problem. Later, Balbiani gave a complete axiomatization of the set of wffs valid in every information structure, using the technique of *copying* introduced by Vakarelov [101]. In fact, in [2], complete axiomatizations of logics with semantics based on various types of structures with relative accessibility relations is presented. One of these is a logic for information structures (cf. [1]). This, as required, is a multi-modal logic with a modal operator $[P]$ for each $P \subseteq \mathcal{A}$. The operator $[P]$ corresponds to the lower approximation with respect to the indiscernibility relation relative to the set P of attributes. Apart from the $S5$–axioms for each modal operator, the axiom $[P]\alpha \vee [Q]\alpha \rightarrow [P \cup Q]\alpha$ is considered. The canonical model obtained for this system only satisfies the condition $R_{B \cup C} \subseteq R_B \cap R_C$. Such a model is called *decreasing*. Using the method of copying, one obtains from a decreasing model, a model that satisfies condition (I2) (viz. $R_{B \cup C} = R_B \cap R_C$) and preserves satisfiability as well.

Logics with Attribute Expressions. The logics discussed so far lack an important aspect related to the study of information systems. The language of these logics cannot refer to attributes or attribute values which are essential parts of an information systems. In this section, we survey logics with attribute expressions.

Decision Logic. Decision logic (DL) is the simplest logic with this feature. It is a propositional logic, the language of which contains a set $\mathcal{A}$ of attribute constants and for each $a \in \mathcal{A}$, a set Val_a of attribute value constants. Using these constants, atomic wffs are formed which are of the form (a, v), $a \in \mathcal{A}$, $v \in Val_a$, and are called *descriptors*. The wffs of DL are formed in the standard way using the descriptors and Boolean connectives $\neg, \wedge$. Semantics of DL is directly based on the deterministic information systems. The satisfiability of the wffs in a deterministic information system $\mathcal{S} := (U, \mathcal{A}, \bigcup_{a \in \mathcal{A}} Val_a, f)$ at an object $x \in U$ is defined in the natural way. For instance,

$$\mathcal{S}, x \models (a, v) \text{ if and only if } f(x, a) = v.$$

A sound and complete deductive system for DL can be given consisting of the following axioms in addition to the propositional logic axioms.

1. $(a, v) \wedge (a, u) \leftrightarrow \bot$, for any $a \in \mathcal{A}$, $u, v \in Val_a$ and $v \neq u$.
2. $\bigvee_{v \in Val_a} (a, v)$, for every $a \in \mathcal{A}$.

The proof of completeness theorem is very simple. We need to show that every consistent set is satisfiable. Consider the deterministic information system $(W, \mathcal{A}, \{Val_a\}_{a\in\mathcal{A}}, f)$, where W is the set all maximal consistent sets. Moreover, $f(w,a) = v$ if and only if $(a,v) \in w$. Axioms 1 and 2 guarantee that f is a total function. Now, one can prove $\alpha \in w$ if and only if α is satisfiable in the above information system. This gives us the desired result.

Dynamic Information Logic. Orłowska ([75]), defines a logic DIL (dynamic information system logic) with models based on dynamic information systems $\mathcal{DS} := (U, \mathcal{A}, \{Val_a\}_{a\in\mathcal{A}}, T, <, f)$ (cf. Definition 7), in order to deal with the temporal aspect of information. In the language of DIL, atomic statements are descriptors of decision logic, together with an object constant x - so these are triples (x,a,v), and are intended to express: "object x assumes value v for attribute a". There are modal operators to reflect the relations $<$ and $<^{-1}$. So the set of wffs is defined following the scheme

$$(x,a,v) \mid \neg\alpha \mid \alpha \wedge \beta \mid [<]\alpha \mid [<^{-1}]\alpha.$$

The truth of all statements of the language is evaluated in a model based on a dynamic information system, with respect to moments of time, i.e. members of the set T.

A DIL-model is a tuple $\mathcal{M} := (\mathcal{S}, m)$ where $\mathcal{S}$ is a dynamic information system, and m a meaning function which assigns objects, attributes and values from U, A, Val to the respective constants. The satisfiability of a formula α in a model $\mathcal{M}$ at a moment $t(\in T)$ of time is defined inductively as follows:

$$\mathcal{M},\ t \models (x,a,v) \text{ if and only if } f(m(x), t, m(a)) = m(v).$$

For the Boolean cases, we have the usual definitions. For the modal case,

$$\mathcal{M},\ t \models [R]\alpha \text{ if and only if for all } t' \in T, \text{ if } (t,t') \in R, \text{ then } \mathcal{M},\ t' \models \alpha,$$

$R \in \{<, <^{-1}\}$. A sound and complete deductive system of DIL can be given consisting of the axioms of linear time temporal logic along with the axiom,

$$(x,a,v) \wedge (x,a,u) \wedge (x',a,v) \to (x',a,u),$$

which says that the values of attributes are uniquely assigned to objects.

Logic NIL. The logic NIL proposed by Orłowska and Pawlak [81] is an extension of the description logic by enriching the language with modal operators $\Box, \Box_1, \Box_2$ corresponding to similarity relation, inclusion relation and converse of inclusion relation. Wffs are built, as usual, out of the atomic wffs (descriptors) and the connectives. A NIL-model $\mathcal{M} := (U, S, R, m)$ consists of a NIL-structure (U, S, R) (cf. Sect. 2.1), along with a meaning function m from the set of all descriptors to the set $\mathcal{P}(U)$. Satisfiability relation is defined in the usual way.

A sound and complete deductive system for NIL was proposed in [81], consisting of the following three axioms in addition to the KTB axioms for $\Box$ and $S4$ axioms for $\Box_1$ and $\Box_2$.

1. $\alpha \rightarrow \Box_1 \neg \Box_2 \neg \alpha$.
2. $\alpha \rightarrow \Box_2 \neg \Box_1 \neg \alpha$.
3. $\Box \alpha \rightarrow \Box_2 \Box \Box_1 \alpha$.

Note that due to Proposition 3, we also obtain completeness with respect to class of all standard NIL-structures.

Rauszer's Logic for Multi-agent Systems. Rauszer [92] describes a logic, that takes into account a (finite) collection of *agents* and their *knowledge bases.* We denote the logic as $\mathcal{L}_{\mathcal{MA}}$. The language of $\mathcal{L}_{\mathcal{MA}}$ has 'agent constants' along with two special constants 0,1. Binary operations $+, \cdot$ are provided to build the set $\mathcal{T}$ of *terms* from these constants. Wffs of one kind are obtained from terms, and are of the form $s \Rightarrow t$, $s, t \in \mathcal{T}$, where $\Rightarrow$ is a binary relational symbol. $s \Rightarrow t$ is to reflect that "the classification ability of agent t is at least as good as that of agent s".

Furthermore, there are attribute as well as attribute-value constants. Descriptors formed by these constants constitute atomic propositions, and using connectives $\wedge, \neg$ and modal operators I_t, $t \in \mathcal{T}$ (representing 'partial knowledge' of each agent), give wffs of another kind.

$\mathcal{L}_{\mathcal{MA}}$-models are not approximation spaces, but what could be called 'partition spaces' on information systems. Informally put, a model consists of an information system $\mathcal{S} := (U, A, Val, f)$, and a family of *partitions* $\{E_t\}_{t \in \mathcal{T}}$ on the domain U – each corresponding to the knowledge base of an agent. The family is shown to have a lattice structure, and the ordering involved gives the interpretation of the relational symbol $\Rightarrow$. Wffs built out of descriptors are interpreted in the standard way, in the information system $\mathcal{S}$. The partial knowledge operator I_t for a term t reflects the lower approximation operator with respect to the partition E_t on U. An axiomatization of $\mathcal{L}_{\mathcal{MA}}$ is presented, to give soundness and completeness results.

Let us note the following facts about the logics discussed in this section so far. Logics DL and DIL do not have modal operators for indiscernibility or any other relations induced by information systems. So DL can talk about the attribute values of the objects and DIL can express the changes in attribute values of the objects with time, but the language of these logics is not strong enough to talk about (changes in) set approximations. On the other hand, although the language of NIL and Rauszer's logic has both descriptors and modalities for relations induced by information systems, but the logics do not connect the descriptors with the modalities. In fact, in these cases, we do not have induced relations corresponding to each subset of attributes as desired by Orłowska in [78]. In Sect. 12, we will propose a logic which will overcome these limitations and which will also have a sound and complete deductive system.

Some other Logics with Attribute Expressions. A class of logics with attribute expressions is also defined in [76, 77]. Models are based on structures of the form $(U, A, \{ind(P)\}_{P \subseteq A})$, where the 'indiscernibility' relation $ind(P)$ for each subset P of the attribute set A, has to satisfy certain conditions. For the models of one of the logics, for example, the following conditions are stipulated for $ind(P)$:

$(U1)$ $ind(P)$ is an equivalence relation on U,
$(U2)$ $ind(P \cup Q) = ind(P) \cap ind(Q)$,
$(U3)$ if $P \subseteq Q$ then $ind(Q) \subseteq ind(P)$, and
$(U4)$ $ind(\emptyset) = U \times U$.

Other logics may be obtained by changing some of $(U1)$–$(U4)$. The language of the logics has a set of variables each representing a set of attributes, as well as constants to represent all one element sets of attributes. Further, the language can express the result of (set-theoretic) operations on sets of attributes. The logics are multimodal – there is a modal operator to reflect the indiscernibility relation for each set of attributes as above. A usual Kripke-style semantics is given, and a number of valid wffs presented. However, as remarked in [76], we do not know of a complete axiomatization for such logics.

In literature, one can find many other generalizations and extensions of DL apart from those discussed above, for instance [24, 25, 110], but these do not discuss any axiomatization at all.

2.3 Boolean Algebra with Operators

In this section, we present some results on Boolean algebras with operators [9] which we will require in Sect. 3. We restrict ourselves to unary operators only.

Definition 14. *A tuple* $\mathfrak{A} := (A, \cap, \sim, 1, \{f_k\}_{k \in \Delta})$ *is said to be a* Boolean algebra with operators *(BAO) if* $(A, \cap, \sim, 1)$ *is a Boolean algebra and each* $f_k : A \to A$ *satisfies (i)* $f_k(1) = 1$ *and (ii)* $f_k(a \cap b) = f_k(a) \cap f_k(b)$. *Moreover, BAO* $\mathfrak{A}$ *is said to be* complete *if* $\bigcap X$ *and* $\bigcup X$ *exist for all* $X \subseteq A$, *where* $\bigcap X$ *and* $\bigcup X$, *respectively, denote the greatest lower bound (g.l.b) and least upper bound (l.u.b) of the set* X.

We will see that as Boolean algebras, every BAO has a concrete set-theoretic representation. The notion of *complex algebra* plays the leading role in such a representation theorem.

Definition 15. *Let us consider a structure of the form* $\mathfrak{F} := (U, \{R_i\}_{i \in \Delta})$, *where* Δ *is an index set and for each* $i \in \Delta$, $R_i \subseteq U \times U$. *The* complex algebra *of* $\mathfrak{F}$ *(notation* $\mathfrak{F}^+$*) is the structure* $(2^U, \cap, {}^c, U, \{m_{R_i}\}_{i \in \Delta})$, *where*

- ^{c} *is the operation of taking the complement of a set relative to* U, *and* $\cap$ *that of taking the intersection of two sets.*
- $m_{R_i} : 2^U \to 2^U$, *defined as*

$$m_{R_i}(X) := \{x \in U : \textit{ for all } y \textit{ such that } (x, y) \in R_i, y \in X\}.$$

Thus, $\mathfrak{F}^+$ is the expansion of the power set algebra over U with operators m_{R_i} : $2^U \to 2^U$, $i \in \Delta$. It is not difficult to see that $\mathfrak{F}^+$ is a BAO.

Definition 16. *A* filter *of a Boolean algebra $\mathfrak{A} := (A, \cap, \sim, 1)$ is a subset $F \subseteq A$ satisfying the following.*

- *$1 \in F$.*
- *If $a, b \in F$, then $a \cap b \in F$.*
- *If $a \in F$ and $a \cap b = a$, then $b \in F$.*

A filter is proper *if it does not contain the smallest element 0. A proper filter is* prime *if $a \cup b \in F$ implies at least one of a and b belongs to F.*

Let $\mathfrak{A} := (A, \cap, \sim, 1, \{f_k\}_{k \in \Delta})$ be a BAO. The *prime filter frame* of $\mathfrak{A}$, denoted as $\mathfrak{A}_+$, is the structure $(\mathcal{F}(\mathfrak{A}), \{\mathfrak{R}_k\}_{k \in \Delta})$, where $\mathcal{F}(\mathfrak{A})$ is the set of all prime filters of the Boolean algebra $(A, \cap, \sim, 1)$ and $\mathfrak{R}_k \subseteq \mathcal{F}(\mathfrak{A}) \times \mathcal{F}(\mathfrak{A})$ such that $(F, G) \in \mathfrak{R}_k$ if and only if $a \in G$ for all $f_k a \in F$.

Theorem 1 (Jónsson-Tarski). *The function $r : A \to 2^{\mathcal{F}(\mathfrak{A})}$ defined by*

$$r(a) := \{F \in \mathcal{F}(\mathfrak{A}) : a \in F\}$$

is a BAO embedding of $\mathfrak{A}$ into the complex algebra $(\mathfrak{A}_+)^+$.

Note that the embedding given in Jónsson-Tarski theorem may not preserve infinite joins and meets. The theorem is improved in [98] and an embedding is given which preserves all infinite joins and meets in a countably infinite collection of subsets of A satisfying some conditions. For the purpose, a special prime filter called *Q-filter*, is used.

Let $\mathfrak{A} := (A, \cap, \sim, 1)$ be a Boolean algebra and $Q := \{Q_n \subseteq A : n \in \mathbb{N}\}$, where each Q_n is non-empty.

Definition 17 [91]. *A prime filter F of $\mathfrak{A}$ is called a Q-*filter, *if it satisfies the following for each $n \in \mathbb{N}$.*

1. *If $Q_n \subseteq F$ and $\bigcap Q_n$ exists then $\bigcap Q_n \in F$.*
2. *If $\bigcup Q_n$ exists and belongs to F then $Q_n \cap F \neq \emptyset$.*

The set of all Q–filters of $\mathfrak{A}$ is denoted by $\mathcal{F}_Q(\mathfrak{A})$. Note that $\mathcal{F}_Q(\mathfrak{A}) \subseteq \mathcal{F}(\mathfrak{A})$. Now consider the BAO $\mathfrak{A} := (A, \cap, \sim, 1, \{f_i\}_{i \in \Delta})$ and Q as above. Instead of the prime filter frame $(\mathcal{F}(\mathfrak{A}), \{\mathfrak{R}_k\}_{k \in \Delta})$, we look at the Q-filter frame $\mathfrak{A}_Q := (\mathcal{F}_Q(\mathfrak{A}), \{\mathfrak{R}^Q{}_i\}_{i \in \Delta})$, where $\mathfrak{R}_i^Q := \mathfrak{R}_i \cap (\mathcal{F}_Q(\mathfrak{A}) \times \mathcal{F}_Q(\mathfrak{A}))$. In other words, $\mathfrak{R}_i^Q \subseteq \mathcal{F}_Q(\mathfrak{A}) \times \mathcal{F}_Q(\mathfrak{A})$ such that $(F, G) \in \mathfrak{R}^Q{}_i$ if and only if $a \in G$ for all $f_i a \in F$.

Theorem 2 [98]. *Let $\{X_n\}_{n \in \mathbb{N}}$ and $\{Y_n\}_{n \in \mathbb{N}}$ be enumerations of the sets $Q_* := \{Q_m \in Q : \bigcap Q_m \in A\}$ and $Q^* := \{Q_m \in Q : \bigcup Q_m \in A\}$ respectively. Moreover, suppose that Q satisfies the following conditions for each $i \in \Delta$:*

(QF1) *for any n, $\bigcap f_i X_n$ exists and satisfies that $\bigcap f_i X_n = f_i \bigcap X_m$,*

(QF2) *for any* $z \in A$ *and* n*, there exists* m *such that* $\{f_i(z \to x) : x \in X_n\} = X_m$*, where* $z \to x :=\sim z \cup x$*,*
(QF3) *for any* $z \in A$ *and* n*, there exists* m *such that* $\{f_i(y \to z) : y \in Y_n\} = Y_m$*.*

Then the function $r' : A \to 2^{\mathcal{F}_Q(\mathfrak{A})}$ *defined by* $r'(a) := \{F \in \mathcal{F}_Q(\mathfrak{A}) : a \in F\}$ *is a BAO embedding of* $\mathfrak{A}$ *into the complex algebra* $(\mathfrak{A}_Q)^+$ *which also preserves all of* $\bigcap X_n$ *and* $\bigcup Y_n$*.*

We note that $r'(a) = r(a) \cap \mathcal{F}_Q(\mathfrak{A})$ for all $a \in A$, where r is the embedding of the Jónsson-Tarski theorem.

3 A Multiple Source Scenario: Multiple Source Approximation System with Distributed Knowledge Base

In this section, we focus on the situations where information is obtained from different agents about the same set of objects. Different agents may consider different sets of attributes to study the same set of objects, or they may assign different attribute values to the objects for the same attribute. As mentioned in Sect. 2, even though rough set theory has been studied in the multi-agent scenario, the issue of counterparts of standard rough set concepts such as approximations of sets, definability of sets is not addressed. Here, we make a study of these concepts. As mentioned in Sect. 1, we will use the terms 'agent' and 'source' synonymously.

Our study is based on *Multiple-source approximation systems with distributed knowledge base* (MSAS^{D}), formally defined as follows.

Definition 18. *A* multiple-source approximation system with distributed knowledge base (MSAS^{D}) *is a tuple* $\mathfrak{F} := (U, \{R_P\}_{P \subseteq N})$*, where* U *is a non-empty set,* N *an initial segment of the set* $\mathbb{N}$ *of positive integers and for each* $P \subseteq N$*,* R_P *is a binary relation on* U *satisfying the following:*

(M1) R_P *is an equivalence relation;*
(M2) $R_P = \bigcap_{i \in P} R_i$*, for each* $P \subseteq N$*.*

$|N|$ is referred to as the *cardinality of* $\mathfrak{F}$ and is denoted by $|\mathfrak{F}|$.

For $i \in N$, we shall write R_i instead of $R_{\{i\}}$. It denotes the knowledge base of ith source of the system. Moreover, for each $P \subseteq N$, R_P represents the strong distributed knowledge base of the group P of sources (cf. [92]). We will simply call it distributed knowledge base of the group P. Note that because of (M2), it suffices to know the 'atomic' relations R_i, $i \in N$, as they generate any R_P, $P \subseteq N$.

MSAS^{D}s are different from *dynamic spaces* considered by Pagliani in [82] (cf. Sect. 2.1). Moreover, MSAS^{D}s are also different from the information structures introduced by Orłowska [78], where we have condition (M1), but (M2) is replaced by (I1), (I2) (cf. Definition 13). Observe that (M2) implies (I1), (I2) and so every $MSAS^D$ is an information structure. In fact, if the relations are indexed over the power set of finite sets, then these two notions coincide. But in general, we can have an information structure which is not a $MSAS^D$.

We begin our study of rough set theory in the multi-agent setting with an investigation of the notion of approximations of sets in $MSAS^D$. Section 3.1 deals with this issue. Notions of strong/ weak lower and upper approximations are proposed and their properties are explored. Different notions of definability of sets coming out of these notions of strong/weak approximations are proposed in Sect. 3.2. It is observed that standard rough set notions of approximations and definability of sets are obtained as a special case of these proposals. In Sect. 3.3, we give some expressions to dependency [84] in this context, which reflect how much the information provided by a $MSAS^D$ depends on that of a source or a group of sources.

The content of this section is based on the articles [45,46,51].

3.1 Notions of Lower and Upper Approximations in $MSAS^D$

Let $\mathfrak{F} := (U, \{R_P\}_{P \subseteq N})$ be a $MSAS^D$, and $P, Q \subseteq N$, $X \subseteq U$. Corresponding to each group of sources in $\mathfrak{F}$, we have an approximation space representing how the group perceives the objects with their distributed knowledge base. Thus we obtain lower and upper approximations of sets corresponding to each group. Note that if $P \subseteq Q$, we have $R_Q \subseteq R_P$ and so $[x]_{R_Q} \subseteq [x]_{R_P}$. In particular, $[x]_{R_Q} \subseteq [x]_{R_i}$ for $i \in Q$. Using this fact, it is not difficult to obtain the proposition below, which shows how these lower and upper approximations are related.

Proposition 4.

1. $\underline{X}_{R_P} \subseteq \underline{X}_{R_Q}$, *if* $P \subseteq Q$.
2. $\overline{X}_{R_Q} \subseteq \overline{X}_{R_P}$, *if* $P \subseteq Q$.
3. $B_{R_Q}(X) \subseteq B_{R_P}(X)$ *for* $P \subseteq Q$, *where* $B_R(X)$ *represents the boundary of the set* X *in the approximation space* (U, R).
4. $\underline{\underline{X}_{R_Q}}_{R_P} \subseteq \underline{X}_{R_P}$.
5. $\underline{\underline{X}_{R_Q}}_{R_P} \subseteq \underline{X}_{R_Q}$.

Proof. We only prove 4. The rest can be done in the same way.

$$\begin{aligned} & x \in \underline{\underline{X}_{R_Q}}_{R_P} \\ & \Rightarrow [x]_{R_P} \subseteq \underline{X}_{R_Q} \subseteq X \\ & \Rightarrow x \in \underline{X}_{R_P}. \end{aligned}$$

□

From 1 and 2, it follows that if an object is a positive or negative element of a set with respect to the distributed knowledge base of a group P, then it remains so with respect to the distributed knowledge base of any other group which contains P. In other words, with the increase of the number of sources, the knowledge about the objects of the domain at least does not decrease. Note that equality in

4 does not hold in general. Moreover, we may not have any set inclusion between $\underline{\underline{X}_{R_Q}}_{R_P}$ and $\underline{\underline{X}_{R_P}}_{R_Q}$. But in a special case when $R_P \subseteq R_Q$, i.e. when the group P has 'finer' knowledge than the group Q, we obtain the following.

Proposition 5.

1. *$R_P \subseteq R_Q$, if and only if $\underline{\underline{X}_{R_Q}}_{R_P} = \underline{X}_{R_Q}$.*
2. *$R_P \subseteq R_Q$ implies $\underline{\underline{X}_{R_Q}}_{R_P} = \underline{\underline{X}_{R_P}}_{R_Q}$.*

Proof. 1. From the property of lower approximation, we have $\underline{\underline{X}_{R_Q}}_{R_P} \subseteq \underline{X}_{R_Q}$. So, we prove the reverse inclusion. Let $x \in \underline{X}_{R_Q}$. Let $y, z \in U$ such that $(x, y) \in R_P$ and $(y, z) \in R_Q$. We need to prove $z \in X$. Since $(x, y) \in R_P$ and $R_P \subseteq R_Q$, we also have $(x, y) \in R_Q$ and hence, using the transitivity of R_Q, we obtain $(x, z) \in R_Q$. This gives $z \in X$ as $x \in \underline{X}_{R_Q}$.
Conversely, let $(x, y) \in R_P$ and $(x, y) \notin R_Q$. Let $X := \{z \in U : (x, z) \in R_Q\}$. Then $x \in \underline{X}_{R_Q}$ but $x \notin \underline{\underline{X}_{R_Q}}_{R_P}$.

2. Since $\underline{X}_{R_P} \subseteq X$, we have $\underline{\underline{X}_{R_P}}_{R_Q} \subseteq \underline{X}_{R_Q} = \underline{\underline{X}_{R_Q}}_{R_P}$ (by 1).
So, it remains to show $\underline{\underline{X}_{R_Q}}_{R_P} \subseteq \underline{\underline{X}_{R_P}}_{R_Q}$. Let $x \in \underline{\underline{X}_{R_Q}}_{R_P}$. Then we obtain

$$[x]_{R_Q} \subseteq X. \tag{1}$$

Let $y, z \in U$ such that $(x, y) \in R_Q$ and $(y, z) \in R_P$. We need to show $z \in X$. Using the transitivity of R_Q and the fact that $R_P \subseteq R_Q$, we obtain $(x, z) \in R_Q$. Therefore, from (1), we obtain $z \in X$. □

Dually, one can obtain results similar to Propositions 4 and 5 for upper approximation. It follows from these propositions that for $i \in P$, $\underline{X}_{R_i} \subseteq \underline{X}_{R_P}$, $\overline{X}_{R_P} \subseteq \overline{X}_{R_i}$, $\underline{\underline{X}_{R_P}}_{R_i} = \underline{X}_{R_P}$ and $\underline{\underline{X}_{R_P}}_{R_i} = \underline{\underline{X}_{R_i}}_{R_P}$.

Strong, Weak Lower and Upper Approximations for *MSAS*^{D}. Let us consider the following example.

Example 3. Suppose we have information regarding the attribute set {transport facilities(Tra), law and order(LO), literacy(Li)} for the cities Calcutta(C), Delhi(D), Mumbai(M), Chennai(Ch), Bangalore(B) and Kanpur(K) from four different agencies M_1, M_2, M_3 and M_4 (Table 2):

Here g, a, p stand for *good*, *average* and *poor*. Let $U := \{C, M, D, Ch, B, K\}$. Each M_i, $1 \leq i \leq 4$, then gives rise to an equivalence relation R_i on U

$R_1 := \{\{C, B\}, \{M, D\}, \{Ch\}, \{K\}\}$;
$R_2 := \{\{C, D\}, \{M, Ch, B\}, \{K\}\}$;
$R_3 := \{\{C\}, \{M, D, B\}, \{Ch, K\}\}$;
$R_4 := \{\{C\}, \{M, B\}, \{D\}, \{Ch, K\}\}$.

We thus have a MSAS^D $(U, \{R_P\}_{P \subseteq N})$, $N = \{1, 2, 3, 4\})$, where $R_P := \bigcap_{i \in P} R_i$. Many questions may be raised now. For example, for a given $X (\subseteq U)$, we may ask the following.

Table 2. Information from different agencies about the cities

	M_1			M_2			M_3			M_4		
	Tra	*LO*	*Li*	*Tra*	*LO*	*Li*	*Tra*	*LO*	*Li*	*Tra*	*LO*	*Li*
C	*a*	*a*	*g*	*a*	*a*	*a*	*a*	*a*	*g*	*g*	*g*	*a*
M	*g*	*g*	*a*	*a*	*a*	*g*	*a*	*g*	*p*	*p*	*a*	*g*
D	*g*	*g*	*a*	*a*	*a*	*a*	*a*	*g*	*p*	*a*	*a*	*g*
Ch	*g*	*p*	*g*	*a*	*a*	*g*	*a*	*p*	*g*	*p*	*p*	*g*
B	*a*	*a*	*g*	*a*	*a*	*g*	*a*	*g*	*p*	*p*	*a*	*g*
K	*p*	*g*	*g*	*p*	*g*	*p*	*a*	*p*	*g*	*p*	*p*	*g*

(Q1) Which cities are considered to be a positive element of X by every agency?

(Q2) Take any particular city, say, Calcutta. Is it the case that Calcutta is a boundary element of X for every agency?

(Q3) Is it the case that if a city is not a boundary element of X for some agency, then it will also not be a boundary element of X with respect to the distributed knowledge base (R_N) of all the agencies?

(Q4) Is it the case that if a city is not a boundary element of X with respect to R_N, then there is some agency, for which it will also not be a boundary element of the set?

(Q5) Is there any agency which considers a city to be a positive element of X, but some other agency considers the same city to be a negative element of X?

(Q6) Is there a city which is a boundary element of X for each agency, but is not a boundary element of the set with respect to R_N?

(Q7) Let $P \subseteq N$, and take the collection S of cities that are considered to be positive elements of X by at least one agency of P. Now with respect to the distributed knowledge base R_P of P, will the cities of S also be positive elements of S itself? This question points to a kind of 'iteration' in knowledge, that will become clear in the sequel.

These kinds of questions motivate us to give the following definitions.
Let $\mathfrak{F} := (U, \{R_P\}_{P\subseteq N})$ be a MSAS^{D}, $X \subseteq U$ and P a non-empty subset of N.

Definition 19. *The* strong lower approximation $\underline{X}_{s(P)_{\mathfrak{F}}}$, weak lower approximation $\underline{X}_{w(P)_{\mathfrak{F}}}$, strong upper approximation $\overline{X}_{s(P)_{\mathfrak{F}}}$, *and* weak upper approximation $\overline{X}_{w(P)_{\mathfrak{F}}}$ *of X with respect to P, respectively, are defined as follows.*

$$\underline{X}_{s(P)_{\mathfrak{F}}} := \bigcap_{i\in P}\underline{X}_{R_i};\ \underline{X}_{w(P)_{\mathfrak{F}}} := \bigcup_{i\in P}\underline{X}_{R_i}.$$
$$\overline{X}_{s(P)_{\mathfrak{F}}} := \bigcap_{i\in P}\overline{X}_{R_i};\ \overline{X}_{w(P)_{\mathfrak{F}}} := \bigcup_{i\in P}\overline{X}_{R_i}.$$

If there is no confusion, we shall omit $\mathfrak{F}$ as the subscript in the above definition. Moreover, if $P = N$, then we simply write $\underline{X}_s$, $\underline{X}_w$, $\overline{X}_s$ and $\overline{X}_w$, instead of writing $\underline{X}_{s(N)}$, $\underline{X}_{w(N)}$, $\overline{X}_{s(N)}$ and $\overline{X}_{w(N)}$ and we simply call them strong/weak lower and upper approximations. We observe that $x \in \underline{X}_{s(P)}$, provided x is a positive element of X for every source in P. On the other hand, $x \in \underline{X}_{w(P)}$, provided x is a positive element of X for some source in P. Similarly, $x \in \overline{X}_{s(P)}$, if x is a possible element of X for every source in P, and $x \in \overline{X}_{w(P)}$, if x is a possible element of X for some source in P. The relationship between the defined sets is:

$$\underline{X}_{s(P)} \subseteq \underline{X}_{w(P)} \subseteq X \subseteq \overline{X}_{s(P)} \subseteq \overline{X}_{w(P)}. \qquad (*)$$

Moreover, like lower and upper approximations, we obtain:

$$\underline{\emptyset}_{w(P)} = \underline{\emptyset}_{s(P)} = \overline{\emptyset}_{s(P)} = \overline{\emptyset}_{w(P)} = \emptyset; \ \underline{U}_{w(P)} = \underline{U}_{s(P)} = \overline{U}_{s(P)} = \overline{U}_{w(P)} = U.$$

If $\mathfrak{F} := (U, \{R\})$ then there is only one source and $\underline{X}_s = \underline{X}_w = \underline{X}_R$, $\overline{X}_s = \overline{X}_w = \overline{X}_R$. So in the special case of a single approximation space, the weak/strong lower and upper approximations are just the standard lower and upper approximations respectively.

So, based on the information provided by a group P of sources of MSAS^{D} $\mathfrak{F}$, the domain U is divided into five disjoint sets (cf. Fig. 1), viz. $\underline{X}_{s(P)}$, $\underline{X}_{w(P)} \setminus \underline{X}_{s(P)}$, $\overline{X}_{s(P)} \setminus \underline{X}_{w(P)}$, $\overline{X}_{w(P)} \setminus \overline{X}_{s(P)}$, and $(\overline{X}_{w(P)})^c$. Moreover, the possibility of an element $x \in U$ to belong to X on the basis of information provided by the group P of sources of $\mathfrak{F}$, reduces as we go from $\underline{X}_{s(P)}$ to $(\overline{X}_{w(P)})^c$. If $x \in \underline{X}_{s(P)}$, then we are certain that x is an element of X. On the other hand, if $x \in (\overline{X}_{w(P)})^c$, then we are certain that x is not an element of X. Let us give names to the elements of the different regions.

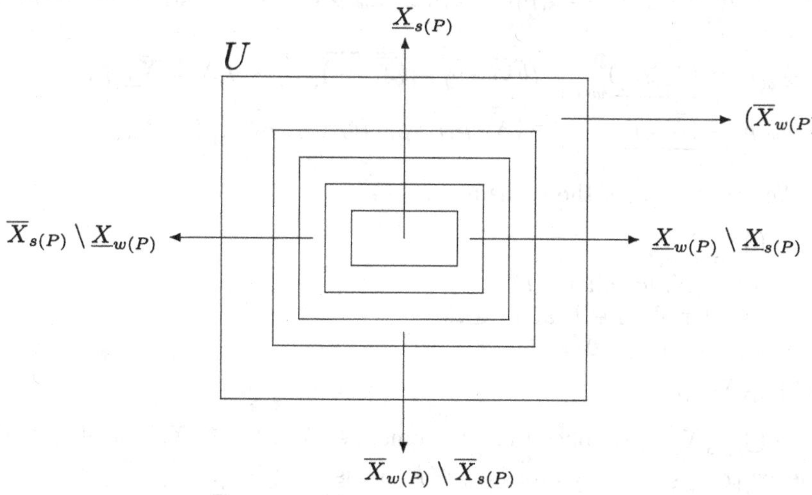

Fig. 1. A partition of U based on the information provided by group P of sources

Definition 20. *$x \in U$ is said to be a*

- certain positive element *of X for P, if $x \in \underline{X}_{s(P)}$,*
- possible positive element *of X for P, if $x \in \underline{X}_{w(P)} \setminus \underline{X}_{s(P)}$,*
- certain negative element *of X for P, if $x \in (\overline{X}_{w(P)})^c$,*
- possible negative element *of X for P, if $x \in \overline{X}_{w(P)} \setminus \overline{X}_{s(P)}$, and*
- certain boundary element *of X for P, if $x \in \overline{X}_{s(P)} \setminus \underline{X}_{w(P)}$.*

So if x is a certain positive element of X for P, then every source belonging to P considers it as a positive element of X. If x is a possible positive element, then there is no source in P which considers it a negative element of X; moreover, there must be at least one source in P which considers it as a positive element, and one for which x is boundary element of X. Similar is the case with negative elements. If x is a certain boundary element of X for P, it is clearly a boundary element of X for each source in P.

Consider a family of subsets $\{X_i\}_{i\in\Delta}$ of U, Δ being the index set.

Proposition 6.

1. *(a)* $\underline{\bigcap_{i\in\Delta} X_i}_{s(P)} = \bigcap_{i\in\Delta} \underline{X_i}_{s(P)}$; *(b)* $\underline{\bigcup_{i\in\Delta} X_i}_{s(P)} \supseteq \bigcup_{i\in\Delta} \underline{X_i}_{s(P)}$;
   (c) $\underline{\bigcap_{i\in\Delta} X_i}_{w(P)} \subseteq \bigcap_{i\in\Delta} \underline{X_i}_{w(P)}$; *(d)* $\underline{\bigcup_{i\in\Delta} X_i}_{w(P)} \supseteq \bigcup_{i\in\Delta} \underline{X_i}_{w(P)}$.
2. *(a)* $\overline{\bigcap_{i\in\Delta} X_i}_{w(P)} \subseteq \bigcap_{i\in\Delta} \overline{X}_{iw(P)}$; *(b)* $\overline{\bigcup_{i\in\Delta} X_i}_{w(P)} = \bigcup_{i\in\Delta} \overline{X}_{iw(P)}$.
   (c) $\overline{\bigcap_{i\in\Delta} X_i}_{s(P)} \subseteq \bigcap_{i\in\Delta} \overline{X}_{is(P)}$; *(d)* $\overline{\bigcup_{i\in\Delta} X_i}_{s(P)} \supseteq \bigcup_{i\in\Delta} \overline{X}_{is(P)}$.
3. *(a)* $\underline{X^c}_{s(P)} = (\overline{X}_{w(P)})^c$; *(b)* $\underline{X^c}_{w(P)} = (\overline{X}_{s(P)})^c$;
   (c) $\overline{X^c}_{s(P)} = (\underline{X}_{w(P)})^c$; *(d)* $\overline{X^c}_{w(P)} = (\underline{X}_{s(P)})^c$.
4. *If $X \subseteq Y$ then $\underline{X}_{s(P)} \subseteq \underline{Y}_{s(P)}$, $\underline{X}_{w(P)} \subseteq \underline{Y}_{w(P)}$, $\overline{X}_{s(P)} \subseteq \overline{Y}_{s(P)}$ and $\overline{X}_{w(P)} \subseteq \overline{Y}_{w(P)}$.*
5. *(a)* $\underline{X}_{w(P)} = \underline{(\underline{X}_{w(P)})}_{w(P)}$; *(b)* $\overline{X}_{s(P)} = \overline{(\overline{X}_{s(P)})}_{s(P)}$; *(c)* $X \subseteq \underline{\overline{X}_{w}}_{s(P)}$;
   (d) $\overline{X}_{w(P)} = \underline{(\overline{X}_{w(P)})}_{w(P)} = \overline{(\overline{X}_{s(P)})}_{w(P)}$; *(e)* $\overline{(\underline{X}_{s(P)})}_{w(P)} \subseteq \underline{X}_{w(P)}$.

Proof. We prove some of the items here.

1(a). $x \in \underline{\bigcap_{i\in\Delta} X_i}_{s(P)}$
$\Leftrightarrow [x]_{R_j} \subseteq \bigcap_{i\in\Delta} X_i$ for all $j \in P$
$\Leftrightarrow [x]_{R_j} \subseteq X_i$ for all $j \in P$ and $i \in \Delta$
$\Leftrightarrow x \in \underline{X_i}_{s(P)}$ for all $i \in \Delta$
$\Leftrightarrow x \in \bigcap_{i\in\Delta} \underline{X_i}_{s(P)}$.

1(b). $x \in \bigcup_{i\in\Delta} \underline{X_i}_{s(P)}$ implies that for some $i \in \Delta$, $[x]_{R_j} \subseteq X_i$ for all $j \in P$. This means $[x]_{R_j} \subseteq \bigcup_{i\in\Delta} X_i$, for all $j \in P$. Thus $x \in \underline{\bigcup_{i\in\Delta} X_i}_{s(P)}$.

3(a). $x \in \underline{X^c}_{s(P)}$
$\Leftrightarrow [x]_{R_i} \subseteq X^c$ for all $i \in P$
$\Leftrightarrow [x]_{R_i} \cap X = \emptyset$ for all $i \in P$

$\Leftrightarrow x \notin \overline{X}_{w(P)}$
$\Leftrightarrow x \in (\overline{X}_{w(P)})^c$.
Thus $\underline{X^c}_{s(P)} = (\overline{X}_{w(P)})^c$. The other part can be done similarly.

5(a). Since $\underline{X}_{w(P)} \subseteq X$, we have

$$\underline{(\underline{X}_{w(P)})}_{w(P)} \subseteq \underline{X}_{w(P)}. \tag{2}$$

Let $x \in \underline{X}_{w(P)}$, then

$$[x]_{R_i} \subseteq X \text{ for some } i \in P. \tag{3}$$

Now, $y \in [x]_{R_i}$ implies $[y]_{R_i} = [x]_{R_i} \subseteq X$ (by (3)) and hence $y \in \underline{X}_{w(P)}$. Thus $[x]_{R_i} \subseteq \underline{X}_{w(P)}$ and so $x \in \underline{(\underline{X}_{w(P)})}_{w(P)}$. This implies

$$\underline{X}_{w(P)} \subseteq \underline{(\underline{X}_{w(P)})}_{w(P)}. \tag{4}$$

From (2) and (4), we have $\underline{X}_{w(P)} = \underline{(\underline{X}_{w(P)})}_{w(P)}$.

5(d). First note that

$$\underline{(\overline{X}_{w(P)})}_{w(P)} \subseteq \overline{X}_{w(P)}. \tag{5}$$

Let $x \in \overline{X}_{w(P)}$. Then $[x]_{R_i} \cap X \neq \emptyset$ for some $i \in P$.
So there exists a z such that

$$z \in [x]_{R_i} \cap X. \tag{6}$$

Now, $y \in [x]_{R_i}$
$\Rightarrow z \in [x]_{R_i} \cap X = [y]_{R_i} \cap X$ (by (6))
$\Rightarrow [y]_{R_i} \cap X \neq \emptyset$
$\Rightarrow y \in \overline{X}_{w(P)}$.

So $[x]_{R_i} \subseteq \overline{X}_{w(P)}$. This implies that $x \in \underline{(\overline{X}_{w(P)})}_{w(P)}$. Thus we have shown

$$\overline{X}_{w(P)} \subseteq \underline{(\overline{X}_{w(P)})}_{w(P)}.$$

This together with (5) gives $\underline{(\overline{X}_{w(P)})}_{w(P)} = \overline{X}_{w(P)}$.

Next, we show $\overline{X}_{w(P)} = \overline{(\overline{X}_{s(P)})}_{w(P)}$. Obviously $\overline{X}_{w(P)} \subseteq \overline{(\overline{X}_{s(P)})}_{w(P)}$ (by 4).
So it remains to show the reverse inclusion. Now,
$x \in \overline{(\overline{X}_{s(P)})}_{w(P)}$
$\Rightarrow [x]_{R_i} \cap \overline{X}_{s(P)} \neq \emptyset$ for some $i \in P$
$\Rightarrow$ there exists a y such that $(x, y) \in R_i$ and $[y]_{R_j} \cap X \neq \emptyset$ for all $j \in P$
$\Rightarrow [y]_{R_i} \cap X \neq \emptyset$
$\Rightarrow [x]_{R_i} \cap X \neq \emptyset$ ($\because (x, y) \in R_i$)
$\Rightarrow x \in \underline{X}_{w(P)}$. □

We observe from Proposition 6 (3) that $\overline{X}_{w(P)}$ is the dual of $\underline{X}_{s(P)}$, while $\overline{X}_{s(P)}$ is the dual of $\underline{X}_{w(P)}$. Moreover, strong lower and weak upper approximations behave like Pawlak's lower and upper approximation with respect to set-theoretic intersection and union. But it is not the case with weak lower and strong upper approximations. In fact, the reverse inclusion in items 1(c) and 1(d) do not hold as shown in Example 4 below. Moreover, from item 5, it follows that, like Pawlak's lower/upper approximations, the weak lower and strong upper approximations are also idempotent. But the strong lower and weak upper approximations do not have this property.

Example 4. Recall the MSAS^{D} of Example 3. For $P := \{2, 3, 4\}$, different lower and upper approximations of some sets are given by the Table 3.

Table 3. Approximations of some sets

Z	$\underline{Z}_{s(P)}$	$\underline{Z}_{w(P)}$	$\overline{Z}_{s(P)}$	$\overline{Z}_{w(P)}$
$\{Ch, M, B\}$	$\emptyset$	$\{Ch, M, B\}$	$\{M, B, Ch\}$	$\{K, M, B, D, Ch\}$
$\{Ch, K\}$	$\{K\}$	$\{Ch, K\}$	$\{Ch, K\}$	$\{M, B, K, Ch\}$
$\{Ch\}$	$\emptyset$	$\emptyset$	$\emptyset$	$\emptyset$
$\{C, K, D\}$	$\{C\}$	$\{C, D, K\}$	$\{C, K, D\}$	$\{C, M, B, D, Ch, K\}$
$\{C, D, B, M\}$	$\{C, D\}$	$\{C, D, B, M, K\}$	$\{C, M, B, D\}$	$\{M, B, Ch, D, C\}$
$\{C, M, D, B, K\}$	$\{C, D\}$	$\{C, B, M, K, D\}$	$\{C, K, M, B, Ch, D\}$	$\{K, M, B, Ch, D, C\}$
$\{C, M, D\}$	$\{C\}$	$\{C, D\}$	$\{C, M, B, D\}$	$\{C, M, B, D, Ch\}$

From the table it follows that for $X_1 := \{Ch, M, B\}$ and $X_2 := \{Ch, K\}$, we have $\underline{X_1}_{w(P)} \cap \underline{X_2}_{w(P)} \not\subseteq \underline{X_1 \cap X_2}_{w(P)}$. Similarly, for $Y_1 := \{C, K, D\}$, $Y_2 := \{C, D, B, M\}$, we have $Ch \in \overline{Y_1 \cup Y_2}_{s(P)}$, but $Ch \notin \overline{Y_1}_{s(P)} \cup \overline{Y_2}_{s(P)}$.
Let $X := \{C, M, D\}$. Then observe that C is a certain positive element of X for the group P. Although D is a positive element of X for both the second and third sources, it is only a possible positive element of X and not a certain positive element for P. Similarly, K and Ch are respectively certain and possible negative elements of X for P. Moreover, M and B are both certain boundary elements of X for the group P.

The next proposition lists some more properties of strong, weak lower and upper approximations.

Proposition 7.

1. $\underline{X}_{s(P \cup Q)} \subseteq \underline{X}_{s(P)} \cup \underline{X}_{s(Q)}$ *(reverse inclusion may not hold).*
2. $\underline{X}_{w(P \cup Q)} = \underline{X}_{w(P)} \cup \underline{X}_{w(Q)}$.
3. $\underline{\underline{X}_{s(P)}}_{R_i} \subseteq \underline{X}_{R_i}$ *(reverse inclusion may not hold).*
4. $\underline{\underline{X}_{w(P)}}_{R_j} \subseteq \underline{X}_{R_j}$ *and* $\underline{X}_{R_i} \subseteq \underline{\underline{X}_{w(P)}}_{R_i}$, *when* $i \in P$.

5. $\underline{\underline{X}_{R_j}}_{w(P)} \subseteq \underline{X}_{R_j}$ *and* $\underline{X}_{R_i} \subseteq \underline{\underline{X}_{R_i}}_{w(P)}$, *when* $i \in P$.
   (Note that $\underline{X}_{R_i} \subseteq \underline{\underline{X}_{R_i}}_{s(P)}$, $i \in P$ *does not hold).*
6. $\underline{\underline{X}_{s(P)}}_{R_i} \subseteq \underline{X}_{s(P)}$ *(reverse inclusion does not hold even if* $i \in P$*).*
7. $\underline{X}_{s(P)} \subseteq \underline{\underline{X}_{w(P)}}_{R_i}$, $i \in P$.

Proof. We only prove 4.
$x \in \underline{X}_{R_i}$
$\Rightarrow [y]_{R_i} \subseteq X$ for all $y \in [x]_{R_i}$
$\Rightarrow y \in \underline{X}_{w(P)}$ for all $y \in [x]_{R_i}$ as $i \in P$
$\Rightarrow [x]_{R_i} \subseteq \underline{X}_{w(P)}$
$\Rightarrow x \in \underline{\underline{X}_{w(P)}}_{R_i}$.
Thus $\underline{X}_{R_i} \subseteq \underline{\underline{X}_{w(P)}}_{R_i}$. Since, $\underline{X}_{w(P)} \subseteq X$, we obtain $\underline{\underline{X}_{w(P)}}_{R_i} \subseteq \underline{X}_{R_i}$. Thus $\underline{\underline{X}_{w(P)}}_{R_i} \subseteq \underline{X}_{R_i}$. □

From Item 1 of Proposition 7, it follows that if x is a certain positive element of X for a group P, then it remains so for every subset of the group. Items 3–7 give us results about 'iterations' of knowledge.

Example 5. Let us recall the MSAS^{D} of Example 4. Note that for $P := \{2,3,4\}$ and $X := \{Ch\}$, we have $Ch \in \underline{X}_{R_P}$, but $\underline{X}_{w(P)} = \emptyset$. This shows that the lower approximation of a set with respect to distributed knowledge base R_P of a group of sources may be different from strong/weak lower approximation of the set. The next proposition shows how the strong/weak lower and upper approximations corresponding to a group P of sources are related with the lower and upper approximations with respect to their distributed knowledge base R_P.

Proposition 8.

1. $\underline{X}_{w(P)} \subseteq \underline{X}_{R_P} \subseteq \overline{X}_{R_P} \subseteq \overline{X}_{s(P)}$.
2. $\underline{X}_{w(P)} \setminus \underline{X}_{s(P)} \subseteq \underline{X}_{R_P}$.
3. $\overline{X}_{R_P} \setminus \underline{X}_{R_P} \subseteq \overline{X}_{s(P)} \setminus \underline{X}_{w(P)}$.
4. $\underline{X}_{s(P)} = \underline{\underline{X}_{s(P)}}_{R_P}$.
5. $\underline{X}_{w(P)} = \underline{\underline{X}_{w(P)}}_{R_P}$ *(but* $\underline{X}_{w(P)} \subseteq \underline{\underline{X}_{s(P)}}_{R_P}$ *does not hold).*
6. $\underline{\underline{X}_{w(P)}}_{R_P} \subseteq \underline{X}_{R(P)}$ *(reverse inclusion does not hold).*

Proof. We only prove 1 and 4.
1. It is enough to show $\underline{X}_{w(P)} \subseteq \underline{X}_{R_P}$ and $\overline{X}_{R_P} \subseteq \overline{X}_{s(P)}$.
Now, $x \in \underline{X}_{w(P)}$

$\Rightarrow [x]_{R_i} \subseteq X$ for some $i \in P$
$\Rightarrow [x]_{R_P} \subseteq X$, $(\because [x]_{R_P} \subseteq [x]_{R_i}$ for all $i \in P)$
$\Rightarrow x \in \underline{X}_{R_P}$.

Again, $x \in \overline{X}_{R_P}$

$\Rightarrow [x]_{R_P} \cap X \neq \emptyset$
$\Rightarrow [x]_{R_i} \cap X \neq \emptyset$ for all $i \in P$ ($\because [x]_{R_P} \subseteq [x]_{R_i}$ for all $i \in P$)
$\Rightarrow x \in \overline{X}_{s(P)}$.

4. $x \in \underline{X}_{s(P)}$
$\Rightarrow [x]_{R_i} \subseteq X$ for all $i \in P$
$\Rightarrow$ for all $y \in [x]_{R_P}$, $[y]_{R_j} \subseteq X$, for all $j \in P$ ($\because [x]_{R_j} = [y]_{R_j}$ for all $j \in P$)
$\Rightarrow$ for all $y \in [x]_{R_P}$, $y \in \underline{X}_{s(P)}$
$\Rightarrow [x]_{R_P} \subseteq \underline{X}_{s(P)}$
$\Rightarrow x \in \underline{\underline{X}_{s(P)}}_{R_P}$.
We also have $\underline{\underline{X}_{s(P)}}_{R_P} \subseteq \underline{X}_{s(P)}$ from the property of lower approximation. Thus we obtain $\underline{\underline{X}_{s(P)}}_{R_P} = \underline{X}_{s(P)}$. □

From 2, it follows that if x is a possible positive element of a set X for a group P, then x is also a positive element of X with respect to the distributed knowledge base R_P. 3 says that if an object is a boundary element of a set with respect to R_P, then it will also be a certain boundary element of the set for P.

Let us return to the Example 3. (Q1) and (Q2) can be rephrased respectively as:
What are the elements of the set $\underline{X}_s$?
Does $C \in \overline{X}_s \setminus \underline{X}_w$?
(Q3)–(Q7) reduce to checking respectively the following set inclusions:

(i) $\overline{X}_{R_N} \setminus \underline{X}_{R_N} \subseteq \overline{X}_s \setminus \underline{X}_w$,
(ii) $\overline{X}_s \setminus \underline{X}_w \subseteq \overline{X}_{R_N} \setminus \underline{X}_{R_N}$,
(iii) $\underline{X}_w \not\subseteq \overline{X}_w \cap (\overline{X}_w \setminus \overline{X}_s)^c$,
(iv) $\overline{X}_s \setminus \underline{X}_w \not\subseteq \overline{X}_{R_N} \setminus \underline{X}_{R_N}$ and
(v) $\underline{X}_{w(P)} \subseteq \underline{\underline{X}_{w(P)}}_{R_P}$.

Note that for $X := \{C, M, D\}$, $\underline{X}_s = \emptyset$ and $\overline{X}_s \setminus \underline{X}_w = \{B\}$ and hence in this case answer to Q2 is no. From Proposition 8 (1), it follows that we always have the inclusion of (i), whatever MSAS^{D} and set is considered, and thus answer to Q3 is yes. The reverse of inclusion in (i), i.e. the inclusion in (ii) may not always hold. For instance, for the MSAS^{D} of Example 3, if $Y := \{B\}$, then $B \in \overline{Y}_s \setminus \underline{Y}_w$, but $B \notin \overline{Y}_{R_N} \setminus \underline{Y}_{R_N}$. Thus answer to Q4 is no. Moreover, if $X := \{B\}$, then answer to Q6 is yes. Since we always have $\underline{Y}_w \subseteq \overline{Y}s$, the answer to Q5 is no. From Proposition 8 (5), it follows that the answer to Q7 is yes.

***MSAS^{D}* and Tolerance Approximation Spaces.** Recall the notion of tolerance approximation space (cf. Sect. 2.1). In this section, we give the relationship between the strong lower, weak upper approximations and the lower, upper approximations for tolerance approximation spaces.

Let us consider a MSAS^{D} $\mathfrak{F} := (U, \{R_P\}_{P \subseteq N})$ and the relation $R_{\mathfrak{F}} := \bigcup_{i \in N} R_i$. $R_{\mathfrak{F}}$ is a tolerance relation and so a tolerance approximation space $(U, R_{\mathfrak{F}})$ is obtained. On the other hand, we have

Proposition 9. *Let (U, R) be a tolerance approximation space with finite U. Then there exists a* MSAS^D $\mathfrak{F}$ *such that* $R_{\mathfrak{F}} = R$.

Proof. For each $x, y \in U$ such that $(x, y) \in R$, consider the set $A_{xy} = \{(x, y)\} \subseteq U \times U$. Let $A_1, A_2, \ldots, A_n$ be an enumeration of all such A_{xy}.
Define equivalence relations $R_1, R_2, \ldots, R_n$ on U such that

$$U/R_i = \{\{x, y\} : (x, y) \in A_i\} \cup \{\{z\} : z \neq x, z \neq y\}.$$

Obviously, $R = R_1 \cup R_2 \cup \ldots \cup R_n$. Let $N = \{1, 2, \ldots, n\}$ and consider the MSAS^D $\mathfrak{F} := (U, \{R_P\}_{P \subseteq N})$, where $R_P := \bigcap_{i \in N} R_i$. As already observed $R_{\mathfrak{F}} = R$. □

Observation 1. In the tolerance approximation space $(U, R_{\mathfrak{F}})$, we have for any $X \subseteq U$, the standard definitions:

$$\underline{X}_{R_{\mathfrak{F}}} := \{x : R_{\mathfrak{F}}(x) \subseteq X\}; \quad \overline{X}_{R_{\mathfrak{F}}} := \{x : R_{\mathfrak{F}}(x) \cap X \neq \emptyset\},$$

where $R_{\mathfrak{F}}(x) = \{y : (x, y) \in R_{\mathfrak{F}}\}$. It is not difficult to show that

$$\underline{X}_{R_{\mathfrak{F}}} = \underline{X}_s; \quad \overline{X}_{R_{\mathfrak{F}}} = \overline{X}_w.$$

From Proposition 9 and Observation 1, it is clear that the strong lower and weak upper approximations corresponding to MSAS^D with finite domains behave like the lower and upper approximations corresponding to tolerance approximation spaces with finite domain. In fact, in Sect. 4.5, we will see that this restriction of 'finite domain' can be removed.

3.2 Different Notions of Definability

In this section, we move to the issue of definability of sets in the case of MSAS^D. In a MSAS^D, it could happen that a set is definable with respect to information provided by one source but not with respect to information provided by another source. So we need to consider different notions of definability. Thus we have the following definitions.
Let $\mathfrak{F} := (U, \{R_P\}_{P \subseteq N})$, and $X \subseteq U$. Let P be a non-empty subset of N.

Definition 21. *X is said to be*

- P-lower definable *in* $\mathfrak{F}$, *if* $\underline{X}_{w(P)} = \underline{X}_{s(P)}$,
- P-upper definable *in* $\mathfrak{F}$, *if* $\overline{X}_{w(P)} = \overline{X}_{s(P)}$,
- P-strong definable *in* $\mathfrak{F}$, *if* $\overline{X}_{w(P)} = \underline{X}_{s(P)}$, i.e. every element of U is either certain positive or certain negative for P,
- P-weak definable *in* $\mathfrak{F}$, *if* $\overline{X}_{s(P)} = \underline{X}_{w(P)}$, i.e. X does not have any certain boundary element for P.

Thus, X is P-lower (upper) definable if and only if each source in N agrees on the positive (negative) elements of X. More formally, we have the following proposition.

Proposition 10.

1. *X is P-lower definable in $\mathfrak{F}$, if and only if $\underline{X}_{R_i} = \underline{X}_{R_j}$, for each $i, j \in P$, i.e. the sets of positive elements in all the approximation spaces (U, R_i) of $\mathfrak{F}$, $i \in P$, are identical.*
2. *X is P-upper definable in $\mathfrak{F}$, if and only if the sets of negative elements in all the approximation spaces (U, R_i), $i \in P$ of $\mathfrak{F}$ are identical.*
3. *X is both P-lower and P-upper definable in $\mathfrak{F}$, if and only if the sets of boundary elements in all the approximation spaces (U, R_i), $i \in P$ of $\mathfrak{F}$ are the same.*

Proof. Let us first prove Item 1. Let X be P-lower definable, i.e. $\underline{X}_{w(P)} = \underline{X}_{s(P)}$.

Therefore, $x \in \underline{X}_{R_j}$
$\Leftrightarrow [x]_{R_j} \subseteq X$
$\Leftrightarrow x \in \underline{X}_{w(P)} = \underline{X}_{s(P)}$
$\Leftrightarrow [x]_{R_i} \subseteq X$
$\Leftrightarrow x \in \underline{X}_{R_i}$.

Conversely, suppose $\underline{X}_{R_i} = \underline{X}_{R_j}$ for all $i, j \in N$. Then,

$$\underline{X}_{w(P)} = \bigcup_{i \in N} \underline{X}_{R_i} = \underline{X}_{R_i} = \bigcup_{i \in N} \underline{X}_{R_i} = \underline{X}_{s(P)}.$$

This completes the proof of Item 1. Similarly, one can prove Item 2. Moreover, Item 3 follows from Items 1 and 2. □

Example 6. Consider the MSAS^{D} of Example 4. Observe that the set $\{Ch, K\}$ is Q-strong definable, where $Q := \{1, 3, 4\}$. Moreover, for $P := \{2, 3, 4\}$, the set $U \setminus \{Ch\}$ is P-upper definable, but it is none of weak, strong or lower P-definable. In fact, it is even not definable in any of the approximation spaces corresponding to sources 2, 3 and 4. Similarly, $\{Ch, M, B\}$ is P-weak definable but it is none of strong, upper or lower P-definable. Here is the exact relationship among these notions of definability.

Proposition 11. *The following are equivalent.*
1. *X is P-strong definable.*
2. *X is P-weak, lower and upper definable.*

The following proposition lists a few more properties of lower and upper definable sets.

Proposition 12.

1. *X is P-upper definable, if and only if X^c is P-lower definable.*
2. *If X and Y are P-upper definable then so are $\overline{X}_s, \underline{\overline{X^c}}_w, \underline{\overline{X^c}}_s, \overline{X}_s \cup \overline{Y}_s, \overline{X}_s \cap \overline{Y}_s$.*
3. *If X and Y are P-lower definable then so are $\underline{X}_w, \overline{X^c}_s, \overline{X^c}_w, \underline{X}_w \cap \underline{Y}_w, \underline{X}_w \cup \underline{Y}_w$.*
4. *An arbitrary union (intersection) of P-upper (P-lower) definable sets is also P-upper (P-lower) definable.*

Proof. 1. Follows from Proposition 6(3).
2. $x \in \overline{(\overline{X}_{s(P)})}_{w(P)}$
$\Rightarrow [x]_i \cap \overline{X}_{s(P)} \neq \emptyset$ for some i
$\Rightarrow$ there exists a $y \in [x]_i \cap \overline{X}_{s(P)}$
$\Rightarrow [y]_i \cap X \neq \emptyset$
$\Rightarrow [x]_i \cap X \neq \emptyset$
$\Rightarrow x \in \overline{X}_{w(P)}$
$\Rightarrow x \in \overline{X}_{s(P)}$ ($\because \overline{X}_{s(P)} = \overline{X}_{w(P)}$).
Therefore, $\overline{(\overline{X}_{s(P)})}_{w(P)} \subseteq \overline{X}_{s(P)}$.
Since $\overline{(\overline{X}_{s(P)})}_{s(P)} = \overline{X}_{s(P)}$, we have $\overline{(\overline{X}_{s(P)})}_{w(P)} \subseteq \overline{(\overline{X}_{s(P)})}_{s(P)}$.
Since $\overline{(\overline{X}_{s(P)})}_{s(P)} \subseteq \overline{(\overline{X}_{s(P)})}_{w(P)}$, we obtain $\overline{(\overline{X}_{s(P)})}_{w(P)} = \overline{(\overline{X}_{s(P)})}_{s(P)}$.
To show that $\underline{X^c}_{w(P)}$ is upper definable, first we prove that if X is upper definable, then $\overline{(\underline{X^c}_{w(P)})}_{s(P)} = \underline{X^c}_{w(P)}$. Then it remains to show $\underline{X^c}_{w(P)} = \overline{(\underline{X^c}_{w(P)})}_{w(P)}$.
Obviously $\underline{X^c}_{w(P)} \subseteq \overline{(\underline{X^c}_{w(P)})}_{s(P)}$, so we only prove the reverse.

$x \in \overline{(\underline{X^c}_{w(P)})}_{s(P)}$
$\Rightarrow [x]_{R_i} \cap \underline{X^c}_{w(P)}$ for all $i \in N$
$\Rightarrow$ for all i there exists $y_i \in [x]_{R_i} \cap \underline{X^c}_{w(P)} = [x]_{R_i} \cap \underline{X^c}_{s(P)}$
($\because \underline{X^c}_{w(P)} = \underline{X^c}_{s(P)}$, by (1))
$\Rightarrow [y_i]_{R_j} \subseteq X^c$ for all j
$\Rightarrow [x]_{R_i} = [y_i]_{R_i} \subseteq X^c$
$\Rightarrow x \in \underline{X^c}_{w(P)}$

Next, we prove $\underline{X^c}_{w(P)} = \overline{(\underline{X^c}_{w(P)})}_{w(P)}$. Obviously $\underline{X^c}_{w(P)} \subseteq \overline{(\underline{X^c}_{w(P)})}_{w(P)}$.
Now,

$x \in \overline{(\underline{X^c}_{w(P)})}_{w(P)}$
$\Rightarrow [x]_{R_i} \cap \underline{X^c}_{w(P)} \neq \emptyset$ for some i
$\Rightarrow$ there exists $y \in [x]_{R_i} \cap \underline{X^c}_{w(P)}$
$\Rightarrow [y]_{R_k} \subseteq X^c$ for some k
$\Rightarrow [y]_{R_k} \cap X = \emptyset$
$\Rightarrow y \notin \overline{X}_{s(P)} = \overline{X}_{w(P)}$
$\Rightarrow [y]_{R_l} \cap X = \emptyset$ for all l
$\Rightarrow [y]_{R_i} \cap X = \emptyset$
$\Rightarrow [x]_{R_i} \cap X = \emptyset$
$\Rightarrow [x]_{R_i} \subseteq X^c$
$\Rightarrow x \in \underline{X^c}_{w(P)}$

Next, we will show that $\overline{X}_{s(P)} \cup \overline{Y}_{s(P)}$ is upper definable.
Since $\overline{(\overline{X}_{s(P)} \cup \overline{Y}_{s(P)})}_{s(P)} \subseteq \overline{(\overline{X}_{s(P)} \cup \overline{Y}_{s(P)})}_{w(P)}$, we only need to prove the reverse inclusion.

Here $x \in \overline{(\overline{X}_{s(P)} \cup \overline{Y}_{s(P)})}_{w(P)}$
$\Rightarrow [x]_{R_i} \cap (\overline{X}_{s(P)} \cup \overline{Y}_{s(P)}) \neq \emptyset$ for some $i \in N$
$\Rightarrow [x]_{R_i} \cap \overline{X}_{s(P)} \neq \emptyset$ or, $[x]_{R_i} \cap \overline{Y}_{s(P)} \neq \emptyset$
$\Rightarrow$ there exists $z \in [x]_{R_i}$ such that either $[z]_{R_j} \cap X \neq \emptyset$ for all j, or, $[z]_{R_j} \cap Y \neq \emptyset$ for all j
$\Rightarrow$ there exists $z \in [x]_{R_i}$ such that either $[z]_{R_i} \cap X \neq \emptyset$ or, $[z]_{R_i} \cap Y \neq \emptyset$
$\Rightarrow [x]_{R_i} \cap X \neq \emptyset$ or, $[x]_{R_i} \cap Y \neq \emptyset$
$\Rightarrow x \in \overline{X}_{w(P)} = \overline{X}_{s(P)}$, or, $x \in \overline{Y}_{w(P)} = \overline{Y}_{s(P)}$.
$\Rightarrow x \in \overline{X}_{s(P)} \cup \overline{Y}_{s(P)}$
$\Rightarrow x \in \overline{(\overline{X}_{s(P)} \cup \overline{Y}_{s(P)})}_{s(P)}$

Thus we have $\overline{(\overline{X}_{s(P)} \cup \overline{Y}_{s(P)})}_{w(P)} \subseteq \overline{(\overline{X}_{s(P)} \cup \overline{Y}_{s(P)})}_{s(P)}$.
In a similar way one can show that $\underline{X^c}_{s(P)}, \overline{X}_{s(P)} \cap \overline{Y}_{s(P)}$ are upper definable.
3. Can be proved dually.
4. Let $\{X_i\}_{i \in \Delta}$ be a family of P-lower definable sets. So $\underline{X_i}_{s(P)} = \underline{X_i}_{w(P)}$, for all $i \in \Delta$. We need to prove $\underline{\bigcap_{i \in \Delta} X_i}_{s(P)} = \underline{\bigcap_{i \in \Delta} X_i}_{w(P)}$.
Obviously, $\underline{\bigcap_{i \in \Delta} X_i}_{s(P)} \subseteq \underline{\bigcap_{i \in \Delta} X_i}_{w(P)}$. So we need to prove the reverse inclusion. Here,

$$\begin{aligned} \underline{\bigcap_{i \in \Delta} X_i}_{w(P)} &\subseteq \bigcap_{i \in \Delta} \underline{X_i}_{w(P)} && \text{(by Proposition 6(4))} \\ &= \bigcap_{i \in \Delta} \underline{X_i}_{s(P)} && (\because X_i \text{ are P-lower definable}) \\ &= \underline{\bigcap_{i \in \Delta} X_i}_{s(P)} && \text{(by Proposition 6).} \end{aligned}$$

One can similarly prove that an arbitrary union of P-upper definable set is also P-upper definable. □

We observe that the collection of upper (lower) definable sets is not closed under intersection (union) – cf. Example 7 below.

Example 7. Let us consider a MSAS^D $\mathfrak{F} := (U, \{R_P\}_{P \subseteq \{1,2\}})$, where $U := \{a, b, c, d\}$, $U/R_1 := \{\{a,c\},\{b\},\{d\}\}$ and $U/R_2 := \{\{a,b\},\{c,d\}\}$. The subsets $Y_1 := \{a\}$, $Y_2 := \{c\}$ of U are lower definable, but their union, i.e. the set $Y_1 \cup Y_2 = \{a,c\}$, is not lower definable. Similarly, the subsets $Z_1 := \{a,b,d\}$, $Z_2 := \{b,c,d\}$ are upper definable, but the set $Z_1 \cap Z_2 = \{b,d\}$ is not upper definable.

For strong definable sets, we have the following.

Proposition 13.

1. *$\emptyset, U$ are both P-strong definable.*
2. *If $X(\subseteq U)$ is P-strong definable then X^c is also P-strong definable.*
3. *An arbitrary union and intersection of P-strong definable subsets of U are also P-strong definable.*

Proof. 1. Trivial.
2. Follows from Proposition 6(3).
3. Let $\{X_i : i \in \Delta\}$ be an arbitrary family of P-strong definable sets.
Let $Y := \bigcup_{i\in\Delta} X_i$ and $Z := \bigcap_{i\in\Delta} X_i$. We have to show $\underline{Y}_{s(P)} = \overline{Y}_{w(P)}$ and $\underline{Z}_{s(P)} = \overline{Z}_{w(P)}$.
Obviously, $\underline{Y}_{s(P)} \subseteq \overline{Y}_{w(P)}$ and $\underline{Z}_{s(P)} \subseteq \overline{Z}_{w(P)}$. Now, using Proposition 6(2), and the fact that each X_i is strong definable, we obtain

$$\overline{Y}_{w(P)} = \bigcup_{i\in\Delta} \overline{X_i}_{w(P)} = \bigcup_{i\in\Delta} \underline{X_i}_{s(P)} \subseteq \underline{\bigcup_{i\in\Delta} X_i}_{s(P)} = \underline{Y}_{s(P)},$$

$$\overline{Z}_{w(P)} \subseteq \bigcap_{i\in\Delta} \overline{X_i}_{w(P)} = \bigcap_{i\in\Delta} \underline{X_i}_{s(P)} = \underline{\bigcap_{i\in\Delta} X_i}_{s(P)} = \underline{Z}_{s(P)}.$$

□

From this Proposition, it may be concluded that the collection of all P-strong definable sets forms a complete field of sets [89]. We do not have such nice properties for P-weak definable sets. It is clear that, if X is definable in an approximation space (U, R_i), $i \in P$, of $\mathfrak{F}$, it is P-weak definable in $\mathfrak{F}$. But the union or intersection of two P-definable sets may not be P-definable, as shown in Example 8.

Example 8. Consider the MSAS^D of Example 7. The subsets $Y_1 := \{a, b\}$ and $Y_2 := \{a, c\}$ of U are weak definable, but the set $Y_1 \cap Y_2 = \{a\}$ is not weak definable. Similarly, the subsets $Z_1 := \{b\}$, $Z_2 := \{c, d\}$ are weak definable, but the set $Z_1 \cup Z_2 = \{b, c, d\}$ is not weak definable.

3.3 Dependency of Information

Dependency of knowledge bases represented by equivalence relations is a very important aspect of rough set theory. Given two equivalence relations R, Q over the same domain U, representing two knowledge bases, we say Q depends on R if and only if $R \subseteq Q$, i.e. $[x]_R \subseteq [x]_Q$ for all $x \in U$. Usually a knowledge base does not depend wholly but partially on other knowledge bases. A measure of this is given [84] by the expression $\frac{|POS_R(Q)|}{|U|}$, assuming U is finite, where $POS_R(Q) = \bigcup_{X\in U/Q} \underline{X}_R$. In this case, we say Q depends in a degree $\frac{|POS_R(Q)|}{|U|} = k$, $(0 \leq k \leq 1)$ on R, and denote it as $R \Rightarrow_k Q$. Note that this measure behaves like a function: given any two equivalence relations Q, R over the same finite domain U, we always have a $k \in [0, 1]$ such that $R \Rightarrow_k Q$.

Let us closely observe the set $POS_R(Q)$. It is not difficult to show that this set is actually the set $\{x \in U : [x]_R \subseteq [x]_Q\}$. In fact, $POS_R(Q)$ consists of precisely all those objects x such that: *if Q considers x to be a positive or negative element of a set, then R considers it as a positive or negative element of the set accordingly; in other words, if R is not able to decide whether x is an element of a set X or not (i.e. x is a boundary element of X with respect to R), then Q is also not able to do so.*

In this section, our aim is to come up with some dependency functions which determine how much the information provided by a MSAS^{D} depends on the information provided by an individual source or a group of sources. Motivated by the set $POS_R(Q)$ as described above, we find that the following sets emerge. Let $\mathfrak{F} := (U, \{R_P\}_{P \subseteq N})$ and $P \subseteq N$.

Definition 22.

$A_1(P) := \{x \in U : [x]_{R_P} \subseteq [x]_{R_N}\}$.
$A_2(P) := \{x \in U : [x]_{R_{N \setminus P}} \not\subseteq [x]_{R_N}\}$.
$A_3(P) := \{x \in U : [x]_{R_Q} \subseteq [x]_{R_P}$ *for some* $Q \subseteq N \setminus P\}$.
$A_4(P) := \{x \in U :$ *for all* $i \in P$ *there exists* $j \in N \setminus P$ *such that* $[x]_{R_j} \subseteq [x]_{R_i}\}$.
$A_5(P) := \{x \in U : [x]_{R_{N \setminus P}} \subseteq [x]_{R_N}\}$.
$A_6(P) := \{x \in U :$ *for all* $j \in N \setminus P$ *there exists* $i \in P$ *such that* $[x]_{R_i} \subseteq [x]_{R_j}\}$.
$A_7(P) := \{x \in U :$ *for all* $Q \subseteq N \setminus P$ *there exists* $i \in P$ *such that* $[x]_{R_i} \subseteq [x]_{R_Q}\}$.

Now, we note that $A_1(P)$ is actually the set $POS_{R_P}(R_N)$ and so it consists of precisely those objects x such that if x is a boundary element of a set X with respect to the distributed knowledge base of the group P, then x will also be the boundary element of X with respect to the distributed knowledge base of all the sources, i.e. for all $X \subseteq U$, $x \notin \overline{X}_{R_N} \setminus \underline{X}_{R_N}$ implies $x \notin \overline{X}_{R_P} \setminus \underline{X}_{R_P}$.

Similarly, $A_2(P)$ consists of precisely those objects x for which there exists a set X such that x is a boundary element of X with respect to the distributed knowledge base of the group $N \setminus P$, but x is not a boundary element of X with respect to the distributed knowledge base of all the sources. In other words, $A_2(P)$ collects all those objects x such that if we dismiss the sources belonging to P from the system, then we lose some information about these objects – information whether these are positive or negative elements of some set.

$A_3(P)$ consists of precisely those objects x such that if x is not a boundary element of some set X with respect to R_P, then it is also not a boundary element of X with respect to the distributed knowledge base of some group consisting of elements which are not in P – in other words, we have *replacement of the group* P in the system regarding the information about these objects in the sense that if the distributed knowledge base of P can determine that x is positive or negative element of a set, then we have some other group not involving the sources of P which can also do so. We get the following proposition giving the interpretation of all the sets defined above.

Proposition 14.

1. *$x \in A_1(P)$ if and only if for all $X \subseteq U$, $x \in \overline{X}_{R_P} \setminus \underline{X}_{R_P}$ implies $x \in \overline{X}_{R_N} \setminus \underline{X}_{R_N}$.*
2. *$x \in A_2(P)$ if and only if there exists a $X \subseteq U$ such that, $x \in (\overline{X}_{R_{N \setminus P}} \setminus \underline{X}_{R_{N \setminus P}}) \cap (\overline{X}_{R_N} \setminus \underline{X}_{R_N})^c$.*

3. $x \in A_3(P)$ *if and only if for all* $X \subseteq U$, $x \notin \overline{X}_{R_P} \setminus \underline{X}_{R_P}$ *implies* $x \notin \overline{X}_{R_Q} \setminus \underline{X}_{R_Q}$ *for some* $Q \subseteq N \setminus P$.
4. $x \in A_4(P)$ *if and only if for all* $i \in P$ *and for all* $X \subseteq U$, $x \notin \overline{X}_{R_i} \setminus \underline{X}_{R_i}$ *implies* $x \notin \overline{X}_{R_j} \setminus \underline{X}_{R_j}$ *for some* $j \in N \setminus P$.
5. $x \in A_5(P)$ *if and only if for all* $X \subseteq U$, $x \notin \overline{X}_{R_N} \setminus \underline{X}_{R_N}$ *implies* $x \notin \overline{X}_{R_{N\setminus P}} \setminus \underline{X}_{R_{N\setminus P}}$.
6. $x \in A_6(P)$ *if and only if for all* $j \in N \setminus P$ *and for all* $X \subseteq U$, $x \notin \overline{X}_{R_j} \setminus \underline{X}_{R_j}$ *implies* $x \notin \overline{X}_{R_i} \setminus \underline{X}_{R_i}$ *for some* $i \in P$.
7. $x \in A_7(P)$ *if and only if for all* $Q \subseteq N \setminus P$ *and for all* $X \subseteq U$, $x \notin \overline{X}_{R_j} \setminus \underline{X}_{R_j}$ *implies* $x \notin \overline{X}_{R_i} \setminus \underline{X}_{R_i}$ *for some* $i \in P$.

Proof. We only prove 1. Let $x \in A_1(S)$ and $X \subseteq U$ such that $x \in \overline{X}_{R_S} \setminus \underline{X}_{R_S}$. We have to show $x \in \overline{X}_{R_N} \setminus \underline{X}_{R_N}$. Since $[x]_{R_S} \subseteq [x]_{R_N}$, $[x]_{R_S} \cap X \neq \emptyset$ gives $[x]_{R_N} \cap X \neq \emptyset$ and $[x]_{R_S} \nsubseteq X$ gives $[x]_{R_N} \nsubseteq X$. Thus $x \in \overline{X}_{R_S} \setminus \underline{X}_{R_S}$ gives $x \in \overline{X}_{R_N} \setminus \underline{X}_{R_N}$.

Conversely, for all $X \subseteq U$, let $x \in \overline{X}_{R_S} \setminus \underline{X}_{R_S}$ imply $x \in \overline{X}_{R_N} \setminus \underline{X}_{R_N}$. We have to show $x \in A_1(S)$, i.e. $[x]_{R_S} \subseteq [x]_{R_N}$. If possible, let it not be the case. Then there exists y such that $(x, y) \in R_S$, but $(x, y) \notin R_N$. Let $X = [x]_{R_N}$. Then $x \in \overline{X}_{R_S} \setminus \underline{X}_{R_S}$ but $x \notin \overline{X}_{R_N} \setminus \underline{X}_{R_N}$, a contradiction. Thus $x \in A_1(S)$. □

The proposition given below expresses some of the sets given in Definition 22 in terms of weak lower approximations.

Proposition 15.

1. $A_4(P) = \bigcap_{i \in P}(\bigcup_{X \in U/R_i} \underline{X}_{w(N\setminus P)})$.
2. $A_6(P) = \bigcap_{i \in N\setminus P}(\bigcup_{X \in U/R_i} \underline{X}_{w(P)})$.
3. $A_7(P) = \bigcap_{Q \subseteq N\setminus P}(\bigcup_{X \in U/R_Q} \underline{X}_{w(P)})$.

Proof. We only prove 1.

$x \in A_4(P)$
$\Leftrightarrow$ for all $i \in P$, there exists $j \in N \setminus P$ such that $[x]_{R_j} \subseteq [x]_{R_i}$
$\Leftrightarrow$ for all $i \in P$, $x \in \underline{[x]_{R_i}}_{w(N\setminus P)}$
$\Leftrightarrow$ for all $i \in P$, $x \in \underline{X}_{w(N\setminus P)}$ for some $X \in U/R_i$
$\Leftrightarrow$ for all $i \in P$, $x \in \bigcup_{X \in U/R_i} \underline{X}_{w(N\setminus P)}$
$\Leftrightarrow x \in \bigcap_{i \in P}(\bigcup_{X \in U/R_i} \underline{X}_{w(N\setminus P)})$. □

Now, just as in standard rough set theory, we define some dependency functions $D : 2^N \rightarrow [0, 1]$ for the MSAS^D $\mathfrak{F} := (U, \{R_P\}_{P \subseteq N})$, using the sets given in Definition 22. Note that $D(P) = k$ will signify that information provided by the MSAS^D depends in a degree k on the information of the group P.

Definition 23. *Let* $\mathfrak{F} := (U, \{R_P\}_{P \subseteq N})$ *be a* MSAS^{D} *with finite* U*. For each* $i \in \{1, 2, \ldots, 7\}$*, we define a dependency function* $D_i : 2^N \to [0, 1]$ *as follows:*

$$D_i(P) := \frac{|A_i(P)|}{|U|} \text{ for } i \in \{1, 2, 6, 7\}, \text{ and}$$

$$D_i(P) := 1 - \frac{|A_i(P)|}{|U|} \text{ for } i \in \{3, 4, 5\}.$$

Example 9. Let us consider the MSAS^{D} $\mathfrak{F}$ of Example 3. Let $P := \{M_1, M_2\}$ and $Q := \{M_3, M_4\}$. Then we have,

$D_1(P) = 1$ and $D_1(Q) = \frac{1}{3}$; $D_2(P) = \frac{2}{3}$ and $D_2(Q) = 0$;
$D_3(P) = \frac{2}{3}$ and $D_2(Q) = 0$; $D_4(P) = \frac{5}{6}$ and $D_4(Q) = \frac{5}{6}$;
$D_5(P) = \frac{2}{3}$ and $D_5(Q) = 0$; $D_6(P) = \frac{1}{2}$ and $D_2(Q) = \frac{1}{6}$;
$D_7(P) = \frac{1}{3}$ and $D_7(Q) = \frac{1}{6}$.

We note the following facts about this example.

(i) According to every dependency function except D_4, information provided by $\mathfrak{F}$ depends on P more than Q.
(ii) According to the dependency function D_4, information provided by $\mathfrak{F}$ depends equally on P and Q. Moreover, $A_4(P) = \{K\}$ and $A_4(Q) = \{C\}$. Therefore, every city except K (C) is such that it is a positive or negative element of some city with respect to the knowledge base of some source belonging to P (Q), but it is a boundary element of the set for every source not belonging to P (Q).
(iii) D_1: the information provided by $\mathfrak{F}$ totally depends on P. So if any object is a positive or negative element of a set with respect to the distributed knowledge base of all sources, then it is also so with respect to the distributed knowledge base of the group P.
(iv) D_6: the information provided by $\mathfrak{F}$ is neither totally dependent nor independent of P. So there are objects (in fact exactly three) such that if any of these is a positive or negative element of some set with respect to the knowledge base of some source not belonging to P, then it is also not a boundary element of the set with respect to the knowledge base of some source belonging to P.
(v) D_5: the information provided by $\mathfrak{F}$ is totally independent of Q. So if any object is a positive or negative element of a set with respect to the distributed knowledge base of all the sources, then it remains so with respect to the distributed knowledge base of the group $N \setminus Q$.

We end this section with the following proposition which lists some of the properties of dependency functions.

Proposition 16.

1. *If* $Q \subseteq P$*, then* $D_i(Q) \leq D_i(P)$*,* $i \in \{1, 2, \ldots, 7\}$.

2. *If* $R_P \subseteq R_Q$*, then* $D_i(P) \geq D_i(Q)$*,* $i \in \{1,3\}$*.*
3. *If* $R_{N\setminus P} \subseteq R_{N\setminus Q}$*, then* $D_i(P) \leq D_i(Q)$*,* $i \in \{2,5\}$*.*
4. $D_i(P \cup Q) \geq D_i(Q)$*,* $i \in \{1,2,\ldots,7\}$*.*
5. *If* $D_7(P) = 1$*, then* $D_6(P) = D_1(P) = 1$*.*
6. *If* $D_6(P) = 1$*, then* $D_1(P) = 1$*.*
7. *If* $D_1(P) = 0$*, then* $D_7(P) = D_6(P) = 0$*.*
8. *If* $D_6(P) = 0$*, then* $D_7(P) = 0$*.*
9. $D_1(P) = 1$ *if and only if* $D_2(N \setminus P) = 0$*.*
10. $D_1(P) = 0$ *if and only if* $D_2(N \setminus P) = 1$*.*

Proof. 1. Note that for $Q \subseteq P$, we have $A_i(Q) \subseteq A_i(P)$ for $i \in \{1,2,6,7\}$ and $A_i(P) \subseteq A_i(Q)$ for $i \in \{3,4,5\}$. This gives $D_i(Q) \leq D_i(P)$ for $i \in \{1,2,\ldots,7\}$.

2 follows from the fact that when $R_P \subseteq R_Q$, then $A_3(P) \subseteq A_3(Q)$ and $A_1(Q) \subseteq A_1(P)$.

3 follows from the fact that when $R_{N\setminus P} \subseteq R_{N\setminus Q}$, then $A_2(P) \subseteq A_2(Q)$ and $A_5(Q) \subseteq A_5(P)$.

4 follows from (1).

5–8 follow from the fact that $A_7(P) \subseteq A_6(P) \subseteq A_1(P)$.

9 follows from the fact that $A_1(P) = U$ if and only if $A_2(N \setminus P) = \emptyset$.

10 follows from the fact that$A_1(P) = \emptyset$ if and only if $A_2(N \setminus P) = U$. □

4 A Logic for MSAS^{D}

In this section, we present a quantified propositional modal logic LMSAS^{D} to facilitate formal reasoning with rough sets in MSAS^{D}s. Sections 4.1 and 4.2 present the syntax and semantics of LMSAS^{D} respectively. In Sect. 4.3, we will consider the semantics under *standard interpretations*, which is a structure of a particular interest to us. In Sect. 4.4, we shall see how the language of LMSAS^{D} may be used to express the properties of rough sets in the multiple-source situation. Some fragments of the logic will be explored in Sect. 4.5 and in the process, we will also obtain the relationship of LMSAS^{D} with the modal system KTB and epistemic logic $S5_n^D$ with distributed knowledge.

The content of this section is based on the articles [45,52].

4.1 Syntax

The alphabet of the language $\mathfrak{L}$ of the logic LMSAS^{D} of MSAS^{D} contains (i) a non-empty countable set Var of variables, (ii) a (possibly empty) countable set $Con := \{c_i : i \in \mathbb{N}\}$ of constants, (iii) a non-empty countable set PV of propositional variables, (iv) a function symbol $*$ and (v) the propositional constants $\top, \bot$.

The set T of *terms* of the language is defined recursively as:

$$x \mid c \mid t_1 * t_2,$$

where $x \in Var$, $c \in Con$ and $t_1, t_2 \in T$. Using the standard Boolean logical connectives $\neg$ (negation) and $\wedge$ (conjunction), the global modal connective A, unary modal connectives $[t]$ (necessity) for each term $t \in T$, and the universal quantifier $\forall$, wffs of LMSAS^{D} are defined recursively as:

$$\top \mid \bot \mid p \mid \neg\alpha \mid \alpha \wedge \beta \mid A\alpha \mid [t]\alpha \mid \forall x\alpha,$$

where $p \in PV$, $t \in T$, $x \in Var$, and α, β are wffs. Let $\mathcal{F}$ and $\overline{\mathcal{F}}$ denote the set of all wffs and closed wffs of LMSAS^{D} respectively.

In what follows, we use the standard basic definitions and notations related to first-order logic (cf. e.g. [40, 72]). For instance, $\alpha(t/x)$ is the wff obtained from α by replacing every free occurrence of x with the term t. Henceforth, for a set Γ of wffs of LMSAS^{D}, $Con(\Gamma), Var(\Gamma), FV(\Gamma)$ and $T(\Gamma)$ will denote the set of constants, variables, free variables and terms respectively, used in the wffs of Γ. Similarly, for a given term t, $B(t)$ will denote the set of all variables and constants used in the term t.

4.2 Semantics

We require the following definition to give the semantics. Recall that MSAS^{D} is the tuple $\mathfrak{F} := (U, \{R_P\}_{P \subseteq N})$ (cf. Definition 18), where we are considering the distributed knowledge base of all possible groups of agents. However, in the rest of this section, we restrict our attention to finite groups of agents. Let $X \subseteq_f Y$ denote that X is a *finite* subset of Y.

Definition 24. *A* MSAS^{D}-structure *is a tuple* $\mathfrak{F} := (U, \{R_P\}_{P \subseteq_f N})$, *where* U, N, R_P *are as in Definition 18.*

Note that, in the definition of MSAS^{D}-structure, one can replace the condition (M2) with the conditions

(I1) $R_\emptyset = U \times U$ and
(I2) $R_{P \cup Q} = R_P \cap R_Q$.

Definition 25. *A tuple* $\mathfrak{M} := (\mathfrak{F}, V, I)$ *is called an interpretation, where* $\mathfrak{F} := (U, \{R_P\}_{P \subseteq_f N})$, R_P *is a binary relation for each* $P \subseteq_f N$, $V : PV \rightarrow 2^U$ *and* $I : Con \rightarrow N$. $\mathfrak{F}$ *is called a* frame. *An interpretation* $\mathfrak{M} := (\mathfrak{F}, V, I)$ *is a* MSAS^{D}*-interpretation if* $\mathfrak{F}$ *is a* MSAS^{D}*-structure.*

An assignment for an interpretation $\mathfrak{M}$ *is a map* $v : Var \rightarrow N$. v *is extended to a map* $\tilde{v} : T \rightarrow 2^N$ *as follows:*

- $\tilde{v}(x) = \{v(x)\}$, $x \in Var$;
- $\tilde{v}(c) = \{I(c)\}$, $c \in Con$; *and*
- $\tilde{v}(t * t') = \tilde{v}(t) \cup \tilde{v}(t')$, $t, t' \in T$.

Thus each term represents a finite set of sources. Variables and constants represent the singleton set and hence point to individual sources. The function symbol $*$ represents the union of two sets. Note that for t, t' such that $B(t) = B(t')$, we obtain $\tilde{v}(t) = \tilde{v}(t')$. In fact, we have for any term s, $\tilde{v}(s) = \bigcup_{r \in B(s)} \tilde{v}(r)$.

Let $\mathfrak{M}$ be an interpretation. As in classical first-order logic, we have, for a variable x, the notion of *x-equivalence* of two assignments v, v' for $\mathfrak{M}$. We now proceed to define *satisfiability* in an interpretation $\mathfrak{M} := (\mathfrak{F}, V, I)$ of a wff α, under an assignment v, at an object w of the domain U, denoted $\mathfrak{M}, v, w \models \alpha$. We omit the cases of Boolean wffs.

Definition 26.

- $\mathfrak{M}, v, w \models A\alpha$, *if and only if for all* w' *in* U *such that* $wR_{\emptyset}w'$, $\mathfrak{M}, v, w' \models \alpha$.
- $\mathfrak{M}, v, w \models [t]\alpha$, *if and only if for all* w' *in* U *such that* $wR_{\tilde{v}(t)}w'$, $\mathfrak{M}, v, w' \models \alpha$.
- $\mathfrak{M}, v, w \models \forall x\alpha$, *if and only if for every assignment* v' *x-equivalent to* v, $\mathfrak{M}, v', w \models \alpha$.

Given an interpretation $\mathfrak{M}$ and an assignment v of it, we use $[\![\alpha]\!]_{\mathfrak{M},v}$, where $\alpha \in \mathcal{F}$, to denote the set $\{w : \mathfrak{M}, v, w \models \alpha\}$. The set $\bigcap_v [\![\alpha]\!]_{\mathfrak{M},v}$ is denoted by $[\![\alpha]\!]_{\mathfrak{M}}$.

α is said to be *valid in* $\mathfrak{M}$, denoted as $\mathfrak{M} \models \alpha$, if and only if $[\![\alpha]\!]_{\mathfrak{M}} = W$. Since we are interested in MSAS^{D}-interpretations, we shall simply say α *is valid*, if α is valid in all MSAS^{D}-interpretations and write $\models_M \alpha$. α is satisfiable in an interpretation $\mathfrak{M} := (\mathfrak{F}, V, I)$ if $\mathfrak{M}, v, w \models \alpha$ for some v and some w. α is satisfiable, if it is satisfiable in some MSAS^{D}-interpretation.

Note that for all terms t_1, t_2, t_3 and all wffs α, the wffs $[t_1 * t_2]\alpha \leftrightarrow [t_2 * t_1]\alpha$ and $[t_1 * (t_2 * t_3)]\alpha \leftrightarrow [(t_1 * t_2) * t_3]\alpha$ are valid in all interpretations. Therefore, henceforth, we shall avoid using brackets while writing terms.

Remark 1. From Definition 26, it is clear that relation $R_{\tilde{v}(t)}$ is used to evaluate the modal operator $[t]$. Further, $\tilde{v}(t)$ is always a finite subset of N. Thus, it follows that in the evaluation of a LMSAS^{D} wff in an interpretation based on a MSAS^{D} $\mathfrak{F} := (U, \{R_P\}_{P \subseteq N})$, the relation R_P for infinite P does not play any role. Moreover, given a MSAS^{D}-structure $\mathfrak{F} := (U, \{R_P\}_{P \subseteq_f N})$, one can obtain a MSAS^{D} $\mathfrak{F}' := (U, \{R'_P\}_{P \subseteq N})$ such that $R'_P = R_P$ for finite P – take $R'_P := \bigcap_{p \in P} R_{\{p\}}$. From this fact it follows that a LMSAS^{D} wff is valid in the class of all interpretations based on MSAS^{D}s if and only if it is valid in the class of all interpretations based on MSAS^{D}-structures.

Let us weaken the condition (I2) defining information structures (cf. Definition 13) and consider the following.

Definition 27. $\mathfrak{F} := (U, \{R_P\}_{P \subseteq_f N})$ *is called a* semi MSAS^{D}*-structure, if it satisfies M1 and the weaker condition* $R_{P \cup Q} \subseteq R_P \cap R_Q$, $P, Q \subseteq_f N$.

We note that the notion of semi MSAS^{D}-structure is a special case of that of a decreasing frame and frame for logic $\mathcal{L}2$ defined in [1] and [2] respectively. In Sect. 5.1, we shall see that a wff of LMSAS^{D} is satisfiable in the class of interpretations based on MSAS^{D}-structures if and only if it is satisfiable in the class of interpretations based on semi MSAS^{D}-structures. Thus in order to prove the completeness of an axiomatic system with respect to the class of interpretations based on MSAS^{D}-structures, it is enough to prove the same with respect to the class of interpretations based on semi MSAS^{D}-structures.

4.3 Standard Interpretation

The following structure is of particular interest to us. Let $\Gamma \subseteq \mathcal{F}$ with finite $Con(\Gamma)$.

Definition 28. *A* MSAS^{D}*-interpretation* $\mathfrak{M}^{\Gamma} := (\mathfrak{F}, V, I_{\Gamma})$ *is a* Γ−standard interpretation, *when* $|\mathfrak{F}| \geq k$, *k being the largest integer such that* $c_k \in Con(\Gamma)$ *and* $I_{\Gamma} : Con \rightarrow N$ *such that* $I_{\Gamma}(c_i) = i$, *for each* $c_i \in Con(\Gamma)$.

We also require to introduce the following special sets of standard interpretations.

Definition 29. *Let* $\Gamma \subseteq \mathcal{F}$ *with finite* $Con(\Gamma)$, *and k the largest integer such that* $c_k \in Con(\Gamma)$. *Consider any integer* $m \geq k$. *We define*

- $\mathcal{C}_{\Gamma}^{m} := \{\mathfrak{M}^{\Gamma} = (\mathfrak{F}, V, I_{\Gamma}) : \mathfrak{F}$ *is a* MSAS^{D}*-structure with* $|\mathfrak{F}| = m\}$;
- $\mathcal{C}_{\Gamma}^{\mathbb{N}} := \{\mathfrak{M}^{\Gamma} = (\mathfrak{F}, V, I_{\Gamma}) : \mathfrak{F}$ *is a* MSAS^{D}*-structure with* $|\mathfrak{F}| = |\mathbb{N}|\}$; *and*
- $\mathcal{C}_{\Gamma} := \mathcal{C}_{\Gamma}^{\mathbb{N}} \cup \bigcup_{m \geq k} \mathcal{C}_{\Gamma}^{m}$. *If* $\Gamma := \{\alpha\}$, *we simply write* $\mathcal{C}_{\alpha}$.

It will be seen that validity of any wff $\alpha \in \mathcal{F}$ with respect to all MSAS^{D}-interpretations is effectively the same as that with respect to only interpretations in $\mathcal{C}_{\alpha}$.

Theorem 3. $\models_M \alpha$, *if and only if* α *is valid in all interpretations in* $\mathcal{C}_{\alpha}$.

From Remark 1, and Theorem 3, it follows that a LMSAS^{D} wff α is valid in the class of all interpretations based on MSAS^{D}s if and only if α is valid in all interpretations in $\mathcal{C}_{\alpha}$.

Before we establish Theorem 3, we observe in the following proposition conditions under which two interpretations, assignment pairs satisfy the same set of wffs.

Proposition 17. *Let* $\alpha \in \mathcal{F}$ *and* $\mathfrak{M}_i := (\mathfrak{F}_i, V_i, I_i)$ *where* $\mathfrak{F}_i := (W, \{R_P^i\}_{P \subseteq_f N_i})$, $i \in \{1, 2\}$, *be two* MSAS^{D}*-interpretations. Let* v_1 *and* v_2 *be assignments for the interpretations* $\mathfrak{M}_1$ *and* $\mathfrak{M}_2$ *respectively, such that the following hold:*

(a) $R^1_{\bar{v}_1(t)} = R^2_{\bar{v}_2(t)}$, *for all* $t \in Con(\{\alpha\}) \cup FV(\{\alpha\})$.
(b) $\{R_i^1\}_{i \in N_1} = \{R_i^2\}_{i \in N_2}$.
(c) $V_1(p) = V_2(p)$ *for all propositional variables p occurring in* α.

Then $\mathfrak{M}_1, v_1, w \models \alpha$ *if and only if* $\mathfrak{M}_2, v_2, w \models \alpha$, *for all* $w \in U$.

The proof is by induction on the complexity of α.
As a consequence, we obtain the following result easily. It will be used in Sect. 6.

Corollary 1. *Consider* $\alpha \in \mathcal{F}$, *two* MSAS^{D}*-interpretations* $\mathfrak{M}_1 := (\mathfrak{F}, V_1, I_1)$, $\mathfrak{M}_2 := (\mathfrak{F}, V_2, I_2)$ *and assignments* v_1, v_2 *such that (i)* $I_1(c) = I_2(c)$ *for all* $c \in Con(\{\alpha\})$, *(ii)* $v_1(x) = v_2(x)$ *for all* $x \in FV(\{\alpha\})$ *and (iii)* $V_1(p) = V_2(p)$ *for all propositional variables in* α. *Then* $\mathfrak{M}_1, v_1, w \models \alpha$ *if and only if* $\mathfrak{M}_2, v_2, w \models \alpha$, *for all* $w \in U$.

The following leads us to Theorem 3.

Proposition 18. *Let $\alpha \in \mathcal{F}$ be such that*
(i) $Con(\{\alpha\}) := \{c_{i_1}, c_{i_2}, \ldots, c_{i_n}\}$, $i_1 < i_2 < \cdots < i_n$,
(ii) $FV(\{\alpha\}) := \{x_{j_1}, x_{j_2}, \ldots, x_{j_r}\}$, $j_1 < j_2 < \cdots < j_r$.
If α is satisfiable in a MSAS^{D}-interpretation, then α is also satisfiable in an interpretation in $\mathcal{C}_\alpha$ with the same domain, under all assignments v_α, where $v_\alpha(x_{j_k}) = i_n + k, k \in \{1, 2, \ldots, r\}$.

Proof. Let α be satisfiable in $\mathfrak{M} := (\mathfrak{F}, V, I)$, $\mathfrak{F} := (U, \{R_P\}_{P \subseteq_f N})$, under the assignment v. Let us consider any α–standard interpretation $\mathfrak{M}^\alpha := (\mathfrak{F}', V, I_\alpha)$, where $\mathfrak{F}' := (U, \{R'_P\}_{P \subseteq_f M})$, that satisfies the following:

$\{R_i\}_{i \in N} = \{R'_i\}_{i \in M}$;
$R'_{i_j} = R_{\tilde{v}(c_{i_j})}$, $j = 1, 2, \ldots, n$; and
$R'_{i_n + k} = R_{\tilde{v}(x_{j_k})}$, $k = 1, 2, \ldots, r$.

Note that in order to fulfil the last two conditions, $\mathfrak{F}'$ is required to have relations at the positions i_j, $j = 1, 2, \ldots, n$ and $i_n + k$, $k = 1, 2, \ldots, r$. Moreover, due to the first condition, we want at least one occurrence of the relations from the set $R := \{R_i\}_{i \in N} \setminus \{R_{\tilde{v}(t)},\ t \in \{c_{i_j} : j = 1, 2, \ldots, n\} \cup \{x_{j_k} : k = 1, 2, \ldots, r\}\}$ (if any) in $\mathfrak{F}'$. Thus, we must have $|\mathfrak{F}'| \geq i_n + r + |R|$. In fact, for each cardinality m, $i_n + r + |R| \leq m \leq |\mathbb{N}|$, we can obtain such a $\mathfrak{F}'$ with $|\mathfrak{F}'| = m$.

From Proposition 17, it follows that α is satisfiable in $\mathfrak{M}^\alpha$ under all assignments v_α. □

4.4 Interpretation in Terms of Rough Sets

Given a MSAS^{D}-interpretation $\mathfrak{M} := (\mathfrak{F}, V, I)$, $\mathfrak{F} := (U, \{R_P\}_{P \subseteq_f N})$, and assignment v of $\mathfrak{M}$, recall that $\llbracket \alpha \rrbracket_{\mathfrak{M},v} := \{w \in U : \mathfrak{M}, v, w \models \alpha\}$. The following illustrates how the quantifiers and modalities are used to express lower/upper approximations.

Proposition 19.

1. $\llbracket \langle t \rangle \alpha \rrbracket_{\mathfrak{M},v} = \overline{(\llbracket \alpha \rrbracket_{\mathfrak{M},v})}_{R_{\tilde{v}(t)}}$; $\llbracket [t] \alpha \rrbracket_{\mathfrak{M},v} = \underline{(\llbracket \alpha \rrbracket_{\mathfrak{M},v})}_{R_{\tilde{v}(t)}}$.
2. $\llbracket [c_{i_1} * c_{i_2} * \cdots * c_{i_m}] \alpha \rrbracket_{\mathfrak{M},v} = \underline{(\llbracket \alpha \rrbracket_{\mathfrak{M},v})}_{R_P}$, *where* $P := \{\tilde{v}(c_{i_j}) : j = 1, 2, \ldots, m\}$.

For α which does not have a free occurrence of x,

3. $\llbracket \forall x [x] \alpha \rrbracket_{\mathfrak{M},v} = \underline{(\llbracket \alpha \rrbracket_{\mathfrak{M},v})}_s$; $\llbracket \exists x [x] \alpha \rrbracket_{\mathfrak{M},v} = \underline{(\llbracket \alpha \rrbracket_{\mathfrak{M},v})}_w$.
4. $\llbracket \forall x \langle x \rangle \alpha \rrbracket_{\mathfrak{M},v} = \overline{(\llbracket \alpha \rrbracket_{\mathfrak{M},v})}_s$; $\llbracket \exists x \langle x \rangle \alpha \rrbracket_{\mathfrak{M},v} = \overline{(\llbracket \alpha \rrbracket_{\mathfrak{M},v})}_w$.

From 2, it follows that in standard interpretations, $[c_{i_1} * c_{i_2} * \cdots * c_{i_m}]$ corresponds to the lower approximation with respect to the distributed knowledge base of the group of sources $\{i_1, i_2, \ldots, i_m\}$. This also shows that LMSAS^{D} can express the approximation of sets with respect to the distributed knowledge base of any finite group of sources. Note that 3 and 4 may not hold if we remove the restriction on α, as shown by Example 10.

Example 10. Let us consider the MSAS^{D}-interpretation $\mathfrak{M} := (\mathfrak{F}, V, I), \mathfrak{F} := (W, \{R_P\}_{P \subseteq \{1,2\}})$, where $W := \{w_1, w_2, w_3\}$, $W/R_1 := \{\{w_1, w_2\}, \{w_3\}\}$, $W/R_2 := \{\{w_1\}, \{w_2, w_3\}\}$, $V(p) := \{w_1, w_2\}$. Let v be an assignment such that $v(x) = 2$. Then we have $w_1 \in [\![\forall x[x][x]p]\!]_{\mathfrak{M},v}$. But $w_1 \notin \underline{([\![[x]p]\!]_{\mathfrak{M},v})}_{R_1}$ and hence $w_1 \notin \underline{([\![\alpha]\!]_{\mathfrak{M},v})}_{s}$.

The constant symbols in the language of LMSAS^{D} are used to express properties with respect to particular approximation spaces in MSAS^{D}s. For instance, given a MSAS^{D} $\mathfrak{F} := (U, \{R_P\}_{P \subseteq \{1,2,3\}})$, suppose one has the query: is $\underline{X}_{R_{\{1,3\}}} \subseteq \underline{X}_s$? This is equivalent to checking whether $[\![\alpha]\!]_{\mathfrak{M}^\alpha} = U$, where α is $[c_1 * c_3]p \rightarrow \forall x[x]p$, $\mathfrak{M}^\alpha := (\mathfrak{F}, V, I_\alpha)$ and $V(p) = X$. The assignment does not play any role in the satisfiability of α as it is a closed wff.

The properties given in Proposition 6 turn into valid wffs of LMSAS^{D}:

Proposition 20. *The following are valid in all* MSAS^{D}*-interpretations.*

1. (a) $\forall x[x](\alpha \wedge \beta) \leftrightarrow \forall x[x]\alpha \wedge \forall x[x]\beta$; (b) $\exists x\langle x\rangle(\alpha \vee \beta) \leftrightarrow \exists x\langle x\rangle\alpha \vee \exists x\langle x\rangle\beta$.
2. (a) $\exists x\langle x\rangle(\alpha \wedge \beta) \rightarrow \exists x\langle x\rangle\alpha \wedge \exists x\langle x\rangle\beta$; (b) $\exists x[x](\alpha \wedge \beta) \rightarrow \exists x[x]\alpha \wedge \exists x[x]\beta$.
3. (a) $\forall x[x]\neg\alpha \leftrightarrow \neg\exists x\langle x\rangle\alpha$; (b) $\exists x[x]\neg\alpha \leftrightarrow \neg\forall x\langle x\rangle\alpha$.
4. (a) $\exists x[x]\alpha \leftrightarrow \exists x[x]\exists y[y]\alpha$; (b) $\exists x\langle x\rangle\forall y[y]\alpha \rightarrow \exists x[x]\alpha$.

The wffs of LMSAS^{D} can express many properties related to MSAS^{D} other than those related to strong/weak lower and upper approximations, as we see below.

Proposition 21.

1. $\neg([c_1]p \vee [c_1]\neg p) \wedge \exists x([c_1 * x]p \vee [c_1 * x]\neg p)$ *is satisfiable.*
2. $\models_M \forall x\forall y([x]p \rightarrow [x * y]p)$.
3. $[c_1 * c_2]p \rightarrow \exists x[x]p$ *is satisfiable.*
4. $\forall x\forall y\exists z([x * y]p \vee [x * y]\neg p \rightarrow [z]p \vee [z]\neg p)$ *is satisfiable.*
5. $\models_M \forall x\forall y([x]p \wedge [y]p \rightarrow [x * y]p)$.

Satisfiability of the wff in 1 in a standard interpretation at the object w corresponds to the following situation. w is an undecidable/boundary element of the set represented by p with respect to the knowledge base of the first source, but there is a source in the system such that with respect to the distributed knowledge base of these two sources, w becomes a decidable (i.e. positive or negative) element of the set. Validity of the wff in 2 corresponds to the property $R_{\{i,j\}} \subseteq R_i$ of the MSAS^{D}, which shows that the distributed knowledge of any two sources is always at least as precise as that of an individual source. If the wff in 3 is valid in a standard interpretation, then it will mean that if an object is a positive element of the set represented by p with respect to the distributed knowledge base of the group consisting of the first two sources, then the object also belongs to the weak lower approximation of the set. Similarly, if the wff in 4 is valid in a MSAS^{D}-interpretation, it will imply that if any group consisting of two sources considers an object x to be a positive/negative element of the set represented by p, then we will have a source in the system which also considers it to be so. Validity of the wff in 5 guarantees that if an object is in the strong lower approximation of a set with respect to a group consisting of two sources, then it is also a positive element of the set with respect to the distributed knowledge base of the group. Such a wff can be given for any finite group of sources.

We end this section with the observation how the questions of the kind posed in Example 3 – rephrased in Page 173 – can now be looked upon as questions of satisfiability of some LMSAS^D-wffs in particular interpretations. More explicitly, for this example, $\mathfrak{F} := (U, \{R_P\}_{P \subseteq N})$, $N = \{1,2,3,4\}$, and we consider a standard interpretation $\mathfrak{M} := (\mathfrak{F}, V, I_\Gamma)$ with $V(p) := X$ and $c_i \in Con(\Gamma)$, $i = 1,2,3$. Then Q1-Q7 are equivalent to the following, for any assignment v. Let $t := c_1 * c_2 * c_3 * c_4$.

Q1. What are the objects $w \in U$ such that $\mathfrak{M}, v, w \models \forall x[x]p$?
Q2. Is $\mathfrak{M}, v, Cal \models \forall x\langle x\rangle p \wedge \neg\exists x[x]p$?
Q3. Is $\mathfrak{M} \models (\langle t\rangle p \wedge \neg[t]p) \rightarrow (\forall x\langle x\rangle p \wedge \neg\exists x[x]p$?
Q4. Is $\mathfrak{M} \models (\forall x\langle x\rangle p \wedge \neg\exists x[x]p) \rightarrow (\langle t\rangle p \wedge \neg[t]p)$?
Q5. Is $\mathfrak{M} \models (\neg\exists x\langle x\rangle p \vee (\exists x\langle x\rangle p \wedge \forall x[x]p)) \rightarrow \neg\exists x[x]p$?
Q6. Is $\mathfrak{M} \models (\forall x\langle x\rangle p \wedge \neg\exists x[x]p) \rightarrow (\langle t\rangle p \wedge \neg[t]p)$?
Q7. Is $\mathfrak{M} \models \exists x[x]p \rightarrow [t]\exists x[x]p$?

4.5 Relationship of LMSAS^D with other Logics

In this section, we discuss some fragments of LMSAS^D, and show that the modal system KTB, epistemic logic $S5_n^D$ and hence $S5$ are embedded in LMSAS^D.

We assume that the modal language and the language of LMSAS^D are based on the same set of propositional variables, and adopt the following notations and definitions.

Λ: the set of modal wffs with L as the modal operator for necessity;
$\mathcal{K}$: the set of all MSAS^D-structures;
$\mathcal{B}$: the set of all tolerance Kripke frames, i.e. frames with reflexive and symmetric relations.

A logic L_1 is *embeddable* into a logic L_2 ($L_1 \rightharpoonup L_2$), provided there is a translation $\star$ of wffs of L_1 into L_2, such that $\vdash_{L_1} \alpha$ if and only if $\vdash_{L_2} \alpha^\star$ for any wff α of L_1.

Logic for Strong Lower and Weak Upper Approximations. Choose and fix a variable, say x, and define a connective $\Box$, for any LMSAS^D wff α:

$$\Box\alpha := \forall x[x]\alpha.$$

Let us consider the set $\mathcal{F}_s$ of wffs defined by the scheme: $p \in PV|\neg\alpha|\alpha \wedge \beta|\Box\alpha$. $\mathcal{F}_s$ consists of only closed wffs and $Con(\mathcal{F}_s) = \emptyset$. So the satisfiability of the wffs of $\mathcal{F}_s$ in an interpretation $(\mathfrak{F}, V, I)$ will be independent of I and assignment v. Let $\mathfrak{M} := (\mathfrak{F}, V, I)$ be a MSAS^D-interpretation. Then from Proposition 19, we have $[\![\Box\alpha]\!]_{\mathfrak{M}} = \underline{([\![\alpha]\!]_{\mathfrak{M}})}_s$ and $[\![\Diamond\alpha]\!]_{\mathfrak{M}} = \overline{([\![\alpha]\!]_{\mathfrak{M}})}_w$ as x is not free in any α.

It follows that the set of wffs $\underline{\mathcal{F}}_s$ can express the properties of strong lower and weak upper approximations.

Let us consider the bijection T_1 from Λ to $\underline{\mathcal{F}}_s$ which fixes propositional variables and takes $L\alpha$ to $\Box T_1(\alpha)$. Further, consider the mapping $\Psi : \mathcal{K} \to \mathcal{B}$ which maps the MSAS^{D}-structure $\mathfrak{F} := (U, \{R_P\}_{P \subseteq_f N})$ to $(U, R_{\mathfrak{F}})$, where $R_{\mathfrak{F}} := \bigcup_{i \in N} R_i$.

The following proposition guarantees that if $T_1(\alpha)$ is satisfiable, then α is also satisfiable in some tolerance frame.

Proposition 22. *Let us consider a* MSAS^{D}*-interpretation* $\mathfrak{M} = (\mathfrak{F}, V, I)$ *and an assignment* v*. Then for all* $\alpha \in \Lambda$ *and* $w \in U$*,*

$$(\Psi(\mathfrak{F}), V), w \models \alpha \text{ if and only if } \mathfrak{M}, v, w \models T_1(\alpha).$$

The proof is by induction on the complexity of α.
Propositions 9, 22 and the finite model property of KTB lead us to

Proposition 23. $\alpha \in \Lambda$ *is satisfiable in a tolerance frame if and only if* $T_1(\alpha)$ *is satisfiable.*

As T_1 is bijective, it follows from Proposition 23 that the logic for strong lower and weak upper approximations is the modal system KTB. Moreover, T_1 gives us

Proposition 24. $KTB \rightharpoonup$ LMSAS^{D}.

Logic for Weak Lower and Strong Upper Approximations. Define the set $\underline{\mathcal{F}}_w$ of wffs for the weak lower approximation by taking $\Box_1\alpha$ as $\exists x[x]\alpha$. By Proposition 19, for any $\mathfrak{M}$, $[\![\Box_1\alpha]\!]_{\mathfrak{M},v} = \underline{([\![\alpha]\!]_{\mathfrak{M},v})}_w$ and $[\![\Diamond_1\alpha]\!]_{\mathfrak{M}} = \overline{([\![\alpha]\!]_{\mathfrak{M}})}_s$.

Proposition 25. *The following holds in* LMSAS^{D}*:*

(N) *If* $\models_M \alpha$*, then* $\models_M \Box_1\alpha$*;*
(M) $\models_M \Box_1(\alpha \wedge \beta) \to (\Box_1\alpha \wedge \Box_1\beta)$*;*
(S4) $\models_M \Diamond_1\Diamond_1\alpha \to \Diamond_1\alpha$*;*
(T) $\models_M \Box_1\alpha \to \alpha$.

(M) and (S4) follow from 2(b) and 5(b) of Proposition 6 respectively, and (T) follows from the inclusion (*) given on Page 168. The axiom K of modal logic is not valid:

Example 11. Consider the MSAS^{D} $\mathfrak{F}$ of Example 10, and V such that $V(p) := \{w_1, w_2\}$, $V(q) := \{w_2\}$. Then for any v, $\mathfrak{M}, v, w_2 \not\models \exists x[x](p \to q) \to (\exists x[x]p \to \exists x[x]q)$.

Let us consider the bijection T_2 from Λ to $\underline{\mathcal{F}}_w$, which fixes propositional variables and takes $L\alpha$ to $\Box_1 T_2(\alpha)$. From Proposition 25, we obtain

Proposition 26. *If* $\alpha \in \Lambda$ *is a theorem of the classical modal system MTS4* [15]*, then* $T_2(\alpha)$ *is a valid wff of* LMSAS^{D}.

Moreover, we also have,

Proposition 27. *If* $T_2(\alpha)$ *is a valid wff of* LMSAS^{D}, *then* $\alpha \in \Lambda$ *is a theorem of S5.*

Proof. Every equivalence frame is also a MSAS^{D} having only one relation; $T_2(\alpha) \in \mathcal{F}_w$ is valid in an equivalence frame (U, R), if and only if $\alpha \in \Lambda$ is valid in (U, R). □

From Propositions 26 and 27, it follows that the logic for weak lower and strong upper approximation lies between the classical modal system MTS4 and modal system S5. However, the exact axiomatization of this system remains a question.

LMSAS^{D} and Epistemic Logic. Let us consider the n-agent epistemic logic $S5_n^D$ [22] with knowledge operators K_i, $i \in N = \{1, \ldots, n\}$ and distributed knowledge operators D_G for groups G of agents. Let T_3 be the translation from the wffs of $S5_n^D$ to those of LMSAS^{D} which takes $K_i\alpha$ and $D_{\{i_1, i_2, \ldots, i_m\}}\alpha$ to $[c_i]T_3(\alpha)$ and $[c_{i_1} * \cdots * c_{i_m}]T_3(\alpha)$ respectively.

Let $\Gamma := \{T_3(\alpha) : \alpha \text{ is a wff of } S5_n^D\}$. Note that $Con(\Gamma) = \{c_1, c_2, \ldots, c_n\}$.

Define a surjection Φ from the set of all $\Gamma-$standard interpretations to the set of all structures of the form $(U, \{R_i\}_{i \in N}, V)$, where each R_i is an equivalence relation on U and $V : PV \rightarrow 2^U$, as follows: for $\mathfrak{M} := (\mathfrak{F}, V, I_\Gamma)$, $\mathfrak{F} := (U, \{R_P\}_{P \subseteq_f M})$,

$$\Phi(\mathfrak{M}) := (U, \{R_i\}_{i \in N}, V).$$

It is clear that

Proposition 28. $\mathfrak{M}, v, w \models T_3(\alpha)$ *if and only if* $\Phi(\mathfrak{M}), w \models \alpha$, *for all assignments* v *and* $w \in U$.

Using Theorem 3, Proposition 28 and the fact that Φ is surjective, we obtain

Proposition 29. α *is valid wff of* $S5_n^D$ *if and only if* $\models_M T_3(\alpha)$.
Hence $S5_n^D \rightharpoonup \text{LMSAS}^{\text{D}}$, *and in particular,* $S5 \rightharpoonup \text{LMSAS}^{\text{D}}$.

It may be noted that epistemic logics will not suffice for our purpose. The semantics for these logics considers a finite and fixed number of agents, thus giving a finite and fixed number of modalities in the language. But in the case of LMSAS^{D}, the number of sources is not fixed, and could also be countably infinite. So it is not possible here to refer to all/some sources using only the connectives $\wedge$, $\vee$, and the quantifiers $\forall$, $\exists$ are used to achieve the task.

5 Axiomatization of LMSAS^{D}

In this section we present the axiomatization of LMSAS^{D} with respect to MSAS^{D}-interpretations proposed in [52].

Consider the following deductive system, where t, t' are terms in T and $\Box \in \{[t]; t \in T\} \cup \{A\}$, $\Diamond\alpha := \neg\Box\neg\alpha$. Note that the connective A is interpreted as the global modal operator.

Axiom schema:

1. All axioms of classical propositional logic.
2. $\forall x\alpha \rightarrow \alpha(t/x)$, where $t \in Con \cup Var$ is free for x in α.
3. $\forall x(\alpha \rightarrow \beta) \rightarrow (\alpha \rightarrow \forall x\beta)$, where the variable x is not free in α.
4. $[t]\alpha \rightarrow [t']\alpha$ when $B(t) \subseteq B(t')$.
5. $A\alpha \rightarrow [t]\alpha$.
6. $\Box(\alpha \rightarrow \beta) \rightarrow (\Box\alpha \rightarrow \Box\beta)$.
7. $\Box\alpha \rightarrow \alpha$.
8. $\Diamond\Box\alpha \rightarrow \alpha$.
9. $\Box\alpha \rightarrow \Box\Box\alpha$.

Rules of inference:

$$\forall.\ \frac{\beta}{\forall x\beta} \qquad MP.\ \frac{\alpha \quad \alpha \rightarrow \beta}{\beta} \qquad N.\ \frac{\alpha}{\Box\alpha}$$

So this is the quantification of a propositional modal system having 'indexed' modalities, obeying axioms of the system $S5$. We note that this is different from both propositional quantification of modal logic [12], and modal predicate logic [59], as none of them have feature of quantification over modalities. Let us denote the deductive system consisting of the above axioms and rules by Λ. The notion of *theoremhood* is defined in the usual way and we write $\vdash_\Lambda \alpha$, if α is a theorem. The Barcan-like wff $\forall x\Box\alpha \rightarrow \Box\forall x\alpha$, where in the case when $\Box = [t]$, $x \notin B(t)$, is not taken as an axiom as it can be deduced. (We thank Yde Venema for indicating that this should be the case). Moreover, if the condition $x \notin B(t)$ is removed, then the wff may not remain valid: consider, e.g. the wff $\forall x[x][x]p \rightarrow [x]\forall x[x]p$, $\mathfrak{F}$ as the MSAS^{D} of Example 10, and the interpretation $\mathfrak{M} := (\mathfrak{F}, V, I)$ with assignment v, where $V(p) = \{w_1, w_2\}$, $v(x) = 1$.

Soundness of LMSAS^{D} with respect to the semantics of MSAS^{D}-interpretations is obtained in a standard manner.

5.1 Completeness

The proof of the completeness theorem for LMSAS^{D} is in the line of modal lower predicate calculus [40]. Note that if we could obtain the completeness for a wff α with respect to the class of all MSAS^{D}-interpretations, then due to Theorem 3, this would also give us completeness for α with respect to the class of all α-standard interpretations.

As in the case of first order logic, we need the $\forall$−property for sets of wffs: a set Γ of wffs of a language $\mathcal{L}$ is said to have the $\forall$−*property* in $\mathcal{L}$ if and only if for every wff α and every variable x from $\mathcal{L}$, there is some variable y in $\mathcal{L}$, not occurring in α, such that $\alpha(y/x) \rightarrow \forall x\alpha \in \Gamma$.

Furthermore, we require to extend the language by adding infinitely many new variables. Let n be an integer and consider the set $Con_n := \{c_1, c_2, \ldots, c_n\}$ of constants. Let $\mathfrak{L}^n$ be the language obtained from $\mathfrak{L}$ by taking the constants from the set Con_n only; $\mathfrak{L}^{n+}$ the language obtained from $\mathfrak{L}^n$ by adding infinitely many new variables; Var^+ the set of all variables in $\mathfrak{L}^{n+}$, and T^{n+}, the set of all terms in $\mathfrak{L}^{n+}$.

With the usual techniques of modal predicate calculus, we have

Proposition 30. *Any $\Lambda-$ consistent set of wffs of $\mathfrak{L}^n$ has a $\Lambda-$maximal consistent extension in $\mathfrak{L}^{n+}$ with the $\forall-$property in $\mathfrak{L}^{n+}$.*

Let us now describe the *canonical interpretation* $\mathfrak{M}_n{}^C := (\mathfrak{F}_n{}^C, V_n{}^C, I_n{}^C)$ for LMSAS$^{\rm D}$ and assignment $v_n{}^C$ required for the proof of the completeness theorem.

Definition 30 (Canonical interpretation).

- $U_n^C := \{w : w$ *is* $\Lambda-$*maximal consistent in* $\mathfrak{L}^{n+}$ *and has the* $\forall-$*property in* $\mathfrak{L}^{n+}\}$;
- $V_n^C : PV \to 2^{U_n^C}$ *is such that* $V_n^C(p) := \{w \in U_n^C : p \in w\}$, *for* $p \in PV$;
- $I_n^C : Con \to \mathbb{N}$ *is such that* $I_n^C(c_i) := i,\ c_i \in Con_n$;
- $v_n^C : Var^+ \to \mathbb{N}$ *is such that* $v_n^C(x_i) := n+i$, *where* $x_1, x_2, \ldots$ *is an enumeration of the variables in* $\mathfrak{L}^{n+}$.
- *Finally, we get* $\mathfrak{F}_n^C := (U_n^C, \{R_n^C{}_P\}_{P\subseteq_f \mathbb{N}})$, *the canonical* MSAS$^{\rm D}$, *by defining* $R_n^C{}_P$ *in the standard way as:*
  - $wR_n^C{}_P w'$, $P \neq \emptyset$, *if and only if for every wff* $[t]\alpha$ *of* $\mathcal{L}^{n+}$ *with* $\tilde{v}_n^C(t) = P$, $[t]\alpha \in w$ *implies* $\alpha \in w'$, *where* $w, w' \in U_n^C$.
  - $wR_n^C{}_\emptyset w'$, *if and only if for every wff* $A\alpha$ *of* $\mathcal{L}^{n+}$, $A\alpha \in w$ *implies* $\alpha \in w'$, *where* $w, w' \in U_n^C$.

Note 1. (i) For each $P \subseteq_f \mathbb{N}$, there exists $t \in T^{n+}$ such that $\tilde{v}_n^C(t) = P$.
(ii) For $t, t' \in T^{n+}$, $\tilde{v}_n^C(t) = \tilde{v}_n^C(t')$ implies $B(t) = B(t')$.

This follows from the fact that for each $j \in \mathbb{N}$, there exists $t \in Con_n \cup Var^+$ such that $\tilde{v}_n^C(t) = j$ and $\tilde{v}_n^C|_{Var^+ \cup Con_n}$ is injective.

Proposition 31 (Truth Lemma). *$\beta \in w$ if and only if $\mathfrak{M}_n^C, v_n^C, w \models \beta$, for any wff β of $\mathfrak{L}^{n+}$ and $w \in U_n^C$.*

The proof, as usual, is by induction on the complexity of the wff β.

Recall Definition 27. We have the following.

Proposition 32. *$\mathfrak{F}_n^C$ is a semi* MSAS$^{\rm D}$*-structure.*

Proof. The equivalence of each of the relation $R_n^C{}_P$ follows from Axioms 7–9. Moreover, the condition $R_n^C{}_{P\cup Q} \subseteq R_n^C{}_P \cap R_n^C{}_Q$ follows from Axiom 4. We omit the detail proof as it is very standard in modal logic. □

Let $\alpha \in \mathcal{F}$ be $\Lambda-$consistent, and k the largest integer such that $c_k \in Con(\{\alpha\})$. From Propositions 30, 31 and 32, we then obtain

Proposition 33. *If α is $\Lambda-$consistent, then there exists an interpretation $\mathfrak{M} = (\mathfrak{F}, V, I)$ based on a semi* MSAS$^{\rm D}$ *$\mathfrak{F} = (U, \{R_P\}_{P\subseteq_f N})$, an object $w \in U$ and an assignment v such that $\mathfrak{M}, v, w \models \alpha$.*

Note that the interpretation obtained in Proposition 33 may not have the properties (I1) and (I2). In order to get these, some more work needs to be done. In fact, we will take standard modal logic approach where one transforms models (interpretations) into desirable form without affecting satisfiability. Moreover, for this purpose, we shall make use of two standard techniques of modal logic: copying, and generated sub-model. Let us first state the following theorem that gives us the completeness result.

Theorem 4. *Let $\alpha \in \mathcal{F}$ and $\mathfrak{M} = (\mathfrak{F}, V, I)$ be an interpretation based on a semi* MSAS^{D}*-structure $\mathfrak{F} = (U, \{R_P\}_{P \subseteq_f N})$ such that $\mathfrak{M}, v, w \models \alpha$ for some assignment v and $w \in U$. Then there exists a* MSAS^{D}*-interpretation $\mathfrak{M}' = (\mathfrak{F}', V', I)$ with $|\mathfrak{F}'| = |\mathfrak{F}|$ and $w' \in U'$ (domain of $\mathfrak{F}'$) such that $\mathfrak{M}', v, w' \models \alpha$. Moreover, if $|U| = n$ and $|\mathfrak{F}| = m$, then $|U'| \leq n \times 2^{2^m \times mn}$.*

The theorem is proved in two steps. First, we obtain the property (I2) following the technique of copying. In the second step, we add the property (I1). The first part follows the proof scheme of Balbiani [1]. For the sake of completeness, we give the main definitions and results.

Definition 31. *Let $\mathfrak{F} = (U, \{R_P\}_{P \subseteq_f N})$ and $\mathfrak{F}' = (U', \{R'_P\}_{P \subseteq_f N})$ be two frames and Ψ be a set of mappings from U to U'. Ψ is said to be a copying from $\mathfrak{F}$ to $\mathfrak{F}'$ provided the following is satisfied.*

- **(C1)** *For every mapping $f, g \in \Psi$ and for every object $x, y \in U$, $f(x) = g(y)$ implies $x = y$.*
- **(C2)** *For every object $x' \in U'$, there exists a mapping $f \in \Psi$ and an object $x \in U$ such that $f(x) = x'$.*
- **(C3)** *For every $P \subseteq_f N$, for every mapping $f \in \Psi$ and for every object $x, y \in U$, $(x, y) \in R_P$ implies there exists a mapping $g \in \Psi$ such that $(f(x), g(y)) \in R'_P$.*
- **(C4)** *For every $P \subseteq_f N$, for every mapping $f, g \in \Psi$ and for every object $x, y \in U$, $(f(x), g(y)) \in R'_P$ implies $(x, y) \in R_P$.*

Proposition 34. *Let $\mathfrak{M} := (\mathfrak{F}, V, I)$ and $\mathfrak{M}' := (\mathfrak{F}', V', I)$ be two interpretations such that the following hold.*

1. *$|\mathfrak{F}| = |\mathfrak{F}'| = |N|$, where N is an initial segment of $\mathbb{N}$.*
2. *There exists a copying Ψ from $\mathfrak{F}$ into $\mathfrak{F}'$.*
3. *$V'(p) = \{f(x) : f \in \Psi \text{ and } x \in V(p)\}$, for all $p \in PV$.*

Then for every wff α, for every mapping $f \in \Psi$, for every $x \in U$ and for every assignment $v : Var \to N$, $\mathfrak{M}, v, x \models \alpha$ if and only if $\mathfrak{M}', v, f(x) \models \alpha$.

The proof is by induction on the complexity of α.

Let us consider the frame $\mathfrak{F} := (U, \{R_P\}_{P \subseteq_f N})$ of Theorem 4, and the following definitions, where $P \subseteq_f N$, 2^N_f denotes the collection of all finite subsets of N, and for any $P, Q \subseteq_f U$, $P + Q := (P \setminus Q) \cup (Q \setminus P)$.
$\Omega := \{f : f : 2^N_f \times N \to 2^U\}$; $U^* := U \times \Omega$.

$\pi(P) : U \times U \to 2^U$ is such that $\pi(P)(x,y) := [x]_{R_P} + [y]_{R_P}$. Note that $\pi(P)(x,y) = \emptyset$, if and only if $(x,y) \in R_P$.
R^*_P is a relation on U^* such that $((x,f),(y,g)) \in R^*_P$, if and only if
a1. $(x,y) \in R_P$;
a2. for all $i \in P$ and for every $Q \subseteq_f N$, if $i \in Q$, then $f(Q,i) + g(Q,i) = \emptyset$ and
a3. for every $Q \subseteq_f N$, $\sum_{i \in Q}(f(Q,i) + g(Q,i)) = \pi(Q)(x,y)$.
Consider the frame $\mathfrak{F}^* := (U^*, \{R^*_P\}_{P \subseteq_f N})$. We then have

Proposition 35.

1. *For each* $P \subseteq_f N$, R^*_P *is an equivalence relation.*
2. $R^*_{P \cup Q} = R^*_P \cap R^*_Q$.
3. $R^*_P \subseteq R^*_\emptyset$.
4. *Let* $P \subseteq_f N$, *and* $f \in \Omega$. *Then for every* $x, y \in U$ *such that* $(x,y) \in R_P$, *there exists* $g \in \Omega$ *such that* $((x,f),(y,g)) \in R^*_P$.
5. *There exists a copying from* $\mathfrak{F}$ *into* $\mathfrak{F}^*$.
6. *If* $|U| = n$ *and* $|\mathfrak{F}| = |N| = m$, m, n *being finite, then* $|U^*| = n \times 2^{2^m \times mn}$.

Proof. 4. Let ϕ be a mapping from $2^N_f \times 2^N_f$ to N such that for all $S, Q \subseteq_f N$, if $Q \not\subseteq S$, then $\phi(S,Q) \in Q$ and $\phi(S,Q) \notin S$. For every $Q \subseteq_f N$ and $i \in N$, let $g(Q,i)$ be the subset of U obtained as follows:
If $i \notin Q$, then $g(Q,i) = \emptyset$.
If $i \in Q$ and $i \in P$, then $g(Q,i) = f(Q,i)$.
If $i \in Q$ and $i \notin P$, then either $i \neq \phi(P,Q)$ – in this case $g(Q,i) = \emptyset$ or $i = \phi(P,Q)$ – in this case $g(Q,i) = \sum_{j \in Q \setminus P} f(Q,j) + \pi(Q)(x,y)$.
One can then show that $((x,f),(y,g)) \in R^*_P$. □

Now take the valuation V of Theorem 4 and let $V^* : PV \to 2^{U^*}$ be such that $V^*(p) := \{f^*(x) : f^* \in \Omega^* \text{ and } x \in V(p)\}$. Then from Proposition 34 and the fact that $\mathfrak{M}, v, w \models \alpha$, we obtain $\mathfrak{M}^*, v, f(w) \models \alpha$, where $f \in \Omega^*$ and $\mathfrak{M}^* := (\mathfrak{F}^*, V^*, I)$.

In order to incorporate property (I1), consider $\mathfrak{F}^{*g} := (U^{*g}, \{R^{*g}_P\}_{P \subseteq_f N})$, the interpretation generated by $f(w)$ using the equivalence relation $R^*_\emptyset$ as follows: U^{*g} is the equivalence class of $f(w)$ with respect to the equivalence relation $R^*_\emptyset$; R^{*g}_P, V^{*g} are the restrictions of R^*_P and V^* to U^{*g} respectively (that is: $R^{*g}_P := R^*_P \cap (U^{*g} \times U^{*g})$, and for each $p \in PV$, $V^{*g}(p) := V^*(p) \cap U^{*g}$). Then one can show that

Proposition 36. $\mathfrak{F}^{*g}$ *is a* MSAS^D*-structure.*

Using the property of generated sub-models and that $\mathfrak{M}^*, v, f(w) \models \alpha$, we obtain $\mathfrak{M}^{*g}, v, f(w) \models \alpha$. This gives us Theorem 4, and finally, the completeness theorem.

Theorem 5 (Completeness). $\models_M \alpha$ *implies* $\vdash_\Lambda \alpha$.

6 Some Decidability Results Related to LMSAS^{D}

The content of this section is based on the articles [45,52].

From Theorem 3, it follows that given a wff $\alpha \in \mathcal{F}$, if we could decide whether there exists an interpretation in $\mathcal{C}_\alpha$ where α is satisfiable, then we would be able to decide whether α is satisfiable in any MSAS^{D}-interpretation or not. This question remains open, but we have the following result.

Theorem 6. *Given a wff $\alpha \in \mathcal{F}$, an integer n and $I : Con \rightarrow \{1, 2, \ldots, n\}$, we can decide whether there exists a* MSAS^{D}*-interpretation $\mathfrak{M} := (\mathfrak{F}, V, I)$ with $|\mathfrak{F}| = n$ such that α is satisfiable in $\mathfrak{M}$.*

Suppose α is satisfiable in a MSAS^{D}-interpretation $\mathfrak{M} := (\mathfrak{F}, V, I)$ with $|\mathfrak{F}| = n$. Then, we will show that without affecting the satisfiability of α, $\mathfrak{M}$ can be transformed into a MSAS^{D}-interpretation $\mathfrak{M}' := (\mathfrak{F}', V, I)$, with $|U'| \leq n \times 2^{2^D \times Dn}$, where U' is the domain of $\mathfrak{F}'$, $D = 2^{|\Delta| \times n^m}$, and Δ is the set of all sub-wffs of α. This result will be the main part of the proof of Theorem 6, and will be proved using a very standard technique of modal logic, known as filtration [9]. So, let us move to provide a detailed proof of this theorem.

Let $\Sigma \subseteq \mathcal{F}$ be a sub-wff closed set. Let $N := \{1, 2, \ldots, n\}$, and consider the interpretation $\mathfrak{M} := (\mathfrak{F}, V, I)$, where $\mathfrak{F} := (U, \{R_P\}_{P \subseteq_f N})$ is a MSAS^{D}-structure and $V : PV \rightarrow 2^U$, I, n are the same as in Theorem 6. Consider the binary relation $\equiv_\Sigma$ on U defined as:
$w \equiv_\Sigma w'$, if and only if for all $\beta \in \Sigma$ and all assignments v for $\mathfrak{M}$, $\mathfrak{M}, v, w \models \beta$ if and only if $\mathfrak{M}, v, w' \models \beta$.

Definition 32 (Filtration interpretation).

- *$U^{f_\Sigma} := \{[w] : w \in U\}$, $[w]$ is the equivalence class of w with respect to the equivalence relation $\equiv_\Sigma$;*
- *$R_P^{f_\Sigma} \subseteq U^{f_\Sigma} \times U^{f_\Sigma}$ is defined as:*
  $([w], [w']) \in R_P^{f_\Sigma}$ if and only if there exist $w_1 \in [w]$ and $w_2 \in [w']$ such that $(w_1, w_2) \in R_P$;
  ${R_P^{f_\Sigma}}^*$ is the transitive closure of $R_P^{f_\Sigma}$;
- *$V^{f_\Sigma} : PV \rightarrow 2^{U^{f_\Sigma}}$ is defined as: $V^{f_\Sigma}(p) := \{[w] \in U^{f_\Sigma} : w \in V(p)\}$.*

Let $\mathfrak{M}^{f_\Sigma}$ and ${\mathfrak{M}^{f_\Sigma}}^*$ denote the interpretations $((U^{f_\Sigma}, \{{R_P}^{f_\Sigma}\}_{P \subseteq_f N}), V^{f_\Sigma}, I)$ and $((U^{f_\Sigma}, \{{R_P}^{{f_\Sigma}^*}\}_{P \subseteq_f N}), V^{f_\Sigma}, I)$ respectively. We shall see that, although the interpretation $\mathfrak{M}^{f_\Sigma}$ preserves satisfaction (cf. Proposition 40), the relation $R_P^{f_\Sigma}$ may not be an equivalence. So, we consider the interpretation ${\mathfrak{M}^{f_\Sigma}}^*$ instead of $\mathfrak{M}^{f_\Sigma}$, and proceed to get the desired decidability result viz. Theorem 6. We begin with the following propositions listing some of the properties of $R_P^{f_\Sigma}$ and ${R_P^{f_\Sigma}}^*$.

Proposition 37.

1. *For each* $P \subseteq_f N$, $R_P^{f_\Sigma}$ *is reflexive and symmetric.*
2. $R_\emptyset^{f_\Sigma} = U^{f_\Sigma} \times U^{f_\Sigma}$.
3. *If* $(w,u) \in R_P$ *then* $([w],[u]) \in R_P^{f_\Sigma}$;
4. *If* $([w],[u]) \in R_P^{f_\Sigma}$, *then for all* $\langle t \rangle \alpha \in \Sigma$ *and for all assignment* v *with* $\tilde{v}(t) = P$ *and* $\mathfrak{M}, v, u \models \alpha$, *we have* $\mathfrak{M}, v, w \models \langle t \rangle \alpha$.

Proof. We will only prove 4. Since $([w],[u]) \in R_P^{f_\Sigma}$, there exists $w' \in [w]$ and $u' \in [u]$ such that $(w',u') \in R_P$. Since $\langle t \rangle \alpha \in \Sigma$, so $\alpha \in \Sigma$. Therefore, $\mathfrak{M}, v, u' \models \alpha$ ($\because u \equiv_\Sigma u'$ and $\mathfrak{M}, v, u \models \alpha$). But $(w',u') \in R_P$ and hence $\mathfrak{M}, v, w' \models \langle t \rangle \alpha$. This gives $\mathfrak{M}, v, w \models \langle t \rangle \alpha$ as $\langle t \rangle \alpha \in \Sigma$ and $w \equiv_\Sigma w'$. □

Proposition 38.

1. *For each* $P \subseteq_f N$, ${R_P^{f_\Sigma}}^*$ *is an equivalence relation.*
2. ${R_\emptyset^{f_\Sigma}}^* = U^{f_\Sigma} \times U^{f_\Sigma}$.
3. ${R_{P \cup Q}^{f_\Sigma}}^* \subseteq {R_P^{f_\Sigma}}^* \cap {R_Q^{f_\Sigma}}^*$

Proof. We will only prove 3. Let $([w],[u]) \in {R_{P \cup Q}^{f_\Sigma}}^*$. Then there exists $[w_i]$, $i = 1,2,\ldots,k$ such that $([w],[w_1]) \in R_{P \cup Q}^{f_\Sigma}$, $([w_k],[u]) \in R_{P \cup Q}^{f_\Sigma}$ and $([w_i],[w_{i+1}]) \in R_{P \cup Q}^{f_\Sigma}$, $i \in \{1,2,\ldots,k-1\}$. This implies that there exists $w \in [w], u' \in [u]$ and $w_i' \in [w_i]$, $i = 1,2,\ldots,k$ such that $(w,w_1) \in R_{P \cup Q}$, $(w_k,u) \in R_{P \cup Q}$ and $(w_i,w_{i+1}) \in R_{P \cup Q}$, $i \in \{1,2,\ldots,k-1\}$. Since $R_{P \cup Q} \subseteq R_P \cap R_Q$, we obtain $(w,w_1) \in R_P$, $(w_k,u) \in R_P$, $(w_i,w_{i+1}) \in R_P$ and $(w,w_1) \in R_Q$, $(w_k,u) \in R_Q$, $(w_i,w_{i+1}) \in R_Q$, $i \in \{1,2,\ldots,k-1\}$. This gives $(w,u) \in R_P \cap R_Q$ as R_P and R_Q are transitive. Thus, we obtain $([w],[u]) \in {R_P^{f_\Sigma}}^* \cap {R_Q^{f_\Sigma}}^*$. □

Note 2. Thus the interpretation $\mathfrak{M}^{{f_\Sigma}^*}$ is based on the semi MSAS^{D}-structure $(U^{f_\Sigma}, \{{R_P^{f_\Sigma}}^*\}_{P \subseteq_f N})$.

Next, we show that $\mathfrak{M}^{{f_\Sigma}^*}$ preserves satisfaction in the following sense.

Proposition 39. *For all wffs* $\alpha \in \Sigma$, *all assignment* v *for* $\mathfrak{M}$ *and all objects* $w \in U$, $\mathfrak{M}, v, w \models \alpha$ *if and only if* $\mathfrak{M}^{{f_\Sigma}^*}, v, [w] \models \alpha$.

To prove it, we need the Propositions 40–42. The proofs are routine.

Proposition 40. *For all wffs* $\alpha \in \Sigma$, *all assignment* v *for* $\mathfrak{M}$ *and all objects* $w \in U$, $\mathfrak{M}, v, w \models \alpha$ *if and only if* $\mathfrak{M}^{f_\Sigma}, v, [w] \models \alpha$.

Proof. Proof is by induction on the complexity of α. □

Proposition 41. *For all wffs* $[t]\alpha \in \Sigma$, *all assignment* v *for* $\mathfrak{M}$ *and all objects* $[w],[u] \in U^{f_\Sigma}$ *such that* $([w],[u]) \in R_{\tilde{v}(t)}^{f_\Sigma}$ *and* $\mathfrak{M}^{f_\Sigma}, v, [w] \models [t]\alpha$, *we have* $\mathfrak{M}^{f_\Sigma}, v, [u] \models [t]\alpha$.

Proof. Since $([w],[u]) \in R^{f_\Sigma}_{\tilde{v}(t)}$, there exists $w' \in [w]$ and $u' \in [u]$ such that $(w',u') \in R_{\tilde{v}(t)}$. Now,
$\mathfrak{M}^{f_\Sigma}, v, [w] \models [t]\alpha$
$\Rightarrow \mathfrak{M}, v, w \models [t]\alpha$ (by Proposition 40)
$\Rightarrow \mathfrak{M}, v, w' \models [t]\alpha \quad (\because w' \in [w])$
$\Rightarrow \mathfrak{M}, v, w' \models [t][t]\alpha$ ($\because [t]\beta \rightarrow [t][t]\alpha$ is valid in all MSAS^{D}-interpretation)
$\Rightarrow \mathfrak{M}, v, u' \models [t]\alpha \quad (\because (w',u') \in R_{\tilde{v}(t)})$
$\Rightarrow \mathfrak{M}, v, u \models [t]\alpha \quad (\because u' \in [u])$
$\Rightarrow \mathfrak{M}^{f_\Sigma}, v, [u] \models [t]\alpha$ (by Proposition 40). □

Proposition 42. *For all wffs $\alpha \in \Sigma$, all assignment v for $\mathfrak{M}$ and all objects $w \in U$, $\mathfrak{M}^{f_\Sigma}, v, [w] \models \alpha$ if and only if ${\mathfrak{M}^{f_\Sigma}}^*, v, [w] \models \alpha$.*

Proof. Proof is by induction on the complexity of α. We only provide the proof of the case when α is of the form $[t]\beta$.
Let $\mathfrak{M}^{f_\Sigma}, v, [w] \models [t]\beta$. We have to show ${\mathfrak{M}^{f_\Sigma}}^*, v, [w] \models [t]\beta$. Let $([w],[u]) \in {R^{f_\Sigma}_{\tilde{v}(t)}}^*$. Then there exists $w_1, w_2, \ldots, w_k$ such that $([w],[w_1]) \in R^{f_\Sigma}_{\tilde{v}(t)}$, $([w_k],[u]) \in R^{f_\Sigma}_{\tilde{v}(t)}$ and $([w_i],[w_{i+1}]) \in R^{f_\Sigma}_{\tilde{v}(t)}$, $i \in \{1,2,\ldots,k-1\}$. Since $\mathfrak{M}^{f_\Sigma}, v, [w] \models [t]\beta$, by Proposition 41, we have $\mathfrak{M}^{f_\Sigma}, v, [w_k] \models [t]\beta$. This implies $\mathfrak{M}^{f_\Sigma}, v, [u] \models \beta$ as $([w_k],[u]) \in R^{f_\Sigma}_{\tilde{v}(t)}$. Thus by induction hypothesis, we obtain ${\mathfrak{M}^{f_\Sigma}}^*, v, [u] \models \beta$ and hence ${\mathfrak{M}^{f_\Sigma}}^*, v, [w] \models [t]\beta$. The other direction is obvious. □

Combining Propositions 40 and 42, we obtain the proof of Proposition 39. Thus, we have the following result which will lead us to the finite model property (Proposition 44).

Proposition 43. *Let $\alpha \in \mathcal{F}$ and consider n and I given in Theorem 6. Let Δ be the set of all sub-wffs of α. If α is satisfiable in a* MSAS^{D}*-interpretation $\mathfrak{M} := (\mathfrak{F}, V, I)$ with $|\mathfrak{F}| = n$, then it is satisfiable in an interpretation $\mathfrak{M}' := (\mathfrak{F}', V, I)$ based on a semi* MSAS^{D}*-structure $\mathfrak{F}' := (U', \{R'_P\}_{P \subseteq_f N})$ with $|U'| \leq 2^{|\Delta| \times n^m}$, where $|Var\{\alpha\}| = m$.*

Proof. Due to Note 2 and Proposition 39, it is enough to show that $|U^{f_\Delta}| \leq 2^{|\Delta| \times n^m}$. Let $Var(\Delta) := \{x_1, x_2, \ldots, x_m\}$. Let Asg denote the collection of all assignments for $\mathfrak{M}$. Consider the equivalence relation $\approx$ defined on Asg as follows:

$$v_1 \approx v_2 \text{ if and only if } v_1(x_j) = v_2(x_j),\ j = 1, 2, \ldots, m.$$

Let $Asg/\approx$ denote the quotient set, as usual. Note that $|Asg/\approx| \leq n^m$. Now define a map $g : U^{f_\Delta} \rightarrow 2^{\Sigma \times Asg/\approx}$ as:

$$g([w]) := \{(\beta,[v]) \in \Delta \times Asg/\approx : {\mathfrak{M}^{f_\Sigma}}^*, v, w \models \beta\}.$$

Clearly g is injective and hence U^{f_Δ} contains at most $2^{|\Delta| \times n^m}$ elements. □

Let $D = 2^{|\Delta| \times n^m}$. Then using Theorems 4 and 43, we obtain,

Proposition 44 (Finite model property). *Let $\alpha \in \mathcal{F}$ and consider n and I given in Theorem 6. Let Δ be the set of all sub-wffs of α. If α is satisfiable in a* MSAS^{D}*-interpretation $\mathfrak{M} := (\mathfrak{F}, V, I)$ with $|\mathfrak{F}| = n$, then it is satisfiable in a* MSAS^{D}*-interpretation $\mathfrak{M}' := (\mathfrak{F}', V, I)$, where $|U'| \leq n \times 2^{2^D \times Dn}$ (U' is the domain of $\mathfrak{F}'$).*

Proof of Theorem 6. Since there are only $n^{|FV(\{\alpha\})|}$ distinct mappings from $FV(\{\alpha\})$ to N, from Propositions 1 and 44, we obtain the desired result. □

Since there are only $n^{|Con(\{\alpha\})|}$ distinct mappings from $Con(\{\alpha\})$ to N, from Proposition 1 and Theorem 6, we obtain,

Theorem 7. *Given a wff $\alpha \in \mathcal{F}$ and an integer n, we can decide whether there exists a* MSAS^{D}*-interpretation $\mathfrak{M} := (\mathfrak{F}, V, I)$ with $|\mathfrak{F}| = n$ such that α is satisfiable in $\mathfrak{M}$.*

In practical problems, MSAS^{D}s with finite domains would be of particular relevance. We have the following decidable problem related to such approximation systems.

Theorem 8. *Given an integer t and a wff $\alpha \in \mathcal{F}$, we can decide whether there exists an interpretation in $\mathcal{C}_\alpha$ with a domain of cardinality t, in which α is satisfiable.*

Proof. We shall construct a finite set F of MSAS^{D}-structures with domain of cardinality t and an assignment v, such that if α is satisfiable in any interpretation in $\mathcal{C}_\alpha$ with domain of cardinality t, then it must be satisfiable in an interpretation in $\mathcal{C}_\alpha$ based on some MSAS^{D}-structure from F under the assignment v.

Let $c_{i_1}, c_{i_2}, \ldots, c_{i_n}$ and $x_{j_1}, x_{j_2}, \ldots, x_{j_r}$ give respectively the complete list of constants and free variables of α, with $i_1 < i_2 < \ldots < i_n$, $j_1 < j_2 < \ldots < j_r$. Now consider any set U with $|U| = t$. The set F consists of all MSAS^{D}-structures $(U, \{R_P\}_{P \subseteq_f N})$ with cardinality $|N| \geq i_n + r$, that satisfy conditions (1)–(4) below.

Suppose m is the number of distinct equivalence relations on U. These relations may be assigned to the positions $i_1, \ldots, i_n, i_n + 1, \ldots, i_n + r \in N$ in $m^{n+r} = g(\text{say})$ different ways. Let $A := \{A_1^{l_1}, A_2^{l_2}, \ldots, A_g^{l_g}\}$ be the set of all distinct assignments of the m relations to these positions. The superscript l_s on $A_s^{l_s}$ indicates that l_s of the $n + r$ relations assigned by $A_s^{l_s}$, are distinct.

(1) The relations $R_{i_1}, \ldots, R_{i_n}, R_{i_n+1}, \ldots, R_{i_n+r}$ in the MSAS^{D}-structure $(U, \{R_P\}_{P \subseteq_f N})$, $|N| \geq i_n + r$, must be determined by an assignment in A. Consider any such MSAS^{D}-structure, corresponding to the member $A_s^{l_s}$, say, in A.
(2) The cardinality $|N|$ must be such that $|N| \leq i_n + r + (m - l_s)$. If $m > l_s$, at the positions $i_n + r + 1, \ldots, |N|$, we must have relations that are distinct from each other, as well as from each of $R_{i_1}, \ldots, R_{i_n}, R_{i_n+1}, \ldots, R_{i_n+r}$.
(3) For $j < i_n + r$ with $j \notin \{i_1, \ldots, i_n, i_n + 1, \ldots, i_n + r\}$, $R_j = R_l$, for some l in $\{i_1, \ldots, i_n, i_n + 1, \ldots, i_n + r, i_n + r + 1, \ldots, |N|\}$.

Furthermore,

(4) no two MSAS^{D}s-structure corresponding to the same $A_s^{l_s}$, have the same *set* of relations.

For a fixed $A_s^{l_s}$, the number of corresponding MSAS^{D}-structures in F is

$$C_0^{m-l_s} + C_1^{m-l_s} + \ldots + C_{m-l_s-1}^{m-l_s} + C_{m-l_s}^{m-l_s} = d_s \text{ (say)}.$$

As F is just the union of the collections of MSAS^{D}-structures corresponding to all the assignments in A (satisfying (2)–(4)), $|F| = \sum_{s=1}^{g} d_s$, making F finite.

Let $P(\alpha)$ be the set of all propositional variables occurring in α and V_1, V_2, ..., V_k be the list of all distinct mappings from $P(\alpha)$ to 2^U. Let v be an assignment such that $v(x_{j_q}) = i_n + q$, $q = 1, 2, \ldots, r$. From Proposition 17, it follows that if α is satisfiable in an interpretation with domain of cardinality t, then it must be satisfiable in one of the $\alpha-$standard interpretations based on a MSAS^{D}-structure in F with the valuation from the set $\{V_1, V_2, \ldots, V_k\}$ under the assignment v. So by checking the satisfiability under v in all $\alpha-$ standard interpretations (finite in number) based on the MSAS^{D}-structures of F with the valuation from the set $\{V_1, V_2, \ldots, V_k\}$, we can decide whether α is satisfiable in an interpretation with domain of cardinality t. □

Let us illustrate the construction of the set F in Theorem 8 for the wff $\alpha := [c_1]p \wedge ([c_3]p \rightarrow (\forall x[x]q \wedge [x_2]p))$, and $t = 3$. For α, $i_n = 3, r = 1$. Take any U with cardinality 3. Then $m = 5$; denote by $\rho_1, \rho_2, \rho_3, \rho_4, \rho_5$ the equivalence relations on U. MSAS^{D}-structures $(U, \{R_P\}_{P \subseteq_f N})$ in F would have cardinality $|N| \geq 4$, and the relations R_1, R_3, R_4 would be determined by assignments from $A := \{A_1^{l_1}, \ldots, A_g^{l_g}\}$, $g = 5^3$. Let us consider A_s^2. By condition (2), the MSAS^{D}-structures corresponding to A_s^2 would have $|N| \leq 7$. Suppose $|N| = 6$, and A_s^2 defines $R_1 := \rho_5$, $R_3 := \rho_2$, $R_4 := \rho_5$. We consider all MSAS^{D}-structures $(U, \{R_P\}_{P \subseteq_f N})$ such that (i) $R_1 = \rho_5, R_3 = \rho_2, R_4 = \rho_5$ (ii) R_5, R_6 are distinct and are taken from the set $\{\rho_1, \rho_3, \rho_4\}$ (condition-2); R_2 can be any of the relations $R_1, R_3 - R_6$ (condition-3). Now, for example, both $R_1 = \rho_5, R_2 = \rho_3, R_3 = \rho_2, R_4 = \rho_5, R_5 = \rho_3, R_6 = \rho_4$ and $R_1 = \rho_5, R_2 = \rho_2, R_3 = \rho_2, R_4 = \rho_5, R_5 = \rho_4, R_6 = \rho_3$ satisfy the above. However, due to condition (4), we choose any one of them. A_s^2 would thus give $8(= d_s)$ MSAS^{D}-structures. Finally, F is the union of the collections of MSAS^{D}-structures so obtained, corresponding to each assignment $A_s^{l_s}$ in A.

From Proposition 18 and Theorem 8, we obtain,

Corollary 2. *Given an integer t and a wff $\alpha \in \mathcal{F}$, we can decide whether there exists a* MSAS^{D} *interpretation with domain of cardinality t, in which α is satisfiable.*

The assumption of finiteness of the domain in this result, entails that there is a finite number of distinct equivalence relations on the domain. However, it should be noted that in the class of interpretations where decidability is being checked, there is no restriction on the *cardinality of the* MSAS^{D}*-structures.* This is in contrast to the assumption of Theorem 7 – the class of interpretations considered there, must have a given finite cardinality.

6.1 A Decidable Fragment

Let us now present a decidable fragment of LMSAS^{D}. Following the notation of first order logic [14], a wff α is said to be Σ_0^0, Σ_1^0 or Σ_2^0 according as it is of the form β, $\exists x_1 \exists x_2 \cdots \exists x_n \beta$, or $\exists x_1 \exists x_2 \cdots \exists x_n \forall y_1 \forall y_2 \cdots \forall y_m \beta$ respectively, where β does not involve any quantifier. The following is immediate.

Proposition 45. *Consider a Σ_0^0 wff $\alpha \in \mathcal{F}$. Let $\mathfrak{M} := (\mathfrak{F}, V, I)$ and $\mathfrak{M}' := (\mathfrak{F}', V, I')$ be two* MSAS^{D}*-interpretations with the same domain U and v, v' two assignments such that $R_{\tilde{v}(t)} = R_{\tilde{v'}(t)}$ for all $t \in Con(\{\alpha\}) \cup Var(\{\alpha\})$. Then for all $w \in U$,*

$$\mathfrak{M}, v, w \models \alpha \text{ if and only if } \mathfrak{M}', v', w \models \alpha.$$

The following proposition is crucial to obtain the decidable fragment.

Proposition 46. *Let α be a Σ_2^0 wff with n and m existential and universal quantifiers respectively. Let $z_1, z_2, \ldots, z_r$ be the complete list of free variables occurring in α. If α is satisfiable in an $\alpha-$standard interpretation $\mathfrak{M} := (\mathfrak{F}, V, I_\alpha)$, then it is also satisfiable in a* MSAS^{D}*-interpretation $\mathfrak{M}' := (\mathfrak{F}', V, I_\alpha)$, where each relation occurring in $\mathfrak{F}'$ also occurs in $\mathfrak{F}$ and $|\mathfrak{F}'| = k+n+r$, k being the largest integer such that $c_k \in Con(\{\alpha\})$.*

Proof. Let $\alpha := \exists x_1 \exists x_2 \cdots \exists x_n \forall y_1 \forall y_2 \cdots \forall y_m \beta$, where β does not contain any quantifier, such that it is satisfiable in the $\alpha-$standard interpretation $\mathfrak{M} := (\mathfrak{F}, V, I_\alpha)$, $\mathfrak{F} := (U, \{R_P\}_{P \subseteq_f N})$, under the assignment v. So for some $w \in U$, $\mathfrak{M}, v, w \models \alpha$. Then there exists an assignment v_1 such that $v_1(x) = v(x)$ for $x \notin \{x_1, x_2, \ldots, x_n\}$ and

$$\mathfrak{M}, v_1, w \models \forall y_1 \forall y_2 \cdots \forall y_m \beta \tag{7}$$

Consider the $\alpha-$standard interpretation $\mathfrak{M}' := (\mathfrak{F}', V, I_\alpha)$, where $\mathfrak{F}' := (U, \{R'_P\}_{P \subseteq_f M})$, $M := \{1, 2, \ldots, k+n+r\}$ and
(i) $R'_j = R_j$, $j \in \{1, 2, \ldots, k\}$, (ii) $R'_{k+l} = R_{v_1(x_l)}$, $l \in \{1, 2, \ldots, n\}$, and
(iii) $R'_{k+n+l} = R_{v_1(z_l)}$, $l \in \{1, 2, \ldots, r\}$.
Let v'_1 be an assignment for $\mathfrak{M}'$ such that
$v'_1(x_i) = k+i$, $x_i \in \{x_1, x_2, \ldots, x_n\}$ and $v'_1(z_i) = k+n+i$, $z_i \in \{z_1, z_2, \ldots, z_r\}$.
We shall prove that

$$\mathfrak{M}', v'_1, w \models \forall y_1 \forall y_2 \cdots \forall y_m \beta. \tag{8}$$

If possible, let (8) not hold. Then there exists an assignment v'_2 such that $v'_2(y) = v'_1(y)$ for $y \notin \{y_1, y_2, \ldots, y_m\}$ and

$$\mathfrak{M}', v'_2, w \not\models \beta. \tag{9}$$

Let us consider an assignment v_2 for the interpretation $\mathfrak{M}$ such that the following are satisfied. (i) $v_2(y) = v_1(y)$, if $y \notin \{y_1, y_2, \ldots, y_m\}$. (ii) For $y \in \{y_1, y_2, \ldots, y_m\}$, $v_2(y) = j$, if $v'_2(y) = j \in \{1, 2, \ldots, k\}$ and $v_2(y) = v_1(x_j)$,

if $v_2'(y) = k+j \in \{k+1, k+2, \ldots, k+n\}$. (iii) $v_2(y) = v_1(z_j)$, if $v_2'(y) = k+n+j \in \{k+n+1, k+n+2, \ldots, k+n+r\}$.
Note that $R'_{\tilde{v_2'(t)}} = R_{\tilde{v_2}(t)}$ for all $t \in Con(\{\beta\}) \cup Var(\{\beta\})$.
In fact, for $t = x_i$, $i \in \{1, 2, \ldots, n\}$, $R'_{\tilde{v_2'(t)}} = R'_{\tilde{v_1'(t)}} = R'_{k+i} = R_{\tilde{v_1}(x_i)} = R_{\tilde{v_2}(x_i)}$.
For $t = z_i$, $i \in \{1, 2, \ldots, r\}$, $R'_{\tilde{v_2'(t)}} = R'_{\tilde{v_1'(t)}} = R'_{k+n+i} = R_{\tilde{v_1}(z_i)} = R_{\tilde{v_2}(z_i)}$.
For $t = y_i \in \{y_1, y_2, \ldots, y_m\}$ with $v_2'(y) = j \in \{1, 2, \ldots, k\}$, $R'_{\tilde{v_2'(t)}} = R'_j = R_j = R_{\tilde{v_2}(t)}$.
For $t = y_i \in \{y_1, y_2, \ldots, y_m\}$ such that $v_2'(y) = k+j \in \{k+1, k+2, \ldots, k+n\}$, $R'_{\tilde{v_2'(t)}} = R'_{k+j} = R_{\tilde{v_1}(x_j)} = R_{\tilde{v_2}(t)}$.
For $t = y_i \in \{y_1, y_2, \ldots, y_m\}$ with $v_2'(y) = k+n+j \in \{k+n+1, k+n+2, \ldots, k+n+r\}$,
$R'_{\tilde{v_2'(t)}} = R'_{k+n+j} = R_{\tilde{v_1}(z_j)} = R_{\tilde{v_2}(t)}$.
Therefore, from Proposition 45 and (9), we obtain $\mathfrak{M}, v_2, w \not\models \beta$. But since $v_2(y) = v_1(y)$ for all $y \notin \{y_1, y_2, \ldots, y_m\}$, $\mathfrak{M}, v_1, w \not\models \forall y_1 \forall y_2 \ldots \forall y_m \beta$. This contradicts (7). Therefore, (8) holds. □

From Theorems 3, 6 and Proposition 46, we obtain the following theorem.

Theorem 9. *Satisfiability of a Σ_2^0 wff can be decided.*

Observe that for the satisfiability of a Σ_2^0 wff $\exists x_1 \exists x_2 \cdots \exists x_n \forall y_1 \forall y_2 \cdots \forall y_m \beta$, the choice of the relations attached with the modalities for x_i's does not vary with the relations attached with the modalities for y_i's. This fact is used to obtain Proposition 46. But the situation changes for other wffs. For instance, in case of the wff $\forall y \exists x \langle y \rangle \langle x \rangle p$, p being a propositional variable, we require different relations to be attached with the modality $\langle x \rangle$ depending on the relation attached with the modality $\langle y \rangle$. So Proposition 46 fails for this wff, as shown by the following example.

Example 12. Let $\alpha := \forall y \exists x \langle y \rangle \langle x \rangle p$ and $(\mathfrak{F}, V, I_\alpha)$ the α-standard interpretation where $\mathfrak{F} := (U, \{R_P\}_{P \subseteq_f \mathbb{N}})$, $U := \{w, w'\} \cup \{w_i : i \in \mathbb{N}\}$, $V(p) := \{w'\}$ and

$U/R_1 := \{\{w, w_1\}, \{w'\}, \{w_j : j \geq 2\}\}$, while
$U/R_i := \{\{w, w_i\}, \{w_{i-1}, w'\}, \{w_j : j \neq i, i-1\}\}$, for $i \geq 2$.

Note that for all assignments v, we have $\mathfrak{M}, v, w \models \alpha$. Moreover, if the relation R_i is attached with the modality $\langle y \rangle$, then we must attach the relation R_{i+1} with the modality $\langle x \rangle$. Thus, if any of the relations is removed, we lose satisfiability of the wff in the resultant MSAS^{D}. Formally, $\mathfrak{M}', v, w \not\models \alpha$ for any α-standard interpretation $\mathfrak{M}' := (\mathfrak{F}', V, I_\alpha)$, where $\mathfrak{F}' := (U, \{R'_P\}_{P \subseteq_f M})$ with $\{R'_i\}_{i \in M} \subsetneq \{R_i\}_{i \in \mathbb{N}}$.

7 Algebraic Semantics for LMSAS^{D}

In this section, we present an algebraic semantics for LMSAS^{D}, which appeared in [47].

Let us recall the Definition 14 of Boolean algebra with operators (*BAO*). In this section, we are interested only in those (complete) *BAOs* where $\Delta = 2^{\mathbb{N}}_f$ (the set of all finite subsets of $\mathbb{N}$) and each f_k, f_l, $l, k \subseteq_f \mathbb{N}$ satisfies the following additional conditions:

(B1) $f_k a \leq f_k f_k a$;
(B2) $f_k a \leq a$;
(B3) $a \leq f_k g_k a$, where $g_k := \sim f_k \sim$;
(B4) $f_k a \cup f_l a \leq f_{k \cup l} a$ and
(B5) $f_\emptyset a \leq f_k a$.

The class of all complete *BAOs* which satisfies (B1)–(B5) is denoted by $\mathfrak{C}$. We shall obtain completeness of LMSAS^{D} with respect to the class $\mathfrak{C}$.

Definition 33. *Let* $\mathfrak{A} := (A, \cap, \sim, 1, \{f_k\}_{k \subseteq_f \mathbb{N}})$ *be a BAO satisfying (B1)–(B5). By an* assignment *in* $\mathfrak{A}$, *we mean a function* $\theta : PV \to A$. θ *can be extended uniquely to a meaning function* $\tilde{\theta} : \overline{\mathcal{F}} \to A$ *as follows:*

$\tilde{\theta}(\neg\alpha) = \sim \tilde{\theta}(\alpha)$;
$\tilde{\theta}(\alpha \wedge \beta) = \tilde{\theta}(\alpha) \wedge \tilde{\theta}(\beta)$;
$\tilde{\theta}([t]\alpha) = f_k(\tilde{\theta}(\alpha))$, $k = \{i \in \mathbb{N} : c_i \in B(t)\}$;
$\tilde{\theta}(A\alpha) = f_\emptyset \tilde{\theta}(\alpha)$ *and*
$\tilde{\theta}(\forall x \alpha) = \bigcap\{\tilde{\theta}(\alpha(c_j/x)) : j \in \mathbb{N}\}$, *provided the g.l.b. exists.*

We define $\tilde{\theta}(\alpha) := \tilde{\theta}(cl(\alpha))$ *for* $\alpha \in \mathcal{F}$, *where* $cl(\alpha)$ *denotes the closure of the wff* α.

Note that in order to define the natural translation corresponding to all possible assignments from closed LMSAS^{D} wffs to elements of *BAO*, we only require the existence of joins and meets of the set of the form $\{\tilde{\theta}(\alpha(c_j/x)) : j \in \mathbb{N}\}$, where α is an LMSAS^{D} wff with only one free variable x and θ is an assignment. This motivates us to define a realization for LMSAS^{D} in the line of realization of first order formalized languages [89].

Definition 34. *A BAO* $\mathfrak{A} := (A, \cap, \sim, 1, \{f_k\}_{k \subseteq_f \mathbb{N}})$ *satisfying (B1)–(B5) is said to be a* realization for LMSAS^{D}, *if for every assignment* $\theta : PV \to A$, $\tilde{\theta}(\alpha)$ *exists for all* $\alpha \in \overline{\mathcal{F}}$.

Note that every complete *BAO* satisfying (B1)–(B5) is a realization for LMSAS^{D}. But not all realizations for LMSAS^{D} are complete *BAO*. For instance, if a *BAO* $\mathfrak{A}$ satisfying (B1)–(B5), has only one distinct function symbol, then each set $\{\tilde{\theta}(\alpha(c_j/x)) : j \in \mathbb{N}\}$ will be singleton and thus $\mathfrak{A}$ becomes a realization which may not necessarily be complete.

Consider the complex algebra of a MSAS^{D}-structure (cf. Definition 15). We obtain the following.

Proposition 47. *Every complex algebra of a semi* MSAS^{D}*-structure is a complete BAO satisfying (B1)–(B5).*

Proof. Let $\mathfrak{F} := (U, \{R_P\}_{P \subseteq_f N})$ be a semi MSAS^{D}-structure. Since $\mathfrak{F}^+$ is a power set algebra, we only need to prove (B1)–(B5).

(B1): $x \in m_{R_k}(X)$
$\Rightarrow$ for all y such that $(x, y) \in R_k$, $y \in X$
$\Rightarrow$ for all y and z such that $(x, y) \in R_k$ and $(y, z) \in R_k$, $z \in X$ ($\because$ R_k is transitive)
$\Rightarrow x \in m_{R_k} m_{R_k}(X)$.

(B2): $x \in m_{R_k}(X)$
$\Rightarrow$ for all y such that $(x, y) \in R_k$, $y \in X$
$\Rightarrow x \in X$ ($\because$ R_k is reflexive).

(B3): $x \in \sim m_{R_k} \sim m_{R_k}(X)$
$\Rightarrow$ there exists y such that $(x, y) \in R_k$ and $y \in m_{R_k}(X)$
$\Rightarrow$ there exists y such $(x, y) \in R_k$ and for all z with $(y, z) \in R_k$, $z \in X$
$\Rightarrow x \in X$ ($\because$ R_k is symmetric).

(B4): $x \in m_{R_k}(X)$
$\Rightarrow$ for all y such that $(x, y) \in R_k$, $y \in X$
$\Rightarrow$ for all y such that $(x, y) \in R_{k \cup l}$, $y \in X$ ($\because$ $R_{k \cup l} \subseteq R_k$)
$\Rightarrow x \in m_{R_{k \cup l}}(X)$.
Therefore, $m_{R_k}(X) \subseteq m_{R_{k \cup l}}(X)$. Similarly, we obtain $m_{R_l}(X) \subseteq m_{R_{k \cup l}}(X)$ and hence $m_{R_k}(X) \cup m_{R_l}(X) \subseteq m_{R_{k \cup l}}(X)$.

(B5): $x \in m_{R_\emptyset}(X)$
$\Rightarrow$ for all y such that $(x, y) \in R_\emptyset$, $y \in X$
$\Rightarrow$ for all y such that $(x, y) \in R_k$, $y \in X$ ($\because$ $R_\emptyset \subseteq R_k$)
$\Rightarrow x \in m_{R_k}(X)$. □

Let us denote the class of all realizations of LMSAS^{D} and complex algebras of semi MSAS^{D}-structures by $\mathfrak{R}$ and $\mathfrak{Cm}$ respectively. So we have $\mathfrak{Cm} \subseteq \mathfrak{C} \subseteq \mathfrak{R}$.

Definition 35. *Let* $\mathfrak{A} := (A, \cap, \sim, 1, \{f_k\}_{k \subseteq_f \mathbb{N}})$ *be a realization for* LMSAS^{D}. *Then we write* $\mathfrak{A} \Vdash \alpha \approx \beta$ *if and only if for every assignment* $\theta : PV \to A$, $\tilde{\theta}(\alpha) = \tilde{\theta}(\beta)$. *We simply write* $\mathfrak{R} \Vdash \alpha$ *if* $\mathfrak{A} \Vdash \alpha \approx \top$ *for all* $\mathfrak{A} \in \mathfrak{R}$. *Similarly we write* $\mathfrak{C} \Vdash \alpha$ *and* $\mathfrak{Cm} \Vdash \alpha$ *according as* $\mathfrak{A} \Vdash \alpha \approx \top$ *for all* $\mathfrak{A} \in \mathfrak{C}$ *or* $\mathfrak{A} \in \mathfrak{Cm}$ *respectively.*

Proposition 48 (Soundness). *If* $\vdash_\Lambda \alpha$ *then* $\mathfrak{R} \Vdash \alpha$ *and hence* $\mathfrak{C} \Vdash \alpha$ *and* $\mathfrak{Cm} \Vdash \alpha$.

Proposition 49 (Completeness). *For* $\alpha \in \mathcal{F}$, *if* $\mathfrak{C} \Vdash \alpha$, *then* $\vdash_\Lambda \alpha$.

We begin our journey to prove the above completeness theorem with the Lindenbaum algebra $\mathfrak{Ln}$ for LMSAS^{D}. In fact, giving exactly the same argument as in the modal logic case, one can easily show that $\mathfrak{Ln} := (\overline{\mathcal{F}}/\equiv, \cap, \sim, 1, \{f_k\}_{k \subseteq_f \mathbb{N}})$, where $1 = [\top]$, and $f_k([\alpha]) := [[t]\alpha]$, t being a term consisting of only constants with $k = \{i : i \in B(t)\}$, is a BAO. Moreover, axioms 7–9, 4, 5 give us the properties (B1)–(B5). $\mathfrak{Ln}$ is, in fact, a realization for LMSAS^{D}. But in order to prove this, we need a few more definitions and results.

Let $p_1, p_2, \ldots$ be an enumeration of the propositional variables and $\theta' : PV \to \overline{\mathcal{F}}/\equiv$ be an assignment. Let $\alpha_1, \alpha_2, \ldots$ be countably many distinct wffs such that $\theta'(p_i) := [\alpha_i]$. For a given wff α, α^* denotes the wff obtained from α by uniform replacement of propositional variables p_i's by α_i's. By induction on the complexity of α, we obtain

Proposition 50. *The wff $(\alpha(c_j/x))^*$ is same as the wff $\alpha^*(c_j/x)$, $j \in \mathbb{N}$.*

Proof. By induction on the complexity of α. We only consider the case when α is of the form $[t]\beta$ or $\forall y\beta$.

Let α is of the form $[t]\beta$. Then
$(([t]\beta)(c_j/x))^*$
$= ([t']\beta(c_j/x))^*$, where t' is obtained from t by replacing x with c_j
$= [t'](\beta(c_j/x))^*$
$= [t']\beta^*(c_j/x)$ (induction hypothesis)
$= ([t]\beta)^*(c_j/x)$.

Next, suppose α is of the form $\forall y\beta$. If x is different from y, then
$((\forall y\beta)(c_j/x))^*$
$= \forall y(\beta(c_j/x))^*$
$= \forall y\beta^*(c_j/x)$ (induction hypothesis)
$= (\forall y\beta)^*(c_j/x)$.
If α is of the form $\forall x\beta$, then,
$((\forall x\beta)(c_j/x))^* = (\forall x\beta)^* = \forall x\beta^* = (\forall x\beta^*)(c_j/x) = (\forall x\beta)^*(c_j/x)$. □

Proposition 51. *Consider $\overline{\mathcal{F}}/\equiv$. Then for any $\alpha \in \mathcal{F}$ which has only x as free variable, $\bigcap_j[\alpha(c_j/x)]$ exists and is given by $[\forall x\alpha]$.*

Proof. It follows from the facts that (i) $\vdash_\Lambda \forall x\alpha \to \alpha(c_i/x)$, for all $i \in \mathbb{N}$ and (ii) $\vdash_\Lambda \delta \to \alpha(c_i/x)$ for all $i \in \mathbb{N}$, implies $\vdash_\Lambda \delta \to \forall x\alpha$. □

We use Propositions 50 and 51 to get

Proposition 52. $\tilde{\theta}'(\alpha) = [\alpha^*]$, *for all $\alpha \in \overline{\mathcal{F}}$.*

Proof. By induction on the complexity of α. We only provide the proof for the case when α is of the form $\forall x\beta$ or $[t]\beta$.
Here, $\tilde{\theta}'(\forall x\beta)$
$= \bigcap_j \tilde{\theta}'(\beta(c_j/x))$
$= \bigcap_j[(\beta(c_j/x))^*]$ (by induction hypothesis)
$= \bigcap_j[\beta^*(c_j/x)]$ (by Proposition 50)
$= [\forall x\beta^*]$ (by Proposition 51)
$= [(\forall x\beta)^*]$.

Now, let us consider the case when α is of the form $[t]\beta$. Let $B(t) = \{c_{i_1}, c_{i_2}, \ldots, c_{i_m}\}$ and $k = \{i : c_i \in B(t)\}$.

Now, $\tilde{\theta}'([t]\beta)$
$= f_k\tilde{\theta}'(\beta)$
$= f_k[\beta^*]$ (by induction hypothesis)
$= [[t]\beta^*]$,
$= [([t]\beta)^*]$. □

Thus, we obtain

Proposition 53. *$\mathfrak{Ln} := (\overline{\mathcal{F}}/\equiv, \cap, \sim, 1, \{f_k\}_{k\subseteq_f \mathrm{N}})$ is a realization for* LMSAS^{D}.

Due to Proposition 53, we obtain the completeness theorem with respect to the class of all realizations. But, as mentioned earlier, we want the completeness with respect to the class $\mathfrak{C}$. It can be shown, as in the propositional logic case, that the Lindenbaum algebra $\mathfrak{Ln}$ defined above is *not* a complete BAO and so we need to do some more work in order to get the completeness theorem with respect to the class $\mathfrak{C}$. Note that we would achieve our goal if we could embed any LMSAS^{D} realization $\mathfrak{A} := (A, \cap, \sim, 1, \{f_k\}_{k\subseteq_f \mathrm{N}})$ into some complex algebra. At this point one may think of the BAO consisting of all subsets of the set of all prime filters of the BAO $\mathfrak{A}$, as described in the Jónsson-Tarski theorem (cf. Theorem 1). But the embedding given in this theorem may not preserve infinite joins and meets. This problem could be overcome if we consider the BAO consisting of all subsets of the set of all Q-filters (cf. Definition 17) instead of ultra-filters. Here, Q is a countably infinite collection of subsets of A satisfying certain conditions and the embedding obtained in this case preserves all the infinite joins and meets *in* Q. Since this embedding may not preserve *all* existing joins and meets, the question again arises whether even this embedding will be able to give us the desired result? The answer is yes.

Let us note the following trivial fact about the Q-filter frames (cf. Sect. 2.3).

Proposition 54. *If $\mathfrak{A} := (A, \cap, \sim, 1, \{f_k\}_{k\subseteq_f \mathrm{N}})$ be a BAO satisfying (B1)–(B5), then $\mathfrak{A}_Q$ is a semi* MSAS^{D}*-structure.*

Let us consider the Lindenbaum algebra $\mathfrak{Ln}$ and the canonical assignment θ^c which maps propositional variables to its class, i.e. $\theta^c(p) = [p]$. For each wff α with a single free variable x, let us define the set $Q_\alpha := \{\tilde{\theta}^c(\alpha(c_j/x)) : j \in \mathrm{N}\}$ and let $Q := \{Q_\alpha : \alpha$ has the single free variable $x\}$. Note that Q is countable. Take an enumeration $\{X_n\}_{n\in\mathrm{N}}$ and $\{Y_n\}_{n\in\mathrm{N}}$ of the set $Q_* := \{Q_\alpha \in Q : \bigcap Q_\alpha \in \overline{\mathcal{F}}/\equiv\}$ and $Q^* := \{Q_\alpha \in Q : \bigcup Q_\alpha \in \overline{\mathcal{F}}/\equiv\}$ respectively. Then it is not difficult to obtain:

Proposition 55. *Q satisfies the condition (QF1)–(QF3) of Theorem 2.*

Proof. (QF1): Consider a $Q_\alpha = \{\tilde{\theta}^c(\alpha(c_j/x)) : j \in \mathrm{N}\}$. We will show

$$f_k \bigcap_j \tilde{\theta}^c(\alpha(c_j/x)) = \bigcap_j f_k\tilde{\theta}^c(\alpha(c_j/x)).$$

Let t be a term consisting of only constants such that $k = \{i : c_i \in B(t)\}$. Then

$f_k \bigcap_j \tilde{\theta^c}(\alpha(c_j/x))$
$= f_k[\forall x\alpha]$ (by Propositions 51 and 52)
$= [[t]\forall x\alpha]$
$= [\forall x[t]\alpha]$ ($\because \vdash_\Lambda [t]\forall x\alpha \leftrightarrow \forall x[t]\alpha$ as $x \notin B(t)$)
$= \bigcap_j \tilde{\theta^c}([t]\alpha(c_j/x))$
$= \bigcap_j f_k\tilde{\theta^c}(\alpha(c_j/x))$.

(QF2): Let $k \subseteq_f \mathbb{N}$, $[\beta] \in \overline{\mathfrak{F}}/_\equiv$ and consider the set $\{\tilde{\theta^c}(\alpha(c_j/x)) : j \in \mathbb{N}\}$, where α has a single free variable x.

$\{f_k([\beta] \to \tilde{\theta^c}(\alpha(c_j/x))); j \in \mathbb{N}\}$
$= \{f_k(\tilde{\theta^c}(\beta) \to \tilde{\theta^c}(\alpha(c_j/x))); j \in \mathbb{N}\}$ (by Proposition 52)
$= \{f_k(\tilde{\theta^c}(\beta \to \alpha(c_j/x))); j \in \mathbb{N}\}$
$= \{\tilde{\theta^c}([t](\beta \to \alpha(c_j/x))); j \in \mathbb{N}\}$, where t is a term such that $k = \{i : c_i \in B(t)\}$
$= \{\tilde{\theta^c}(([t](\beta \to \alpha))(c_j/x)); j \in \mathbb{N}\} \in Q$ ($\because B(t)$ does not have any variables and β is closed wff).
(QF3) can be proved similarly. □

Therefore, from Theorem 2, we obtain,

Proposition 56. *There exists a BAO embedding r of $\mathfrak{Ln}$ into $(\mathfrak{Ln}_Q)^+$ such that $r(\bigcap_j \tilde{\theta^c}(\alpha(c_j/x))) = \bigcap_j r(\tilde{\theta^c}(\alpha(c_j/x)))$.*

We note that by Proposition 54, $\mathfrak{Ln}_Q$ is a semi MSAS^D-structure and hence by Proposition 47, $(\mathfrak{Ln}_Q)^+$ is a complete *BAO* satisfying (B1)–(B5).

Proposition 57. *Consider the assignment γ in the BAO $(\mathfrak{Ln}_Q)^+ \in \mathfrak{Cm}$ defined as $\gamma(p) := r([p])$, $p \in PV$. Then $\tilde{\gamma}(\alpha) = r([\alpha])$ for all $\alpha \in \overline{\mathcal{F}}$.*

Proof. Proof is by induction on the complexity of the wff α. Basis case is given, so we only need to prove the induction case. Here, we only consider the case when α is of the form $\forall x\beta$ or $[t]\beta$. Let us first consider α of the form $\forall x\beta$.

Here, $\tilde{\gamma}(\forall x\beta)$
$= \bigcap_j \tilde{\gamma}(\beta(c_j/x))$
$= \bigcap_j r([\beta(c_j/x)])$ (induction hypothesis)
$= \bigcap_j r(\tilde{\theta^c}(\beta(c_j/x)))$ (by Proposition 52)
$= r(\bigcap_j \tilde{\theta^c}(\beta(c_j/x)))$ (by Proposition 56)
$= r(\tilde{\theta^c}(\forall x\beta))$
$= r([\forall x\beta])$ (by Proposition 52). □

Proposition 58.

(i) For $\alpha \in \overline{\mathcal{F}}$, $\mathfrak{Cm} \Vdash \alpha$ implies $\vdash_\Lambda \alpha$.
(ii) For $\alpha \in \mathcal{F}$, $\mathfrak{Cm} \Vdash \alpha$ implies $\vdash_\Lambda \alpha$.

Proof. (i) If possible, let $\nvdash_{\Lambda} \alpha$. Then $[\alpha] \neq 1$. Now, consider the algebra $(\mathfrak{Ln}_Q)^+ \in \mathfrak{Cm}$ and the assignment γ defined in Proposition 57. Since $[\alpha] \neq 1$, we have $r([\alpha]) \neq 1$. Therefore, $\tilde{\gamma}(\alpha) \neq 1$, a contradiction.
(ii) If possible, let $\nvdash_{\Lambda} \alpha$. Then $\nvdash_{\Lambda} cl(\alpha)$ and hence by (i), we obtain a $\mathfrak{A} \in \mathfrak{Cm}$ and an assignment θ in $\mathfrak{Cm}$ such that $\tilde{\theta^c}(cl(\alpha)) \neq 1$ and thus we obtain $\tilde{\theta^c}(\alpha) \neq 1$. □

So Proposition 49 follows from Proposition 58 and the fact that $\mathfrak{Cm} \subseteq \mathfrak{C}$.

We end this section with the remark that the soundness and completeness theorems establish a strong connection between the MSAS$^{\rm D}$s and the class $\mathfrak{C}$ of algebras. It follows that the operators f_i, f_s and f_w are the counterparts of the lower, strong lower and weak lower approximations respectively, where $f_s(a) := \bigcap_{i\in \mathrm{N}} f_i(a)$ and $f_w(a) := \bigcup_{i\in \mathrm{N}} f_i(a)$. Thus one could study the properties of MSAS$^{\rm D}$s involving the different notions of lower and upper approximations in the algebras of the class $\mathfrak{C}$ using these operators, and conversely. For instance, the properties $\underline{X^c}_s = (\overline{X}_w)^c$ and $\underline{(\overline{X}_w)}_w = \overline{(\overline{X}_s)}_w$ of MSAS$^{\rm D}$ correspond to the properties $f_s(\sim a) =\sim g_w a$ and $f_w g_w a = g_w g_s a$ of the algebras of $\mathfrak{C}$ respectively, where $g_s(a) := \bigcap_{i\in \mathrm{N}} g_i(a)$ and $g_w(a) := \bigcup_{i\in \mathrm{N}} g_i(a)$.

8 A Temporal Dimension: Dynamic Spaces

In classical rough set theory, most of the concepts discussed are static, in the sense that time does not play any role. In this section we study rough sets in the situation where the knowledge base changes with time. In terms of information systems, this change may be due to a variation in the set of attributes with time, or objects taking different attribute values at different time points, as in the case of dynamic information systems (cf. Definition 7). It is not difficult to see that addition or deletion of attributes may be required with inflow of information. On the other hand, availability of more information may also warrant enlarging the set of attribute values, due to say, a finer classification of categories. The knowledge base may also need to be updated due to lack of, or imprecise information about the attributes as in case of incomplete and non-deterministic information systems respectively.

In any case, to incorporate new information, one needs to update the old knowledge base. Thus, a sequence of information and corresponding updates determines a sequence giving the knowledge base at different time points. Our focus is on the behavior of rough sets in such a scenario, viz. when the knowledge base evolves with time. Objects of the domain may belong to positive, possible, boundary or negative regions of a rough set with respect to the current knowledge base, and may transit between these regions as there is an update with time. Our interest is in questions that arise in this context, and in expressing these questions in a formal logical framework. We consider the general situation where the relation on the domain representing the knowledge base is not necessarily an equivalence. Thus, we consider Pagliani's notion of dynamic spaces (cf. Sect. 2.1) restricted to finite families of relations.

We do not address the possible reasons behind the changes in the knowledge base (as described earlier) in this section. However, we shall try to address this issue in Sect. 12.

Definition 36. *A* dynamic space *is a finite sequence* $\mathfrak{F} := \mathcal{F}_1, \mathcal{F}_2, \ldots, \mathcal{F}_N$, *where* $\mathcal{F}_i := (W, P_i)$ *and* $P_i \subseteq W \times W$, $i = 1, 2, \ldots, N$.

$|N|$ is referred to as the *cardinality of* $\mathfrak{F}$ and is denoted by $|\mathfrak{F}|$. The elements of $\{1, 2, \ldots, N\}$ are the time points. Moreover, the relation P_t, $1 \leq t \leq N$ represents the information about the domain W of the objects at the time point t.

A dynamic space consisting of equivalence relations will be called a *dynamic approximation space* (DAS). Observe that a dynamic space is a finite collection of Kripke frames (on the same domain), and so we get different classes of dynamic spaces according as the relations in all the frames involved are of a certain type. The dynamic space consisting of Kripke I-frames, $I \in \{K, K4, T, B, S4, KTB, KB4, S5\}$, is called a *dynamic I space*, following standard modal logic nomenclature (cf. Table 4). Thus, the class of dynamic K spaces is actually the class of all dynamic spaces and the class of dynamic $S5$ spaces is just the class of DASs. As DASs are specially relevant to the classical rough sets, we devote Sect. 8.1 in exploring their properties. So, the knowledge base here is represented by a partition. We find that a flow of information results in different patterns of changes of the partition. For example, the gain of information with time may be such that hitherto indistinguishable objects become distinguishable. It is also possible that a stage t is reached after which the partition does not change - e.g. if the attributes deleted are dispensable, leading eventually to a reduct (cf. Definition 5) of the attribute set at stage t. A time point may also be reached where we can distinguish every object from another. Moreover, in such a situation, gain in information about additional attributes will not have any affect on the partition. Considering this variety of changes, different types of DASs are defined. The section also relates dynamic information systems with DASs.

Table 4. Different classes of dynamic spaces

Type of dynamic space	Relation type
Dynamic K space	Any binary relation
Dynamic $K4$ space	Transitive
Dynamic T space	Reflexive
Dynamic B space	Symmetric
Dynamic $S4$ space	Reflexive and transitive
Dynamic KTB space	Reflexive and symmetric
Dynamic $KB4$ space	Symmetric and transitive
Dynamic $S5$ space	Equivalence

In Sect. 8.2, we return to the general structure of a dynamic space and formulate a logic which can express the notions of approximations of sets (cf. Definitions 2 and 12) relative to time. Note that the temporal logic DIL defined for dynamic information systems [75] (cf. Sect. 2.2) can express the changes in attribute values of the objects with time, but, as mentioned in Sect. 2.2, due to the absence of modal operators for lower/upper approximations, it cannot express approximations of sets. Thus, we consider a language $\mathcal{L}$ having temporal operators as well as modal operators for 'necessity' and 'possibility'. Satisfiability of the wffs of $\mathcal{L}$ is defined in structures based on dynamic spaces. It is shown that the different dynamic I spaces as well as the different types of DASs mentioned earlier can be characterized by $\mathcal{L}$-wffs.

The content of this section is based on the article [7].

8.1 Dynamic Approximation Space

From the definition of dynamic information system (cf. Definition 7), it follows that corresponding to each time point, there is a deterministic information system: for $t \in T$, take the deterministic information system $\mathcal{DS}_t := (W, A, \{Val_a\}_{a \in A}, f_t)$, where $f_t(x, a) = f(x, t, a)$ for all $x \in W$ and $a \in A$. Thus given a dynamic information system $\mathcal{DS}$, we obtain a family of approximation spaces $\{\mathcal{C}(\mathcal{DS}_t)\}_{t \in T}$, where $\mathcal{C}$ is the map from the set of all deterministic information systems to the set of all approximation spaces defined by $\mathcal{C}(\mathcal{S}) := (W, Ind_{\mathcal{S}}(A))$, for $\mathcal{S} = (W, A, \{Val_a\}_{a \in A}, f)$. In particular, we have

Observation 2. With every finite time dynamic information system $\mathcal{DS}$, one can associate a unique DAS such that the ith approximation space in the collection is obtained from the information system corresponding to the ith point in T. Following Vakarelov [100], we call such a DAS *standard*, and denote it by $\mathfrak{F}_{\mathcal{DS}}$.

Using the fact that $\mathcal{C}$ is onto, it is easy to see that given a family $\{(W, P_i)\}_{i \in I}$ of approximation spaces with a linear order $<$ on the index set I, there is a dynamic information system $\mathcal{DS} := (W, A, \{Val_a\}_{a \in A}, I, <, f)$, with the set I indexing time and such that $\mathcal{C}(\mathcal{DS}_i) = (W, P_i)$, $i \in I$.
Thus, given a DAS $\mathfrak{F}$, we obtain a finite time dynamic information system $\mathcal{DS}$ such that $\mathfrak{F}_{\mathcal{DS}} = \mathfrak{F}$. In the line of Proposition 3, we have now arrived at

Proposition 59. *Every DAS is a standard DAS and conversely.*

As remarked in the Introduction, there may be various kinds of changes in the partition of a domain with time, resulting in different DASs. For instance, a DAS could reflect refinement of partitions on some domain, with the inflow of information. Further, in the ideal situation, the refinement (which may occur after a stage) would lead to the finest possible partition – that with singleton sets. We define the following types of DASs.

Let $\mathfrak{F} := \mathcal{F}_1, \mathcal{F}_2, \ldots, \mathcal{F}_N$, where $\mathcal{F}_i := (W, P_i)$, $i = 1, \ldots, N$.

Definition 37. $\mathfrak{F}$ *is*

- monotonic refined, *if* $P_{i+1} \subseteq P_i$ *for all* $i \in \{1, 2, \ldots, N-1\}$, $N \geq 2$;
- eventually monotonic refined, *if there exists a* j, $1 \leq j < N$ *such that* $P_{i+1} \subseteq P_i$ *for all* $i \in \{j, j+1, \ldots, N-1\}$;
- eventually discrete, *if* $\mathfrak{F}$ *has a discrete approximation space* $\mathcal{F}_i$ *(that is,* P_i *is the identity relation), and all* $\mathcal{F}_j$, $j > i$, *are discrete;*
- saturated, *if there exists* $i < N$ *such that* $P_i = P_j$ *for all* $j \geq i$; *and*
- ideal, *if* $\mathfrak{F}$ *is monotonic refined and has a discrete approximation space.*

$\mathfrak{F}$ is clearly saturated, if it is eventually discrete, or ideal. The only technical restriction is that in the former case, the $\mathcal{F}_i$ in the definition has $i < N$, and in the latter, the discrete approximation space in $\mathfrak{F}$ is some $\mathcal{F}_i$ with $i < N$. Let us observe some simple instances in terms of information systems, where we obtain some of the DASs defined above.

Example 13. Let $T := \{t_1, t_2, \ldots, t_n\}$ be the time points with the ordering $t_1 < t_2 < \ldots < t_n$. In contrast to the case of a dynamic information system, we may have the attribute set varying with time. Suppose no deletion of attributes takes place. Let $\mathcal{S}_{t_i} := (W, A_{t_i}, \{Val_a\}_{a \in A_{t_i}}, f_{t_i})$ be an information system such that

- $A_{t_i} \subseteq A_{t_{i+1}}$, $i = 1, 2, \ldots n-1$, (∗)
- if $a \in A_{t_i} \cap A_{t_j}$, we have $f_{t_i}(x, a) = f_{t_j}(x, a)$, for all $x \in W$.

The second condition asserts that if a is found amongst the attributes considered at two different time points, the value assigned for it to any object remains unchanged. So we have the collection of information systems $\mathfrak{S} := \{\mathcal{S}_{t_1}, \ldots, \mathcal{S}_{t_n}\}$. As in Observation 2, consider $\mathfrak{F}_{\mathfrak{S}}$. It is easy to see that $\mathfrak{F}_{\mathfrak{S}}$ is monotonic refined. The ideal situation for $\mathfrak{F}$ would of course require that we reach a time point, say t_k, such that for any two objects $x, y \in W$, there exists an attribute $a \in A_{t_k}$ with $f_{t_k}(x, a) \neq f_{t_k}(y, a)$. However, it could also be the case that there exists a j, $2 \leq j < n-1$ such that $A_{t_i} \subseteq A_{t_{i+1}}$, $i = j, j+1, \ldots n-1$, and then $\mathfrak{F}_{\mathfrak{S}}$ would be eventually monotonic refined.

We note that if we do not have the condition (∗) of Example 13, that is, we are also deleting attributes with time (for instance, when some attribute is not considered important enough), then $\mathfrak{F}_{\mathfrak{S}}$ may not be monotonic refined. In fact, there may be coarsening or refinement, or no pattern may be followed at all. If we disallow addition, but allow deletion of attributes, then there would be progressive coarsening of the partitions. In this case, if $\mathfrak{F}_{\mathfrak{S}}$ is saturated after time t_i (say), we know that the attributes deleted from A_{t_i} are *dispensable* [84]. In fact, one may eventually reach a *reduct* of A_{t_i}, that is, a minimal attribute set preserving the classification due to A_{t_i}.

Example 14. Let $\mathcal{DS} := (W, A, \{Val_a\}_{a \in A}, T, <, f)$ be a dynamic information system. Gain in information may result in a finer classification of categories in the attribute value set, making it change with time. Now suppose the categorization obeys the following for any $x, y \in W$, $a \in A$, $t, t' \in T$:

if $f(x, t, a) \neq f(y, t, a)$, then for $t < t'$, $f(x, t', a) \neq f(y, t', a)$.

For instance, at time t there may be categories of possessing 'between 2–4 moons' and 'between 5–7 moons', but in a refined categorization one may be able to determine if the objects have exactly 2, 3 or 4 moons. Clearly, $\mathfrak{F}_{\mathcal{DS}}$ is then monotonic refined. As before, if we reach a stage with an attribute such that all objects take different values for it, then $\mathfrak{F}_{\mathcal{DS}}$ would become ideal. But it could also be the case that $\mathfrak{F}_{\mathcal{DS}}$ is not of any of the defined types.

Observation 3.

- Suppose $\mathfrak{F}$ is monotonic refined and $x \in W$ is a positive/negative element of $X \subseteq W$ in (W, R_i), for some $i = 1, 2, \ldots, N$. Then x remains so in all the future approximation spaces, that is, in all (W, R_j), $j = i, i+1, \ldots, N$. Hence if a subset X is definable in some approximation space of $\mathfrak{F}$, it will remain so in all the future approximation spaces.
- Let $\mathfrak{F}$ be eventually discrete. Then every object of W finally becomes a 'definite' element, that is, an approximation space is eventually reached where that element is either positive or negative.

In the sequel, we shall see how these properties, among others, may be 'formally' expressed.

8.2 Logics for Dynamic Spaces

In this section we present a language $\mathcal{L}$ which can express the approximations of sets relative to time. The language $\mathcal{L}$ contains a unary modal connective $\Box$ (for *necessity*), unary temporal connectives $\oplus$ (*next*), $\ominus$ (*previous*) and the binary temporal connectives $\mathcal{U}$ (*until*), $\mathcal{S}$ (*since*). So wffs of $\mathcal{L}$ are given as:

$$\bot \mid \top \mid p \in PV \mid \neg\alpha \mid \alpha \wedge \beta \mid \Box\alpha \mid \ominus\alpha \mid \oplus\alpha \mid \alpha\mathcal{U}\beta \mid \alpha\mathcal{S}\beta,$$

where $\bot, \top$ are the logical constants for *false* and *true* respectively, and PV denotes the (countable) set of all propositional variables. Apart from the usual derived connectives $\vee, \rightarrow, \leftrightarrow, \Diamond$ (for *possibility*), there are the following:

$F\alpha := \top\mathcal{U}\alpha$ (*some time in the future*);
$G\alpha := \neg F\neg\alpha$ (*always in the future*);
$P\alpha := \top\mathcal{S}\alpha$ (*some time in the past*);
$H\alpha := \neg P\neg\alpha$ (*always in the past*).

Semantics. Let $\mathfrak{F} := \mathcal{F}_1, \mathcal{F}_2, \ldots, \mathfrak{F}_N$ be a dynamic space, where $\mathcal{F}_i := (W, P_i)$. A *valuation function* V on $\mathfrak{F}$ is a map from the set PV of propositional variables to 2^W. $\mathfrak{M} := (\mathfrak{F}, V)$ is called a *model*. Let $T_{\mathfrak{F}} := \{1, 2, \ldots, N\}$ be the set of time points.

Definition 38. *The satisfiability of a $\mathcal{L}$-wff α in a model $\mathfrak{M}$ at $t \in T_{\mathfrak{F}}$ and $w \in W$, denoted as $\mathfrak{M}, t, w \models \alpha$, is defined inductively:*

- $\mathfrak{M}, t, w \models \top$ *and* $\mathfrak{M}, t, w \not\models \bot$;

- *For each propositional variable* p, $\mathfrak{M}, t, w \models p$, *if and only if* $w \in V(p)$;
- *The standard definitions for the Boolean cases;*
- $\mathfrak{M}, t, w \models \Box\alpha$, *if and only if* $\mathfrak{M}, t, w' \models \alpha$ *for all* w' *such that* $(w, w') \in P_t$;
- $\mathfrak{M}, t, w \models \oplus\alpha$, *if and only if* $t < N$ *and* $\mathfrak{M}, t+1, w \models \alpha$;
- $\mathfrak{M}, t, w \models \ominus\alpha$, *if and only if* $t > 1$ *and* $\mathfrak{M}, t-1, w \models \alpha$;
- $\mathfrak{M}, t, w \models \alpha\mathcal{U}\beta$, *if and only if there exists* r *with* $t \leq r \leq N$ *such that* $\mathfrak{M}, r, w \models \beta$, *and for all* k *such that* $t \leq k < r$, $\mathfrak{M}, k, w \models \alpha$;
- $\mathfrak{M}, t, w \models \alpha\mathcal{S}\beta$, *if and only if there exists* r *with* $1 \leq r \leq t$ *such that* $\mathfrak{M}, r, w \models \beta$, *and for all* k *such that* $r < k \leq t$, $\mathfrak{M}, k, w \models \alpha$.

Remark 2. Conditions of satisfiability of the derived connectives F, G, P and H are then obtained as follows:

- $\mathfrak{M}, t, w \models F\alpha$, if and only if there exists a r with $t \leq r \leq N$ and $\mathfrak{M}, r, w \models \alpha$;
- $\mathfrak{M}, t, w \models G\alpha$, if and only if for all r with $t \leq r \leq N$, $\mathfrak{M}, r, w \models \alpha$;
- $\mathfrak{M}, t, w \models P\alpha$, if and only if there exists a r with $1 \leq r \leq t$ and $\mathfrak{M}, r, w \models \alpha$;
- $\mathfrak{M}, t, w \models H\alpha$, if and only if for all r with $1 \leq r \leq t$, $\mathfrak{M}, r, w \models \alpha$.

A $\mathcal{L}$-wff α is said to be *satisfiable* in $\mathfrak{F}$, if there exists some valuation function V on $\mathfrak{F}$ such that $\mathfrak{M}, 1, w \models \alpha$ for some $w \in W$, where $\mathfrak{M} := (\mathfrak{F}, V)$. α is satisfiable in a given a class $\mathfrak{G}$ of dynamic spaces if α is satisfiable in some $\mathfrak{F} \in \mathfrak{G}$. The notion of validity is then defined in the usual way: α is *valid* in $\mathfrak{F}$ if for all valuations V, and for all $w \in W$, $\mathfrak{M}, 1, w \models \alpha$, where $\mathfrak{M} := (\mathfrak{F}, V)$. Moreover, α is valid in a given a class $\mathfrak{G}$ of dynamic spaces if α is valid in all $\mathfrak{F} \in \mathfrak{G}$.

Remark 3. One can also define a notion of satisfiability by requiring the satisfiability of the wff at some time point, at some object. That is, we may call a wff α satisfiable in $\mathfrak{F}$ if there exists some valuation function V on $\mathfrak{F}$ such that $\mathfrak{M}, t, w \models \alpha$ for some $w \in W$ and some $t \in T_{\mathfrak{F}}$. But in this article, we shall consider only the notions of satisfiability and validity given above.

α is said to be a *propositional wff*, in brief $\alpha \in PF$, if α involves only Boolean connectives.

Remark 4. The interpretation of propositional variables is *rigid* here, in the sense that it is not time dependent. Accordingly, the truth value of any propositional wff α remains the same at every object, irrespective of time points. More formally, if $\mathfrak{M} := (\mathfrak{F}, V)$ and $w \in W$, we have $\mathfrak{M}, t, w \models \alpha$ if and only if $\mathfrak{M}, t', w \models \alpha$, for any $t, t' \in T_{\mathfrak{F}}$.

A natural generalization would be the case where the interpretation of propositional variables also changes with time. This would mean that, if $\mathbf{N} := \{1, 2, \ldots, N\}$, $\mathcal{F}_i := (W, P_i)$, $i \in \mathbf{N}$, and $\mathfrak{F} := \mathcal{F}_1, \mathcal{F}_2, \ldots, \mathcal{F}_N$, the valuation function V on $\mathfrak{F}$ would be a map from PV to $2^{W \times \mathbf{N}}$. In order to make the presentation simple, our discussions in this article are confined to rigid interpretations of propositional variables only. But we must remark that all the results obtained in this section and in Sect. 10 can be easily proved for this generalized semantics as well.

Remark 5. In temporal logics, the time frame is usually considered to be unbounded in the future. But in the current article, it is taken to be finite both in the past and future. Such a consideration is very relevant to some problems in computer science and artificial intelligence [13] and more particularly, in rough set theory. For instance, in the situation of the knowledge base (information system) evolving through time by means of updates, for all practical purposes, we consider only finitely many knowledge bases (information systems) corresponding to the different stages of updates (cf. Sect. 12), and hence this gives rise to a finite underlying time frame.

Definition 39. *Each $I \in \{K, K4, T, B, S4, KTB, KB4, S5\}$ determines a logic, denoted $L(T, I)$, consisting of all $\mathcal{L}$-wffs valid in the class of dynamic I spaces. Moreover, for $N \in \mathbb{N}$, we use $L_N(T, I)$ to denote the logic consisting of all $\mathcal{L}$-wffs valid in the class of dynamic I spaces $\mathfrak{F}$ with $|\mathfrak{F}| = N$.*

All the valid wffs of the modal system I, $I \in \{K, K4, T, B, S4, KTB, KB4, S5\}$, are valid in the class of all dynamic I spaces. Some valid temporal wffs are given in the following proposition. As intended, the usual axioms of linear temporal logic [71] are valid here, with modifications to account for finiteness both in the past and future (the last expressed by wffs 2(b) and 2(c) of the proposition below).

Proposition 60. *The following are valid in all dynamic spaces.*

1. (a) $\alpha\mathcal{U}\beta \leftrightarrow \beta \vee (\alpha \wedge \oplus(\alpha\mathcal{U}\beta))$;
   (b) $\alpha\mathcal{S}\beta \leftrightarrow \beta \vee (\alpha \wedge \ominus(\alpha\mathcal{S}\beta))$;
2. (a)$\alpha\mathcal{U} \perp \rightarrow \perp$; (b) $P \neg \ominus \top$; (c) $F \neg \oplus \top$;
3. (a) $G\alpha \rightarrow \alpha$; (b) $G\alpha \rightarrow G\neg \oplus \neg\alpha$;
4. $G(\alpha \rightarrow \beta) \rightarrow (G\alpha \rightarrow G\beta)$; *similarly for* H;
5. (a) $\alpha \rightarrow \neg \oplus \neg \ominus \alpha$; (b) $\alpha \rightarrow \neg \ominus \neg \oplus \alpha$;
6. $G(\alpha \rightarrow \oplus\alpha) \rightarrow G(\alpha \rightarrow G\alpha)$;
7. (a)$\ominus\alpha \rightarrow \neg \ominus \neg\alpha$; (b) $\oplus\alpha \rightarrow \neg \oplus \neg\alpha$.

As in standard modal logic, we have the following notion of characterization.

Definition 40. *Let $\mathfrak{G}, \mathfrak{K}$ be two classes of dynamic spaces. Let α be a $\mathcal{L}$-wff and Γ a set of $\mathcal{L}$-wffs.*

(1) We say that α characterizes *$\mathfrak{G}$ if for all dynamic spaces $\mathfrak{F}$, $\mathfrak{F} \in \mathfrak{G}$ if and only if α is valid in $\mathfrak{F}$. Similarly, we say that Γ characterizes $\mathfrak{G}$ if, $\mathfrak{F} \in \mathfrak{G}$ if and only if every $\mathcal{L}$-wff of Γ is valid in $\mathfrak{F}$.*
(2) α characterizes *the class $\mathfrak{G}$ of dynamic spaces* relative *to $\mathfrak{K}$ if for all dynamic spaces $\mathfrak{F} \in \mathfrak{K}$, $\mathfrak{F} \in \mathfrak{G}$ if and only if α is valid in $\mathfrak{F}$. Γ characterizes $\mathfrak{G}$ relative to $\mathfrak{K}$ if for all dynamic spaces $\mathfrak{F} \in \mathfrak{K}$, $\mathfrak{F} \in \mathfrak{G}$ if and only if every $\mathcal{L}$-wff of Γ is valid in $\mathfrak{F}$.*

Observation 4. It is clear that if a set $\Gamma = \{\alpha_1, \alpha_2, \ldots, \alpha_n\}$ of modal wffs characterizes a class $\mathfrak{G}$ of Kripke frames, then the set $\{G\alpha_1, G\alpha_2, \ldots, G\alpha_n\}$ of wffs will characterize the class of dynamic spaces consisting of frames from $\mathfrak{G}$.

Recall the special dynamic approximation spaces of Definition 37. The following proposition gives the wffs characterizing these classes of dynamic spaces relative to the class of DASs.

Proposition 61. *Let p be a propositional variable. Then Table 5 gives different classes of DASs, and corresponding characterizing wffs.*

Proof. In each case, we only prove the converse part, that is, if the given $\mathcal{L}$-wff is valid in the DAS $\mathfrak{F}$, then $\mathfrak{F}$ belongs to the named family.

1. Suppose $\mathfrak{F} = \mathcal{F}_1, \mathcal{F}_2, \ldots, \mathcal{F}_N$ is not monotonic refined where $\mathcal{F}_i := (W, R_i)$. Then there exist R_i and R_{i+1} such that $(a,b) \in R_{i+1}$ but $(a,b) \notin R_i$ for some $a, b \in W$. Let us consider a valuation function V such that $V(p) := W \setminus \{b\}$. Then clearly $\mathfrak{M}, i, a \not\models \Box p \rightarrow \oplus\Box p \vee \neg\oplus\top$, where $\mathfrak{M} := (\mathfrak{F}, V)$, and hence $\mathfrak{M}, 1, a \not\models G(\Box\alpha \rightarrow (\oplus\Box\alpha \vee \neg\oplus\top))$.

2. Let $\mathfrak{F}$ be not eventually monotonic refined. So we must have $R_N \not\subseteq R_{N-1}$. Then there exist $a, b \in W$ such that $(a,b) \in R_N$ but $(a,b) \notin R_{N-1}$. Let us consider a valuation function V such that $V(p) := W \setminus \{b\}$. Then clearly $\mathfrak{M}, N-1, a \not\models \Box p \rightarrow \oplus\Box p \vee \neg\oplus\top$, where $\mathfrak{M} := (\mathfrak{F}, V)$, and hence $\mathfrak{M}, 1, a \not\models F\ominus G(\Box p \rightarrow \oplus\Box p \vee \neg\oplus\top)$.

3. Suppose $\mathfrak{F}$ is not saturated. Then $R_{N-1} \neq R_N$. So either

(i) there exist $a, b \in W$ such that $(a,b) \in R_{N-1}$ and $(a,b) \notin R_N$, or
(ii) there exist $a, b \in W$ such that $(a,b) \notin R_{N-1}$ and $(a,b) \in R_N$.

Take a valuation V such that $V(p) := W \setminus \{b\}$ and consider $\mathfrak{M} := (\mathfrak{F}, V)$. Then in both cases we obtain $\mathfrak{M}, N-1, a \not\models (\Box p \leftrightarrow \oplus\Box p)$. Hence $\mathfrak{M}, 1, a \not\models F(G\ominus(\Box p \leftrightarrow \oplus\Box p)$.

4. Since $G(\Box\alpha \rightarrow (\oplus\Box\alpha \vee \neg\oplus\top))$ is valid in $\mathfrak{F}$, it must be monotonic refined. If possible let $\mathfrak{F}$ be not discrete. Then we must have $a, b \in W$ such that $b \in [a]_{R_i}$ for all i. Let us consider a valuation function V such that $V(p) := W \setminus \{b\}$. Then $\mathfrak{M}, i, a \not\models \Box p$ for all i, where $\mathfrak{M} := (\mathfrak{F}, V)$. We also have $\mathfrak{M}, i, a \not\models \Box\neg p$ for all i. Thus $\mathfrak{M}, 1, a \not\models F(\Box p \vee \Box\neg p)$.

5. This can be proved similarly. □

Table 5. Characterizing wffs for different classes of DASs

	Class of DAS	Characterizing wff
1.	Monotonic refined	$G\ (\Box p \rightarrow (\oplus\Box p \vee \neg\oplus\top))$
2.	Eventually monotonic refined	$F\ominus G\ (\Box p \rightarrow \oplus\Box p \vee \neg\oplus\top)$
3.	Saturated	$F(G\ominus(\alpha \leftrightarrow \oplus\alpha))$, where α does not involve $\oplus, \ominus, \mathcal{U}, \mathcal{S}$
4.	Ideal	$G\ (\Box p \rightarrow (\oplus\Box p \vee \neg\oplus\top)) \wedge F\ (\Box p \vee \Box\neg p)$
5.	Eventually discrete	$F\ G\ (\Diamond p \rightarrow \Box p)$

We remark that characterizing wffs of the Proposition 61 together with $S5$ axioms will give us the characterizing set of wffs for each of the classes of DASs mentioned.

Interpretation in Terms of Rough Sets. As we have noted earlier, the connection with rough sets is obtained by considering DASs. Given a DAS $\mathfrak{F} := \mathcal{F}_1, \mathcal{F}_2, \ldots, \mathcal{F}_N$, $\mathcal{F}_i := (W, P_i)$, and model $\mathfrak{M} := (\mathfrak{F}, V)$ for some valuation V, let us denote the set $\{w \in W : \mathfrak{M}, i, w \models \alpha\}$ by $[\![\alpha]\!]_{\mathfrak{M},i}$. The following proposition gives the set-theoretic interpretations of some $\mathcal{L}$-connectives and their combinations. 1(a)–(b), 3(a)–(b) and 5(a)–(b) give the interpretations in terms of rough sets.

Proposition 62.

1. *(a)* $[\![\Box\alpha]\!]_{\mathfrak{M},i} = \underline{[\![\alpha]\!]_{\mathfrak{M},i}}_{P_i}$; *(b)* $[\![\Diamond\alpha]\!]_{\mathfrak{M},i} = \overline{[\![\alpha]\!]_{\mathfrak{M},i}}_{P_i}$;
2. *(a)* $[\![\oplus\alpha]\!]_{\mathfrak{M},i} = [\![\alpha]\!]_{\mathfrak{M},i+1}$, $i < N$; *(b)* $[\![\ominus\alpha]\!]_{\mathfrak{M},i} = [\![\alpha]\!]_{\mathfrak{M},i-1}$, $i > 1$;
3. *(a)* $[\![\oplus\Box\alpha]\!]_{\mathfrak{M},i} = \underline{[\![\alpha]\!]_{\mathfrak{M},i+1}}_{P_{i+1}}$, $i < N$; *(b)* $[\![\ominus\Box\alpha]\!]_{\mathfrak{M},i} = \underline{[\![\alpha]\!]_{\mathfrak{M},i-1}}_{P_{i-1}}$, $i > 1$;
4. *(a)* $[\![F\alpha]\!]_{\mathfrak{M},i} = \bigcup_{i\le j\le N} [\![\alpha]\!]_{\mathfrak{M},j}$; *(b)* $[\![G\alpha]\!]_{\mathfrak{M},i} = \bigcap_{i\le j\le N} [\![\alpha]\!]_{\mathfrak{M},j}$;
5. *(a)* $[\![F\Box\alpha]\!]_{\mathfrak{M},i} = \bigcup_{i\le j\le N} \underline{[\![\alpha]\!]_{\mathfrak{M},j}}_{P_j}$; *(b)* $[\![G\Box\alpha]\!]_{\mathfrak{M},i} = \bigcap_{i\le j\le N} \underline{[\![\alpha]\!]_{\mathfrak{M},j}}_{P_j}$;
6. *(a)* $[\![P\Box\alpha]\!]_{\mathfrak{M},i} = \bigcup_{1\le j\le i} \underline{[\![\alpha]\!]_{\mathfrak{M},j}}_{P_j}$; *(b)* $[\![H\Box\alpha]\!]_{\mathfrak{M},i} = \bigcap_{1\le j\le i} \underline{[\![\alpha]\!]_{\mathfrak{M},j}}_{P_j}$.

We can use the above wffs to express the approximations of sets relative to time in generalized approximation spaces as well (cf. Definition 12).

Instances of valid/satisfiable statements, involving a mixture of the different modalities are given in the following straightforward proposition. Some may be derived as consequences of Proposition 61.
Recall the notion of a propositional wff and the result given in Remark 4.

Proposition 63. *Let $\alpha \in PF$.*

1. $(F\,\Box\alpha \vee P\,\Box\alpha) \rightarrow P\,(\neg \ominus \top \wedge G\,\Diamond\alpha)$ *is valid in the class of all dynamic I spaces, $I \in \{T, S4, KTB, S5\}$.*
2. $\Diamond\alpha\ \mathcal{U}\ (\Diamond\alpha \rightarrow \Box\alpha)$ *is valid in the class of all eventually discrete DASs.*
3. *The following are valid in the class of all monotonic refined DASs.*
   (a) $\neg \oplus \neg\, G \ominus (\Box\alpha \rightarrow \oplus\Box\alpha)$
   (b) $(\alpha \rightarrow \Box\alpha) \rightarrow G(\alpha \rightarrow \Box\alpha)$.
   (c) $\Box\alpha \rightarrow G\Box\alpha$, $\neg\Diamond\alpha \rightarrow G\neg\Diamond\alpha$.
4. *In a monotonic refined DAS $\mathfrak{F}$, if $\mathcal{M}_1, w \models_{\mathfrak{F}} \alpha \rightarrow F\Box\alpha$ for all $w \in W$, $\alpha \in PF$, then there exists a j such that $\mathcal{M}_1, w \models_{\mathfrak{F}} \oplus^j (\alpha \rightarrow \Box\alpha)$ for all $w \in W$.*
5. $\Box\alpha \wedge F(\neg\Box\alpha \wedge \Diamond\alpha)$ *is satisfiable, but not valid in the class of all DASs.*
6. $F\,(\Diamond\alpha \rightarrow \Box\alpha)$ *is satisfiable, but not valid in the class of all DASs.*

Let us interpret some of the above wffs in a model $\mathfrak{M} := (\mathfrak{F}, V)$ based on a DAS $\mathfrak{F}$. Note that, as mentioned in Remark 4), the satisfiability of α only depends on V. Let V interpret α to be the set X in the domain of $\mathfrak{F}$.

The wff 1 says that if an object is in the lower approximation of the set X at some time (current, past, or future), then it will always be in the upper approximation of X.

The satisfiability of wff 2 at all objects at a time point guarantees that every object will move to a *certain* region of the set X at some future time, and before that it remains in the upper approximation of X. We note that even if $\mathfrak{F}$ has a single discrete approximation space, this wff would be valid in it.

The wff 3b expresses the fact that if X is definable at the current time then it will remain so at all the future times.

The wff in 3c expresses that if an object is in a certain region of X, it will remain so at all the future times.

Item 4 of the proposition says that if each element of X is an element of the lower approximation of X with respect to some approximation space in $\mathfrak{F}$, then there must be an approximation space in $\mathfrak{F}$ where X becomes definable. A similar result holds if we replace $\Box$ by $\Diamond$.

Wff 5 is satisfied by all those elements which are necessarily inside the set X at the current time but which move to the boundary region of X at some future time.

If $\mathfrak{F}$ is monotonic refined, then satisfiability of wff 6 at all objects at a time point means that the set X becomes definable eventually (even if it is rough at that time point).

We note that Observation 3 presented earlier, has found expression in the language $\mathcal{L}$.

Example 15. Let us consider a DAS $\mathfrak{F} := \mathfrak{F}_1, \ldots, \mathfrak{F}_N$ and suppose 1 is the current time point. Then questions such as the following could be raised.

Q1. Does the partition become finer with time?
Q2. Can a time point be reached after which the partition becomes static?
Q3. For each object, is there a time point at which it becomes distinguishable from all other objects?
Q4. Could two rough sets X and Y be related such that each object gets into the positive region of X at some future time point, and before that it is in the negative region of Y?
Q5. Do we have an object such that it is not even in the possible region of a given rough set X at any time point?
Q6. What are the objects that are currently in the boundary region of X, but will get into one of its 'certain' (viz. positive/negative) regions in the future?
Q7. Is X, which is currently rough, definable at the next time point, and remain so in all future time? In other words, could its boundary region be erased from the next time point onwards?

Let us see how the language $\mathcal{L}$ can be used to phrase these questions. Let p, q be propositional variables and $\mathfrak{M} := (\mathfrak{F}, V)$ a model such that $V(p) := X$ and $V(q) := Y$.

Q1 and Q2 are about whether $\mathfrak{F}$ is monotonic refined, or saturated (respectively). The wffs for these are given in Proposition 61.
Q3 is equivalent to checking the 1-validity of the $\mathcal{L}$-wff $F(\Diamond p \rightarrow \Box p)$.
Q4 is equivalent to checking the 1-validity of $\neg\Diamond q \,\mathcal{U}\, \Box p$ under V.
Q5 is equivalent to checking the 1-satisfiability of $G\neg\Diamond p$ under V.
Condition of Q6 is satisfied by the objects belonging to the set $[\![(\Diamond p \wedge \neg\Box p) \wedge F\,(\Box p \vee \neg\Diamond p)]\!]_{\mathfrak{M},1}$.
Q7 is answered by checking if the $\mathcal{L}$-wff $\Diamond p \wedge \neg\Box p$ is 1-satisfiable and $\oplus G(\Diamond p \rightarrow \Box p)$ is 1-valid.

9 $\mathcal{L}$-Semantics in Perspective

In this section, we present a comparative study of the logics $L(T, I)$ with some of the known logics in literature. The content of this section is based on the article [7,54].

9.1 $\mathcal{L}$-Semantics as a Fibring over a Combination of Temporal and Kripke Frames

We refer to [29] for the following.

Definition 41. *Consider Kripke frames $\mathcal{F}_1 := (W_1, R_1)$ and $\mathcal{F}_2 := (W_2, R_2)$ with binary accessibility relations. The frame $\mathcal{F}_1 \times \mathcal{F}_2 := (W_1 \times W_2, R_1^*, R_2^*)$ is called their* product, *where*
$R_1^* := \{((x, z), (y, z)) : xR_1y,\ z \in W_2\}$, *and*
$R_2^* := \{((z, x), (z, y)) : xR_2y,\ z \in W_1\}$.

Definition 42. *Let $\mathcal{F}_i := (W, R_i)$, $i = 1, \ldots, N$, be a collection of Kripke frames over the same set of possible worlds. Their* fusion *is the frame $\mathcal{F}_1 * \mathcal{F}_2 * \cdots * \mathcal{F}_N := (W, R_1, R_2, \ldots, R_N)$.*

Let $\mathfrak{F} := \mathcal{F}_1, \mathcal{F}_2, \ldots, \mathcal{F}_N$ be a dynamic space, where $\mathcal{F}_i := (W, R_i)$, $i \in \{1, 2, \ldots, N\}$. Consider a finite linear temporal frame $T := (\{1, 2, \ldots, N\}, <)$, where $<$ is the natural linear order on $\{1, 2, \ldots, N\}$, that is $i < j$ if and only if $j = i+1$. Further, let $\mathcal{F}_i^* := T \times \mathcal{F}_i$, $i \in \{1, 2, \ldots, N\}$, and $\mathcal{F} := \mathcal{F}_1^* * \mathcal{F}_2^* * \cdots * \mathcal{F}_N^*$. So $\mathcal{F} = (T \times W, <^*, R_1^*, R_2^*, \ldots, R_N^*)$.

A valuation function V on $\mathcal{F}$, is (as before) a function from the set PV of propositional variables to 2^W. Let us now consider a *selection function* f on $\mathcal{F}$, which is a bijective map from the set T to the set $\{R_1^*, R_2^*, \ldots, R_N^*\}$. Let us call $(\mathcal{F}, f)$ a *$\mathcal{L}$-structure.*

Definition 43. *The satisfiability of a wff α with respect to a 'model' $\mathcal{M} := (\mathcal{F}, f, V)$ at $(t, w) \in T \times W$ (in notation $(t, w) \models_{\mathcal{M}} \alpha$) is defined inductively, as follows. We omit the Boolean cases.*

- *For each propositional variable p, $(t, w) \models_{\mathcal{M}} p$ if and only if $w \in V(p)$;*

- $(t, w) \models_{\mathcal{M}} \Box\alpha$ *if and only if* $(t, w') \models_{\mathcal{M}} \alpha$ *for all* (t, w') *with* $(t, w) f(t) (t, w')$;
- $(t, w) \models_{\mathcal{M}} \oplus \alpha$ *if and only if there exists* (t', w) *such that* $(t, w) <^* (t', w)$ *and* $(t', w) \models_{\mathcal{M}} \alpha$*; A similar definition for* $(t, w) \models_{\mathcal{M}} \ominus \alpha$;
- $(t, w) \models_{\mathcal{M}} \alpha \mathcal{U} \beta$ *if and only if there exist* $(t_1, w), (t_2, w), \ldots, (t_j, w)$ *such that* $t = t_1$, $(t_1, w) <^* (t_2, w) <^* \cdots <^* (t_j, w)$*. Further,* $(t_j, w) \models_{\mathcal{M}} \beta$*, and for all* $k = 1, 2, \ldots, j-1$, $(t_k, w) \models_{\mathcal{M}} \alpha$*; A similar definition for* $(t, w) \models_{\mathcal{M}} \alpha \mathcal{S} \beta$.

Here, the role of the selection function f is to pick out the R_i^* with respect to which the $\Box$ modality is to be evaluated. This is the basic idea of *fibring.*

That the above semantics may be identified with one given in Sect. 8.2, is established by the following proposition.

Proposition 64. *There exists a bijective mapping* g *between the sets of all* $\mathcal{L}$*-structures and dynamic spaces. Moreover, for every wff* α*, valuation function* V, $\mathcal{M} := (\mathcal{F}, f, V)$, $g((\mathcal{F}, f)) := \mathfrak{F}$, $\mathfrak{M} := (\mathfrak{F}, V)$,

$$\mathfrak{M}, i, w \models \alpha \text{ if and only if } (i, w) \models_{\mathcal{M}} \alpha.$$

Proof. Given a $\mathcal{L}$-structure $\mathcal{T} = (\mathcal{F}, f)$ where $\mathcal{F} = (T \times W, <^*, R_1^*, R_2^*, \ldots, R_N^*)$ and $T = \{1, 2, \ldots, N\}$, we define a dynamic space $\mathfrak{F}_{\mathcal{T}}$: the ith Kripke frame in $\mathfrak{F}_{\mathcal{T}}$ is $(W, R'_{f(i)})$, $i = 1, 2, \ldots, N$, where $(w, w') \in R'_{f(i)}$ if and only if $((t, w), (t, w') \in R^*_{f(i)})$ for some t. Let us take $g(\mathcal{T}) := \mathfrak{F}_{\mathcal{T}}$. It is easy to verify that this g is bijective. The other part of the proposition can be proved by a simple induction on the complexity of the wff α. $\Box$

9.2 Multi-modal Logics

We shall use the logic $L_N(T, I)$, $N \in \mathbb{N}$, consisting of all the wffs valid in the class of dynamic I spaces $\mathfrak{F}$ with $|\mathfrak{F}| = N$. Moreover, the following notion of embedding will be very helpful.

Definition 44. *A logic* L_1 *is* embeddable *into a logic* L_2 ($L_1 \rightharpoonup L_2$)*, provided there is a translation* $\star$ *of wffs of* L_1 *into* L_2*, such that* α *is satisfiable in* L_1 *if and only if* α^* *is satisfiable in* L_2*. If* $L_1 \rightharpoonup L_2$*, and* $L_2 \rightharpoonup L_1$*, then we write* $L_1 \rightleftharpoons L_2$.

For an integer N, let us consider the multi-modal version I_N of the modal logic I ($\in \{K, K4, T, B, S4, KTB, KB4, S5\}$) consisting of modal operators $\Box_i$, $i = 1, 2, \ldots, N$. Let us use I_N to denote the the set of all wffs of I_N as well. Since a structure of the form $(W, \{P_1, P_2, \ldots, P_N\})$ can equivalently be written as the dynamic space $\mathfrak{F} := \mathcal{F}_1, \mathcal{F}_2, \ldots, \mathcal{F}_N$, where $\mathfrak{F}_i := (W, P_i)$, and vice versa, the semantics of I_N can be given based on dynamic spaces. Therefore, one can discuss validity of any I_N-wff, or its satisfiability in a given class of dynamic spaces of cardinality N. The logic I_N thus consists of all I_N-wffs which are valid in the class of dynamic I spaces of cardinality N. We have the following theorem.

Theorem 10. *Let* $I \in \{K, K4, T, B, S4, KTB, KB4, S5\}$*, and* $N \in \mathbb{N}$*. Then* $L_N(T, I) \rightleftharpoons I_N$.

Theorem 10 shows that the logic $L_N(T,I)$ is identifiable with the logic I_N. We sketch a proof, giving the involved translations explicitly.

Let $\mathbf{N} := \{1, 2, \ldots, N\}$, and let $\mathcal{L}$, the language of the logics $L(T,I)$, also denote the set of all wffs of $L(T,I)$. Assuming that the wffs of I_N and $\mathcal{L}$ are formed using the same set PV of propositional variables, we define mappings $\Psi_1 : \mathbf{N} \times \mathcal{L} \to I_N$, and $\Psi_2 : I_N \to \mathcal{L}$ as follows.

Definition 45.

1. $\Psi_1 : \mathbf{N} \times \mathcal{L} \to I_N$ *such that*

$$\Psi_1(i,p) := p, \; p \in PV,$$
$$\Psi_1(i,\neg\alpha) := \neg\Psi_1(i,\alpha), \; \textit{and} \; \Psi_1(i,\alpha\wedge\beta) := \Psi_1(i,\alpha)\wedge\Psi_1(i,\beta),$$
$$\Psi_1(i,\Box\alpha) := \Box_i\Psi_1(i,\alpha),$$
$$\Psi_1(i,\oplus\alpha) := \begin{cases} \Psi_1(i+1,\alpha) & \textit{if}\; i<N \\ \bot & \textit{if}\; i=N \end{cases}$$
$$\Psi_1(i,\ominus\alpha) := \begin{cases} \Psi_1(i-1,\alpha) & \textit{if}\; i>1 \\ \bot & \textit{if}\; i=1 \end{cases}$$
$$\Psi_1(i,\alpha\mathcal{U}\beta) := \bigvee_{i\le j\le N}\left(\Psi_1(j,\beta)\wedge\bigwedge_{i\le k\le j-1}\Psi_1(k,\alpha)\right),$$
$$\Psi_1(i,\alpha\mathcal{S}\beta) := \bigvee_{1\le j\le i}\left(\Psi_1(j,\beta)\wedge\bigwedge_{j+1\le k\le i}\Psi_1(k,\alpha)\right).$$

2. $\Psi_2 : I_N \to \mathcal{L}$ *such that*

$$\Psi_2(p) := p, \; p \in PV,$$
Ψ_2 *is homomorphic for Boolean connectives,*
$$\Psi_2(\Box_i\alpha) := P(\neg\ominus\top\wedge\oplus^{i-1}\Box\Psi_2(\alpha)), \; \textit{where} \; \oplus^i := \underbrace{\oplus\oplus\cdots\oplus}_{i-times}.$$

Let us consider a model $\mathfrak{M} := (\mathfrak{F}, V)$, where $|\mathfrak{F}| = N$. Then by an easy induction on the complexity of the wff α, we obtain the following.

Proposition 65.

1. *For* $\alpha \in \mathcal{L}$, $\mathfrak{M}, t, w \models \alpha$ *if and only if* $\mathfrak{M}, w \models \Psi_1(t,\alpha)$.
2. *For* $\alpha \in I_N$, $\mathfrak{M}, w \models \alpha$ *if and only if* $\mathfrak{M}, 1, w \models \Psi_2(\alpha)$.

Proof of Theorem 10. We prove $L_N(T,I) \rightharpoonup I_N$ and $I_N \rightharpoonup L_N(T,I)$. Consider the translation $\star$ from the $\mathcal{L}$-wffs to the wffs of I_N such that $\alpha^\star := \Psi_1(1,\alpha)$. Under this translation, we obtain $L_N(T,I) \rightharpoonup I_N$ as a direct consequence of Proposition 65. To prove $I_N \rightharpoonup L_N(T,I)$, we use the translation Ψ_2 and Proposition 65. □

9.3 First Order Temporal Logics

Let us consider a temporal first order language with equality based on the set of propositional variables PV which, along with the temporal operators $\oplus$, $\ominus$, $\mathcal{U}$, $\mathcal{S}$, has atomic wffs $\top$, $\bot$, unary predicates P_p corresponding to each $p \in PV$, and a binary predicate Q. Let $\mathbb{F}$ denote the wffs of this language.

It is known that when interpreted on models, modal wffs are equivalent to first order wffs in one free variable, and this result is proved using the standard translation ST_x mapping propositional variable p to $P_p(x)$, $\Box\phi$ to $\forall y(Rxy \rightarrow ST_y(\phi))$, y being a fresh variable. Therefore, it indicates that such a connection should also be there between the logics $L(T, I)$ and first order temporal logics. In fact, a connection between $L(T, I)$ and the *monodic packed fragment* of $\mathbb{F}$ can be established by extending the standard translation ST_x to the set of $\mathcal{L}$-wffs such that ST_x is homomorphic for temporal operators. We give the details below, but first let us recall that an interpretation of $\mathbb{F}$ is a structure of the form $(T, W, \{\overline{Q}_t\}_{t\in T}, \{\overline{P}_p\}_{p\in PV})$, where W is a set of objects, T a non-empty set of time points with a suitable ordering, and $\overline{Q}_t \subseteq W \times W$, $\overline{P}_p \subseteq W$, for each $t \in T$ and $p \in PV$. Let us use FOI^N, where $N \in \mathbb{N}$, to denote the class of all interpretations of $\mathbb{F}$ over the initial segment $\{1, 2, \ldots, N\}$ of $\mathbb{N}$ with natural ordering, as the underlying time frame. Let F_N denote the logic consisting of all wffs valid in the class FOI^N of interpretations.

Note that a model $\mathfrak{M} := (\mathfrak{F}, V)$ of $L(T, I)$, where $\mathfrak{F} := \mathcal{F}_1, \mathcal{F}_2, \ldots, \mathcal{F}_N$, and $\mathcal{F}_i := (W, P_i)$, can be viewed as an element $(\mathbf{N}, W, \{P_t\}_{t\in\mathbf{N}}, \{\overline{P}_p\}_{p\in PV})$ of FOI^N, where $\mathbf{N} := \{1, 2, \ldots, N\}$ and $\overline{P}_p := V(p)$. Similarly, interpretations $\mathfrak{M}$ in FOI^N can be viewed as models of $L(T, I)$. Moreover, whether we view $\mathfrak{M}$ as a model of $L(T, I)$ or an interpretation of $\mathbb{F}$, the time domain and structure is the same, and as the translation ST_x is homomorphic for temporal operators, we immediately obtain

Proposition 66. *For any $\mathcal{L}$−wff α,*

$$\mathfrak{M}, t, w \models \alpha \text{ if and only if } \mathfrak{M}, t \models ST_x(\alpha)[w].$$

In order to give the precise connection between $L(T, I)$ and the monodic packed fragment of $\mathbb{F}$, we give the definition of the latter, as presented in [37].

Definition 46. *An $\mathbb{F}$-wff γ is said to be* packing guard *if γ is a conjunction of atomic and existentially-quantified atomic wffs (possibly equalities) such that any two distinct free variables of γ co-occur free in some conjunct of γ. The* monodic packed fragment *of $\mathbb{F}$ consists of the following wffs.*

- *Any atomic $\mathbb{F}$-wff (which can be equality, $\top$, or $\bot$) is monodic packed.*
- *Boolean combinations of monodic packed wffs are monodic packed.*
- *If γ is packing guard, ϕ is a monodic packed wff, every free variable of ϕ is free in γ, and $\overline{y}$ is a tuple of variables, then $\exists\overline{y}(\gamma \wedge \phi)$ is a monodic packed wff.*
- *If x is a variable and ϕ, ψ are monodic packed wffs with free variable at most x, then $\phi\ \mathcal{U}\ \psi$ and $\phi\ \mathcal{S}\ \psi$ are monodic packed wffs.*

– *If x is a variable and ϕ is a monodic packed wff with free variable at most x, then $\oplus\phi$ and $\ominus\phi$ are monodic packed wffs.*

As a consequence of Proposition 66, the translation ST_x will lead us to an embedding of $L_N(T, K)$ into the monodic packed fragment of $\mathbb{F}$, provided we add a conjunct which imposes the condition that the interpretation of propositional variables is rigid. Such a condition may be given by the wff $\overline{p}$ defined as follows:

$$\forall x\Big((p(x) \to Gp(x)) \wedge (\neg p(x) \to G\neg p(x))\Big).$$

Note that $\overline{p}$ is a monodic packed wff. $\overline{p}$ is not unique – there may be other wffs expressing the rigidity condition (e.g. ones involving the operator H).

We have the following theorem.

Theorem 11. *For $I \in \{K,\ K4,\ T,\ B,\ S4,\ KTB,\ KB4,\ S5\}$, and $N \in \mathbb{N}$, $L_N(T, I)$ is embeddable into the monodic packed fragment of F_N.*

Proof. First, let $I = K$. We choose a first order variable x_0 and fix it. Consider the translation Φ_K from $\mathcal{L}$ to the monodic packed fragment of $\mathbb{F}$, defined as follows.

$$\Phi_K(\alpha) := ST_{x_0}(\alpha) \wedge \bigwedge_{p\in\alpha} \overline{p},$$

where $p \in \alpha$ means that the propositional variable p occurs in α. Now using Proposition 66, we obtain the desired result for K. Similar would be the case for the other I's: we would need to add a conjunct in the above translation according to the type of relations in the respective dynamic I spaces. For instance, for $I = B$, the dynamic I spaces consist of symmetric relations, and so the required translation would be

$$\Phi_B(\alpha) := ST_{x_0}(\alpha) \wedge \bigwedge_{p\in\alpha} \overline{p}\ \wedge \forall x\forall y(Qxy \to Qyx).$$

□

Remark 6. We can easily obtain a result similar to Theorem 11 for the semantics discussed in Remark 4, where the interpretation of the propositional variables is not rigid, that is, it also changes with time. For this, we just need to remove the conjunct $\bigwedge_{p\in X} \overline{p}$ from the translations.

Although we have given the semantics of the language $\mathcal{L}$ with a finite time line, it can, in a natural way, be extended to models with $\mathbb{N}$ itself as the underlying time frame. Then Theorem 11 can be extended to this class of time structures as well. That is, if $L_{\mathbb{N}}(T, I)$ $(F_{\mathbb{N}})$ is the logic consisting of $\mathcal{L}$–wffs (respectively $\mathbb{F}$–wffs) valid in the class of dynamic I spaces (respectively, interpretations of $\mathbb{F}$) with $\mathbb{N}$ as the underlying time frame, then we have

Theorem 12. *For $I \in \{K, K4, T, B, S4, KTB, KB4, S5\}$,*

$$L_{\mathbb{N}}(T, I) \rightharpoonup F_{\mathbb{N}}.$$

The proof follows exactly in the lines of that of Theorem 11. We shall use this theorem in Sect. 11.3 to obtain a decidable fragment of the logic $L(T, I)$.

9.4 Finger and Gabbay's Proposal of Combination of Modal Logics

In Sect. 9.1, it is show that the semantics of $L(T, I)$ can be obtained as a combination of temporal and basic modal logic, and thus one would like to see how these logics are related with the existing proposals of combination of modal logics. We consider the logics presented in [10,27].

Finger and Gabbay in [27] introduced a general methodology to combine an arbitrary logical system L with a pure propositional temporal logic T (such as linear temporal logic with '*Since*' and '*Until*'). Let us call the resultant combined logic $T(L)$.

Let us consider the modal system $S5$ and see how the logic $T(S5)$ is related with $L(T, S5)$. A similar argument would work for $I \in \{K, K4, T, B, S4, KTB, KB4\}$. In order to define the wffs of $T(S5)$, $S5$-wffs are divided into two classes: (a) a wff belongs to the set of Boolean combinations, BC_K, if and only if it is built from other wffs by using one of the Boolean connectives $\neg$ or $\wedge$, or any other connective defined only in terms of those; (b) it belongs to the set ML_{S5} of 'monolithic' wffs otherwise. Then the set $L_{T(S5)}$ wffs of $T(S5)$ is defined as follows.

- If $\alpha \in ML_{S5}$, then $\alpha \in L_{T(S5)}$;
- If $\alpha, \beta \in L_{T(S5)}$, then $\neg\alpha$, $\alpha \wedge \beta, \alpha\mathcal{S}\beta, \alpha\mathcal{U}\beta \in L_{T(S5)}$.

Thus, the wffs of $T(S5)$ form a proper subset of $\mathcal{L}$-wffs: only those $\mathcal{L}$-wffs are considered in which temporal operators do not come in the scope of $\Box$. So, for instance, $\Box(\alpha\mathcal{U}\beta)$ is not a wff of $T(S5)$. Note that if we restrict ourselves to only such wffs in $\mathcal{L}$, it would deprive $\mathcal{L}$ of one of its salient features. We would not be able to compute expressions such as $(\underline{X}_{R_1})_{\overline{R_2}}$, that is, where there is an iteration of lower/upper approximation operators corresponding to different relations. In $\mathcal{L}$, the syntactic counterpart of the afore-mentioned expression is the wff $\oplus\Box\ominus\Box p$.

The semantics of $T(S5)$ is based on a structure of the form $\mathcal{M}_t := (T, <, g)$, where $(T, <)$ represents the underlying flow of time and g is a function which associates each time point with a tuple (M_t, w_t). $M_t := (W_t, R_t, V_t)$ is an $S5$ model and $w_t \in W_t$. The satisfiability relation is defined as follows:

1. $\mathcal{M}_T, t \models \alpha, \alpha \in ML_{S5}$, if and only if $M_t, w_t \models \alpha$, where $g(t) := (M_t, w_t)$;
2. $\mathcal{M}_T, t \models \alpha\mathcal{U}\beta$, if and only if there exists $s \in T$ such that $t < s$ and $\mathcal{M}_T, s \models \beta$, and for every $u \in T$, if $t < u < s$ then $\mathcal{M}_T, u \models \alpha$.

The Boolean and '*Since*' cases are defined in the standard way. The difference between this semantics and that of $L(T, S5)$ becomes clear now. In $T(S5)$, for each time point, the object w_t and the model M_t are fixed. So when one moves to a time point t, an $S5$-wff must be evaluated at the object w_t, and across time points, these objects would vary in general. In $L(T, S5)$, however, when one moves across time points using only the $\oplus, \ominus, \mathcal{U}$ or $\mathcal{S}$ operators, the object at which a wff is to be evaluated, remains the same. For instance, the satisfiability of the wff $\oplus\Box p$ at $(t, w) \in T \times W$ in $L(T, S5)$ semantics entails that the object

w is in the lower approximation of the set represented by p with respect to the relation at the $t+1^{th}$ time point. On the other hand, satisfiability of the same wff at the same time point in $T(S5)$ semantics means that the object w_{t+1}, is in the lower approximation of the set represented by p with respect to the relation at the $t+1^{th}$ time point. So, in this case, the satisfiability of the wff $\oplus\Box p$ does not depend on the object w at which it is evaluated.

9.5 Bonanno's Proposal

Now, let us move to the logic proposed by Bonanno [10], which is also very closely related with $L(T, I)$. The language of this logic contains three modal operators: B (belief operator), $\mathbb{I}$ (information operator) and A (global modal operator). It also contains temporal operators $\oplus$ and $\ominus$, but does not have *until* and *since* operators. The wffs are given as: $p \in PV|\alpha|\neg\alpha|\alpha \wedge \beta|B\alpha|\mathbb{I}\alpha|A\alpha$.
The semantics is based on a structure of the form $(T, <, W, \{\mathcal{B}_t\}_{t\in T}, \{\mathcal{I}_t\}_{t\in T})$, called *temporal belief revision frame*, where

- $(T, <)$ is a next-time branching frame, that is, $<\subseteq T \times T$ such that,
  - if $t_1 < t_3$ and $t_2 < t_3$, then $t_1 = t_2$ and
  - if $(t_1, \ldots, t_n)$ is a sequence with $t_i < t_{i+1}$, then $t_n \neq t_1$,
- $\mathcal{B}_t, \mathcal{I}_t \subseteq W \times W$.

The satisfiability conditions for the operators B, $\oplus$ and $\ominus$ are given as follows. Let $(t, w) \in T \times W$.

- $t, w \models B\alpha$ if and only if $t, w' \models \alpha$ for all w' such that $(w, w') \in \mathcal{B}_t$.
- $t, w \models \oplus\alpha$ if and only if $t', w \models \alpha$ for all t' such that $t < t'$. (10)
- $t, w \models \ominus\alpha$ if and only if $t', w \models \alpha$ for all t' such that $t' < t$. (11)

One can see that the satisfiability condition of B is the same as that of the operator $\Box$ of $L(T, I)$. Moreover, the satisfiability conditions of $\oplus$ and $\ominus$ given by (10), (11) and Definition 38 coincide when we consider a finite linear time line.

A deductive system for the set of wffs valid in all temporal belief revision frames has not been obtained, and the problem has been cited as open in [11]. However, a deductive system is proposed for wffs valid in all structures of the form $(T, <, W, \{W_t\}_{t\in T}, \{\mathcal{B}_t\}_{t\in T}, \{\mathcal{I}_t\}_{t\in T})$, called *general temporal belief revision frames*, where corresponding to each time point t, we have the domain $W_t \subseteq W$ and $\mathcal{B}_t, \mathcal{I}_t \subseteq W_t \times W_t$. The satisfiability conditions of $\oplus$ and $\ominus$ are now given as follows.

Let $\hookrightarrow\subseteq (T \times W) \times (T \times W)$ be such that $(t, w) \hookrightarrow (t', w')$ if and only if (1) $w = w'$, (2) $w \in W_t \cap W_{t'}$ and either (3a) $t < t'$, or (3b) $t <^* t'$ and for every $x \in T$, if $t <^* x$ and $x <^* t'$, then $w \notin W_x$, where $<^*$ denotes the transitive closure of $<$.

- $t, w \models \oplus\alpha$ if and only if $t', w \models \alpha$ for all t' such that $(t, w) \hookrightarrow (t', w)$. (12)
- $t, w \models \ominus\alpha$ if and only if $t', w \models \alpha$ for all t' such that $(t', w) \hookrightarrow (t, w)$. (13)

Note that the temporal belief revision frame is obtained as a special case of the general temporal belief revision frame by imposing the condition $W_t = W$ for all t. Moreover, in the case of temporal belief revision frame, the conditions (12) and (13) just reduce to (10) and (11) respectively.

So, in [10] a branching time structure is considered, whereas, in $L(T, I)$ we have a finite linear time line. In Sect. 11.3, we shall consider a fragment $\mathbb{L}_2$ of $L(T, I)$, which properly contains Bonanno's logic without the operators $\mathbb{I}$ and A, and where the operator B is identified with $\Box$. This fragment of $L(T, I)$ will be shown in Sect. 11.3 to be decidable, entailing the decidability of the above fragment of Bonanno's logic with respect to the class of all temporal belief revision frames over finite linear time line.

10 Tableau-Based Proof Method for $L(T, I)$

We now present tableau-based proof procedures for the logics $L(T, I)$, $I \in \{K, K4, T, B, S4, KTB, KB4, S5\}$, corresponding to the classes of dynamic I spaces. Prefixed wffs are used for the purpose. The technique is in the line of the one given by [28] with appropriate modifications. The soundness and completeness theorems for the tableau-based proof procedures of $L(T, I)$ will be proved in Sects. 10.1 and 10.2 respectively. To prove the completeness theorem, we present a systematic procedure P1 to construct an '(I, N)-tableau' for a given wff X (N an integer) such that, if the (I, N)-tableau for X obtained following P1 is not closed, then X cannot have a closed (I, N)-tableau at all. However, P1 may not terminate. In Sect. 10.3, this limitation is overcome: P1 is modified to yield terminating procedures having the above mentioned feature.

The content of this section is based on the article [54].

We begin with some basic definitions.

Let '$\overline{T}$' and '$\overline{F}$' be two new symbols. By a *signed wff*, we mean $\overline{T}X$ or $\overline{F}X$ where X is a $\mathcal{L}$-wff. The complexity of a signed wff $\overline{T}X$ or $\overline{F}X$, is the complexity of X in the usual sense. A *signed sub-wff* of X, or $\overline{T}X$ or $\overline{F}X$, is a signed wff of the form $\overline{T}Y$ or $\overline{F}Y$, where Y is a sub-wff of X. We denote the set of all signed sub-wffs of X (or, $\overline{T}X$ or $\overline{F}X$) by $S(X)$.

$\overline{T}X$ behaves like X and $\overline{F}X$ behaves like $\neg X$. Thus, we have the following.

Definition 47.

(i) $\mathfrak{M}, t, w \models \overline{T}X$ *if and only if* $\mathfrak{M}, t, w \models X$, *and*
(ii) $\mathfrak{M}, t, w \models \overline{F}X$ *if and only if* $\mathfrak{M}, t, w \not\models X$.

An $\alpha-$wff is a wff of one of the forms $\overline{T}(X \wedge Y), \overline{F}(X \vee Y), \overline{F}(X \to Y), \overline{F}(\neg X)$. For each $\alpha-$wff, two components, α_1 and α_2 are defined in Fig. 2(a). Similarly the β, $\oplus$, $\ominus$, ν and π wffs and their components are given by Fig. 2.

Apart from these, we have two more types of wffs which involve '$\mathcal{U}$' and '$\mathcal{S}$' operators, that is, wffs of the form $\overline{T}(X\mathcal{U}Y)$, $\overline{F}(X\mathcal{U}Y)$, $\overline{T}(X\mathcal{S}Y)$, $\overline{F}(X\mathcal{S}Y)$.

From Definitions 38 and 47, it is not difficult to see,

- $\mathfrak{M}, t, w \models \alpha$, if and only if $\mathfrak{M}, t, w \models \alpha_1$ and $\mathfrak{M}, t, w \models \alpha_2$.

(a) α -wff

α	α_1 α_2
$\overline{T}(X \wedge Y)$	$\overline{T}X$ $\overline{T}Y$
$\overline{F}(X \vee Y)$	$\overline{F}X$ $\overline{F}Y$
$\overline{F}(X \rightarrow Y)$	$\overline{T}X$ $\overline{F}Y$
$\overline{F}(\neg X)$	$\overline{T}X$ $\overline{T}X$

(b) β -wff

β	β_1 β_2
$\overline{T}(X \vee Y)$	$\overline{T}X$ $\overline{T}Y$
$\overline{F}(X \wedge Y)$	$\overline{F}X$ $\overline{F}Y$
$\overline{T}(X \rightarrow Y)$	$\overline{F}X$ $\overline{T}Y$
$\overline{T}(\neg X)$	$\overline{F}X$ $\overline{F}X$

(c) $\oplus$ -wff

$\oplus$	$\oplus_0$
$\overline{T}(\oplus X)$	$\overline{T}X$
$\overline{F}(\oplus X)$	$\overline{F}X$

(d) $\ominus$ -wff

$\ominus$	$\ominus_0$
$\overline{T}(\ominus X)$	$\overline{T}X$
$\overline{F}(\ominus X)$	$\overline{F}X$

(e) ν -wff

ν	ν_0
$\overline{T}(\Box X)$	$\overline{T}X$
$\overline{F}(\Diamond X)$	$\overline{F}X$

(f) π -wff

π	π_0
$\overline{T}(\Diamond X)$	$\overline{T}X$
$\overline{F}(\Box X)$	$\overline{F}X$

Fig. 2. Different types of signed wffs and their components.

- $\mathfrak{M}, t, w \models \nu$, if and only if $\mathfrak{M}, t, w' \models \nu_0$ for all w' such that $w, w' \in P_t$.
- $\mathfrak{M}, t, w \models \overline{T}(X\mathcal{U}Y)$, if and only if there exists r with $t \leq r \leq N$ such that $\mathfrak{M}, r, w \models \overline{T}Y$ and for all k such that $t \leq k < r$, $\mathfrak{M}, k, w \models \overline{T}X$.

One can obtain similar results for other types of signed wffs.

Let us recall the notion of a prefix considered in [28] to give the tableau-based proof procedures for modal logics. It is defined as a non-empty string over the set $\mathbb{N}$ of positive integers. Wffs are labeled with the prefix to name the world (object) where each wff is supposed to hold. Satisfiability of a $\mathcal{L}$-wff depends on the object as well as on the time point where the wff is evaluated. Thus, we need to modify the notion of a prefix so that it not only names the object, but also mentions the time point where the wff is supposed to hold. In fact, if our interest is in the dynamic spaces of cardinality N, then we will consider strings over the set $\mathbb{N} \cup \{R_1, R_2, \ldots, R_N\}$. The elements from $\{R_1, R_2, \ldots, R_N\}$ will keep track of the time points. We formally define a N-prefix as follows. For a string σ, let us write $|\sigma|$ and $(\sigma)_i$, $i \leq |\sigma|$, to denote the length of σ and the ith element of σ. The empty string is denoted by ε.

Definition 48. *Let N be an integer. A string σ over the alphabet* $\mathbb{N} \cup \{R_1, R_2, \ldots, R_N\}$ *is called a* N-prefix *if it satisfies the following:*

(i) $(\sigma)_1 \in \{R_1, R_2, \ldots, R_N\}$ *for* $|\sigma| \geq 1$.
(ii) There does not exist a j *such that* $(\sigma)_j, (\sigma)_{j+1} \in \{R_1, R_2, \ldots, R_N\}$.
(iii) If $(\sigma)_i = R_n$ *and* $(\sigma)_j = R_m$, *where* $i < j$ *and for all* k *with* $i < k < j$, $(\sigma)_k \in \mathbb{N}$, *then* $n \neq m$.

We will denote the set of all N-prefixes by $P(N)$. For instance, observe that for $N \geq 3$, R_2135, R_21R_3 and R_2135R_1157 are N-prefixes, but $1R_2135$, $R_21R_3R_1157$ and R_2135R_2157 are not. Note that $\varepsilon \in P(N)$, for all N. We call ε the *empty* N*-prefix*. ε is usually not taken as a prefix, but we do so for making some statements compact.

Notation 13. In the rest of this section $\tau, \sigma, \mu, \sigma_1, \sigma'$ etc. will denote N-prefixes. X, Y, Z etc. will be used to denote the $\mathcal{L}$-wffs as well as the signed wffs and the context should be clear from the occurrence. Furthermore, l, t etc. in a N-prefix will denote the elements from $\mathbb{N}$.

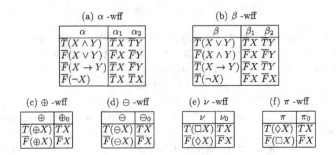

Definition 49. *The* characteristic *of a N-prefix σ, denoted as $char(\sigma)$, is the integer $n \in \{1, 2, \ldots, N\}$ such that $\sigma = \tau R_n$ or $\sigma = \tau R_n l_1 l_2 \cdots l_r$ where $l_i \in \mathbb{N}$.*

Notation 14. For a N-prefix of the form σt, $t \in \mathbb{N}$, we write $(\sigma t)_n^*$ to denote the N-prefix $\sigma t R_n$ if $char(\sigma t) \neq n$, and σt, otherwise.

If a wff is labeled by a N-prefix σ with $char(\sigma) = n$, then it signifies that the wff needs to hold at the object represented by σ at the time point n. Moreover, the N-prefixes σ and σR_l will be used to refer to the same object. Thus, the N-prefixed wffs σX and $\sigma R_l X$ will indicate that the wff X needs to hold at the object represented by σ (or, σR_l), but at the time points $char(\sigma)$ and $char(\sigma R_l)$ respectively. In other words, σ and σR_l represent the same object of the domain, but in the context of different approximation spaces/time points (which are determined by the characteristics of the N-prefixes). Thus σt, $(\sigma t)_k^*$ represent the same object, possibly at different time points: $(\sigma t)_k^*$ indicates the time point k.

Consider the strings $R_1 45 R_2 R_3$ and $R_1 45 R_3$. Using the interpretation of N-prefixes given above, we note that both $R_1 45 R_2 R_3$ and $R_1 45 R_3$ refer to the same object, and also in the context of same time point 3. Thus both $R_1 45 R_2 R_3$ and $R_1 45 R_3$ carry the same information, and therefore, we don't need to include $R_1 45 R_2 R_3$ as a N-prefix. This is why we have imposed the restriction (ii) on N-prefixes. One can similarly see the justification for the Condition (iii) in the definition of N-prefixes.

Let X be a wff. Given an integer N, an (I, N)-tableau for X, $I \in \{K, K4, T, B, S4, KTB, KB4, S5\}$, is the tree each of whose nodes is marked with a N-prefixed signed wff (that is, wff of the form $\sigma \overline{T} X$ or $\sigma \overline{F} X$, where $\sigma \in P(N)$), obtained by enlarging the one node tree $R_1 1 \overline{F} X$ using certain (I, N)-*tableau extension rules*, which will be provided later. An (I, N)-branch (that is, a branch of a (I, N)-tableau) is called *closed* if it contains either of the following:

1. $\sigma \overline{T} X$ and $\sigma \overline{F} X$,
2. $\sigma \overline{T} \perp$,
3. $\sigma \overline{F} \top$,
4. $\sigma \overline{T} \oplus X$, where $char(\sigma) = N$,
5. $\sigma \overline{T} \ominus X$, where $char(\sigma) = 1$.

Clearly, an (I, N)-tableau is going to be interpreted in the dynamic spaces of cardinality N. Now, condition 4 corresponds to the situation where we need a wff of the form $\oplus X$ to be satisfied at the time point N of a dynamic space with cardinality N, which is obviously not possible. Similarly, condition 5 points to an absurdity.

An (I, N)-branch that is not closed is called *open*. An (I, N)-tableau is said to be *closed* if each branch of it is closed.

Definition 50.

(i) A wff X is said to be (I, N)-tableau provable *if we obtain a closed (I, N)-tableau for X.*

(ii) If X is (I,N)-tableau provable for all N, then we say that X is I-tableau provable.

In order to give the full set of tableau extension rules, we need a few more definitions.

Definition 51.

(i) $\sigma \in P(N)$ is said to be used *on an (I,N)-tableau branch if σZ occurs on the branch for some signed wff Z.*
(ii) $\sigma \in P(N)$ is said to be unrestricted *on an (I,N)-tableau branch if σ is not an initial segment (proper or otherwise) of any N-prefix used on the branch.*
(iii) $\sigma' \in P(N)$ is said to be a simple extension *of $\sigma \in P(N)$ if $\sigma' = \sigma n$ for some $n \in \mathbb{N}$.*

Let us pause for a moment here and again recall the tableau-based proof procedures of modal logics given in [28]. The accessibility relation, say R^K, on the set of prefixes for the modal system K is defined such that $(\sigma, \sigma') \in R^K$ if and only if $\sigma' = \sigma t$, $t \in \mathbb{N}$. If we want the relation to be reflexive, then we also include the relation R^r, where $(\sigma, \sigma') \in R^r$ if and only if $\sigma = \sigma'$. Similarly, if we want the symmetry, then we consider the relation R^s, where $(\sigma, \sigma') \in R^s$ if and only if $(\sigma', \sigma) \in R^K$, that is, if and only if $\sigma = \sigma' t$. Therefore, the accessibility relation for the modal system T, B can be given as $R^K \cup R^r$ and $R^K \cup R^s$ respectively.

Let us return to the situation here. A natural question arises: could the relation R^K also be taken as the accessibility relation for the class of dynamic K spaces? Moreover, would the relation R^r be enough to impose reflexivity? The answer to both the questions is negative. In fact, we shall require the following relations to define the accessibility relations for different classes of dynamic spaces.

Definition 52. *Let $n \leq N$. We define the relations $R1, R2, R3, R^4_n, R^{s4}_n, R^5_n, R^{s5}_n, R^6_n, R^{s6}_n, R^7_n, R^{s7}_n$ on the set $P(N) \setminus \{\varepsilon\}$ of N-prefixes as follows.*

- *$(\sigma, \sigma') \in R1$ if and only if $\sigma = \sigma'$;*
- *$(\sigma, \sigma') \in R2$ if and only if $\sigma = \sigma' R_l$ for some $l \in \{1, 2, \ldots, N\}$;*
- *$(\sigma, \sigma') \in R3$ if and only if $\sigma' = \sigma R_l$ for some $l \in \{1, 2, \ldots, N\}$;*
- *$(\sigma, \sigma') \in R^4_n$ if and only if $\sigma' = \sigma t$ for some $t \in \mathbb{N}$, where $char(\sigma) = char(\sigma') = n$;*
- *$(\sigma, \sigma') \in R^{s4}_n$ if and only if $\sigma = \sigma' t$ for some $t \in \mathbb{N}$, where $char(\sigma) = char(\sigma') = n$;*
- *$(\sigma, \sigma') \in R^5_n$ if and only if $\sigma = \tau R_l$ and $\sigma' = \tau R_n t$ for some $l \in \{1, 2, \ldots, N\}$ and $t \in \mathbb{N}$;*
- *$(\sigma, \sigma') \in R^{s5}_n$ if and only if $\sigma' = \tau R_l$ and $\sigma = \tau R_n t$ for some $l \in \{1, 2, \ldots, N\}$ and $t \in \mathbb{N}$;*
- *$(\sigma, \sigma') \in R^6_n$ if and only if $\sigma = \tau R_l$ and $\sigma' = \tau t$ for some $l \in \{1, 2, \ldots, N\}$ and $t \in \mathbb{N}$, where $char(\sigma') = char(\tau) = n$;*

- $(\sigma, \sigma') \in R_n^{s6}$ *if and only if* $\sigma' = \tau R_l$ *and* $\sigma = \tau t$ *for some* $l \in \{1, 2, \ldots, N\}$ *and* $t \in \mathbb{N}$, *where* $char(\sigma) = char(\tau) = n$;
- $(\sigma, \sigma') \in R_n^{7}$ *if and only if* $\sigma' = \sigma R_n t$ *for some* $t \in \mathbb{N}$;
- $(\sigma, \sigma') \in R_n^{s7}$ *if and only if* $\sigma = \sigma' R_n t$ *for some* $t \in \mathbb{N}$.

Since the N-prefixes σ and σR_l will be interpreted to represent the same object, if we want the accessibility relation to be reflexive, then $R1$ will not be sufficient for the purpose and we shall require $R2$ and $R3$ as well. For a similar reason, we will require R_n^5, R_n^6, R_n^7 in addition to R_n^4 to give the accessibility relation for the class of dynamic K spaces. $R_n^{s4}, R_n^{s5}, R_n^{s6}, R_n^{s7}$ will be used to give symmetry.

The following definition gives the accessibility relations for different classes of dynamic spaces using the relations defined above.

Definition 53. *Let* $I \in \{K, K4, T, B, S4, KTB, KB4, S5\}$ *and* $n \leq N$. *We define the* I-accessibility relation for characteristic n, R_n^I, *on the set* $P(N) \setminus \{\varepsilon\}$ *of* N*-prefixes as follows. Let us write* R^* *to denote the transitive closure of the relation* R.

- $R_n^K := R_n^4 \cup R_n^5 \cup R_n^6 \cup R_n^7$.
- $R_n^T := R1 \cup R2 \cup R3 \cup R_n^K$.
- $R_n^{K4} := (R_n^K)^*$.
- $R_n^B := R_n^4 \cup R_n^5 \cup R_n^6 \cup R_n^7 \cup R_n^{s4} \cup R_n^{s5} \cup R_n^{s6} \cup R_n^{s7}$.
- $R_n^{KB4} := (R_n^B)^*$.
- $R_n^{S4} := (R_n^T)^*$.
- $R_n^{KTB} := R1 \cup R2 \cup R3 \cup R_n^B$.
- $R_n^{S5} := (R_n^{KTB})^*$.

Now, we are in a position to give the tableau extension rules for each of the classes of dynamic I spaces, $I \in \{K, K4, T, B, S4, KTB, KB4, S5\}$. Figures 3–4 give the rules for these classes corresponding to the N-prefixed wffs. We call these (I, N)-*tableau extension rules.* Except one, the rules for the *Since* operator are not listed, as these can be given in the lines of those for the *Until* operator.

The rules involving temporal operators, that is, $\oplus, \ominus, \mathcal{U}, \mathcal{S}$-rules, will be called the *temporal rules.*

In order to understand the tableau extension rules, let us consider a wff $\oplus\alpha$. To evaluate $\oplus\alpha$ at the time point t at the object w, we need to evaluate the wff α at the same object w, but at the time point $t+1$. To capture this fact, the $\oplus$ rule is designed such that if the branch contains $\sigma m \oplus \alpha$, $m \in \mathbb{N}$, then using the $\oplus(b)$-rule, we introduce the wff $\sigma m R_{l+1} \alpha$ (assuming $char(\sigma m) = l$). The N-prefix in the latter is used to indicate the same object referred to by σm (recall the comments following Definition 49), along with the fact that we have shifted focus to the time point $l+1$. Similarly, one may explain the rule $\mathcal{U}(d)$. Note that in a model $\mathfrak{M}$ with number of time points N, for any $w \in U$ and $n < N$, we have $\mathfrak{M}, n, w \models \neg(X\mathcal{U}Y)$, if and only if either $\mathfrak{M}, k, w \models \neg Y$ for all k with $n \leq k \leq N$, or there exists s with $n \leq s < N$ such that $\mathfrak{M}, s, w \models \neg X$ and for all k such that $n \leq k \leq s$, $\mathfrak{M}, k, w \models \neg Y$. Observe that the wff $X\mathcal{U}Y$

$\dfrac{\sigma\alpha}{\begin{array}{c}\sigma\alpha_1\\ \sigma\alpha_2\end{array}}$ <u>α-rule</u>	1st $\sigma\beta$ 2nd $\sigma\beta_1 \mid \sigma\beta_2$ <u>β-rule</u>	$\dfrac{\sigma R_n\oplus,\ \ n<N}{\sigma^*_{n+1}\oplus_0}$ <u>$\oplus(a)$-rule</u>
$\dfrac{\sigma R_n t_1 t_2\cdots t_r\oplus,\ \ n<N}{\sigma^*_{n+1}\oplus_0}$ <u>$\oplus(b)$-rule</u>	$\dfrac{\sigma R_n\ominus,\ \ n>1}{\sigma^*_{n-1}\ominus_0}$ <u>$\ominus(a)$-rule</u>	$\dfrac{\sigma R_n t_1 t_2\cdots t_r\ominus,\ \ n>1}{\sigma^*_{n-1}\ominus_0}$ <u>$\ominus(b)$-rule</u>
$\dfrac{\sigma\nu}{\sigma'\nu_0}$ where $(\sigma,\sigma')\in R^I_n$, σ' is used on the branch and $char(\sigma)=char(\sigma')=n$. <u>$\nu$-rule</u>	$\dfrac{\sigma\pi}{\sigma'\pi_0}$ σ' is an unrestricted simple extension of σ. <u>π-rule</u>	$\dfrac{\sigma R_n\overline{T}p}{\sigma\overline{T}p}$ $\dfrac{\sigma R_n\overline{F}p}{\sigma\overline{F}p}$ p is propositional variable. <u>Propositional-rule</u>

$\dfrac{\sigma\overline{F}(X\mathcal{U}Y),\ \ char(\sigma)=N}{\sigma\overline{F}Y}$ <u>$\mathcal{U}$ end point-rule</u>	$\dfrac{\sigma\overline{F}(X\mathcal{S}Y),\ \ char(\sigma)=1}{\sigma\overline{F}Y}$ <u>$\mathcal{S}$ first point-rule</u>

Fig. 3. (I,N)-tableau extension rules.

can be made false at (n,w) in $N-n+1$ possible ways. Accordingly, the rule $\mathcal{U}(d)$ for $\tau\overline{F}(X\mathcal{U}Y)$, $n<N$, where $\tau=\sigma R_n t_1 t_2\cdots t_r$, asks for the introduction of $N-n+1$ branches. We again note that all N-prefixes of the kind τ^*_m in the rule indicate the same object referred to by τ, along with the fact that we have shifted focus to the time point m. The tableau extension rules for the other wffs may be understood in a similar manner.

Given a N-prefix σ, we would like to investigate next what would be the form of any N-prefix μ related to σ by the relation R^I_n. We would then be able to see the possible forms of an accessible N-prefix – this is crucial for the application of the ν rule. Moreover, the information about the form of such N-prefixes will also be used for obtaining important results related with the soundness, completeness

$$\sigma R_n \overline{T}(X\mathcal{U}Y)$$

1st	2nd	$\cdots$	i^{th}, $1 \leq i \leq N-n+1$	$\cdots$	$(N-n)^{th}$	$(N-n+1)^{th}$
$\sigma_n^*\overline{T}Y$	$\sigma_{n+1}^*\overline{T}Y$	$\cdots$	$\sigma_{n+(i-1)}^*\overline{T}Y$	$\cdots\cdots$	$\sigma_{N-1}^*\overline{T}Y$	$\sigma_N^*\overline{T}Y$
	$\sigma_n^*\overline{T}X$	$\cdots$	$\sigma_{n+(i-k)}^*\overline{T}X$	$\cdots\cdots$	$\sigma_{N-k}^*\overline{T}X$	$\sigma_{N+1-k}^*\overline{T}X$
			$2 \leq k \leq i$		$2 \leq k \leq N-n$	$2 \leq k \leq N-n+1$

$\mathcal{U}(a)$-rule

$$\sigma R_n t_1 t_2 \cdots t_r \overline{T}(X\mathcal{U}Y), \quad \tau = \sigma R_n t_1 t_2 \cdots t_r$$

1st	2nd	$\cdots$	i^{th}, $1 \leq i \leq N-n+1$	$\cdots$	$(N-n)^{th}$	$(N-n+1)^{th}$
$\tau_n^*\overline{T}Y$	$\tau_{n+1}^*\overline{T}Y$	$\cdots$	$\tau_{n+(i-1)}^*\overline{T}Y$	$\cdots\cdots$	$\tau_{N-1}^*\overline{T}Y$	$\tau_N^*\overline{T}Y$
	$\tau_n^*\overline{T}X$	$\cdots$	$\tau_{n+(i-k)}^*\overline{T}X$	$\cdots\cdots$	$\tau_{N-k}^*\overline{T}X$	$\tau_{N+1-k}^*\overline{T}X$
			$2 \leq k \leq i$		$2 \leq k \leq N-n$	$2 \leq k \leq N-n+1$

$\mathcal{U}(b)$-rule

$$\sigma R_n \overline{F}(X\mathcal{U}Y),\ n < N$$

1st	2nd	$\cdots$	i^{th}, $1 \leq i \leq N-n$	$\cdots$	$(N-n)^{th}$	$(N-n+1)^{th}$
$\sigma_n^*\overline{F}X$	$\sigma_{n+1}^*\overline{F}X$	$\cdots$	$\sigma_{n+(i-1)}^*\overline{F}X$	$\cdots\cdots$	$\sigma_{N-1}^*\overline{F}X$	$\sigma_N^*\overline{F}Y$
$\sigma_n^*\overline{F}Y$	$\sigma_{n+(2-k)}^*\overline{F}Y$	$\cdots$	$\sigma_{n+(i-k)}^*\overline{F}Y$	$\cdots\cdots$	$\sigma_{N-k}^*\overline{F}Y$	$\sigma_{N-k}^*\overline{F}Y$
	$1 \leq k \leq 2$		$1 \leq k \leq i$		$1 \leq k \leq N-n$	$1 \leq k \leq N-n$

$\mathcal{U}(c)$-rule

$$\sigma R_n t_1 t_2 \cdots t_r \overline{F}(X\mathcal{U}Y),\ n < N \text{ and } \tau = \sigma R_n t_1 t_2 \cdots t_r$$

1st	2nd	$\cdots$	i^{th}, $1 \leq i \leq N-n$	$\cdots$	$(N-n)^{th}$	$(N-n+1)^{th}$
$\tau_n^*\overline{F}X$	$\tau_{n+1}^*\overline{F}X$	$\cdots$	$\tau_{n+(i-1)}^*\overline{F}X$	$\cdots\cdots$	$\tau_{N-1}^*\overline{F}X$	$\tau_N^*\overline{F}Y$
$\tau_n^*\overline{F}Y$	$\tau_{n+(2-k)}^*\overline{F}Y$	$\cdots$	$\tau_{n+(i-k)}^*\overline{F}Y$	$\cdots\cdots$	$\tau_{N-k}^*\overline{F}Y$	$\tau_{N-k}^*\overline{F}Y$
	$1 \leq k \leq 2$		$1 \leq k \leq i$		$1 \leq k \leq N-n$	$1 \leq k \leq N-n$

$\mathcal{U}(d)$-rule

Fig. 4. (I, N)-tableau extension rules, continued.

and termination of the proof procedures. The form of μ depends on I as well as on the form of σ itself. Here we will only give the propositions determining the form for $I = S5$, but one can obtain similar results for other Is. In fact, the argument will be simpler for all the other choices of Is.

Let us first consider the N-prefix of the form $\tau := \tau' R_n t_1 t_2 \cdots t_s,\ s \geq 1$ and the relation R_n^{S5}. We investigate the form of μ for which we have $(\tau, \mu) \in R_n^{S5}$. Since R_n^{S5} is reflexive, we should have $(\tau, \tau) \in R_n^{S5}$. Moreover, due to symmetry and transitivity of R_n^{S5}, we also expect to have $(\tau, \tau' R_n d_1 d_2 \cdots d_k) \in R_n^{S5}$. Furthermore, as mentioned above, σ and σR_l will be interpreted to represent the same object, and thus we should also have $(\tau, \tau R_l)$ and $(\tau, \tau' R_n d_1 d_2 \cdots d_k R_l) \in R_n^{S5}$. In fact, the following proposition shows that these are the only possibilities for μ.

Proposition 67. *Let $\tau = \tau' R_n t_1 t_2 \cdots t_s,\ s \geq 1$ and $(\tau, \mu) \in R_n^{S5}$. Then μ must be in either of the following forms.*

(a) μ is τ'.
(b) μ is $\tau' R_l$, where $R_l \in \{R_1, R_2, \ldots, R_N\}$.
(c) μ is $\tau' R_n d_1 d_2 \cdots d_k$, where $d_i \in \mathbb{N}$.
(d) μ is $\tau' R_n d_1 d_2 \cdots d_k R_l$, where $R_l \in \{R_1, R_2, \ldots, R_N\}$ and $d_i \in \mathbb{N}$.

Proof. The proposition is proved by showing that if

$$\tau R_n^{KTB} \sigma_1 R_n^{KTB} \sigma_2 \cdots R_n^{KTB} \sigma_j,$$

then $\sigma_j = \tau'$ or σ_j is in either of the above-mentioned forms. We use induction on j. Note that since $\tau' R_n t_1 t_2 \cdots t_s \in P(N)$, we have $char(\tau') \neq n$.
Basis case: Let $j = 1$.
Since $(\tau, \sigma_1) \in R_n^{KTB}$ and $R_n^{KTB} = R1 \cup R2 \cup R3 \cup R_n^4 \cup R_n^5 \cup R_n^6 \cup R_n^7 \cup R_n^{s4} \cup R_n^{s5} \cup R_n^{s6} \cup R_n^{s7}$, we must have $(\tau, \sigma_1) \in R$ for some $R \in \{R1, R2, R3, R_n^4, R_n^5, R_n^6, R_n^7, R_n^{s4}, R_n^{s5}, R_n^{s6}, R_n^{s7}\}$. But note that $(\tau, \sigma_1) \notin R$ for $R \in \{R2, R_n^5, R_n^6, R_n^7\}$. Therefore, $\sigma_1 = \tau'$ (in the case when $(\tau, \sigma_1) \in R1$), or σ_1 must be in one of the following forms:

- τR_l (when $(\tau, \sigma_1) \in R3$), or
- $\tau' R_n t_1 t_2 \cdots t_s t_{s+1}$ (when $(\tau, \sigma_1) \in R_n^4$), or
- $\tau' R_n$, where $s = 1$ (when $(\tau, \sigma_1) \in R_n^{s4}$), or
- $\tau' R_n t_1 t_2 \cdots t_{s-1}$, where $s \geq 2$ (when $(\tau, \sigma_1) \in R_n^{s4}$), or
- $\tau' R_l$, where $s = 1$ (when $(\tau, \sigma_1) \in R_n^{s5}$), or
- $\tau' R_n t_1 t_2 \cdots t_{s-1} R_l$, where $s \geq 2$ (when $(\tau, \sigma_1) \in R_n^{s6}$), or
- τ', where $s = 1$ (when $(\tau, \sigma_1) \in R_n^{s7}$).

Thus in each case, we obtain σ_j in the desired form.
Induction case: Suppose (a) $\sigma_j = \tau'$, or σ_j is in one of the following forms:

(b) $\tau' R_l$, or
(c) $\tau' R_n d_1 d_2 \cdots d_k$, or
(d) $\tau' R_n d_1 d_2 \cdots d_k R_l$

Case (a): $\sigma_j = \tau'$
Since $char(\tau') \neq n$ and $(\sigma_j, \sigma_{j+1}) \in R_n^{KTB}$, $\sigma_{j+1} = \tau'$ (when $(\sigma_j, \sigma_{j+1}) \in R1$) or σ_{j+1} must be in either of the following forms:
- $\tau' R_l$ (when $(\sigma_j, \sigma_{j+1}) \in R3$), or
- $\tau' R_n d_1$ (when $(\sigma_j, \sigma_{j+1}) \in R_n^7$).

Case (b): σ_j is of the form $\tau' R_l$.
In this case, $\sigma_{j+1} = \tau'$ (when $(\sigma_j, \sigma_{j+1}) \in R2$), or σ_{j+1} must be in either of the following forms:

- $\tau' R_l$ (when $(\sigma_j, \sigma_{j+1}) \in R1$), or
- $\tau' R_l d_1$, $l = n$ (when $(\sigma_j, \sigma_{j+1}) \in R_n^4$), or
- $\tau' R_n d_1$, (when $(\sigma_j, \sigma_{j+1}) \in R_n^5$).

Case (c): σ_j is of the form $\tau' R_n d_1 d_2 \cdots d_k$.
Then, $\sigma_{j+1} = \tau'$ and $k = 1$ (when $(\sigma_j, \sigma_{j+1}) \in R_n^{s7}$), or σ_{j+1} must be in one of the following forms:

- $\tau' R_n d_1 d_2 \cdots d_k$ (when $(\sigma_j, \sigma_{j+1}) \in R1$), or
- $\tau' R_n d_1 d_2 \cdots d_k R_l$ (when $(\sigma_j, \sigma_{j+1}) \in R3$), or
- $\tau' R_n d_1 d_2 \cdots d_k d_{k+1}$ (when $(\sigma_j, \sigma_{j+1}) \in R_n^4$), or
- $\tau' R_n d_1 d_2 \cdots d_{k-1}$ and $k \geq 2$ (when $(\sigma_j, \sigma_{j+1}) \in R_n^{s4}$), or
- $\tau' R_n$ and $k = 1$ (when $(\sigma_j, \sigma_{j+1}) \in R_n^{s4}$), or
- $\tau' R_l$ and $k = 1$ (when $(\sigma_j, \sigma_{j+1}) \in R_n^{s5}$), or
- $\tau' R_n d_1 d_2 \cdots d_{k-1} R_l$ and $k \geq 2$ (when $(\sigma_j, \sigma_{j+1}) \in R_n^{s6}$).

Case (d): σ_j is of the form $\tau' R_n d_1 d_2 \cdots d_k R_l$.
Then, σ_{j+1} must be in either of the following forms:

- $\tau' R_n d_1 d_2 \cdots d_k R_l$ (when $(\sigma_j, \sigma_{j+1}) \in R1$), or
- $\tau' R_n d_1 d_2 \cdots d_k$ (when $(\sigma_j, \sigma_{j+1}) \in R2$), or
- $\tau' R_n d_1 d_2 \cdots d_k d_{k+1}$ (when $(\sigma_j, \sigma_{j+1}) \in R_n^6$).

Thus, in each case, we obtain σ_{j+1} in the desired form. This completes the proof. □

As a direct consequence of Proposition 67, we obtain the following corollary.

Corollary 15. *Let $\mu, \tau \in P(N)$ be such that (i) $\tau = \tau' R_n t_1 t_2 \cdots t_s$, $s \geq 1$, (ii) $(\tau, \mu) \in R_n^{S5}$ and (iii) $char(\mu) = n$. Then μ must be either $\tau' R_n$, or of the form $\tau' R_n d_1 d_2 \cdots d_k$, $d_i \in \mathbb{N}$.*

We end this section with the two following propositions, similar to Proposition 67.

Proposition 68. *Let $\tau = \tau' R_h t_1 t_2 \cdots t_s R_n \in P(N)$, (so, $s \geq 1$, $h \neq n$) and $(\tau, \mu) \in R_n^{S5}$. Then $\mu = \tau' R_h t_1 t_2 \cdots t_s$ or μ must be in either of the form (a) $\tau' R_h t_1 t_2 \cdots t_s R_l$, or (b) $\tau d_1 d_2 \cdots d_r$, or (c) $\tau d_1 d_2 \cdots d_r R_l$.*

Proposition 69. *Let $\tau = \tau' R_h t_1 t_2 \cdots t_s$, $s \geq 1$, $h \neq n$ and $(\tau, \mu) \in R_n^{S5}$. Then $\mu = \tau$ or μ must be in either of the form (a) τR_l, or (b) $\tau R_n d_1 d_2 \cdots d_r$, or (c) $\tau R_n d_1 d_2 \cdots d_r R_l$.*

10.1 Soundness

In this section, we shall prove the soundness of the tableau-based proof procedures proposed in Sect. 10. We begin with the following definitions. Recall the relations R_n^I given by Definition 53.

Definition 54.

(i) *Let S be a set of N-prefixed wffs and $\mathfrak{F} := \mathcal{F}_1, \mathcal{F}_2, \ldots \mathcal{F}_N$ be a dynamic I space, where $\mathcal{F}_i = (W, P_i)$. By an I-*interpretation *of S in $\mathfrak{F}$, we mean a mapping Int from the set of N-prefixes that occur in S to W such that the following conditions are satisfied:*
  (a) If $(\sigma, \sigma') \in R_n^I$ then $(Int(\sigma), Int(\sigma')) \in P_n$, $n = 1, 2, \ldots, N$.
  (b) $Int(\sigma) = Int(\sigma R_l)$, for all $l \in \mathbb{N}$.

(ii) *A* set S *of N-prefixed wffs is said to be (I, N)-*satisfiable *if there exists some dynamic I space $\mathfrak{F} := \mathcal{F}_1, \mathcal{F}_2, \ldots, \mathcal{F}_N$ of cardinality N, an I-interpretation Int of S in $\mathfrak{F}$, and a valuation function $V : P \rightarrow 2^W$ such that for each $\sigma Z \in S$,*

$$\mathfrak{M}, n, Int(\sigma) \models Z$$

where $char(\sigma) = n$ and $\mathfrak{M} = (\mathfrak{F}, V)$.

(iii) *A* branch *of an (I, N)-tableau is said to be (I, N)-*satisfiable *if the set of N-prefixed wffs on it is (I, N)-satisfiable.*

(iv) *An (I, N)-*tableau *is said to be (I, N)-*satisfiable *if some branch of it is (I, N)-satisfiable.*

Observe that the condition (b) in the definition of interpretation signifies that the N-prefixes σ and σR_l represent the same object.

Example 16. Let us consider a $(S5, 2)$-tableau for the wff $F\Box r \rightarrow \Diamond p \vee \Box q$, $p, q, r \in PV$, given in Fig. 5.

Let S be the set of all 2-prefixed wffs occurring on the right branch of the tableau of Fig. 5. Note that $\{R_1 1, R_1 11, R_1 1 R_2\}$ is set of all 2-prefixes that occur in S. Let us consider the dynamic $S5$ space $\mathfrak{F} := \mathcal{F}_1, \mathcal{F}_2$, where $\mathfrak{F}_i := (W, P_i)$, $W := \{x, y\}$, $W/P_1 := W \times W$ and $W/P_2 := \{\{x\}, \{y\}\}$. Let Int be the function from the set of 2-prefixes occurring in S to W defined as

$$Int(R_1 1) = Int(R_1 1 R_2) = x;$$
$$Int(R_1 11) = y.$$

Note that Int is an interpretation of S in $\mathfrak{F}$ as it satisfies both the defining conditions (a), and (b) of an I-interpretation (cf. Definition 54). In fact, condition (a) is a direct consequence of the fact that $(R_1 1, R_1 11), (R_1 1 R_2, R_1 11) \notin R_2^{S5}$, which, in turn, follows from Proposition 67. Further, using Int and the valuation V which maps p to $\emptyset$, q and r to $\{x\}$, one can show that the right branch of the above tableau is $(S5, 2)$-satisfiable. For instance, for the wff $R_1 1 \overline{F}(F\Box r \rightarrow \Diamond p \vee \Box q)$ lying on the right branch, we have $\mathfrak{M}, 1, Int(R_1 1) \models \overline{F}(F\Box r \rightarrow \Diamond p \vee \Box q)$ as $\mathfrak{M}, 2, x \models \Box r$, but $\mathfrak{M}, 1, x \not\models \Diamond p$ and $\mathfrak{M}, 1, x \not\models \Box q$, $\mathfrak{M} := (\mathfrak{F}, V)$.

1. $R_1 1\overline{F}(F\Box r \to \Diamond p \vee \Box q)$
2. $R_1 1\overline{T}F\Box r$ (applying α-rule on 1)
3. $R_1 1\overline{F}(\Diamond p \vee \Box q)$ (α-rule on 1)
4. $R_1 1\overline{F}\Diamond p$ (α-rule on 3)
5. $R_1 1\overline{F}\Box q$ (α-rule on 3)
6. $R_1 11\overline{F}q$ (π-rule on 5)
7. $R_1 1\overline{F}p$ (ν-rule on 4)
8. $R_1 11\overline{F}p$ (ν-rule on 4)

Left branch:

9. $R_1 1\overline{T}\Box r$ ($\mathcal{U}(b)$-rule on 2)
12. $R_1 1\overline{T}r$ (ν-rule on 9)

Right branch:

10. $R_1 1R_2\overline{T}\Box r$ ($\mathcal{U}(b)$-rule on 2)
11. $R_1 1\overline{T}\top$ ($\mathcal{U}(b)$-rule on 2)
13. $R_1 1R_2\overline{T}r$ (ν-rule on 10)
14. $R_1 1\overline{T}r$ (Propositional-rule on 13)

Fig. 5. A $(S5, 2)$-tableau for $F\Box r \to \Diamond p \vee \Box q$.

Using the definition of a closed branch, it is not difficult to obtain

Proposition 70. *A closed branch of an (I, N)-tableau cannot be (I, N) satisfiable and hence a closed (I, N)-tableau cannot be (I, N)-satisfiable.*

The following proposition will lead us to the desired theorem.

Proposition 71. *Let T be an (I, N)-tableau that is (I, N)-satisfiable. Let T' be the (I, N)-tableau that results from a single (I, N)-tableau rule being applied to T. Then T' is also (I, N)-satisfiable.*

Proof. Let T be (I, N)-satisfiable under Int. By the given condition, some branch of T, say θ is (I, N)-satisfiable. The result follows easily in the case when the (I, N)-rule is applied to a branch other than θ, as in this case, θ is unaltered in T'. So, consider the case when the (I, N)-rule in question is applied to the branch θ. Let

$$\mathfrak{M}, n, Int(\sigma) \models Z, \text{ where } char(\sigma) = n, \ \mathfrak{M} := (\mathfrak{F}, V), \tag{14}$$

for all σZ occurring on θ. Let $\mathfrak{F} := \mathcal{F}_1, \mathcal{F}_2, \ldots, \mathcal{F}_N$, where $\mathcal{F}_i := (W, P_i)$. The proof argument for the cases when the rule $\alpha, \beta, v, \oplus, \ominus, \mathcal{U}$ or $\mathcal{S}$ is applied on θ, is same for all $I \in \{K, K4, T, B, S4, KTB, KB4, S5\}$. The argument when the $\pi-$rule is applied, depends on I. Here we give proof for $S5$. The other cases can be proved similarly.

Let us first consider the case when the α-rule is applied on θ. Let $\sigma\alpha$ occur on θ and $\sigma\alpha_1$ and $\sigma\alpha_2$ be added to the end of θ by the α-rule. Then, since $\mathfrak{M}, n, Int(\sigma) \models \alpha$, where $char(\sigma) = n$, we obtain $\mathfrak{M}, n, Int(\sigma) \models \alpha_1$ and

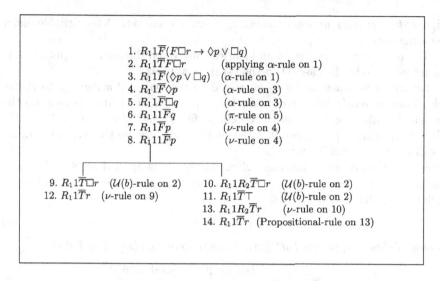

$\mathfrak{M}, n, Int(\sigma) \models \alpha_2$, and hence the enlarged branch is also (I, N)-satisfiable under the same Int.

The result can be proved in an identical way when the β-rule is applied on θ. So, let us move to the case when the ν-rule is applied.

Let $\sigma\nu$ occur on θ and $\sigma'\nu_0$ be added to the end of θ according to the ν-rule. Then $char(\sigma') = char(\sigma) = n$, (say), $(\sigma, \sigma') \in R_n^I$ and σ' is used on the branch. In this case Int is already defined for σ' and $(Int(\sigma), Int(\sigma')) \in P_n$. Thus, $\mathfrak{M}, n, Int(\sigma) \models \nu$ gives $\mathfrak{M}, n, Int(\sigma') \models \nu_0$ Therefore the enlarged branch is still (I, N)-satisfiable under Int.

Now, let us consider the case when the π-rule is applied on θ and $\sigma'\pi_0$ is added to the end of θ. Let $char(\sigma) = n$. Since $\mathfrak{M}, n, Int(\sigma) \models \pi$, there exists Γ such that $(Int(\sigma), \Gamma) \in P_n$, and

$$\mathfrak{M}, n, \Gamma \models \pi_0 \tag{15}$$

Now we define mapping a Int^* with domain $dom(Int) \cup \{\sigma'\}$ as follows.

$$Int^*(\tau) := \begin{cases} Int(\tau) & \text{if } \tau \text{ is used on } \theta \\ \Gamma & \text{for } \tau = \sigma'. \end{cases}$$

Our claim is that Int^* is an interpretation for the set of N-prefixed wffs occurring on the branch obtained by extending θ by the application of the above π rule.

Let τ be any N-prefix used on the branch θ and $(\sigma', \tau) \in R_l^{TS5}$ for some $l \leq N$. Then by Proposition 69, it follows that $l = n$ (otherwise σ' becomes an initial segment of τ). Thus, using the transitivity of R_n^{S5}, we obtain $(\sigma, \tau) \in R_n^{TS5}$ and hence $(Int(\sigma), Int(\tau)) \in P_l$, that is, $(Int^*(\sigma), Int^*(\tau)) \in P_l$. Using this fact, and transitivity of P_l, we obtain $(Int^*(\sigma'), Int^*(\tau)) \in P_l$. Clearly Int^* also satisfies the other condition for interpretation. Thus we have proved our claim. Therefore from (15), we obtain $\mathfrak{M}, n, Int^*(\sigma') \models \pi_0$. This shows that the enlarged branch is (I, N)-satisfiable under Int^*.

Now let us work on the case when the $\oplus(a)$-rule is applied on θ. Let $\sigma R_n \oplus$ occur on θ, $n < N$, and suppose $\sigma^*_{n+1}\oplus_0$ is added to the end of θ. There are two possibilities: σ^*_{n+1} may or may not be used on θ. Let us first consider the case when σ^*_{n+1} is used on θ. In this case Int is defined for σ^*_{n+1}. Moreover, $Int(\sigma) = Int(\sigma^*_{n+1})$. Therefore, from $\mathfrak{M}, n, Int(\sigma R_n) \models \oplus$, we obtain $\mathfrak{M}, n+1, Int(\sigma^*_{n+1}) \models \oplus_0$.

Now, let us suppose σ^*_{n+1} is not used on θ. Then, we must have $\sigma^*_{n+1} = \sigma R_{n+1}$. We define Int^* with domain $dom(Int) \cup \{\sigma R_{n+1}\}$ as follows.

$$Int^*(\tau) := \begin{cases} Int(\tau) & \text{if } \tau \text{ is used on } \theta \\ Int(\sigma) & \text{for } \tau = \sigma R_{n+1}. \end{cases}$$

We claim that Int^* is an interpretation for the set of N-prefixed wffs occurring on the extended branch. Let τ be any N-prefix used on the branch θ and $(\tau, \sigma R_{n+1}) \in R_l^{TS5}$ for some $l \leq N$. This implies $(Int(\tau), Int(\sigma)) \in P_l$ and hence $(Int^*(\tau), Int^*(\sigma R_{n+1})) \in P_l$. The other property of interpretation is trivially satisfied by Int^*. Moreover, it is not difficult to see that $\mathfrak{M}, n+1, Int^*(\sigma R_{n+1}) \models \oplus_0$.

All the remaining cases can be proved in the same fashion. □

Using Proposition 71, we obtain, for any $I \in \{K, K4, T, B, S4, KTB, KB4, S5\}$,

Theorem 16 (Soundness). *If X is I-tableau provable then X is valid in the class of all dynamic I spaces.*

Proof. Suppose X is I-tableau provable, but not valid in the class of all dynamic I spaces. Then there exists some dynamic I space $\mathfrak{F} := \mathcal{F}_1, \mathcal{F}_2, \ldots, \mathcal{F}_N$, for some N where $\mathcal{F}_i := (W, P_i)$, a valuation function $V : PV \to 2^W$ and $\Gamma \in W$ such that

$$\mathfrak{M}, 1, \Gamma \models \overline{F}X \tag{16}$$

for some i, $1 \leq i \leq N$, where $\mathfrak{M} := (\mathfrak{F}, V)$.

Consider the singleton set $S := \{R_1 1\overline{F}X\}$ of N-prefixed wff and interpretation Int of S defined as $Int(R_1 1) := \Gamma$. From (16), it follows that the one node (I, N)-tableau

$$R_1 1\overline{F}X$$

is (I, N)-satisfiable under Int. Then by Proposition 71, so is every subsequent (I, N)-tableau. But, by Proposition 70, an (I, N)-satisfiable (I, N)-tableau cannot be closed, contradicting the assumption that X is I-tableau provable. This completes the proof. □

10.2 Completeness

We now address the completeness of the tableau-based proof procedures proposed in Sect. 10. In order to do that, we first give a systematic procedure to construct an (I, N)-tableau for a given wff X such that if the (I, N)-tableau for X obtained following this procedure is not closed, then X cannot have a closed (I, N)-tableau. Moreover, in that case, the resultant tableau will give enough information to construct a counter model for X.

Let us call a N-prefixed wff *atomic* if it is of the form $\sigma\overline{T}A$ or $\sigma\overline{F}A$ where σ is of the form τl, $l \in \mathbb{N}$, and A is an atomic wff, that is, A is a propositional variable or $\top$ or $\bot$. Note that in [28], a prefixed wff of the form $\sigma\overline{T}A$ and $\sigma\overline{F}A$, A being atomic, is taken to be atomic, but in our case, the N-prefixed wff, say $R_1 12R_2\overline{T}p$, $p \in PV$, is not atomic. This deviation is due to the fact that in [28], no tableau extension rule can be applied on the wff $\sigma\overline{T}A$ or $\sigma\overline{F}A$, but it is not the case with $R_1 12R_2\overline{T}p$. In fact, the propositional rule is applicable on it.

We now describe the systematic procedure to construct an (I, N)-tableau for a given wff X, step by step as follows.

Step-1: We begin by placing $R_1 1\overline{F}X$ at the origin.

Suppose the nth stage of the construction is completed. The tableau constructed so far could be either closed or open. If it is closed then stop the construction. Otherwise, we move to step $n + 1$.

Step-$n+1$: Choose an occurrence of a N-prefixed wff σZ as high up in the tree as possible, satisfying the following.

(i) σZ is not chosen in any of the previous steps;
(ii) σZ is not atomic;
(iii) σZ is not of the form $\sigma \overline{F} \oplus X$, where $char(\sigma) = N$;
(iv) σZ is not of the form $\sigma \overline{F} \ominus X$, where $char(\sigma) = 1$;

If there is no such σZ, then stop the construction. If there is more than one such wff at the same level, then choose the one on the left-most branch. Once the selection is done, extend each open branch θ through the occurrence of σZ using the tableau extension rules as follows.

(a) If σZ is of a form other than $\sigma \nu$ and $\sigma \pi$, extend the branch using the suitable tableau extension rule. For instance, if σZ is of the form $\sigma \alpha$, add $\sigma \alpha_1$ and $\sigma \alpha_2$ to the end of θ following the α-rule. Similarly, if σZ is of the form $\sigma' R_n F(X \mathcal{U} Y)$, where $n < N$, then following the $\mathcal{U}(c)$-rule, split the end of θ into $N - n + 1$ forks and add ${\sigma'}^*_n \overline{F} X, {\sigma'}^*_n \overline{F} Y$ to the end of the first fork, ${\sigma'}^*_n \overline{F} Y, {\sigma'}^*_{n+1} \overline{F} Y, \ldots, {\sigma'}^*_N \overline{F} Y$ to the end of the $(N-n+1)^{th}$ fork and ${\sigma'}^*_{n+(i-1)} \overline{F} X, {\sigma'}^*_n \overline{F} Y, {\sigma'}^*_{n+1} \overline{F} Y, \ldots, {\sigma'}^*_{n+(i-2)} \overline{F} Y, {\sigma'}^*_{n+(i-1)} \overline{F} Y$ to the end of the ith fork, $2 \leq i \leq N - n$.
(b) If σZ is of the form $\sigma \nu$ and $char(\sigma) = n$, then for each N-prefix σ' satisfying (i) $char(\sigma') = n$ (ii) $(\sigma, \sigma') \in R^I_n$ (iii) σ' is used on θ, add $\sigma' \nu_0$ to the end of θ, after which add a fresh occurrence of $\sigma \nu$ to the end of θ.
(c) If σZ is of the form $\sigma \pi$, then add $\sigma k \pi_0$ to the end of θ, where k is the smallest integer such that σk is unrestricted on θ.

The step n+1 is completed once each open branch through the σZ is extended.

This ends the description of our systematic procedure. Let us name this procedure P1. We note that in following P1, it is possible to fall into an endless sequence of steps. For example, procedure P1 to construct a $(B,3)$-tableau for $\oplus \neg \Diamond \Box \Diamond X$ $(X \in PV)$ will never terminate as shown by the following example.

Example 17. The $(B,3)$-tableau for $\oplus \neg \Diamond \Box \Diamond X$ $(X \in PV)$, obtained by following P1 is given in Fig. 6.

To handle this situation in the proof of the completeness theorem, we use *König's lemma* which states that an infinite, finitely generated tree must have an infinite branch. A finitely generated tree is one where each node has only a finite number of immediate successors, and an infinite tree has infinitely many nodes. We note that (I,N)-tableaux are finitely generated as each node has at most N successors.

The open branches of a tableau obtained by following the procedure P1 have certain properties which lead us to the desired counter model. These properties are assembled to give the following definition.

1. $R_1 1\overline{F} \oplus \neg\Diamond\Box\Diamond X$	
2. $R_1 1 R_2 \overline{F} \neg\Diamond\Box\Diamond X$	($\oplus(b)$-rule on 1)
3. $R_1 1 R_2 \overline{T} \Diamond\Box\Diamond X$	(α-rule on 2)
4. $R_1 1 R_2 1 \overline{T} \Box\Diamond X$	(π-rule on 3)
5. $R_1 1 R_2 \overline{T} \Diamond X$	(ν-rule on 4)
6. $R_1 1 R_2 1 \overline{T} \Box\Diamond X$	(4 is added following P1)
7. $R_1 1 R_2 2 \overline{T} X$	(π-rule on 5)
8. $R_1 1 R_2 \overline{T} \Diamond X$	(ν-rule on 6)
9. $R_1 1 R_2 1 \overline{T} \Box\Diamond X$	(6 is added following P1)
10. $R_1 1 R_2 3 \overline{T} X$	(π-rule on 8)
....	
....	

Fig. 6. $(B, 3)$-tableau for $\oplus\neg\Diamond\Box\Diamond X$ obtained by following P1.

Definition 55. *Let S be a set of N-prefixed wffs. S is called (I, N)*-downward saturated *provided the following conditions are satisfied:*

1. *(a) For any X, it cannot be the case that both $\sigma\overline{T}X$ and $\sigma\overline{F}X$ belong to S for some N-prefix σ.*
   (b) $\sigma\overline{T}\bot$, $\sigma\overline{F}\top \notin S$.
   (c) $\sigma\overline{T} \oplus X \notin S$, where $char(\sigma) = N$.
   (d) $\sigma\overline{T} \ominus X \notin S$, where $char(\sigma) = 1$.
2. *If $\sigma\rho \in S$, $\sigma \neq R_l$ and $\rho \in \mathbb{N} \cup \{R_1, R_2, \ldots, R_N\}$, then $\sigma \in S$. (Here $\sigma \in S$ means $\sigma\psi \in S$ for some ψ).*
3. *Let $\sigma X \in S$ and suppose σX is such that some (I, N)-tableau extension rule, other than the ν and π rules, is applicable on it (note that at most one rule can be applicable on σX). Suppose application of that rule on σX results in splitting of the branch in, say, k forks. Then there is a fork such that every N-prefixed wff occurring on it belongs to S. Thus if $\sigma\alpha \in S$, then $\sigma\alpha_1$ and $\sigma\alpha_2$ both belong to S. Similarly, if $\sigma R_n \overline{T}(X\mathcal{U}Y) \in S$ then $\sigma^*_n\overline{T}Y \in S$, or, there exists some i, $2 \leq i \leq N - n + 1$ such that $\sigma^*_{n+(i-1)}\overline{T}Y \in S$ and for all k, $0 \leq k \leq i - 2$, $\sigma^*_{n+k}\overline{T}X \in S$.*
4. *If $\sigma\nu \in S$, where $char(\sigma) = n$, then $\sigma'\nu_0 \in S$ for all σ' occurring in S such that (i) $(\sigma, \sigma') \in R^I_n$ (ii) $char(\sigma') = char(\sigma) = n$.*
5. *If $\sigma\pi \in S$, then $\sigma'\pi_0 \in S$ for some σ' which is a simple extension of σ.*

Observe that the above definition is a natural extension of the one given in [28] keeping in view the presence of tableau extension rules which were not there in [28]. The conditions (1a)–(1d) correspond to the openness of the branch. Conditions (3)–(5) are based on the tableau extension rules. A condition like (2) is not required in [28], but we need it here to handle the ν-rule (as we shall see in Proposition 73 below).

It is not difficult to see the following.

Proposition 72. *The set of all N-prefixed wffs occurring on an open branch θ of an (I, N)-tableau, obtained by following procedure P1, is an (I, N)-downward saturated set.*

Now, we are in a position to give the desired counter model.

Definition 56. *Let S be an (I, N)-downward saturated set. Consider the dynamic I space $\mathfrak{F}^S := \mathcal{F}_1{}^S, \mathcal{F}_2{}^S, \ldots, \mathcal{F}_N{}^S$ where*

$\mathcal{F}_i^S = (W^S, P_i^S)$,
$W^S = \{\sigma : \sigma$ *is not of the form* $\sigma' R_l$ *and* $\sigma Z \in S$ *for some signed wff* $Z\}$,
$(\sigma, \sigma') \in P_n^S$ *if and only if* $(\sigma, \sigma') \in R_n^I$, $n = 1, 2, \ldots, N$. *So,* P_n^S *is the restriction of* R_n^I *on* W^S.

$\mathfrak{F}^S$ *will be called the* dynamic I space generated by S.

Given an (I, N)-downward saturated set S, let $V^S : PV \to 2^{W^S}$ be the valuation defined as,

$$V^S(p) := \{\sigma \in W^S : \sigma\overline{T}p \in S\}.$$

Let Int^S be the interpretation defined by

$$Int^S(\sigma) := \begin{cases} \sigma & \text{if } \sigma \text{ is not of the form } \tau R_l \\ Int^S(\tau) & \text{if } \sigma = \tau R_l. \end{cases}$$

The main result of the section is as follows – it leads us to the completeness theorem.

Proposition 73. *Let S be an (I, N)-downward saturated set. Then for any N-prefixed signed wff σZ,*

$$\sigma Z \in \textit{Simplies}\ \mathfrak{M}^S, n, Int^S(\sigma) \models Z,$$

where $char(\sigma) = n$ and $\mathfrak{M}^S := (\mathfrak{F}^S, V^S)$.

Proof. The proof is by induction on the complexity of Z. Let $char(\sigma) = n$.

Basis case: Note that $\sigma\overline{T}\bot$ and $\sigma\overline{F}\top$ cannot belong to S. Moreover, we always have $\mathfrak{M}^S, n, Int^S(\sigma) \models \overline{T}\top$ and $\mathfrak{M}^S, n, Int^S(\sigma) \models \overline{F}\bot$. Thus it remains only to consider the case when $Z = \overline{T}p$ or $\overline{F}p$, where p is a propositional variable. Now, σ is either of the form $\sigma' t$, or $\sigma' t R_n$, $t \in \mathbb{N}$. First suppose σ is of the form $\sigma' t$. Then $\sigma' t\overline{T}p \in S$ implies $\sigma' t \in V^S(p)$ and hence $\mathfrak{M}^S, n, \sigma' t \models \overline{T}p$. Thus, we obtain $\mathfrak{M}^S, n, Int^S(\sigma' t) \models \overline{T}p$ as $Int^S(\sigma' t) = \sigma'$. Similarly, one can show that if $\sigma' t\overline{F}p \in S$, then $\mathfrak{M}^S, n, \sigma' t \models \overline{F}p$.

Next, suppose $\sigma' t R_n \overline{T}p \in S$. Then using the definition of a downward saturated set, we obtain $\sigma' t\overline{T}p \in S$. Thus, $\mathfrak{M}^S, n, Int^S(\sigma' t) \models \overline{T}p$, following the above argument. This gives $\mathfrak{M}^S, n, Int^S(\sigma' t R_n) \models \overline{T}p$ as $Int^S(\sigma' t R_n) = Int^S(\sigma' t)$. Similarly, we obtain the desired result when $\sigma' t R_n \overline{F}p \in S$.

Induction case: The argument for the cases when Z is a wff other than a ν-wff, is the same for all I. The argument when Z is a ν-wff depends on I. Here we give the proof for $S5$.

Let Z be a α-wff. Then $\sigma Z \in S$ implies $\sigma\alpha_1, \sigma\alpha_2 \in S$. Therefore, by induction hypothesis, we obtain $\mathfrak{M}^S, n, Int^S(\sigma) \models \alpha_i$, $i = 1, 2$. Thus $\mathfrak{M}^S, n, Int^S(\sigma) \models Z$.

Now suppose Z is a ν-wff. Let $\sigma' \in W$ be such that $(Int^S(\sigma), \sigma') \in P_n^S$. We need to show that $\mathfrak{M}^S, n, \sigma' \models \nu_0$. Let $char(\sigma') = k$. Since $\sigma' \in W$, σ' must be of the form $\sigma' = \sigma'' R_k i_1 i_2 \cdots i_l$, $l \geq 1$ and $i_j \in \mathbb{N}$, $i \leq j \leq l$. Therefore, $Int^S(\sigma') = \sigma'$.

First suppose σ is of the form $\tau_1' R_n t_1 t_2 \cdots t_s$, $t_j \in \mathbb{N}$. So, $Int^S(\sigma) = \sigma$. Since $(Int^S(\sigma), \sigma') \in P_n^S$, that is $(\sigma, \sigma') \in P_n^S$, we obtain $(\sigma, \sigma') \in R_n^{S5}$. Therefore, by Proposition 67, either (a) $\sigma' = \tau_1'$ or (b) σ' is of the form $\tau_1' R_n d_1 d_2 \cdots d_r$. Let τ be the following N-prefix.

$$\tau := \begin{cases} \sigma' R_n & \text{if } \sigma' = \tau_1' \\ \sigma' & \text{if } \sigma' \text{ is of the form } \tau_1' R_n d_1 d_2 \cdots d_r. \end{cases}$$

Observe the following facts:

- $(\sigma, \tau) \in R_n^{S5}$;
- $char(\tau) = n$;
- $Int^S(\sigma') = Int^S(\tau) = \sigma'$;
- τ is an initial segment of σ' or σ.

Therefore, using the fact that $\sigma\nu \in S$ and the definition of a downward saturated set, we obtain $\tau\nu_0 \in S$. By the induction hypothesis, $\mathfrak{M}^S, n, Int^S(\tau) \models \nu_0$, that is, $\mathfrak{M}^S, n, \sigma' \models \nu_0$. In the other case when σ is of the form $\tau_1' R_n$, we still obtain a τ satisfying the above-mentioned properties, and hence the desired result.

Now, we move to the case when σZ is of the form $\sigma' R_n \overline{T}(X \mathcal{U} Y)$, where $n < N$. Then either $\sigma' R_n \overline{T} Y \in S$, or there exists i, $2 \leq i \leq N - n + 1$ such that $\sigma'^*_{n+i-1} \overline{T} Y \in S$ and for all k, $0 \leq k \leq i - 2$, $\sigma'^*_{n+k} \overline{T} X \in S$. This implies that either $\mathfrak{M}^S, n, Int^S(\sigma' R_n) \models \overline{T} Y$, or there exists i, $2 \leq i \leq N - n + 1$ such that $\mathfrak{M}^S, n + i - 1, Int^S(\sigma'^*_{n+i-1}) \models \overline{T} Y$ and for all k, $0 \leq k \leq i - 2$, $\mathfrak{M}^S, n + k, Int^S(\sigma'^*_{n+k}) \models \overline{T} X$. Therefore, $\mathfrak{M}^S, n, Int^S(\sigma' R_n) \models \overline{T} Y$, or there exists $j > n$ such that $\mathfrak{M}^S, j, Int^S(\sigma' R_n) \models \overline{T} Y$ and for all s, $n \leq s < j$, $\mathfrak{M}^S, s, Int^S(\sigma' R_n) \models \overline{T} X$ $\quad (\because Int^S(\sigma' R_n) = Int^S(\sigma'^*_d) = Int^S(\sigma'))$. Thus we have $\mathfrak{M}^S, n, Int^S(\sigma' R_n) \models \overline{T}(X \mathcal{U} Y)$.
All the remaining cases can be proved in the same way. □

Now, we have the following completeness theorem. Let $I \in \{K, K4, T, B, S4, KTB, KB4, S5\}$.

Theorem 17 (Completeness). *If X is valid in the class of all dynamic I spaces, then X is I-tableau provable.*

Proof. Suppose X is not I-tableau provable. Then there exists some N such that X is not (I, N)-tableau provable. Let us apply the procedure P1 to construct an (I, N)-tableau for X. There are two possibilities: the procedure will terminate

1. $R_1 1\overline{F} \oplus \neg\Diamond\Box\Diamond X$	
2. $R_1 1 R_2 \overline{F} \neg\Diamond\Box\Diamond X$	($\oplus(b)$-rule on 1)
3. $R_1 1 R_2 \overline{T} \Diamond\Box\Diamond X$	(α-rule on 2)
4. $R_1 1 R_2 1 \overline{T} \Box\Diamond X$	(π-rule on 3)
5. $R_1 1 R_2 \overline{T} \Diamond X$	(ν-rule on 4)
6. $R_1 1 R_2 2 \overline{T} X$	(π-rule on 5)

Fig. 7. $(B, 3)$-tableau for $\oplus\neg\Diamond\Box\Diamond X$, obtained by following P2.

and leave us with an open branch, say θ, or the procedure will not terminate. If the procedure does not terminate, then by König's lemma there still must be an open branch, say θ again. In either case, it is easy to check that the set S of N-prefixed wffs on branch θ will be an (I, N)-downward saturated set. Since $R_1 1\overline{F}X$ is an element of this set, it follows by Proposition 73 that $\mathfrak{M}, 1, Int^S(R_i 1) \models \overline{F}X$. This shows that X is not valid in the class of all dynamic I spaces, a contradiction. Therefore X is I-tableau provable. $\Box$

Remark 7. From Theorems 16 and 17, it follows that the logic $L(T, I)$ is the set of all I-tableau provable wffs, $I \in \{K, K4, T, B, S4, KTB, KB, S5\}$.

10.3 Termination

The systematic procedure P1 to construct the (I, N)-tableaux given in Sect. 10.2, as observed there, may not terminate (cf. Example 17). In this section, we will modify it to obtain terminating procedures such that if X is (I, N)-tableau provable, then the tableau for X obtained by following the modified procedures will be closed. In Sect. 11, we shall see how these terminating procedures lead us to obtain a decidable fragment of $L(T, I)$, $I \in \{K, K4, T, B, S4, KTB, KB4, S5\}$. We treat logics with and without transitivity separately. Our first modification of P1, presented below, will work for logics corresponding to the classes of those dynamic I spaces that do not ask for transitivity of the relations.

Terminating Procedure P2 for $\boldsymbol{I \in \{K, T, B, KTB\}}$. P1 is modified in the following way: if a N-prefixed wff say, $\sigma\psi$, is supposed to be added to the end of a branch, but we find that it already occurs on the same branch, then we do not add it. This means that the systematic ν-rule should be modified so that instead of continuously adding occurrences of $\sigma\nu$ to branch ends we do the following. We remember that whenever a new N-prefix σ' is introduced on a branch containing $\sigma\nu$, where $(\sigma, \sigma') \in R_n^I$, $char(\sigma) = char(\sigma') = n$ and there is no occurrence of $\sigma'\nu_0$ on the branch, then we also add $\sigma'\nu_0$ to the branch end, but we do not add a fresh occurrence of $\sigma\nu$ itself. Let us call this modified procedure P2. Observe that this modification does not cause the loss of any information that may have been used to construct the counter model to prove the completeness theorem in Sect. 10.2.

Example 18. The $(B, 3)$-tableau for $\oplus\neg\Diamond\Box\Diamond X$, $X \in PV$, obtained by following P2 is given in Fig. 7.

1. $R_1 1 \overline{F} \oplus \neg\Diamond\Box\Diamond X$
2. $R_1 1 R_2 \overline{F} \neg\Diamond\Box\Diamond X$ ($\oplus(b)$-rule on 1)
3. $R_1 1 R_2 \overline{T} \Diamond\Box\Diamond X$ (α-rule on 2)
4. $R_1 1 R_2 1 \overline{T} \Box\Diamond X$ (π-rule on 3)
5. $R_1 1 R_2 \overline{T} \Diamond X$ (ν-rule on 4)
6. $R_1 1 R_2 2 \overline{T} X$ (π-rule on 5)
7. $R_1 1 R_2 2 \overline{T} \Diamond X$ (ν-rule on 4 using prefix $R_1 1 R_2 2$)
8. $R_1 1 R_2 21 \overline{T} X$ (π-rule on 7)
9. $R_1 1 R_2 21 \overline{T} \Diamond X$ (ν-rule on 4 using prefix $R_1 1 R_2 21$)

....

....

Fig. 8. $(KB4, 3)$-tableau for $\oplus\neg\Diamond\Box\Diamond X$, obtained by following P2.

In Example 17, we have seen that the procedure P1 of $(B, 3)$-tableau construction for the wff $\oplus\neg\Diamond\Box\Diamond X$ does not terminate. But Example 18 shows that the procedure P2 terminates in this case. In fact, we will prove the following main result of the section.

Theorem 18. *For $I \in \{K, T, KTB, B\}$, the procedure P2 terminates.*

We note, at this point, that P2 may not terminate for $I \in \{K4, S4, KB4, S5\}$. The example below establishes this for $I = KB4$.

Example 19. The $(KB4, 3)$-tableau for $\oplus\neg\Diamond\Box\Diamond X$, $X \in PV$, obtained by following P2 is given in Fig. 8.

Observe that, unlike the $(KB4, 3)$-tableau for $\oplus\neg\Diamond\Box\Diamond X$ in Fig. 8, in the $(B, 3)$-tableau for the same wff (cf. Fig. 7), we cannot apply the ν-rule on 4 using prefix $R_1 1 R_2 2$ as $(R_1 1 R_2 1, R_1 1 R_2 2) \notin R_2^B$.

In order to prove Theorem 18, we shall require the following definitions.

Definition 57.

(i) The w-length *of a N-prefix σ, denoted as $L_w(\sigma)$, is defined as:*
  (a) $L_w(\tau R_n) = 0$
  (b) $L_w(\tau R_n l_1 l_2 \cdots l_r) = r$, $l_i \in \mathbb{N}, r \geq 1$.

(ii) The t-length *of a N-prefix σ, denoted as $L_t(\sigma)$, is the number of elements from the set $\{R_1, R_2, \ldots, R_N\}$ in σ.*

Let us consider an (I, N)-tableau for some wff X constructed following the procedure P2, and let θ be a branch of it.

Notation 19. For a wff $\tau Z \in \theta$, let us write $H_\theta(\tau Z)$ to denote the depth of the occurrence of τZ on θ measured from the root.

The two following lemmas will be used to prove Theorem 18. The proofs are given in Appendices A and B.

Lemma 1. *There exists an integer d such that θ will not have any N-prefixed wff with t-length greater than d.*

Lemma 2. *Let $I \in \{K, T, KTB, B\}$ and m be an integer. Then there is only a finite number of N-prefixes σ on θ with $L_t(\sigma) \leq m$.*

Note that although Lemma 1 holds for all I, Lemma 2 holds only for $I \in \{K, T, KTB, B\}$. The $(KB4, 3)$-tableau for $\oplus\neg\Diamond\Box\Diamond X$ given in Fig. 8, shows that Lemma 2 does not hold for $I = KB4$. In fact, we only have one branch there and it contains all the 3-prefixes from the set $\{R_1 1 R_2 21,\ R_1 1 R_2 211,\ R_1 1 R_2 21111, \ldots\}$. The same wff can also be used to show that Lemma 2 does not hold for $K4, S4, S5$.

Let us now see how the two lemmas can be used to obtain the termination of P2 for $I \in \{K, T, KTB, B\}$.

Proof of Theorem 18. If possible, let there exist a wff X, an integer N and $i \in \{1, 2, \ldots, N\}$ such that the procedure P2 to produce an (I, N)-tableau for X does not terminate. Then by König's lemma, we must be constructing an infinite branch, say θ. But, using Lemmas 1 and 2, we will show that θ cannot be infinite and this will prove the result.

Lemma 1 guarantees the existence of an integer d such that $L_t(\sigma) \leq d$ for all σ occurring on θ. Then, from Lemma 2, it follows that there are only finitely many N-prefixes on θ. Note that for any given N-prefix σ, if σZ occurs on θ, then Z must be a signed sub-wff of X, and there are only finitely many signed sub-wffs of X. Using these facts, it follows that only finitely many N-prefixed wffs can occur on θ. Since θ contains no repetitions, it follows that θ is a finite branch. □

So, our next task is to prove the Lemmas 1 and 2. We shall use the following Proposition to prove Lemma 1.

Definition 58.

*(i) The M-*degree *of a signed wff X, denoted as $D_M(X)$, is the number of occurrences of $\Box, \Diamond$ in X.*
*(ii) The t-*degree *of a signed wff X, denoted as $D_t(X)$, is the number of occurrences of $\oplus$, $\ominus$, $\mathcal{U}$, $\mathcal{S}$ in X.*

Proposition 74. *Let G_{k+1} consist of the N-prefixed wffs of the form $\tau R_n Z$ occurring on θ such that $L_t(\tau R_n) = k+1$ and $\tau R_n Z$ is added on θ applying a single temporal rule on a N-prefixed wff of the form $\tau Z'$ for some Z'. (Note that in this case, we will have $D_t(Z') > D_t(Z)$). Then*

$$\{Z : \sigma Z \in \theta \text{ and } L_t(\sigma) \geq k+1\} \subseteq \bigcup_{\sigma Z \in G_{k+1}} S(Z) \tag{17}$$

where $S(Z)$ is the set of all signed sub-wffs of Z.

Proof. Let $\{\sigma Z \in \theta : L_t(\sigma) \geq k+1\} = \{\delta_1, \delta_2, \ldots\}$, where $H_\theta(\delta_{i+1}) > H_\theta(\delta_i)$, $i = 1, 2, \ldots$. It is enough to show for all n that if δ_n is σZ, then $Z \in \bigcup_{\sigma Y \in G_{k+1}} S(Y)$. We prove it by showing that

(i) δ_1 has this property and
(ii) if $\delta_1, \delta_2, \ldots, \delta_m$ have this property, then δ_{m+1} also has this property.

Let $A := \bigcup_{\sigma Y \in G_{k+1}} S(Y)$, and δ_1 be σZ. We need to show $Z \in A$. Since σZ is the first wff on θ with a N-prefix of t-length greater than k, we must have $\sigma Z \in G_{k+1}$. This implies $Z \in A$.

Next suppose the result holds for $\delta_1, \delta_2, \ldots, \delta_m$. We prove for it for δ_{m+1}. Let δ_{m+1} be σZ and suppose it is obtained by a single rule application on a N-prefixed wff, say $\sigma' Z'$. Here, we only consider the cases when δ_m is obtained by the application of the ν-rule or $\oplus(a)$-rule or $\oplus(b)$-rule on $\sigma' Z'$. The other cases can be done in a similar fashion.

ν-rule: In this case Z' is ν and Z is ν_0. Let $char(\sigma) = char(\sigma') = n$. Since $(\sigma, \sigma') \in R_n^I$, we have $L_t(\sigma) = L_t(\sigma')$ (for $I = S5$, this follows from Propositions 67 and 68 and for other I's, this follows from similar kinds of results). This implies $\sigma' Z' \in \{\delta_1, \delta_2, \ldots, \delta_m\}$ and hence, by induction hypothesis, $Z' \in A$. Therefore, ν_0, that is, Z belongs to A.

$\oplus(a)$-rule: In this case $\sigma' Z'$ must be of the form $\tau R_n \oplus$ and so σZ is $\tau^*_{n+1} \oplus_0$. Therefore, $L_t(\sigma') \geq L_t(\sigma)$ and hence $\tau R_n \oplus \in \{\delta_1, \delta_2, \ldots, \delta_m\}$. This implies $\oplus \in A$ and hence $Z \in A$.

$\oplus(b)$-rule: In this case $\sigma' Z'$ must be of the form $\tau d \oplus, d \in \mathbb{N}$ and so σZ is $\tau d R_{n+1} \oplus_0$, where $char(\tau) = n$. Therefore, $L_t(\sigma') \geq k$. If $L_t(\sigma') \geq k+1$, then we will have $Z \in A$ by an argument similar to the one above. So, let $L_t(\sigma') = k$. In this case, we will have $\sigma Z \in G_{k+1}$ and so $Z \in A$. □

Corollary 20.

(i) $\max\{D_t(Z) : \sigma Z \in \theta$ *for some* σ *with* $L_t(\sigma) = k+1\} = \max\{D_t(Z) : \sigma Z \in G_{k+1}\}$.
(ii) $\max\{D_t(Z) : \sigma Z \in \theta$ *for some* σ *with* $L_t(\sigma) = k\} > \max\{D_t(Z) : \sigma Z \in G_{k+1}\}$.

Proof. Note that we are calculating maximum over finite sets of natural numbers. Let
$L_1 := \max\{D_t(Z) : \sigma Z \in \theta$ for some σ with $L_t(\sigma) = k+1\}$,
$L_2 := \max\{D_t(Z) : \sigma Z \in G_{k+1}\}$ and
$L_3 := \max\{D_t(Z) : \sigma Z \in \theta$ for some σ with $L_t(\sigma) = k\}$.
Let $L_i = D_t(Z_i)$, where $\sigma_i Z_i \in \theta$ and $L_t(\sigma_1) = k+1$, $L_t(\sigma_3) = k$, $\sigma_2 Z_2 \in G_{k+1}$.

(i). From Proposition 74, we obtain $Z_1 \in S(Y)$ for some $\sigma Y \in G_{k+1}$. This gives

$$L_1 = D_t(Z_1) \leq D_t(Y) \leq L_2. \tag{18}$$

Moreover, since $\sigma_2 Z_2 \in G_{k+1}$, we have $L_t(\sigma_2) = k+1$ and hence $L_2 = D_t(Z_2) \leq L_1$. Combining this with (18), we obtain $L_1 = L_2$.

(ii). Since $\sigma_2 Z_2 \in G_{k+1}$, using the definition of G_{k+1}, it follows that σ_2 must be of the form τR_n, $L_t(\tau) = k$, and there exists some Z' such that $\tau Z' \in \theta$ and $D_t(Z') > D_t(Z)$. Therefore, we obtain $L_3 \geq D_t(Z') > D_t(Z) = L_2$. □

Now, we are in a position to prove Lemma 1.

Proof of Lemma 1. From Corollary 20, we obtain,

$$\max\{D_t(Z) : \sigma Z \in \theta \text{ for some } \sigma \text{ with } L_t(\sigma) = k+1\} < \max\{D_t(Z) : \sigma Z \in \theta \text{ for some } \sigma \text{ with } L_t(\sigma) = k\}.$$

Thus it follows that as t-length becomes larger, t-degree becomes smaller. Hence, for some d, any N-prefixed wff on θ whose N-prefix is of t-length d must be of t-degree 0. Therefore no N-prefix of t-length greater than d could have been introduced on θ. Thus we have the result. □

Now we proceed to prove Lemma 2, giving a number of intermediate results. We begin with the following.

Proposition 75.

(1) If $\tau R_n Z \in \theta$ is obtained from $\sigma Z'$ by the application of a single rule, where $\sigma \neq \tau$, then $D_M(Z') \geq D_M(Z)$ and σ is of the form τR_l, or $\tau R_n d_1 d_2 \cdots d_r$, $d_i \in \mathbb{N}$.
(2) If $\tau R_n d_1 d_2 \cdots d_r Z \in \theta$ is obtained from $\sigma Z'$ by the application of a single rule, where $\sigma \neq \tau$, then $D_M(Z') \geq D_M(Z)$ and σ is of the form τR_l, or $\tau R_l t_1 t_2 \cdots t_s$, or $\tau R_l t_1 t_2 \cdots t_s R_m$, $t_i \in \mathbb{N}$.
(3) If $\sigma_1 Z_1, \sigma_2 Z_2, \ldots, \sigma_n Z_n$ is a finite sequence of N-prefixed wffs on θ such that (a) $\sigma_1 Z_1 = \tau R_l Z$, (b) $\sigma_i Z_i$ is obtained from $\sigma_{i+1} Z_{i+1}$, $i = 1, 2, \ldots, n-1$, by the application of a single rule and (c) $\sigma_i \neq \tau$, $i = 1, 2, \ldots, n$, then τ is a proper initial segment of each σ_i, and $D_M(Z_{i+1}) \geq D_M(Z_i)$, $i = 1, 2, \ldots, n-1$.

Proof. We only prove (1). (2) can be shown in a similar way and (3) obtains from (1) and (2). The argument of the proof remains the same for all I when $\tau R_n Z \in \theta$ is obtained from $\sigma Z'$ by the application of a rule other than the ν-rule. Here, we only give the proof for the case when $\tau R_n Z \in \theta$ is obtained from $\sigma Z'$ by the application of the ν-rule. Note that in this case, we have $D_M(Z') \geq D_M(Z)$. Moreover, we also have $(\sigma, \tau R_n) \in R_n^I$ and $char(\sigma) = n$. Let us first consider $I = S5$. Then, we have $(\tau R_n, \sigma) \in R_n^{S5}$, and hence from Proposition 68, we obtain σ of the form τR_n or $\tau R_n d_1 d_2 \cdots d_r$. If $I = S4$, then $(\sigma, \tau R_n) \in R_n^{K4}$ implies there exists $\sigma_1, \sigma_2, \ldots \sigma_r$ such that $\sigma_1 = \sigma$, $\sigma_i \neq \sigma_j$ for $i \neq j$, $\sigma_i \neq \tau R_n$, for all i and $\sigma_1 \; R_n^{K4} \; \sigma_2 \; R_n^{K4} \cdots \sigma_r \; R_n^{K4} \; \tau R_n$. It is not difficult to see that each σ_i is either τ or of the form τR_m, $m \neq n$. Thus, it follows that τ is a proper initial segment of σ as σ is different from τ ($\because char(\sigma) = n = char(\tau R_n) \neq char(\tau)$). One can similarly prove the result for other I's. □

Using this proposition, we obtain

Proposition 76. *Let $\tau R_l Z \in \theta$. Then we must have $\tau Z' \in \theta$ for some Z' with $D_M(Z') \geq D_M(Z)$ and $H_\theta(\tau Z') < H_\theta(\tau R_l Z)$.*

Proof. By the given condition, we must have a finite sequence $\sigma_1 Z_1, \sigma_2 Z_2, \ldots, \sigma_n Z_n$ of N-prefixed wffs on θ such that,

(a) $\sigma_1 Z_1$ is $\tau R_l Z$ and σ_n is $R_i 1$;
(b) $\sigma_k Z_k$ is obtained from $\sigma_{k+1} Z_{k+1}$, $k = 1, 2, \ldots, n-1$ by the application of a single rule and hence $H_\theta(\sigma_{k+1}) < H_\theta(\sigma_k)$.

Therefore, we obtain $\sigma_k = \tau$ for some k, $2 \leq k \leq n$. Otherwise, by Proposition 75, we would have τ a proper initial segment of each σ_m, a contradiction as σ_n is $R_i 1$. Let $\sigma_k = \tau$ and $\sigma_j \neq \tau$ for all j, $1 \leq j < k$. We will show that $\sigma_k Z_k$, that is τZ_k is the desired wff. Obviously, $H_\theta(\tau Z_k) < H_\theta(\tau R_l Z)$. Moreover, by Proposition 75, we obtain $D_M(Z_{m+1}) \geq D_M(Z_m)$, $m = 1, 2, \ldots, k-2$, and hence $D_M(Z_{k-1}) \geq D_M(Z_{k-2}) \geq \cdots \geq D_M(Z_1)$. Since $\sigma_k Z_k$ is obtained from $\sigma_{k-1} Z_{k-1}$ using only one rule, we have $D_M(Z_k) \geq D_M(Z_{k-1})$. Therefore, $D_M(Z_k) \geq D_M(Z_1) = D_M(Z)$ ($\because Z_1 = Z$). □

Notation 21. Let $\tau R_n \in P(N)$. By $A_{\tau R_n}$, we will denote the set of all N-prefixes of the form $\tau R_n d_1 d_2 \cdots d_r$, $r \geq 1$, $d_i \in \mathbb{N}$.

Proposition 77. *Let $I \in \{K, T, KTB, B\}$. Let $\sigma Z \in \theta$ be such that (a) $\sigma \in A_{\tau R_l}$, (b) $L_w(\sigma) \geq 2$, (c) $D_M(Z) > 0$. Then there exists some $\sigma' Z' \in \theta$ such that (i) $\sigma' \in A_{\tau R_l}$, (ii) σ' is a proper initial segment of σ, (iii) $L_w(\sigma') + 1 = L_w(\sigma)$, (iv) $D_M(Z') > D_M(Z)$, and (v) $H_\theta(\sigma' Z') < H_\theta(\sigma Z)$.*

Proof. We prove the result for $I = B$. The other cases can be proved similarly. Let $A = \{\sigma_1 Z_1 \in \theta : \sigma_1 \in A_{\tau R_l}, L_w(\sigma_1) \geq 2, D_M(Z_1) > 0\} = \{\delta_1, \delta_2, \ldots\}$, where $H_\theta(\delta_{i+1}) > H_\theta(\delta_i)$, $i = 1, 2, \ldots$. Note that $\sigma Z \in A$. So, it is enough to show for each δ_j that if δ_j is $\sigma_j Z_j$, then there exists some $\sigma'_j Z'_j \in \theta$ such that (i) $\sigma'_j \in A_{\tau R_l}$ (ii) σ'_j is proper initial segment of σ_j (iii) $L_w(\sigma'_j) + 1 = L_w(\sigma_j)$ (iv) $D_M(Z'_j) > D_M(Z_j)$ and (v) $H_\theta(\sigma'_j Z'_j) < H_\theta(\delta_j)$. We will prove it by showing that,

1. δ_1 has the required property and
2. if $\delta_1, \delta_2, \ldots, \delta_n$ have the required property, then δ_{n+1} also has the required property.

Note that δ_1 must be of the form $\tau R_l t_1 t_2 \cdots t_r Z_1$, $r \geq 2$. Suppose δ_1 is obtained from μY by the application of a single rule. $\mu Y \notin A$ as $H_\theta(\mu Y) < H_\theta(\delta_1)$. Moreover, this single rule must be either the ν-rule or π-rule. For instance, it cannot be the $\oplus(a)$-rule, because otherwise we will obtain $\mu = \tau R_l t_1 t_2 \cdots t_r R_{l-1}$ and $D_M(Y) = D_M(Z_1) > 0$. But due to Proposition 76, this will give us $\tau R_l t_1 t_2 \cdots t_r Z' \in \theta$ for some Z' with $D_M(Z') \geq D_M(Y)$ and $H_\theta(\tau R_l t_1 t_2 \cdots t_r Z') < H_\theta(\mu Y)$. Therefore we obtain $\tau R_l t_1 t_2 \cdots t_r Z' \in A$ with $H_\theta(\tau R_l t_1 t_2 \cdots t_r Z') < H_\theta(\delta_1)$, a contradiction. Similarly, one can show that it cannot be any other rule.

In the case when δ_1 is obtained from μY by the application of the ν-rule, μ must be either $\tau R_l t_1 t_2 \cdots t_{r-1}$, or $\tau R_l t_1 t_2 \cdots t_r t_{r+1}$. But, as $\mu Y \notin A$, we must have $r = 2$ and $\mu = \tau R_1$. Similarly, when δ_1 is obtained by the application of the π-rule, we obtain $r = 2$ and $\mu = \tau R_1$. Thus in both cases, μY has all the desired properties.

Now suppose we have the required property for $\delta_1, \delta_2, \ldots, \delta_n$. We prove it for δ_{n+1}. Let δ_{n+1} be $\sigma_1 Z_1$ and suppose δ_{n+1} is obtained by the application of a single rule on $\sigma_2 Z_2$. We will consider the cases when the applied rule is α or ν or $\oplus(a)$. The other cases can be proved similarly.

α-rule: In this case $\sigma_1 = \sigma_2$ and $\sigma_2 Z_2 \in \{\delta_1, \delta_2, \ldots, \delta_n\}$. Therefore, we have the required result.

ν-rule: Let σ_1 be $\tau R_l t_1 t_2 \cdots t_m t_{m+1}$, $m \geq 1$. Then σ_2 is either $\tau R_l t_1 t_2 \cdots t_m$, or $\tau R_l t_1 t_2 \cdots t_{m+1} t_{m+2}$. If the former, we are done. So let σ_2 be $\tau R_l t_1 t_2 \cdots t_{m+1} t_{m+2}$. Since $H_\theta(\sigma_2 Z_2) < H_\theta(\sigma_1 Z_1)$ and $D_M(Z_2) > D_M(Z_1)$, $\sigma_2 Z_2$ must be one of δ_1, δ_2, $\ldots, \delta_n$. Therefore, $\tau R_l t_1 t_2 \cdots t_{m+1} Z_3 \in \theta$ for some Z_3, such that

(i) $H_\theta(\tau R_l t_1 t_2 \cdots t_{m+1} Z_3) < H_\theta(\sigma_2 Z_2)$ and
(ii) $D_M(Z_3) > D_M(Z_2)$.

Therefore $\tau R_l t_1 t_2 \cdots t_{m+1} Z_3$ must be one of $\delta_1, \delta_2, \ldots, \delta_n$ and hence we have $\tau R_l t_1 t_2 \cdots t_m Z_4 \in \theta$ for some Z_4 satisfying all the required conditions.

$\oplus(a)$-rule: Let σ_1 be $\tau R_l t_1 t_2 \cdots t_m t_{m+1}$. Then σ_2 must be $\tau R_l t_1 t_2 \cdots t_m t_{m+1} R_{l-1}$ and $D_M(Z_1) = D_M(Z_2)$. By Proposition 76, we must have

$$\tau R_l t_1 t_2 \cdots t_m t_{m+1} Z_3 \in \theta$$

for some Z_3, where (i) $D_M(Z_3) \geq D_M(Z_2)$ and (ii) $H_\theta(\tau R_l t_1 t_2 \cdots t_{m+1} Z_3) < H_\theta(\sigma_2 Z_2)$. Therefore $\tau R_l t_1 t_2 \cdots t_{m+1} Z_3$ must be one of $\delta_1, \delta_2, \cdots, \delta_n$ and hence we have the result. □

Proposition 78. *Let $I \in \{K, T, KTB, B\}$. Then for a given $\tau R_n \in P(N)$, there exists an integer d such that θ will not have any N-prefixed wff with a N-prefix $\sigma \in A_{\tau R_n}$ and w-length greater than d.*

Proof. If θ is finite then there is nothing to prove. Let θ be an infinite branch. Let $\tau R_n l_1 l_2 \cdots l_{r+1} Z$ occur on θ, where $D_M(Z) > 0$ and $r \geq 1$. Then by Proposition 77, we must have

$$\tau R_n l_1 l_2 \cdots l_r Z' \in \theta$$

for some Z' such that $D_M(Z') > D_M(Z)$. Thus it follows that the maximum M-degree of all wffs on θ with $\tau R_n l_1 l_2 \cdots l_{r+1}$ as N-prefix is lower than the maximum M-degree of all wffs on θ with $\tau R_n l_1 l_2 \cdots l_r$ as N-prefix. So, for the wffs with N-prefixes from the set $A_{\tau R_n}$, as the w-length becomes larger, the M-degree becomes smaller. This means that there exists an integer d such that if σZ occurs in θ, $\sigma \in A_{\tau R_n}$ and $L_w(\sigma) = d$, then $D_M(Z) = 0$. Hence we have the proposition. □

Note that Propositions 77 and 78 do not hold for $I \in \{K4, S4, KB4, S5\}$. The $(KB4, 3)$-tableau for $\oplus\neg\Diamond\Box\Diamond X$ (cf. Fig. 8) shows this fact for $I = KB4$.

Proposition 78 along with the following proposition gives us Lemma 2.

Proposition 79. *θ can have only a finite number of N-prefixes of any given length.*

Proof. If θ is finite then there is nothing to prove. So, let θ be an infinite branch. If possible, suppose there is an integer k such that there exist infinitely many N-prefixes of length k. Let m be the smallest such integer. Note that there is no N-prefix of length one and it is easy to see that the only N-prefix of length two is the N-prefix $R_i 1$ and it occurs finitely often. Hence $m > 2$. Now, since there is only a finite number of N-prefixes of length $m - 1$, we can have only a finite number of N-prefixes of length m of the form τR_l, $l \in \{1, 2, \ldots, N\}$. Thus we must have an infinite number of N-prefixes of length m of the form τt, $t \in \mathbb{N}$.

Now, the only way a N-prefix of the form τt, $t \in \mathbb{N}$ of length $m > 2$ can be introduced on θ is by the application of the $\pi-$ rule to a N-prefixed wff of the form $\tau\pi$, where τ is of length $m - 1$. Moreover, the $\pi-$rule can be applied to such a N-prefixed wff only once. So if there are infinitely many N-prefixes of length m on θ of the form τt, there must be infinitely many wffs on θ of the form $\sigma\pi$, where σ is of length $m - 1$. This contradicts the choice of m. Thus we have the result. □

Proof of Lemma 2. If θ is finite then there is nothing to prove. So, let θ be an infinite branch. By Propositions 78 and 79, it follows that there is only a finite number of N-prefixes on θ of t-length 1. Next we assume that there are only finitely many N-prefixes of t-length n, $1 \leq n \leq m - 1$ and we prove the result for $n+1$. From our assumption it follows that there are only finitely many N-prefixes of t-length $n + 1$ of the form τR_l on θ. Let $\tau_1 R_{l_1}, \tau_2 R_{l_2}, \ldots, \tau_r R_{l_r}$ be the complete list of such N-prefixes. By Proposition 78, there exist integers D_j, $1 \leq j \leq r$ such that $L_w(\sigma) \leq D_j$ for all $\sigma \in A_{\tau_j R_{l_j}}$ occurring on θ. Let $D := \max\{D_j : 1 \leq j \leq r\}$. Then $L_w(\sigma) \leq D$ for all σ occurring on θ of t-length $n + 1$.

Recall our assumption that there are only finitely many N-prefixes of t-length n occurring in θ. Let $L := \max\{L(\sigma) : \sigma \text{ occurs in } \theta \text{ and } L_t(\sigma) = n\}$, where $L(\sigma)$ denoted the length of σ. Observe that $L(\sigma) \leq L + D + 1$ for all σ occurring on θ of t-length $n + 1$. Now from Proposition 79 it follows that there is only a finite number of N-prefixes on θ of t-length $n + 1$. This completes the proof. □

Terminating Procedure P3 for $I \in \{K4, KB4, S4, S5\}$. We have seen through Example 19 that Lemma 2 does not hold for $I \in \{K4,\ KB4,\ S4,\ S5\}$, and this cost us the termination of P2 (cf. Theorem 18) for these Is. So we modify P2 for these Is as follows.

Recall that $S(X)$ is the set of all signed sub-wffs of X and it is finite, say of size D. We put the constraint on the procedure P2 such that we do not apply any rule on the N-prefixed wffs σZ with $L_w(\sigma) = 2^D + 3$. The reason why we consider the bound to be $2^D + 3$ will be clear in the sequel. Let us call this modified procedure P3.

Note that we will not have any N-prefixed wff σZ with $L_w(\sigma) > 2^D+3$ on the (I, N)-tableau for X obtained by following procedure P3. This fact is enough to get Lemma 2 for a branch θ of an (I, N)-tableau, $I \in \{K4, KB4, S4, S5\}$, obtained by following P3. This together with Lemma 1 (again for P3) guarantees the termination of P3.

Observe that if there are $2^D + 1$ N-prefixes, say $\sigma_1, \sigma_2, \dots, \sigma_{2^D+1}$, occurring in a branch θ, then we must have k, l, $1 \le k < l \le 2^D + 1$ such that $\{Z : \sigma_k Z \in \theta\} = \{Z : \sigma_l Z \in \theta\}$. We will see how this fact is used to show that if there exists some closed (I, N)-tableau for X, then the procedure P3 will generate one such tableau. For this, we prove the following.

Proposition 80. *Let $I \in \{K4, KB4, S4, S5\}$ and B and A be the (I, N)-tableaux for a wff X obtained by following the procedures P2 and P3 respectively. Let B be closed. Then A is also closed.*

Notation 22. The set $\{\sigma : \sigma Z \in \theta$ for some Z with $H_\theta(\sigma Z) \le n\}$ will be denoted by $P_n(\theta)$.

We will use the following result to prove Proposition 80.

Proposition 81 (Branch mapping theorem). *Let $I \in \{K4, KB4, S4, S5\}$ and consider an (I, N)-tableau A for a wff X and a branch θ of it, obtained by following the procedure P3. Consider the complete list of N-prefixes on θ with $w-$ length $2^D + 3$ given by*

$$\zeta_j := \sigma^j R_{n_j} t_1^j t_2^j \cdots t_{r_j}^j t_{r_j+1}^j \cdots t_{s_j}^j t_{s_j+1}^j \cdots t_{2^D+3}^j,\ j \in \{1, 2, \dots m\}.$$

Suppose for each $j \in \{1, 2, \dots m\}$, there exist integers r_j, s_j, $2^D + 3 > s_j > r_j > 1$ such that the N-prefixes $\tau_j := \sigma^j R_{n_j} t_1^j t_2^j \cdots t_{r_j}^j t_{r_j+1}^j \cdots t_{s_j}^j$ and $\tau_j' := \sigma^j R_{n_j} t_1^j t_2^j \cdots t_{r_j}^j$ satisfy

$$\{Z : \tau_j' Z \in \theta\} = \{Z : \tau_j Z \in \theta\}. \tag{19}$$

Let B be the (I, N)-tableau for X obtained by following the procedure P2. Then for each $n \in \mathbb{N}$, there exists some branch θ_n in B and a mapping Ω_n from $P_n(\theta_n) \cup \{\varepsilon\}$, ε being the empty string, to the set of N-prefixes occurring on θ, satisfying the following.

(i) $\Omega_n(\varepsilon) = \varepsilon$,
(ii) $\Omega_n(R_i 1) = R_i 1$,
(iii) $\Omega_n(\mu R_k) = \Omega_n(\mu) R_k$,
(iv) $\Omega_n(\mu R_k d_1 d_2 \cdots d_a) = \Omega_n(\mu) R_k d_1' d_2' \cdots d_{a'}'$, *for some* $d_j' \in \mathbb{N}, j = 1, \dots a'$, $a, a' \ge 1$,
(v) For $\mu Z \in \theta_n$ *with* $H_{\theta_n}(\mu Z) \le n$, $\Omega_n(\mu) Z \in \theta$,
(vi) $\Omega_n(\mu) \notin \{\sigma : \tau_j$ *is a proper initial segment of* σ *for some* $j\}$.

Moreover, if the tableau B is closed, then the branch θ of tableau A will also be closed.

Proof. We only prove the result for $I = S5$. It can be proved for other I in the same way. The first part of the proposition is proved by induction on n.

Basis case: $n = 1$
Note that τ_j cannot be an initial segment of R_i1 for any $j \in \{1, 2, \dots m\}$. We take θ_1 to be any branch and $\Omega(\varepsilon) = \varepsilon$, $\Omega_1(R_i1) = R_i1$.

Induction case: Suppose the result holds for n. We prove it for $n + 1$. If the set $\{\sigma Z : \sigma Z \in \theta_n \text{ and } H_{\theta_n}(\sigma Z) = n + 1\}$ is empty, then we just need to take $\theta_{n+1} := \theta_n$ and $\Omega_{n+1} := \Omega_n$. So, let us consider the case when the above set is non-empty. We will define θ_{n+1} in such a way that it is the same as the branch θ_n up to depth n. Moreover, $\Omega_{n+1}(\sigma) = \Omega_n(\sigma)$ for all σ such that $\sigma Y \in \theta_n$ for some Y with $H_{\theta_n}(\mu Y) \leq n$ and $\Omega_{n+1}(\varepsilon) = \varepsilon$. Let $\mu' Z' \in \theta_n$ with $H_{\theta_n}(\mu' Z') \leq n$ be such that by the application of a rule on this wff, we obtain the wff μZ on θ_n with $H_{\theta_n}(\mu Z) = n + 1$. Depending on the rule applied, we will determine θ_{n+1} as follows. We only consider the cases when the π, ν, $\oplus(a)$ or $\mathcal{U}(a)$ rule is applied. The other cases can be handled similarly.

π-rule: In this case we take θ_{n+1} to be θ_n. So, we need to define Ω_{n+1} on μ. It is done as follows.

We have $Z' = \pi$, $Z = \pi_0$ and $\Omega_n(\mu')$ is defined. Moreover, $\Omega_n(\mu')\pi \in \theta$. Note that μ must be of the form $\mu' d$ for some $d \in \mathbb{N}$. Further, $\Omega_n(\mu')$ will be of the form $\Omega_n(\mu_1)R_k d'_1 d'_2 \cdots d'_{a'}$ or $\Omega_n(\mu_1)R_k$, according as μ' is of the form $\mu_1 R_k d_1 d_2 \cdots d_a$, or $\mu_1 R_k$. Since $\Omega_n(\mu') \neq \zeta_l$, $l \in \{1, 2, \dots, m\}$, the π-rule was applied on $\Omega_n(\mu')\pi$ in θ to obtain, say, $\Omega_n(\mu')j\pi_0$ in θ for some $j \in \mathbb{N}$.

If $\Omega_n(\mu')j \in \{\sigma : \tau_q$ is a proper initial segment of σ for some q, $1 \leq q \leq m\}$, then we must have $\Omega_n(\mu') = \tau_q = \sigma^q R_{n_q} t^q_1 t^q_2 \cdots t^q_{r_q} t^q_{r_q+1} \cdots t^q_{s_q}$ for some $q \in \{1, 2 \cdots, m\}$ ($\because \tau_q$ cannot be a proper initial segment of $\Omega_n(\mu')$). Then μ' and μ must be of the form $\mu_1 R_{n_q} d'_1 d'_2 \cdots d'_{a'}$ and $\mu_1 R_{n_q} d'_1 d'_2 \cdots d'_{a'} d$ respectively and $\Omega_n(\mu_1) = \sigma^q$. Moreover, $\Omega_n(\mu')\pi \in \theta$ and (19) give $\tau'_q \pi \in \theta$ and hence we must also have $\tau'_q m \pi_0 \in \theta$ for some $m \in \mathbb{N}$. So, we define $\Omega_{n+1}(\mu) = \tau'_q m = \Omega_n(\mu_1) R_{n_q} t^q_1 t^q_2 \cdots t^q_{r_q} m = \Omega_{n+1}(\mu_1) R_{n_q} t^q_1 t^q_2 \cdots t^q_{r_q} m$.

On the other hand, if $\Omega_n(\mu')j \notin \{\sigma : \tau_q$ is a proper initial segment of σ for some $q\}$, then we define $\Omega_{n+1}(\mu) = \Omega_n(\mu')j = \Omega_{n+1}(\mu')j$.

ν-rule: Take θ_{n+1} to be θ_n. Note that $char(\mu) = char(\mu') = k$, (say) and $(\mu', \mu) \in R^{S5}_k$. Moreover, we must also have a N-prefixed wff $\mu Y \in \theta_n$ with $H_{\theta_n}(\mu Y) \leq n$. Since $H_{\theta_n}(\mu' Z') \leq n$, both $\Omega_n(\mu)$ and $\Omega_n(\mu')$ are defined and $\Omega_n(\mu)Y, \Omega_n(\mu')Z' \in \theta$. We take $\Omega_{n+1}(\mu) := \Omega_n(\mu)$. So, we only need to show $\Omega_{n+1}(\mu)Z$, that is, $\Omega_n(\mu)Z \in \theta$. Note that when $\mu = \mu'$, then $\Omega_n(\mu) = \Omega_n(\mu')$ and as $\Omega_n(\mu')Z' \in \theta$, we also have $\Omega_n(\mu)Z \in \theta$ (due to application of ν-rule). So let $\mu \neq \mu'$. Let us consider the case when μ' is of the form $\mu_1 R_k d_1 d_2 \cdots d_a$. The case when μ' is of the form $\mu_1 R_k$ can be handled similarly. Since $(\mu', \mu) \in R^{S5}_k$, $\mu \neq \mu'$ and $char(\mu) = char(\mu') = k$, by Proposition 67, μ must be in the form $\mu_1 R_k l_1 l_2 \cdots l_b$, or $\mu_1 R_k$. Accordingly,

$\Omega_n(\mu)$ is either $\Omega_n(\mu_1)R_k l'_1 l'_2 \cdots l'_{b'}$, or, $\Omega_n(\mu_1)R_k$. Moreover, $\Omega_n(\mu')$ is of the form $\Omega_n(\mu_1)R_k d'_1 d'_2 \cdots d'_{a'}$. Thus, we obtain $(\Omega_n(\mu'), \Omega_n(\mu)) \in R_k^{S5}$ and this gives $\Omega_n(\mu)Z \in \theta$ (due to the application of the ν-rule on $\Omega_n(\mu')Z' \in \theta$).

$\oplus(a)$-rule: Take $\theta_{n+1} := \theta_n$. μ' is of the form $\mu_1 R_k$, and μ is $(\mu_1)^*_{k+1}$. Note that, we must have a N-prefixed wff $\mu_1 Y \in \theta_n$ for some Y such that $H_{\theta_n}(\mu_1 Y) \leq n$. Thus, $\Omega_n(\mu_1)$ and $\Omega_n(\mu')$ are defined and $\Omega_n(\mu') = \Omega_n(\mu_1)R_k$, $\Omega_n(\mu')Z' \in \theta$, $(\Omega_n(\mu_1))^*_{k+1}Z \in \theta$. Moreover, $char(\mu_1) = char(\Omega_n(\mu_1))$. Note that if $\Omega_n(\mu)$ is defined (which is the case when $\mu Y' \in \theta_n$ for some Y' with $H_{\theta_n}(\mu Y') \leq n$), then we must have $\Omega_n(\mu) = (\Omega_n(\mu_1))^*_{k+1}$ and hence $\Omega_{n+1}(\mu) = (\Omega_n(\mu_1))^*_{k+1}$. Otherwise, we take $\Omega_{n+1}(\mu) = (\Omega_n(\mu_1))^*_{k+1}$.

β-rule: So $\mu = \mu'$. Since $H_{\theta_n}(\mu\beta) \leq n$, $\Omega_n(\mu)$ is defined and $\Omega_n(\mu)\beta \in \theta$. So, we take $\Omega_{n+1}(\mu) = \Omega_n(\mu)$. Let j^{th} branch, $j \in \{1,2\}$, be selected to obtain θ when the β-rule is applied on $\Omega_n(\mu')Z'$. Then, we also take θ_{n+1} to be a branch obtained by choosing the jth-branch when the β-rule is applied on $\mu'Z'$.

$\mathcal{U}(a)$-rule: Let $\mu'Z'$ be $\sigma R_k \overline{T}(Z_1 \mathcal{U} Z_2)$. Then we also have $\Omega_n(\sigma)R_k\overline{T}(Z_1\mathcal{U}Z_2) \in \theta$. Suppose θ is obtained by selecting the jth branch, $1 \leq j \leq N-(k+1)$, when the $\mathcal{U}(a)$-rule is applied on $\Omega_n(\sigma)R_k\overline{T}(Z_1\mathcal{U}Z_2)$. Then, we also take θ_{n+1} to be a branch obtained by selecting the jth-branch when the $\mathcal{U}(a)$-rule is applied on $\sigma R_k\overline{T}(Z_1\mathcal{U}Z_2)$. Let $\mu_2 Y \in \theta_{n+1}$ be such that $H_{\theta_{n+1}}(\mu_2 Y) = n+1$. Note that μ_2 will be of the form σ^*_l, $k \leq l \leq N$. Moreover, if $\Omega_n(\mu_2)$ is defined, then we must have $\Omega_n(\mu_2) = (\Omega_n(\sigma))^*_l$ and hence $\Omega_{n+1}(\mu_2) = (\Omega_n(\sigma))^*_l$. Otherwise, we take $\Omega_{n+1}(\mu_2) = (\Omega_n(\sigma))^*_l$, where $\Omega_{n+1}(\sigma) = \Omega_n(\sigma)$. One can now easily verify the desired conditions.

Now, we prove the second part of the proposition. Suppose the tableau B is closed. Then B must be a finite tableau and hence each branch of B must be finite. Let n be the smallest integer such that $P_{n+1}(\theta')$ is empty set for each branch θ' of B.

By the first part of the proposition, we must have a branch θ_n of B and mapping Ω_n satisfying the conditions (1)–(6) listed above. But using these conditions and the fact that the branch θ_n is closed, one can easily check that the branch θ is also closed. □

We can thus see that Proposition 81 actually gives rise to a mapping from the branches of tableau A to the branches of tableau B, preserving openness. This fact leads us to the proof of Proposition 80. In fact, we shall again use this technique of branch mapping in Sect. 11.3, to prove the decidability of a fragment of the logic $L(T, I)$.

Proof of Proposition 80. We show that every branch of A is closed. Let us consider a branch θ of A, and let $\zeta_j := \sigma^j R_{n_j} t_1^j t_2^j \cdots t_{2^D+3}^j$, $j \in \{1, 2, \ldots, m\}$, be the complete list of N-prefixes on θ with $w-length$ 2^D+3. Since the cardinality of the set of all subsets of $S(X)$ is 2^D, for each $j \in \{1, 2, \ldots, m\}$, we obtain r_j, s_j with $1 < r_j < s_j < 2^D + 3$ such that

$$\{Z : \sigma_j R_{n_j} t_1^j t_2^j \cdots t_{r_j}^j Z \in \theta\} = \{Z : \sigma_j R_{n_j} t_1^j t_2^j \cdots t_{r_j}^j \cdots t_{s_j}^j Z \in \theta\}.$$

Thus from Proposition 81 and the fact that the tableau B is closed, we obtain θ as a closed branch. □

11 Decidability Results Relative to Different Classes of Dynamic Spaces

A logic L is decidable, if given a L–wff X, we can decide whether $X \in L$ or not. The general decidability result of the logics $L(T, I)$ is still open, but Sects. 11.1 and 11.2 present decidability results with respect to certain classes of dynamic I spaces. Further, two decidable fragments of $L(T, I)$ are provided in Sect. 11.3.

The content of this section is based on the article [55].

11.1 Decidability for the Class of Dynamic I Spaces with Fixed Cardinality

Throughout this section, we assume that $I \in \{K, K4, T, BA, S4, KTB, KB4, S5\}$. We begin with the decidability of the logic $L_N(T, I)$, consisting of all $\mathcal{L}$-wffs which are valid in the class of all dynamic I spaces $\mathfrak{F}$ with $|\mathfrak{F}| = N$, where N is a given positive integer. Recall that $|\mathfrak{F}|$, termed the *cardinality of the dynamic I space* $\mathfrak{F}$, denotes the number of constituent Kripke frames in $\mathfrak{F}$. We note that although the semantics of the language $\mathcal{L}$ is given with respect to a finite time line, it can be extended in a natural way, to the models with $\mathbb{N}$ as the underlying time frame. Therefore, we may extend the definition of the dynamic I spaces so that $|\mathfrak{F}|$ is not necessarily a finite positive integer, but possibly $|\mathbb{N}|$ as well. Thus we also bring into our consideration the logic $L_{\mathbb{N}}(T, I)$, which consists of all $\mathcal{L}$-wffs which are valid in the class of all dynamic I spaces $\mathfrak{F}$ with $|\mathfrak{F}| = |\mathbb{N}|$.

Remark 8. The above-mentioned extension of dynamic I spaces will only be considered in this section (i.e. Sect. 11.1), and in the discussion on the first decidable fragment in Sect. 11.3. In the rest of the article, by a dynamic I space, we shall mean a dynamic I space of finite cardinality only.

Theorem 23. *The logics $L_N(T, I)$ and $L_{\mathbb{N}}(T, I)$ are decidable.*

Proof. In Sect. 10.3, we have given terminating procedures to construct an (I, N)-tableau for a given wff X such that if X has a closed (I, N)-tableau, then the procedures generate such a tableau. Using this fact along with soundness and completeness theorems of the tableau based proof procedures (cf. Theorems 16 and 17), we obtain the decidability of $L_N(T, I)$. However, the decidability of $L_N(T, I)$ is also a direct consequence of the connections of $L_N(T, I)$, and $L_{\mathbb{N}}(T, I)$ with multi-modal logic and first order temporal logic, detailed in Sects. 9.2 and 9.3. It is seen that $L_N(T, I)$ is identifiable with the multi-modal logic I_N (cf. Theorem 10) which is decidable, thus also giving the same complexity to the decidability problems of the two logics. It is also observed that $L_N(T, I)$, and $L_{\mathbb{N}}(T, I)$ are embeddable into the monodic packed fragment of a set $\mathbb{F}$ of wffs in a first order temporal language with equality (cf. Theorems 11 and 12). Since the monodic packed fragment of the first order temporal logic with equality is decidable [37], we obtain the decidability of $L_{\mathbb{N}}(T, I)$ (in fact decidability of $L_N(T, I)$ as well). □

Remark 9. We recall that the monodic packed fragment of first order temporal logic with equality is known to be decidable in the class of interpretations with underlying time frame $\mathbb{N}$, and for any given $N \in \mathbb{N}$, this result can easily be proved for the class of interpretations with underlying time frame having N number of time points. But the decidability of this fragment is not known to the authors for the class of interpretations with finite but *varying* number of time points. Therefore, we cannot see that the decidability of the logic $L(T, I)$ of dynamic I spaces can be proved using its connection with first order temporal logic.

Decidability of the logics $L_N(T, I)$, $N \in \mathbb{N}$, also gives the semi-decidability of the logics $L(T, I)$.

Theorem 24. *$L(T, I)$ is semi-decidable.*

The decidability of $L_{\mathbb{N}}(T, I)$ will help us to determine a decidable fragment of $L(T, I)$, as we shall see in Sect. 11.3.

11.2 Decidability for the Class of Dynamic *I* Spaces with Domain of Fixed Cardinality

We now prove the following result.

Theorem 25. *Given a $\mathcal{L}$-wff X and an integer n, we can decide the validity of X in the class of all dynamic I spaces with domain of cardinality n.*

The decidability of linear temporal logic (LTL) with finite time line can be proved using a result which roughly says that, given any model $\mathfrak{M}$ of an LTL wff α, if two time points satisfy exactly the same sub-wffs of α, then all intermediate time points can be removed to get a smaller model $\mathfrak{M}'$ for α. A natural question then is, whether this result also holds true for $L(T, I)$. In fact, it is not the case, as in the case of $L(T, I)$, apart from time points, we also need to take into account the objects of the models. But we shall consider a generalization of the above approach here.

Let $sub(X)$ denote the set of all sub-wffs of the wff X. Further, we will use $[t, w]_{\mathfrak{M}, X}$ to denote the set $\{Y \in sub(X) : \mathfrak{M}, t, w \models Y\}$.

Let $\mathfrak{F} := \mathcal{F}_1, \mathcal{F}_2, \ldots, \mathcal{F}_N$ be a dynamic space and $i, j \in \{1, 2, \ldots, N\}$ be such that $i < j$. Let $D := N - (j - i) \geq 1$. Consider the monotonic increasing and bijective function $f : \{1, 2, \ldots, D\} \to \{1, 2, \ldots, i-1\} \cup \{j, j+1, \ldots, N\}$ defined by,

$$f(k) := \begin{cases} k & \text{if } k < i \\ k + j - i & \text{if } k \geq i. \end{cases}$$

Note that (i) $f(k) < i$, if and only if $k < i$, $k \in \mathbb{N}$, (ii) $i \in \{1, 2, \ldots, D\}$, and (iii) $f(i) = j$.

Let $\mathfrak{F}^{i,j}$ be the dynamic space given by $\mathcal{F}'_1, \mathcal{F}'_2, \ldots, \mathcal{F}'_D$, where $\mathcal{F}'_k := \mathcal{F}_{f(k)}$. Note that $|\mathfrak{F}^{i,j}| < |\mathfrak{F}|$. Consider a valuation V and models $\mathfrak{M} := (\mathfrak{F}, V)$ and $\mathfrak{M}^{i,j} := (\mathfrak{F}^{i,j}, V)$. Then, we have the following result – the proof is given in Appendix D.

Proposition 82. *Let $[i, w]_{\mathfrak{M},X} = [j, w]_{\mathfrak{M},X}$, for all $w \in U$. Then, for all w and for all $k \in \{1, 2, \ldots, D\}$, we have*

$$[k, w]_{\mathfrak{M}^{i,j},X} = [f(k), w]_{\mathfrak{M},X}.$$

Proof. The proof is by induction on the complexity of the wff $Y \in sub(X)$. We only consider the case when Y is of the form $Z_1 \mathcal{U} Z_2$. Let us assume $\mathfrak{M}, f(k), w \models Z_1 \mathcal{U} Z_2$ and prove $\mathfrak{M}^{i,j}, k, w \models Z_1 \mathcal{U} Z_2$. The other direction can be proved giving a similar argument.

So, in this case, we have r, $f(k) \leq r \leq N$ such that

$$\mathfrak{M}, r, w \models Z_2, \text{ and} \tag{20}$$
$$\mathfrak{M}, t, w \models Z_1 \text{ for all } t \text{ with } f(k) \leq t < r. \tag{21}$$

Case 1: $f(k) < i$
Then, $k = f(k) < i$.
Subcase 1.1: $r = i$
From (20), and the given condition, we obtain $\mathfrak{M}, j, w \models Z_2$ and hence

$$\mathfrak{M}, f(i), w \models Z_2 \ (\because f(i) = j).$$

Therefore, using induction hypothesis, we obtain $\mathfrak{M}^{i,j}, i, w \models Z_2$.

Now, consider t, $f(k) = k \leq t < i = r$. By (21), $\mathfrak{M}, f(t), w \models Z_1$. Since $f(t) = t$, by induction hypothesis, we obtain $\mathfrak{M}^{i,j}, t, w \models Z_1$.

Subcase 1.2: $r > i$

In this case, from (20), (21) and the fact that $f(k) < i < r$, we obtain $\mathfrak{M}, i, w \models Z_1 \mathcal{U} Z_2$. Therefore, by the given condition, we obtain $\mathfrak{M}, j, w \models Z_1 \mathcal{U} Z_2$. This means, there exists l, $j \leq l \leq N$ such that

$$\mathfrak{M}, l, w \models Z_2, \text{ and} \tag{22}$$
$$\mathfrak{M}, m, w \models Z_1 \text{ for all } m \text{ with } j \leq m < l. \tag{23}$$

Now, using the property of f, we obtain l' with $l' \geq i > k$ and $f(l') = l$. Therefore, from (22) and induction hypothesis, we obtain $\mathfrak{M}^{i,j}, l', w \models Z_2$.

Let us consider t, $k \leq t < l'$. We need to show $\mathfrak{M}^{i,j}, t, w \models Z_1$. Note that $f(k) \leq f(t) < f(l') = l$. Moreover, either $f(t) \geq j$, or $f(t) < i$. In either case, we obtain $\mathfrak{M}, f(t), w \models Z_1$, using (23), or (21). Thus, we obtain the desired result by the induction hypothesis.

Subcase 1.3: $r < i$
In this case, $f(r) = r$. Therefore, (20) and the induction hypothesis give

$$\mathfrak{M}^{i,j}, r, w \models Z_2.$$

Moreover, for any t, $k \leq t < r$, we obtain $f(k) = k \leq f(t) = t < r$ and hence using (21) and the induction hypothesis, we get $\mathfrak{M}^{i,j}, t, w \models Z_1$.

Case 2: $f(k) \geq j$
In this case, $r \geq j$ and hence there exists $r' \geq i$ such that $f(r') = r$. Therefore, using (20) and the induction hypothesis, we obtain $\mathfrak{M}^{i,j}, r', w \models Z_2$. Since $f(r') = r \geq f(k)$, $r' \geq k$.

Now, consider any t with $k \leq t < r'$. Then, $f(k) \leq f(t) < f(r') = r$. Using (21) and the induction hypothesis, we get $\mathfrak{M}^{i,j}, t, w \models Z_1$.

As a consequence of this proposition, we obtain

Proposition 83. *Let X be a $\mathcal{L}$-wff that is satisfiable in a dynamic I space $\mathfrak{F} := \mathcal{F}_1, \mathcal{F}_2, \ldots, \mathcal{F}_l$, where $\mathcal{F}_i := (U, P_i)$, $i = 1, 2, \ldots, l$. Let $|U| = n$. Then there exists some dynamic I space $\mathfrak{F}^*$ with $|\mathfrak{F}^*| \leq 1 + 2^{|sub(X)|n}$ such that X is satisfiable in $\mathfrak{F}^*$.*

Proof. By the given condition, there exists some valuation function V on $\mathfrak{F}$ and a $w \in U$ such that $\mathfrak{M}, 1, w \models X$, where $\mathfrak{M} := (\mathfrak{F}, V)$.

Let $U := \{w_1, w_2, \ldots, w_n\}$ and $B_{kr} := [k, w_r]_{\mathfrak{M},X}$, $k \in \{1, 2, \ldots, n\}$.
With each $k \in \{1, 2, \ldots, n\}$, we associate an element

$$(B_{k1}, B_{k2}, \ldots, B_{kn})$$

of the set $B^n := \underbrace{B \times B \times \cdots \times B}_{n\text{-times}}$, where $B = 2^{sub(X)}$. Then we observe that

$[k, s]_{\mathfrak{M},X} \doteq [t, s]_{\mathfrak{M},X}$ for all $s \in U$, if and only if k and t associate to the same element of B^n.
Note that $|B^n| \leq 2^{|sub(X)|n}$.
If $|\mathfrak{F}| \leq 1 + 2^{|sub(X)|n}$, then we have nothing to prove. So let $|\mathfrak{F}| > 1 + 2^{|sub(X)|n}$. Then we must have $a, b \in \{2, 3, \ldots, l\}$ such that

$$[a, s]_{\mathfrak{M},X} = [b, s]_{\mathfrak{M},X} \text{ for all } s \in U.$$

Therefore, using Proposition 82 we obtain a $\mathfrak{F}_1$ such that $|\mathfrak{F}_1| < |\mathfrak{F}|$ and

$$\mathfrak{M}, 1, w \models X.$$

If $|\mathfrak{F}_1| \leq 1 + 2^{|sub(X)|n}$, we are done; otherwise we again apply the above arguments on $\mathfrak{F}_1$.
Proceeding in this way, we get the desired result. □

As a direct consequence of Proposition 83, we obtain Theorem 25.

11.3 Decidable Fragments of $L(T, I)$

Now, we present two decidable fragments of $L(T, I)$, $I \in \{ K, K4, T, B, S4, KTB, KB4, S5\}$.

First Decidable Fragment. As in Theorem 23, let us again consider the extension of the definition of the dynamic I spaces so that for a dynamic I space $\mathfrak{F}$, the cardinality of $\mathfrak{F}$, $|\mathfrak{F}|$, is not necessarily a finite integer, but possibly $|\mathbb{N}|$. Following the standard convention, we write $|\mathfrak{F}| < \infty$ to mean that $|\mathfrak{F}|$ is finite.

We consider the following fragments of the set of all $\mathcal{L}$–wffs.

Definition 59.

(i) $\mathbb{L}_0$ *is the set of all wffs* γ *such that if* $\neg\gamma_1$ *is a sub-wff of* γ, *then* γ_1 *does not have any sub-wff of the form* $\oplus\delta$, *or* $\delta_1 \mathcal{U} \delta_2$.
(ii) $\neg\mathbb{L}_0 := \{\neg X : X \in \mathbb{L}_0\}$.

Note that $\mathbb{L}_0$ and $\neg\mathbb{L}_0$ are not closed under negation. We prove that the satisfiability problem for the fragment $\mathbb{L}_0$ with respect to the class of all dynamic I spaces $\mathfrak{F}$ with $|\mathfrak{F}| < \infty$, is decidable. This is done by showing that a wff X of $\mathbb{L}_0$ is satisfiable in this class of dynamic I spaces, if and only if it is satisfiable in the class of dynamic I spaces with cardinality $|\mathbb{N}|$.

As a consequence, we obtain $\neg\mathbb{L}_0$ as a decidable fragment of $L(T, I)$.

Definition 60. *Given a model* $\mathfrak{M} := (\mathfrak{F}, V)$, $\mathfrak{F} := \mathcal{F}_1, \mathcal{F}_2, \ldots, \mathcal{F}_N$, *by an* extension of $\mathfrak{M}$, *we mean a model* $\mathfrak{M}^* := (\mathfrak{F}^*, V)$, $\mathfrak{F}^* := \mathcal{F}_1^*, \mathcal{F}_2^*, \ldots$, *where* $|\mathfrak{F}^*| > N$, *and* $\mathcal{F}_i^* = \mathcal{F}_i$ *for* $1 \le i \le N$.

Observe that cardinality of $\mathfrak{F}^*$ in Definition 60 could be finite or infinite.

Proposition 84. *Let* $X \in \mathbb{L}_0$. *Let* $\mathfrak{M}$ *be a model and* $\mathfrak{M}^*$ *be an extension of* $\mathfrak{M}$. *Then*

$$\mathfrak{M}, t, w \models X \text{ implies } \mathfrak{M}^*, t, w \models X.$$

Proof. The proof is by induction on the complexity of X. We only consider the case when X is of the form $\neg Y$. Note that Y cannot be of the form $\oplus Y_1$ or $Y_1 \mathcal{U} Y_2$, as $X \in \mathbb{L}_0$. Suppose Y is of the form $\Box Y_1$, and the constituent frames of $\mathfrak{M}$ are (U, P_i). Then

$$\begin{aligned}
&\mathfrak{M}, t, w \models \neg\Box Y_1 \\
&\Rightarrow \mathfrak{M}, t, w' \models \neg Y_1 \text{ for some } w' \text{ with } (w, w') \in P_t \\
&\Rightarrow \mathfrak{M}^*, t, w' \models \neg Y_1 \text{ (by induction hypothesis)} \\
&\Rightarrow \mathfrak{M}^*, t, w \models \neg\Box Y_1.
\end{aligned}$$

The other cases, when Y is of the form $Y_1 \wedge Y_2$, $\ominus Y_1$, $Y_1 \mathcal{S} Y_2$ can be done in a similar way. □

From Proposition 84, it follows that if a wff $X \in \neg\mathbb{L}_0$ is not satisfiable in a model $\mathfrak{M}$, then we cannot make it satisfiable in any model obtained by extending $\mathfrak{M}$. This result is obtained due to the constraint put on the wffs of the form $X\mathcal{U}Y$ and $\oplus X$ by the defining conditions of $\mathbb{L}_0$; it is not true in general. For instance, consider the model $\mathfrak{M} := (\mathfrak{F}, V)$, $\mathfrak{F} := \mathcal{F}_1, \mathcal{F}_2$, and its extension $\mathfrak{M}^* := (\mathfrak{F}^*, V)$, $\mathfrak{F}^* := \mathcal{F}_1, \mathcal{F}_2, \mathcal{F}_3$, where $\mathcal{F}_t := (\{w, w'\}, P_t)$, $t = 1, 2, 3$, $[w]_{P_i} := \{w\}$, $[w']_{P_i} := \{w'\}$, $i = 1, 2$, $[w]_{P_3} := \{w, w'\}$, and $V(p) := \{w'\}$, $V(q) := \{w\}$. Then $\mathfrak{M}, t, u \models \neg(\top\ \mathcal{U}\ (\Diamond p \wedge \Diamond q))$ as $\mathfrak{M}, t, u \models \neg(\Diamond p \wedge \Diamond q)$, for $t = 1, 2$ and $u \in \{w, w'\}$. On the other hand, $\mathfrak{M}^*, 1, w \models \top\ \mathcal{U}\ (\Diamond p \wedge \Diamond q)$ as $\mathfrak{M}^*, 3, w \models \Diamond p \wedge \Diamond q$.

Proposition 85. *Let $X \in \mathbb{L}_0$, and $\mathfrak{M}^* := (\mathfrak{F}^*, V)$ be a model with $|\mathfrak{F}^*| = |\mathbb{N}|$. If $\mathfrak{M}^*, t, w \models X$, then there exists some model $\mathfrak{M} := (\mathfrak{F}, V)$ of which $\mathfrak{M}^*$ is an extension such that $t \leq |\mathfrak{F}| < \infty$ and $\mathfrak{M}, t, w \models X$.*

Proof. The proof is again by induction on the complexity of X. Let us just consider the cases when X is of the form (i) $\neg Y$, and (ii) $Y_1\mathcal{U}Y_2$.

(i) Since $X \in \mathbb{L}_0$, Y cannot be of the form $\oplus Y_1$, or $Y_1\mathcal{U}Y_2$. Suppose Y is of the form $\Box Y_1$, and the constituent frames of $\mathfrak{M}^*$ are (U, P_i). Then

$$\begin{aligned}
&\mathfrak{M}^*, t, w \models \neg\Box Y_1\\
&\Rightarrow \mathfrak{M}^*, t, w' \models \neg Y_1 \text{ for some } w' \text{ with } (w, w') \in P_t\\
&\Rightarrow \mathfrak{M}, t, w' \models \neg Y_1\\
&\quad \text{(by induction hypothesis, where } \mathfrak{M} \text{ is as described in proposition)}\\
&\Rightarrow \mathfrak{M}, t, w \models \neg\Box Y_1.
\end{aligned}$$

(ii) $\mathfrak{M}^*, t, w \models Y_1\mathcal{U}Y_2$ implies that there exists r with $t \leq r$ such that $\mathfrak{M}^*, r, w \models Y_2$, and for all k with $t \leq k < r$, $\mathfrak{M}^*, k, w \models Y_1$. Therefore, by induction hypothesis, there exist models $\mathfrak{M}_r, \mathfrak{M}_k$, $t \leq k < r$, such that

$$\mathfrak{M}_r, r, w \models Y_2, \tag{24}$$
$$\mathfrak{M}_k, k, w \models Y_1,\ t \leq k < r. \tag{25}$$

Let $\mathfrak{F}_r$, $\mathfrak{F}_k$ be the constituent dynamic I spaces of the models $\mathfrak{M}_r$ and $\mathfrak{M}_k$, $t \leq k < r$, respectively, and let us consider the model $\mathfrak{M} := (\mathfrak{F}_l, V)$, where $|\mathfrak{F}_l| := \max\{|\mathfrak{F}_r|, |\mathfrak{F}_k|, t \leq k < r\}$. Then from (24), (25) and Proposition 84, we obtain $\mathfrak{M}, r, w \models Y_2$, and for all k with $t \leq k < r$, $\mathfrak{M}, k, w \models Y_1$. This gives $\mathfrak{M}, t, w \models Y_1\mathcal{U}Y_2$.
All the other cases can be proved in a similar way. □

Proposition 86.

1. *A wff $X \in \mathbb{L}_0$ is satisfiable in the class of dynamic I spaces with finite cardinality, if and only if X is satisfiable in the class of dynamic I spaces with cardinality $|\mathbb{N}|$.*
2. *A wff $X \in \neg\mathbb{L}_0$ is valid in the class of dynamic I spaces with finite cardinality, if and only if X is valid in the class of dynamic I spaces with cardinality $|\mathbb{N}|$.*

Proof. Item 1 is a direct consequence of Propositions 84 and 85. Item 2 obtains from Item 1. □

Now using the decidability of the logic $L_{\mathbb{N}}(T, I)$ (cf. Theorem 23) and Proposition 86, we obtain the following decidability result.

Theorem 26.

1. *Given a wff $X \in \mathbb{L}_0$, we can decide the satisfiability of X in the class of all dynamic I spaces $\mathfrak{F}$ with $|\mathfrak{F}| < \infty$.*
2. *Given a wff $X \in \neg\mathbb{L}_0$, we can decide the validity of X in the class of all dynamic I spaces $\mathfrak{F}$ with $|\mathfrak{F}| < \infty$.*

Second Decidable Fragment. Given a wff X, we would like to know which are the signed wffs that can occur in an (I, N)-tableau for X. If Z is a signed wff occurring in some (I, N)-tableau for X, then obviously $Z \in S(X)$. The notion of *ingredients* of a signed wff given below determines such a set more precisely.

Let $C \in \{\overline{T}, \overline{F}\}$. We shall use $\hat{C}$ to denote $\overline{T}$ or $\overline{F}$ according as C is $\overline{F}$ or $\overline{T}$.

Definition 61. *The* ingredients *of a signed wff CX, denoted as $Ing(CX)$, is a set of signed wffs defined inductively as follows.*

$$\begin{aligned}
&Ing(CX) := \{CX\}, \text{ for } X \in PV,\\
&Ing(C\neg X) := \{C\neg X\} \cup Ing(\hat{C}X),\\
&Ing(C(X\sharp Y)) := \{C(X\sharp Y)\} \cup Ing(CX) \cup Ing(CY),\ \sharp \in \{\vee, \wedge, \mathcal{U}, \mathcal{S}\},\\
&Ing(C(X \to Y)) := \{C(X \to Y)\} \cup Ing(\hat{C}X) \cup Ing(CY),\\
&Ing(C\sharp X) := \{C\sharp X\} \cup Ing(CX),\ \sharp \in \{\Box, \Diamond, \oplus, \ominus\}.
\end{aligned}$$

We ask readers to compare the above definition of ingredients with different types of signed wffs, and their components given in Fig. 2.

Example 20. Let us determine the ingredients of a signed wff $\overline{F}(\Box p \to \oplus q)$, where $p, q \in PV$.

$$\begin{aligned}
&Ing(\overline{F}(\Box p \to \oplus q))\\
&= \{\overline{F}(\Box p \to \oplus q)\} \cup \underbrace{Ing(\overline{T}\Box p)}_{=\{\overline{T}\Box p\} \cup \underbrace{Ing(\overline{T}p)}_{=\overline{T}p}} \cup \underbrace{Ing(\overline{F} \oplus q)}_{=\{\overline{F}\oplus q\} \cup \underbrace{Ing(\overline{F}q)}_{\{\overline{F}q\}}}\\
&= \{\overline{F}(\Box p \to \oplus q), \overline{T}\Box p, \overline{T}p, \overline{F}q\}.
\end{aligned}$$

Obviously $Ing(CX) \subseteq S(CX)$. Moreover, only the signed wffs from the set $Ing(\overline{F}X)$ will occur in an (I, N)-tableau for X. Note that $Ing(\overline{F}X)$ also gives us some clue about the rules which can be applied on a wff in the construction of an (I, N)-tableau for X. For instance, if $\overline{T}\Box Y \notin Ing(\overline{F}X)$, then $\overline{T}\Box Y$ will not occur in any (I, N)-tableau for X, and hence we cannot have any application of ν-rule on $\overline{T}\Box Y$ in any (I, N)-tableau for X.

Let us consider the two following fragments of the set of all $\mathcal{L}$-wffs.

Definition 62.

(i) $\mathbb{L}_1$ *is the set of all wffs* γ *such that if* $\gamma_1 \mathcal{U} \gamma_2$ *is a sub-wff of* γ*, then* γ_1 *and* γ_2 *do not involve any temporal operator.*
(ii) $\mathbb{L}_2$ *is the set consisting of all those* $\gamma \in \mathbb{L}_1$ *such that* $Ing(\overline{F}\gamma)$ *does not contain any wff of the form*
 - $\overline{T}\Box\delta$*, or* $\overline{F}\Diamond\delta$*, where* δ *involves the operator* $\mathcal{U}$*,*
 - $\overline{T}(\delta_1 \mathcal{S} \delta_2)$*, or* $\overline{F}(\delta_1 \mathcal{S} \delta_2)$*, where* δ_1 *or* δ_2 *involves the operator* $\mathcal{U}$*.*

Let $\gamma \in \mathbb{L}_1$ be such that in any of its sub-wffs of the form $\Box\delta$, $\Diamond\delta$, δ does not involve the operator $\mathcal{U}$, and in any of its sub-wffs of the form $\delta_1 \mathcal{S} \delta_2$, neither of δ_1, δ_2 involves $\mathcal{U}$. Then obviously $\gamma, \neg\gamma \in \mathbb{L}_2$. Note that $\mathbb{L}_2$ can have wffs where $\mathcal{U}$ occurs in the scope of $\Box$ and $\Diamond$. For instance, $\Box(\gamma_1 \mathcal{U} \gamma_2) \in \mathbb{L}_2$, but $\Diamond(\gamma_1 \mathcal{U} \gamma_2) \notin \mathbb{L}_2$, where γ_1, γ_2 are any wffs formed using Boolean connectives and $\Box$, $\Diamond$. Moreover, $\mathbb{L}_2$ is not closed under negation: $\Box(p\mathcal{U}q) \in \mathbb{L}_2$, but $\neg\Box(p\mathcal{U}q) \notin \mathbb{L}_2$, where $p, q \in PV$.

Remark 10. If $X \in \mathbb{L}_2$, then there is no question of the ν-rule, $\mathcal{S}$-rule or $\mathcal{U}$-rule application in an (I, N)-tableau for X, on any wff of the form σCY, where Y has an *until* sub-wff, that is, a sub-wff with operator $\mathcal{U}$. This also means that if n is the total number of $\mathcal{U}$s occurring in X and $\sigma_1 Z_1, \sigma_2 Z_2, \ldots \sigma_m Z_m$ is the complete list of *until* wffs on any branch θ, then we must have $m \leq n$.

We prove the decidability of the fragment $\mathbb{L}_2$, using the tableau-based proof procedures. We shall write P_I to mean procedure P2 or P3 according as $I \in \{K, T, KTB, B\}$ or $I \in \{K4, KB4, S4, S5\}$.

Let us consider a branch θ of an (I, N)-tableau and suppose $\sigma R_n \overline{T} X \mathcal{U} Y \in \theta$, $n < N$. From the tableau extension rule $\mathcal{U}(a)$, it is clear that when this rule is applied on $\sigma R_n \overline{T} X \mathcal{U} Y$, it results in splitting of the original branch into $N-n+1$ branches and θ is obtained by selecting one of these. Suppose θ is obtained by selecting the jth branch, $1 \leq j \leq N - n + 1$. Then, we obtain the integer $m = n + (j - 1)$, $n \leq m \leq N$ such that $(\sigma)^*_m \overline{T} Y \in \theta$ and $(\sigma)^*_l \overline{T} X \in \theta$ for all l, $n \leq l < m$. Informally, this m signifies that Y holds at the object represented by σ at the time point m in the future of current time point n. Furthermore, X holds at the object represented by σ at all the time points between m and n, including n. Thus, we have the following notion of fulfillment of an *until* wff in a branch.

Definition 63. *An occurrence of the* N*-prefixed wff* $\sigma R_n \overline{T} X \mathcal{U} Y$ *(or,* $\sigma R_n l_1\, l_2 \cdots l_r\, \overline{T} X \mathcal{U} Y$*,* $\sigma R_n \overline{F} X \mathcal{U} Y$*,* $\sigma R_n l_1 l_2 \cdots l_r \overline{F} X \mathcal{U} Y$*) in a branch* θ *of an* (I, N)*-tableau is said to be* fulfilled *at the* m^{th} *point in* θ*,* $n \leq m \leq N$*, if* θ *is the result of the selection of the* $(m - n + 1)^{th}$ *branch when the* $\mathcal{U}(a)$*-rule (respectively* $\mathcal{U}(b)$*,*$\mathcal{U}(c)$*,*$\mathcal{U}(d)$ *rule) is applied on* $\sigma R_n \overline{T} X \mathcal{U} Y$ *(respectively* $\sigma R_n l_1 l_2 \cdots l_r \overline{T} X \mathcal{U} Y$*,* $\sigma R_n \overline{F} X \mathcal{U} Y$*,* $\sigma R_n l_1 l_2 \cdots l_r \overline{F} X \mathcal{U} Y$*).*

We need one more definition before moving to the main result of the section. Given a branch θ of an (I, N)-tableau for a wff X, we would like to determine a suitable upper bound of the characteristics of N-prefixes which can occur with a given wff on θ. Moreover, we want this bound to be independent of N. The following will help us to achieve the goal.

Definition 64. *The $\oplus$-degree of a wff $Z \in \mathbb{L}_1$, denoted as $D_\oplus(X)$, is defined inductively as follows:*

(i) $D_\oplus(p) = D_\oplus(X\mathcal{U}Y) = 1$, *where p is propositional variable,*
(ii) $D_\oplus(\Box X) = D_\oplus(\Diamond X) = D_\oplus(\ominus X) = D_\oplus(X)$,
(iii) $D_\oplus(\oplus X) = 1 + D_\oplus(X)$,
(iv) $D_\oplus(X \vee Y) = D_\oplus(X \wedge Y) = D_\oplus(X\mathcal{S}Y) = max\{D_\oplus(X), D_\oplus(Y)\}$.

The following proposition gives the desired upper bound for particular types of wffs. The proof is provided in Appendix E.

Proposition 87. *Let $X \in \mathbb{L}_1$ be a wff with $D_\oplus(X) = t$. Consider a branch θ of an (I, N)-tableau for X. Let $\sigma CZ \in \theta$, $C \in \{\overline{T}, \overline{F}\}$, be such that Z is not a proper sub-wff of any wff of the form $\gamma_1\mathcal{U}\gamma_2$ belonging to $Ing(\overline{F}X)$. Then $char(\sigma) \leq t$. Moreover, if Z is of the form $\oplus Z'$, then $char(\sigma) \leq t-1$.*

Proof. There exist N-prefixed wffs $\sigma_1C_1Z_1, \sigma_2C_2Z_2, \ldots, \sigma_nC_nZ_n$, $C_i \in \{\overline{T}, \overline{F}\}$, occurring on the branch θ such that the following hold.

1. $\sigma_1C_1Z_1 = R_11\overline{F}X$ and $\sigma_nC_nZ_n = \sigma CZ$.
2. $\sigma_{i+1}C_{i+1}Z_{i+1}$ is obtained from $\sigma_iC_iZ_i$ by the application of a single rule, say r_i-rule, $i \in \{1, 2, \ldots, n-1\}$.
3. If Z_i is not a propositional variable, then Z_{i+1} is a proper sub-wff of Z_i. Moreover, if Z_i is a propositional variable, then Z_{i+1} is also the same propositional variable.

Since Z_n is not a proper sub-wff of any wff of the form $\gamma_1\mathcal{U}\gamma_2$ belonging to $Ing(\overline{F}X)$, from 3, it follows that Z_i, $i \in \{1, 2, \ldots, n\}$ is also of the same kind. Hence none of the r_i can be a $\mathcal{U}$-rule. From this fact it follows that if $char(\sigma_{i+1}) > char(\sigma_i)$, then r_i must be a $\oplus$-rule. Moreover, in this case $D_\oplus(Z_i) > D_\oplus(Z_{i+1})$. Moreover, note that if $D_\oplus(Z_j) = 1$, then we will have $D_\oplus(Z_k) = 1$ for all $n \leq k \leq j$. In that case none of r_k, $n-1 \leq k \leq j$ can be a $\oplus$-rule. From these facts it follows that at most $t-1$ of these r_i can be a $\oplus$-rule and hence $char(\sigma) \leq t$. Moreover, if Z is of the form $\oplus Z'$, then at most $t-2$ of these r_i can be a $\oplus$-rule as $D_\oplus(Z_n) \geq 2$ and so in that case we obtain $char(\sigma) \leq t-1$. □

Proposition 88 (Branch mapping theorem). *Let us consider an $(I, N+1)$-tableau A for a wff $X \in \mathbb{L}_1$ obtained by following the procedure P_I and a branch θ of it. Let $D_\oplus(X) = t$ and $N > t+1$. Suppose there exists an integer m, $t < m < N$ such that no until wff in θ is fulfilled at the mth point. Choose and fix such a m.*

Let B be an (I, N)-tableau for X obtained by following P_I. Then, for each $n \in \mathbb{N}$, there exists some branch θ_n of B and a mapping g_n from $P_n(\theta_n)$ (cf. Notation 22) to the set of all $N+1$-prefixes occurring on θ such that the following holds:

- $g_n(\tau R_l) = \begin{cases} g_n(\tau)R_{l+1} & \mathit{if}\ l \in \{m, \ldots, N\} \\ g_n(\tau)R_l & \mathit{if}\ l \in \{1, 2, \ldots, m-1\} \end{cases}$
- $g_n(\tau l) = g_n(\tau)l'$, *where* $l, l' \in \mathbb{N}$;
- *For* $\tau Z \in \theta_n$ *with* $H_{\theta_n}(\tau Z) \leq n$, $g_n(\tau)Z \in \theta$.

Moreover, if the tableau B is closed, then the branch θ *of tableau A is also closed.*

Proof. Let us prove the first part of the proposition by induction on n. Note that $m \geq 2$.

Basis case: $n = 1$
In this case any branch will work as θ_1. Moreover, we take $g_1(R_1 1) = R_1 1$.

Induction case: Suppose the result holds for n and we prove for $n+1$. If the set $\{\sigma Z : \sigma Z \in \theta_n \text{ and } H_{\theta_n}(\sigma Z) = n+1\}$ is empty, then we take $\theta_{n+1} := \theta_n$ and $g_{n+1} := g_n$ to get the desired result. So, let us consider the case when the above set is non-empty. We will define θ_{n+1} in such a way that it is the same as the branch θ_n up to depth n. Moreover, $g_{n+1}(\sigma) = g_n(\sigma)$ for all σ such that $\sigma Y \in \theta_n$ for some Y with $H_{\theta_n}(\mu Y) \leq n$. Let $\mu' Z' \in \theta_n$ with $H_{\theta_n}(\mu' Z') \leq n$ be such that by the application of a rule on this wff, we obtain wff μZ on θ_n with $H_{\theta_n}(\mu Z) = n+1$. Depending on the rule applied, we will determine θ_{n+1} as follows. We will only consider the cases when the $\mathcal{U}(a)$, or $\oplus(a)$, or π, or ν, rule is applied. The other cases can be handled similarly.

$\mathcal{U}(a)$-rule: So, $\mu' Z'$ must be of the form $\sigma R_u \overline{T} \delta_1 \mathcal{U} \delta_2$. Since $X \in \mathbb{L}_1$, $\delta_1 \mathcal{U} \delta_2$ cannot be a proper sub-wff of any wff of the form $\delta_3 \mathcal{U} \delta_4$ belonging to $Ing(\overline{F}X)$, and thus by Proposition 87, we obtain

$$char(\sigma R_u) = u \leq t < m. \tag{26}$$

Moreover, by the given condition, $g_n(\sigma R_u)$ is defined and is equal to $g_n(\sigma)R_u$ as $u < m$. We also have $g_n(\sigma)R_u \overline{T} \delta_1 \mathcal{U} \delta_2 \in \theta$.

Suppose, when the $\mathcal{U}(a)$-rule is applied on $g_n(\sigma)R_u \overline{T} \delta_1 \mathcal{U} \delta_2 \in \theta$, the branch θ is obtained by taking the jth-branch, $1 \leq j \leq (N+1) - u + 1$, that is, $g_n(\sigma)R_u \overline{T} \delta_1 \mathcal{U} \delta_2 \in \theta$ is fulfilled in θ at $\zeta = (u + j - 1)^{th}$ point. Note that by the given condition,

$$\zeta \neq m. \tag{27}$$

Moreover, we also have

$$\left.\begin{array}{l} (g_n(\sigma))^*_b \overline{T} \delta_1 \in \theta, \text{ for all } b, u \leq b \leq u + (j-2) = \zeta - 1, \text{ and} \\ (g_n(\sigma))^*_{u+(j-1)} \overline{T} \delta_2 = (g_n(\sigma))^*_\zeta \overline{T} \delta_2 \in \theta \end{array}\right\} \tag{28}$$

We have two possibilities for ζ: $\zeta > m$, or, $\zeta < m$ (by (27) $\zeta \neq m$).
First consider $\zeta > m$.
So, we obtain,

$$\zeta > m > t \geq u. \tag{29}$$

In this case, we choose θ_{n+1} to be a branch obtained by selecting the $(\zeta - u)^{th} = (j-1)^{th}$ branch when the $\mathcal{U}(a)$-rule is applied on $\sigma R_u \overline{T} \delta_1 \mathcal{U} \delta_2 \in \theta_n$.

Let $\mu_1 Z_1 \in \theta_{n+1}$ with $H_{\theta_{n+1}}(\mu_1 Z_1) = n+1$. So, we need to define g_{n+1} on μ_1 if $g_n(\mu_1)$ is not defined.

Let $char(\mu_1) = k$. Note that $\mu_1 Z_1$ is either of the following.

- $(\sigma)_k^* \overline{T} \delta_1$, $u \leq k \leq u + (\zeta - u - 2) = \zeta - 2$, or
- $(\sigma)_k^* \overline{T} \delta_2$, $k = \zeta - 1$.

Writing explicitly, $\mu_1 Z_1$ is obtained as either of the following.

(i) $\sigma \overline{T} \delta_1$, where $char(\sigma) = k$ and $u < k \leq \zeta - 2$, or,
(ii) $\sigma R_k \overline{T} \delta_1$, where $char(\sigma) \neq k$ and $u \leq k \leq \zeta - 2$, or,
(iii) $\sigma \overline{T} \delta_2$, where $char(\sigma) = \zeta - 1 = k$, or,
(iv) $\sigma R_{\zeta-1} \overline{T} \delta_2$, where $char(\sigma) \neq \zeta - 1 = k$.

In case (i), $g_{n+1}(\mu_1)$ is already defined as $\mu_1 = \sigma$. So, we only need to show $g_{n+1}(\sigma)\overline{T}\delta_1 \in \theta$, that is $g_n(\sigma)\overline{T}\delta_1 \in \theta$ ($\because g_{n+1}(\sigma) = g_n(\sigma)$).

If $char(\sigma) = k < m$, then $char(g_n(\sigma)) = k$. Therefore, using (28) for $b = k$, we obtain $g_n(\sigma)\overline{T}\delta_1 = (g_n(\sigma))_k^* \overline{T}\delta_1 \in \theta$. Similarly, If $char(\sigma) = k \geq m$, then $char(g_n(\sigma)) = k+1$. Then again using (28) for $b = k+1$, we obtain $g_n(\sigma)\overline{T}\delta_1 = (g_n(\sigma))_{k+1}^* \overline{T}\delta_1 \in \theta$.

In case (ii), if $g_n(\sigma R_k)$ is not defined, then we take $g_{n+1}(\mu_1)$, that is, $g_{n+1}(\sigma R_k)$ to be $g_{n+1}(\sigma) R_k$ or $g_{n+1}(\sigma) R_{k+1}$ according as $k < m$, or $k \geq m$. Now, using (28), one can easily verify that $g_{n+1}(\mu_1)\overline{T}\delta_1 \in \theta$. One can similarly get the results for the cases (iii) and (iv).

Now, let us consider the case when $\zeta < m$. In this case, we choose θ_{n+1} to be a branch obtained by selecting the $(\zeta - u + 1)^{th} = j^{th}$ branch when the $\mathcal{U}(a)$-rule is applied on $\sigma R_u \overline{T} \delta_1 \mathcal{U} \delta_2 \in \theta_n$, and hence $\sigma R_u \overline{T} \delta_1 \mathcal{U} \delta_2$ is fulfilled at the ζth point in θ_{n+1}. Let $\mu_1 Z_1 \in \theta_{n+1}$ with $H_{\theta_{n+1}}(\mu_1 Z_1) = n+1$. Therefore, $\mu_1 Z_1$ must be either

(*i′*) $\sigma \overline{T} \delta_1$, where $char(\sigma) = k$ and $u < k \leq \zeta - 1$, or,
(*ii′*) $\sigma R_k \overline{T} \delta_1$, where $char(\sigma) \neq k$ and $u \leq k \leq \zeta - 1$, or,
(*iii′*) $\sigma \overline{T} \delta_2$, where $char(\sigma) = \zeta$, or,
(*iv′*) $\sigma R_\zeta \overline{T} \delta_2$, where $char(\sigma) \neq \zeta$.

If $\mu_1 = \sigma$, then $g_{n+1}(\mu_1)$ is already defined. Moreover, if μ_1 is σR_l, $a \leq l \leq \zeta < m$, and $g_n(\mu_1)$ is not defined, then we take $g_{n+1}(\mu_1) := g_{n+1}(\sigma) R_l$. Now in each of the above cases, one can show that $g_{n+1}(\mu_1) Z_1 \in \theta$. As above, we need to use (28) and also the fact that $char(\sigma) = char(g_{n+1}(\sigma))$.

$\oplus(a)$-rule: Take $\theta_{n+1} := \theta_n$. Then $\mu' Z'$ must be of the form $\sigma R_l \oplus \delta$, where $l + 1 \leq t < m$ (by Proposition 87). Moreover, $g_n(\sigma R_l)$ is defined and is given as $g_n(\sigma) R_l$. Further, μ will be σ or σR_{l+1} according as $char(\sigma) = l+1$ or $char(\sigma) \neq l+1$. Therefore, if μ is σ, then $g_{n+1}(\mu)$ is already defined. If μ is σR_{l+1}, and $g_n(\sigma R_{l+1})$ is not defined, then we take $g_{n+1}(\mu)$ to be $g_{n+1}(\sigma) R_{l+1}$. Now, it remains to show that $g_{n+1}(\mu)\delta \in \theta$. But it follows from the facts that $g_{n+1}(\sigma) R_l \oplus \delta \in \theta$ and $char(\sigma) = char(g_n(\sigma))$.

π-rule: Take $\theta_{n+1} := \theta_n$. $\mu' Z'$ and μZ must be of the form $\sigma\pi$ and $\sigma l\pi_0$ respectively. Moreover, $g_n(\sigma)$ is defined and $g_n(\sigma)\pi \in \theta$. Suppose, by the application of the π-rule on $g_n(\sigma)\pi$, $g_n(\sigma)l'\pi_0$ is introduced on θ. We define $g_{n+1}(\sigma l)$ to be $g_n(\sigma)l'$.

ν-rule: Take $\theta_{n+1} := \theta_n$ and $g_{n+1} := g_n$. $char(\mu) = char(\mu') = a$ (say) and $(\mu', \mu) \in R^I_a$. We must have $\mu Z^0 \in \theta'$ with $H_{\theta'}(\mu Z') \leq n$ for some Z^0. Moreover, $g_n(\mu')$ and $g_n(\mu)$ are both defined and $g_n(\mu')Z', g_n(\mu)Z^0 \in \theta$. We need to show $g_n(\mu)Z \in \theta$. Obviously, $char(g_n(\mu)) = char(g_n(\mu'))$. It is not difficult to see that $(g_n(\mu'), g_n(\mu)) \in R^I_{char(g_n(\mu))}$. Thus, we obtain $g_n(\mu)Z \in \theta$.

Let us now prove the second part of the proposition. As the tableau B is closed, each branch of B must be finite. Let n be the smallest integer such that $P_{n+1}(\theta')$ is empty for each branch θ' of B. Now, by the first part of the proposition, we obtain a branch θ_n of B and a mapping g_n. Now using g_n and the fact that θ_n is closed, we get the closure for θ. For instance, if θ_n is closed as $\sigma Z \in \theta_n$, where $char(\sigma) = N$, then $g_n(\sigma)Z \in \theta$ and this will give the closure of θ as $char(g_n(\sigma)) = N + 1$. □

Let $X \in \mathbb{L}_2$ be such that $D_\oplus(X) = t$ and n be the number of *until* operators occurring in X. Let $N > t + n + 1$. Then we have the following.

Proposition 89. *If the (I, N)-tableau B for X obtained by following P_I is closed, then so is the $(I, N + 1)$-tableau A for X obtained by following P_I.*

Proof. Consider an arbitrary branch θ of A, we prove it is closed. Note that, as mentioned in Remark 10, if θ has k N-prefixed *until* wffs, then we must have $k \leq n$. Thus, due to the choice of N, we must have an integer m, $t < m < N$, such that no *until* wff in θ is fulfilled at the mth point. Then using Proposition 88, we obtain the closure of θ. □

Remark 11. For $N > t + n + 1$, where $D_\oplus(X) = t$ and n is the number of *until* operators occurring in $X \in \mathbb{L}_2$, the above proof shows that the requirement of the existence of m as stated in Proposition 88 is met for each branch of the tableau A. Thus, Proposition 88, just as Proposition 81, determines a mapping from the branches of tableau A to the branches of tableau B, preserving openness.

Now using Proposition 89 and the fact that the tableau construction procedure P_I terminates, we have the following decidability result.

Theorem 27. *Given a wff $X \in \mathbb{L}_2$, we can decide whether X is I-tableau provable or not, and hence we can decide the validity of X in the class of all dynamic I spaces.*

Proof. Let X in $\mathbb{L}_2$ be such that $D_\oplus(X) = t$ and n be the number of *until* operators occurring in X. Let $N = t + n + 2$. Then from Proposition 89, it follows that in order to check whether X is I-tableau provable or not, we only need to check whether X is (I, K)-tableau provable or not for each $K \leq N$. Thus, we obtain the desired decidability. □

So, we obtain $\mathbb{L}_2$ as a decidable fragment of the logics $L(T, I)$. It should be remarked that, as $\mathbb{L}_2$ is not closed under negation, Theorem 27 does not give the decidability of the satisfiability problem for the elements of $\mathbb{L}_2$ in general, but obviously we get this decidability for the set $\{X : \neg X \in \mathbb{L}_2\}$.

12 Information and Information Update

The language $\mathcal{L}$ is proposed in Sect. 8 to study the temporal dimension of rough set theory where the knowledge base changes with time. $\mathcal{L}$ can study the effects of an inflow of information on the partitions, but the flow of information itself does not appear in the picture. In this section, we are interested in a dynamic logic for information systems, where we can express the flow of information as well as its effect on the approximation of sets. In order to achieve our aim, we need to concentrate on the following points:

(a) search for a suitable logic for information systems;
(b) formally define the notion of *information* and *information update* for information systems;
(c) look for a suitable *dynamic extension* of the language of the logic for information systems, wherein the flow of information may be expressed; and finally
(d) find a set of *reduction axioms* which would give us a complete axiomatization of the dynamic logics. The reduction axioms would also help to see the effect of the gain in information.

Let us first consider (a) and recall Orłowska's definition of an information structure (cf. Definition 13). We find that this structure is good enough to talk about the indiscernibility relations with respect to different sets of attributes, but an information system is more than that. It actually provides information about what value an object takes for an attribute, and this is used to *generate* the indiscernibility relations. This is not the case in an information structure. Thus it seems that, in order to define a suitable logic for information systems, one needs to bring the attribute, attribute value pairs explicitly into the syntax. So, we stipulate the following desirable properties for a suitable logic of information systems.

(I) *The language should have attribute and attribute value constants.*
(II) *The semantics should be based on a structure having relative accessibility relations with the power set of the set A of attributes as the parameter set. These relations in the structure would be present syntactically as modalities.*
(III) *The relationship between the induced relations (indiscernibility, similarity etc.) with the attributes and attribute values should be reflected syntactically in the relationship between the modalities and the pairs of attribute, attribute value constants.*
(IV) *The logic should have a sound and complete deductive system.*

The necessity of having attribute, attribute value pairs in the language of a logic for information systems has been felt by researchers and this is why in literature one can find many proposals for logics with this feature. But a logic for information systems with all of the above features was not obtained till the proposal of such a logic in [49]. Section 12.2 presents the logics LIS and LDIS based on the proposal of [49]. Fragments $\mathrm{LIS}_\mathfrak{f}$ and $\mathrm{LDIS}_\mathfrak{f}$ turn out to be logics for information systems (which could be deterministic or incomplete) and deterministic information systems respectively. These logics have all the four properties stipulated above. Moreover, in addition to expressing the properties of indiscernibility and similarity relations corresponding to any finite set of attributes, these can also express notions related to dependencies in data and data reduction (cf. Definitions 5 and 6). The presence of modal operators for both indiscernibility and similarity relations, enables us to compare rough set 'approximations' of sets with respect to these relations, as will be clear in the sequel. Sound and complete deductive systems for the logics LIS, LDIS, $\mathrm{LIS}_\mathfrak{f}$ and $\mathrm{LDIS}_\mathfrak{f}$ are given in Sect. 12.3. Decidability of $\mathrm{LIS}_\mathfrak{f}$ and $\mathrm{LDIS}_\mathfrak{f}$ are presented in Sect. 12.4.

In Sect. 12.1, we will come to the issue (b) of formal notions of information and information update for the information systems. In Sect. 12.5, we will see that the presence of attribute and attribute value constants in the language of LIS, LDIS, $\mathrm{LIS}_\mathfrak{f}$ and $\mathrm{LDIS}_\mathfrak{f}$ makes it possible to extend these logics to define dynamic logics DLIS, DLDIS, $\mathrm{DLIS}_\mathfrak{f}$ and $\mathrm{DLDIS}_\mathfrak{f}$, where one can express the notion of information given in Sect. 12.1. As $\mathrm{LIS}_\mathfrak{f}$ and $\mathrm{LDIS}_\mathfrak{f}$ are logics for information systems and deterministic information systems respectively, we obtain $\mathrm{DLIS}_\mathfrak{f}$ and $\mathrm{DLDIS}_\mathfrak{f}$ as the dynamic logics for information systems and deterministic information systems. A set of reduction axioms is provided in Sect. 12.6, giving us sound and complete deductive systems for these dynamic logics. These dynamic logics are related with LIS, LDIS, $\mathrm{LIS}_\mathfrak{f}$ and $\mathrm{LDIS}_\mathfrak{f}$ in the same way as dynamic epistemic logic (DEL) is related with epistemic logic [22,104]. These dynamic logics are contrasted with DEL in Sect. 12.7.

The content of this section is based on the articles [49,53].

12.1 Information and Information Update of Information Systems

Recall the definitions and results involving information systems in Sect. 2.1. Let us consider the incomplete information system (IIS) $\mathcal{K}_1$ of Table 6 below, which provides information about six patients $P1 - P6$ regarding the attributes 'Temperature(T)', 'Headache(H)' and 'Nausea(N)'. This table is a modified form of the one given in [33].

We note that one may have a situation where some of the current information turns out to be incorrect and may get corrected later. Moreover, with new information, information gaps may also be filled. For example, suppose we are provided with the following.

I_1: The information regarding the temperature and nausea of the patient $P4$ is not correct. In fact, $P4$ has nausea and temperature, very high or high.

Table 6. IIS $\mathcal{K}_1$

	T	H	N
$P1$	High	$*$	No
$P2$	Very high	$*$	Yes
$P3$	$*$	No	No
$P4$	No	No	No
$P5$	No	Yes	No
$P6$	High	No	Yes

I_2: According to new information, if a patient has nausea and high or very high temperature, then she must have headache.

The information may appear in some order. For instance, it may happen that I_2 arrives after I_1. Let us denote it by writing $I_1; I_2$. Depending on the order of the information, we obtain different updated ISs. For instance, when we update $\mathcal{K}_1$ with the information I_1, we obtain two ISs $\mathcal{K}_2$ and $\mathcal{K}_3$ (cf. Table 7), one by taking the temperature of P_4 to be high and the other by taking it to be very high. By updating $\mathcal{K}_2$ and $\mathcal{K}_3$ with I_2, we obtain ISs $\mathcal{K}_4$ and $\mathcal{K}_5$ respectively. Thus, the update of $\mathcal{K}_1$ with $I_1; I_2$, i.e. first with I_1 and then with I_2, gives the ISs $\mathcal{K}_4$ and $\mathcal{K}_5$.

Let $X := \{P2, P4, P6\}$ be the set of patients infected with flu and let Q be the attribute set $\{T, H, N\}$. One may ask questions of the following kind.

Table 7. ISs obtained from $\mathcal{K}_1$ by updates

(a) IIS $\mathcal{K}_2$

	T	H	N
P1	high	$*$	no
P2	very high	$*$	yes
P3	$*$	no	no
P4	high	no	yes
P5	no	yes	no
P6	high	no	yes

(b) IIS $\mathcal{K}_3$

	T	H	N
P1	high	$*$	no
P2	very high	$*$	yes
P3	$*$	no	no
P4	very high	no	yes
P5	no	yes	no
P6	high	no	yes

(c) IIS $\mathcal{K}_4$

	T	H	N
P1	high	$*$	no
P2	very high	yes	yes
P3	$*$	no	no
P4	high	yes	yes
P5	no	yes	no
P6	high	yes	yes

(d) IIS $\mathcal{K}_5$

	T	H	N
P1	high	$*$	no
P2	very high	yes	yes
P3	$*$	no	no
P4	very high	yes	yes
P5	no	yes	no
P6	high	yes	yes

Q1 Which are the patients who were initially considered to be positive elements of X with respect to $Ind(Q)$, but after the arrival of information I_1, are not considered so anymore?

Q2 Is P5 a positive element of X with respect to $Ind(Q)$, in all completions (cf. Definition 10) of the ISs obtained by updating with $I_1; I_2$?

Q3 Does the set X become definable with respect to $Ind(Q)$, after getting the information $I_1; I_2$?

Q4 Do we have a patient who does not have headache and nausea, but is infected with the flu? Does the situation remain so even after getting the information $I_1; I_2$?

Q5 Is it the case that some patient was thought not to have nausea, but later with the information $I_1; I_2$, it is found that she does have nausea?

We now formally present the notions of information and information update.

Let us fix the set A of attributes and $Val := \bigcup_{a\in A} Val_a$ of attribute values. In any IS $\mathcal{K} := (W, A, Val \cup \{*\}, f)$, the lower (and upper) approximations with respect to the indiscernibility and similarity relations may be regarded as unary operations on 2^W. Let us denote these operators by $\underline{Ind_{\mathcal{K}}(B)}$ and $\underline{Sim_{\mathcal{K}}(B)}$, for $B \subseteq A$, that is $\underline{Ind_{\mathcal{K}}(B)}(X) := \underline{X}_{Ind_{\mathcal{K}}(B)}$ and $\underline{Sim_{\mathcal{K}}(B)}(X) := \underline{X}_{Sim_{\mathcal{K}}(B)}$. Moreover, for $a \in A, v \in Val_a \cup \{*\}$, the descriptor (a, v) may be regarded as a nullary operation (constant), standing for the element $\{x \in W : f(x,a) = v\}$ of 2^W. Consider the operations of intersection ($\cap$) and complementation (^{c}) on 2^W also. Then it is clear that the structure $\mathfrak{A}_{\mathcal{K}} := (2^W, \cap, ^c, \{\underline{Ind_{\mathcal{K}}(B)} : B \subseteq A\}, \{\underline{Sim_{\mathcal{K}}(B)} : B \subseteq A\}, \{(a,v) : a \in A, v \in Val_a \cup \{*\}\})$ is an algebra of similarity type $\mathfrak{H}_{(A,Val)} := (H, \rho)$, where $H := \{\sqcap, \sim\} \cup \{I_B : B \subseteq A\} \cup \{S_B : B \subseteq A\} \cup \{(a,v) : a \in A, v \in Val_a \cup \{*\}\}$, and $\rho(\sqcap) = 2$, $\rho(\sim) = \rho(I_B) = \rho(S_B) = 1$, $\rho((a,v)) = 0$, for each $a \in A, v \in Val_a \cup \{*\}$.

Definition 65. *Let Var be a countable set of variables and let us denote the set of $\mathfrak{H}_{(A,Val)}$-terms over Var by $T_{\mathfrak{H}_{(A,Val)}}$, i.e. $T_{\mathfrak{H}_{(A,Val)}}$ is the smallest set containing $\{(a,v) : a \in A, v \in Val_a \cup \{*\}\} \cup Var$ such that for $t, s \in T_{\mathfrak{H}_{(A,Val)}}$, $t \sqcap s, \sim t, I_B(t), S_B(t) \in T_{\mathfrak{H}_{(A,Val)}}$.*

Next consider an assignment $\theta : Var \to 2^W$ on the algebra $\mathfrak{A}_{\mathcal{K}}$. θ is extended to the meaning function $\theta_{\mathcal{K}} : T_{\mathfrak{H}_{(A,Val)}} \to 2^W$ inductively, in the routine way:

$$
\begin{aligned}
\theta_{\mathcal{K}}(x) &:= \theta(x) for x \in Var,\\
\theta_{\mathcal{K}}(\sim t) &:= W \setminus \theta_{\mathcal{K}}(t),\\
\theta_{\mathcal{K}}(t \sqcap s) &:= \theta_{\mathcal{K}}(t) \cap \theta_{\mathcal{K}}(s),\\
\theta_{\mathcal{K}}(I_B t) &:= \underline{\theta_{\mathcal{K}}(t)}_{Ind_{\mathcal{K}}(B)},\\
\theta_{\mathcal{K}}(S_B t) &:= \underline{\theta_{\mathcal{K}}(t)}_{Sim_{\mathcal{K}}(B)},\\
\theta_{\mathcal{K}}((a,v)) &:= \{x \in W : f(x,a) = v\}.
\end{aligned}
$$

One may refer to any standard text on Universal Algebra for the above.

We are now ready to present the set $Inf_{(A,Val)}$ of *information*.

Definition 66.

1. *An* atomic information *with respect to* A *and* Val *is a tuple* $\mathcal{I} := (t, a, v)$, *where* $t \in \mathcal{T}_{\mathfrak{H}(A,Val)}$, $a \in A$ *and* $v \in Val_a$;
2. *The set* $Inf_{(A,Val)}$ *of* information *consists of the finite strings formed using the atomic information with respect to* A *and* Val *and the symbols ';' and '∨' following the rules:*
   - *if* $\mathcal{I}$ *is an atomic information then* $\mathcal{I} \in Inf_{(A,Val)}$,
   - *if* $\mathcal{I}, \mathcal{J} \in Inf_{(A,Val)}$, *then* $\mathcal{I};\mathcal{J}, \mathcal{I} \vee \mathcal{J} \in Inf_{(A,Val)}$,
   - *nothing else is in* $Inf_{(A,Val)}$.

Note that v is not $*$ in the atomic information. So information is strictly gained, not lost.

Let us now define *information update.*

Definition 67. *Let* $\mathcal{K} := (W, A, \bigcup_{a \in A} Val_a \cup \{*\}, f)$ *be an IS,* $\theta : Var \to 2^W$ *and* $\mathcal{I} \in Inf_{(A,Val)}$. *The* update of $\mathcal{K}$ with respect to information $\mathcal{I}$ under the assignment θ, *denoted by* $U(\mathcal{K}, \mathcal{I}, \theta)$, *is defined inductively as follows.*

- *If* $\mathcal{I} := (t, a, v)$ *is atomic, then* $U(\mathcal{K}, \mathcal{I}, \theta) := \{(W, A, \bigcup_{a \in A} Val_a \cup \{*\}, f_U)\}$, *where*
$$f_U(x, b) := \begin{cases} v & \text{if } x \in \theta_{\mathcal{K}}(t) \text{ and } a = b \\ f(x, b) & \text{otherwise.} \end{cases}$$
- *If* $\mathcal{I}$ *is* $\mathcal{I}_1;\mathcal{I}_2$, *then* $U(\mathcal{K}, \mathcal{I}, \theta) := \bigcup\{U(\mathcal{J}, \mathcal{I}_2, \theta) : \mathcal{J} \in U(\mathcal{K}, \mathcal{I}_1, \theta)\}$.
- *If* $\mathcal{I}$ *is* $\mathcal{I}_1 \vee \mathcal{I}_2$, *then* $U(\mathcal{K}, \mathcal{I}, \theta) := U(\mathcal{K}, \mathcal{I}_1, \theta) \cup U(\mathcal{K}, \mathcal{I}_2, \theta)$.

Observation 5.

(i) $U(\mathcal{K}, \mathcal{I}, \theta) \neq \emptyset$. Moreover, in the case when $\mathcal{I}$ is atomic, $U(\mathcal{K}, \mathcal{I}, \theta)$ is a singleton.
(ii) $\mathcal{K}' \in U(\mathcal{K}, \mathcal{I}, \theta)$ is also an IS.
(iii) If $\mathcal{I} := (t, a, v)$, then the value of $x \in \theta_{\mathcal{K}}(t)$ under a in $\mathcal{K}$ is *replaced* by the new value v.

From the definition of update it follows that the interpretation of the information $\mathcal{I};\mathcal{J}$ is that the IS is first updated with $\mathcal{I}$ and then the resultant ISs are updated with $\mathcal{J}$. In other words, $\mathcal{I};\mathcal{J}$ interprets the situation where we have first got the information $\mathcal{I}$ and then $\mathcal{J}$. On the other hand, $\mathcal{I} \vee \mathcal{J}$ represents uncertainty in information, i.e. it is known that either $\mathcal{I}$ or $\mathcal{J}$, but it is not known which one.

Let $\mathcal{I}_i := (t_i, a_i, v_i)$ be atomic information for $i = 1, 2, \ldots, n$. Note that $\mathcal{I}_1;\mathcal{I}_2;\ldots;\mathcal{I}_n$ means that $\mathcal{I}_{i+1}$ is obtained after $\mathcal{I}_i$, $i = 1, 2, \ldots, n-1$. How would one update an IS, say $\mathcal{K}$ under an assignment θ, when all the $\mathcal{I}_i$'s are obtained at the same time? Information could then be *inconsistent* in the sense that

(*) there exist $i, j \in \{1, 2, \ldots, n\}$, $i \neq j$, such that $\theta_{\mathcal{K}}(t_i) \cap \theta_{\mathcal{K}}(t_j) \neq \emptyset$, $a_i = a_j$ and $v_i \neq v_j$.

In other words, we may have information which requires the assignment of different values for an attribute to the same object. One may opt not to update the IS at all in such a case, or wish to update information for only those objects which do not belong to $\theta_{\mathcal{K}}(t_i) \cap \theta_{\mathcal{K}}(t_j)$, where i, j satisfy (*). In the current work, we are not interested in inconsistent information of this kind.

If the simultaneous information $\mathcal{I}_i$, $i = 1, 2, \ldots, n$, are not inconsistent as above, then the update of the IS with these will be given by $\mathcal{I}_1; \mathcal{I}_2; \cdots ; \mathcal{I}_n$. In fact, the ordering of the information will not matter in this case.

Let us recall the information I_1 of example given in the beginning of the section. Note that it consists of two simultaneous information, (i) $P4$ has nausea and (2) $P2$ has high or very high temperature. Thus, since these are not inconsistent, we can express this information as,

$\mathcal{I}_1$: $(x, N, yes); ((x, T, high) \vee (x, T, very\ high))$, x is a variable.

Similarly, I_2 can be expressed as

$\mathcal{I}_2$: $(((T, high) \sqcup (T, very\ high)) \sqcap (N, yes), H, yes)$.

With the above formalism, we can express the observations made earlier about the changes that the IIS $\mathcal{K}_1$ under goes after incorporating the information I_1 and I_2. Let θ be the assignment such that $\theta(x) := P4$. Then one can easily verify that

$$\begin{aligned} U(\mathcal{K}_1, \mathcal{I}_1, \theta) &= \{\mathcal{K}_2, \mathcal{K}_3\}, \\ U(\mathcal{K}_2, \mathcal{I}_2, \theta) &= \{\mathcal{K}_4\}, \\ U(\mathcal{K}_3, \mathcal{I}_2, \theta) &= \{\mathcal{K}_5\} \text{ and so,} \\ U(\mathcal{K}_1, \mathcal{I}_1; \mathcal{I}_2, \theta) &= \{\mathcal{K}_4, \mathcal{K}_5\}. \end{aligned}$$

In Sect. 12.5, we shall see how the language of a dynamic logic can express all of the above.

12.2 The Logics for IS and DIS

In this section, we shall present the logics for information systems and deterministic information systems. The language $\mathbb{L}$ of the logics contains (i) a non-empty countable set $\mathcal{AC}$ of attribute constants, (ii) for each $a \in \mathcal{AC}$, a non-empty finite set $\mathcal{VC}_a$ of attribute value constants (iii) a non-empty countable set PV of propositional variables (iv) the propositional constants $\top, \bot$ and (iv) a special symbol $*$. Atomic wffs are the propositional variables, p from PV, and *descriptors* [84], i.e. (a, v), for each $a \in \mathcal{AC}$, $v \in \mathcal{VC}_a \cup \{*\}$. The set of all descriptors is denoted as $\mathcal{D}$.

Using the Boolean logical connectives $\neg$ (negation) and $\wedge$ (conjunction) and unary modal connectives $[I(B)]$ and $[S(B)]$ for each $B \subseteq \mathcal{AC}$, wffs of $\mathbb{L}$ are then defined recursively as:

$$\top \mid \bot \mid (a, v) \mid p \mid \neg\alpha \mid \alpha \wedge \beta \mid [I(B)]\alpha \mid [S(B)]\alpha.$$

Let $\mathbb{F}$ denote the set of all wffs of $\mathbb{L}$, and $\mathbb{F}_f$, the set of all wffs which do not involve any modal operator $[L(B)]$, $L \in \{I, S\}$, where B is an infinite subset of $\mathcal{AC}$.

Let $B \subseteq_f A$ and $C \subseteq_i A$ denote that B is a finite and C an infinite subset of A respectively.

Semantics. $\mathbb{L}$-semantics is based on a structure of the form $(W, \{R_{I(B)}\}_{B\subseteq\mathcal{AC}}, \{R_{S(B)}\}_{B\subseteq\mathcal{AC}})$ equipped with meaning functions for the descriptors and the propositional variables. Formally,

Definition 68. *A* model *is a tuple* $\mathfrak{M} := (W, \{R_{I(B)}\}_{B\subseteq\mathcal{AC}}, \{R_{S(B)}\}_{B\subseteq\mathcal{AC}}, m, V)$*, where*

- W *is a non-empty set;*
- $R_{L(B)} \subseteq W \times W,\ L \in \{I, S\}$*;*
- $m : \mathcal{D} \to 2^W$*;*
- $V : PV \to 2^W$*.*

We now proceed to define *satisfiability* of a wff α in a model $\mathfrak{M}$ at an object w of the domain W. In brief, this is denoted as $\mathfrak{M}, w \models \alpha$.

Definition 69.

$\mathfrak{M}, w \models \top$ *and* $\mathfrak{M}, w \not\models \bot$.
$\mathfrak{M}, w \models (a, v)$ *if and only if* $w \in m(a, v)$*, for* $(a, v) \in \mathcal{D}$.
$\mathfrak{M}, w \models p$*, if and only if* $w \in V(p)$*, for* $p \in PV$.
$\mathfrak{M}, w \models \neg\alpha$*, if and only if* $\mathfrak{M}, w \not\models \alpha$.
$\mathfrak{M}, w \models \alpha \wedge \beta$*, if and only if* $\mathfrak{M}, w \models \alpha$ *and* $\mathfrak{M}, w \models \beta$.
$\mathfrak{M}, w \models [L(B)]\alpha,\ L \in \{I, S\}$*, if and only if for all* w' *in* W *with* $(w, w') \in R_{L(B)}$, $\mathfrak{M}, w' \models \alpha$.

For any wff α *in* $\mathbb{F}$ *and model* $\mathfrak{M}$*, let* $[\![\alpha]\!]_{\mathfrak{M}} := \{w \in W : \mathfrak{M}, w \models \alpha\}$.
α *is* valid in $\mathfrak{M}$*, denoted* $\mathfrak{M} \models \alpha$*, if and only if* $[\![\alpha]\!]_{\mathfrak{M}} = W$.
α *is* valid*, if* $\mathfrak{M} \models \alpha$ *for every model* $\mathfrak{M}$*. It will be denoted by* $\models \alpha$.

Notation 28. For $b \in \mathcal{AC}$, we will simply write $R_{L(b)}$ and $[L(b)]$ instead of writing $R_{L(\{b\})}$ and $[L(\{b\})]$ respectively, $L \in \{I, S\}$.

For obtaining a logic for information systems, we require to impose some properties on the structure defined above.

Definition 70. *An* IS-structure *is defined as a tuple* $\mathfrak{F} := (W, \{R_{I(B)}\}_{B\subseteq\mathcal{AC}}, \{R_{S(B)}\}_{B\subseteq\mathcal{AC}}, m)$*, where* W*,* R_B *and* m *are as in Definition 68, and the following are satisfied.*

(IS1) *For each* $a \in \mathcal{AC}$, $\bigcup\{m(a, v) : v \in \mathcal{VC}_a \cup \{*\}\} = W$.
(IS2) *For each* $a \in \mathcal{AC}$, $m(a, v) \cap m(a, v') = \emptyset$*, for* $v \neq v'$.
(IS3) *For each* $B \subseteq_i \mathcal{AC}$, $R_{I(B)}$ *is an equivalence relation.*
(IS4) *For each* $B \subseteq_i \mathcal{AC}$, $R_{S(B)}$ *is a symmetric and reflexive relation.*
(IS5) *For each* $B \subseteq_i \mathcal{AC}$, $R_{I(B)} \subseteq R_{S(B)}$.
(IS6) $R_{I(\emptyset)} = R_{S(\emptyset)} = W \times W$.
(IS7) $R_{L(B)} \subseteq R_{L(C)}$ *for* $C \subseteq B \subseteq \mathcal{AC}$ *and* $L \in \{I, S\}$.
(IS8) *For* $B \subseteq \mathcal{AC}$ *and* $b \in \mathcal{AC}$, $R_{L(B)} \cap R_{L(b)} \subseteq R_{L(B\cup\{b\})}$, $L \in \{I, S\}$.
(IS9) *For* $b \in \mathcal{AC}$, $(w, w') \in R_{I(b)}$ *if and only if there exists* $v \in \mathcal{VC}_b \cup \{*\}$ *such that* $w, w' \in m(b, v)$.

(IS10) *For $b \in \mathcal{AC}$, $(w, w') \in R_{S(b)}$ if and only if any of the following holds.*
 (a) $w, w' \in m(b, v)$ for some $v \in \mathcal{VC}_b$.
 (b) $w \in m(b, *)$.
 (c) $w' \in m(b, *)$.

In the special case of *deterministic* information systems, we have

Definition 71. *An* IS-*structure* $\mathfrak{F} := (W, \{R_{I(B)}\}_{B \subseteq \mathcal{AC}}, \{R_{S(B)}\}_{B \subseteq \mathcal{AC}}, m)$ *will be called a* DIS-structure *if it satisfies:*

(DIS1) *For all $a \in \mathcal{AC}$, $m(a, *) = \emptyset$.*
(DIS2) $R_{I(B)} = R_{S(B)}$, $B \subseteq_i \mathcal{AC}$.

Models based on IS and DIS structures will be called IS *and* DIS *models* respectively.

Observation 6.

(a) In case of IS-models, the modal operators $[L(\emptyset)]$, $L \in \{I, S\}$ are interpreted as the global modal operator [9].
(b) $\mathfrak{M}, w \models [I(B)]\alpha$ and $\mathfrak{M}, w \models [S(B)]\alpha$, if and only if w is in the lower approximation of the set $[\![\alpha]\!]_{\mathfrak{M}}$ with respect to the relations $R_{I(B)}$ and $R_{S(B)}$ respectively.

By (IS7) and (IS8), we obtain $R_{L(B)} \cap R_{L(b)} = R_{L(B \cup \{b\})}$, and so we have

$$R_{L(B)} = \bigcap_{b \in B} R_{L(b)} \tag{30}$$

for finite $B, L \in \{I, S\}$. Note that the conditions (IS3)–(IS5) are for infinite subsets. But one can obtain these for finite B from the remaining conditions. Moreover, in the definition of IS-structure, one can replace the condition $(IS8)$ by the conditions (IS8(a)) and (IS8(b)), where

(IS8(a)) for $B \subseteq \mathcal{AC}$ and $b \in \mathcal{AC}$, if $(w, w') \in R_{I(B)}$ and there exists $v \in \mathcal{VC}_b \cup \{*\}$ such that $w, w' \in m(b, v)$, then $(w, w') \in R_{I(B \cup \{b\})}$,
(IS8(b)) for $B \subseteq \mathcal{AC}$ and $b \in \mathcal{AC}$, if $(w, w') \in R_{S(B)}$ and either
 (a) $w, w' \in m(b, v)$ for some $v \in \mathcal{VC}_b$, or
 (b) $w \in m(b, *)$, or
 (c) $w' \in m(b, *)$,
 then $(w, w') \in R_{S(B \cup \{b\})}$.

(IS8(a)),(IS8(b)) are useful for getting the axiomatization of the logics for IS-structures, as we shall see in Sect. 12.3.

For DIS-structures, from (DIS1), (IS9), (IS10) and (30), we obtain $R_{I(B)} = R_{S(B)}$ for finite B, but the equality may not hold for infinite B. Thus we need (DIS2), and obtain $R_{I(B)} = R_{S(B)}$ for all $B \subseteq \mathcal{AC}$.

A DIS-structure is thus effectively of the form $(W, \{R_{I(B)}\}_{B \subseteq \mathcal{AC}}, m)$.

Given an IS $\mathcal{S} := (W, \mathcal{AC}, \bigcup_{a \in \mathcal{AC}} \mathcal{VC}_a \cup \{*\}, f)$, the structure

$$\mathfrak{F}_{\mathcal{S}} := (W, \{Ind_{\mathcal{S}}(B)\}_{B \subseteq \mathcal{AC}}, \{Sim_{\mathcal{S}}(B)\}_{B \subseteq \mathcal{AC}}, m^{\mathcal{S}}),$$

where $m^{\mathcal{S}}(a, v) := \{w \in W : f(w, a) = v\}$, is an IS-structure. We shall call it the *standard* IS-*structure* generated by $\mathcal{S}$, following Vakarelov [100].
Note that we can have an IS-structure $\mathfrak{F} := (W, \{R_{I(B)}\}_{B \subseteq \mathcal{AC}}, \{R_{S(B)}\}_{B \subseteq \mathcal{AC}}, m)$ which does not satisfy the equality

$$R_{L(B)} = \bigcap_{b \in B} R_{L(b)},\ L \in \{I, S\}. \tag{31}$$

Such an IS-structure cannot be standard and hence all IS-structures are not standard. But for an IS-structure $(W, \{R_{I(B)}\}_{B \subseteq \mathcal{AC}}, \{R_{S(B)}\}_{B \subseteq \mathcal{AC}}, m)$, the IS $\mathcal{S} := (W, \mathcal{AC}, \bigcup_{a \in \mathcal{AC}} \mathcal{VC}_a \cup \{*\}, f)$, where $f(x, a) = v$ if and only if $x \in m(a, v)$, is such that (i) $m^{\mathcal{S}} = m$, (ii) $Ind_{\mathcal{S}}(B) = R_{I(B)}$, provided $R_{I(B)} = \bigcap_{b \in B} R_{I(b)}$ and (iii) $Sim_{\mathcal{S}}(B) = R_{S(B)}$, provided $R_{S(B)} = \bigcap_{b \in B} R_{S(b)}$ for each $B \subseteq \mathcal{AC}$. Thus it follows that an IS-structure $(W, \{R_{I(B)}\}_{B \subseteq \mathcal{AC}}, \{R_{S(B)}\}_{B \subseteq \mathcal{AC}}, m)$ satisfying $R_{L(B)} = \bigcap_{b \in B} R_{L(b)},\ L \in \{I, S\}$ for all $B \subseteq \mathcal{AC}$ is a standard IS-structure.

In a manner identical to the case of IS-structures, we define the *standard* DIS-*structure* generated by a DIS, and the arguments given above also hold for these structures. Let *standard* IS-*models* (SIS-models) and *standard* DIS-*models* ($SDIS$-models) be the models based on standard IS-structures and standard DIS-structures respectively. For $k \in \{IS, SIS, DIS, SDIS\}$, we shall write $\models_k \alpha$, if α is valid in all k-models.
Let $\alpha \in \mathbb{F}_f$. As in any IS-structure, any finite set B of attributes satisfies (31), we have

Proposition 90.

(i) $\models_{IS} \alpha$ *if and only if* $\models_{SIS} \alpha$ *for all* $\alpha \in \mathbb{F}_f$.
(ii) $\models_{DIS} \alpha$ *if and only if* $\models_{SDIS} \alpha$ *for all* $\alpha \in \mathbb{F}_f$.

Proof. We only prove (i). Same arguments will also hold for (ii). One direction is obvious. To prove the other direction, it is enough to show that if a wff α is satisfiable in an IS-models, then it is also satisfiable in some SIS-model. Let $\mathfrak{M} := (\mathfrak{F}, V)$ be an IS-model such that

$$\mathfrak{M}, w \models \alpha, \tag{32}$$

for some w. Let $\mathfrak{F} := (W, \{R_{I(B)}\}_{B \subseteq \mathcal{AC}}, \{R_{S(B)}\}_{B \subseteq \mathcal{AC}}, m)$. Consider the SIS-structure $\mathfrak{F}' := (W, \{R'_{I(B)}\}_{B \subseteq \mathcal{AC}}, \{R'_{S(B)}\}_{B \subseteq \mathcal{AC}}, m)$, where

$$R'_{L(B)} = \bigcap_{b \in B} R_{L(\{b\})}.$$

Obviously, $R'_{L(B)} = R_{L(B)}$ for finite B. Then one can easily show, using (32), that $\mathfrak{M}', w \models \alpha$, where $\mathfrak{M}' := (\mathfrak{F}', V)$. □

Let $\emptyset \neq B := \{b_1, b_2, \ldots, b_n\} \subseteq_f \mathcal{AC}$. Let $\mathcal{D}_B$ be the set of all B-*basic* wffs [84], i.e. wffs of the form $(b_1, v_1) \wedge (b_2, v_2) \wedge \ldots \wedge (b_n, v_n)$, $v_i \in \mathcal{VC}_{b_i} \cup \{*\}, i = 1, 2, \ldots n$. In the case when $B = \emptyset$, we take $\mathcal{D}_B := \{\top\}$. Let us consider an IS-model $\mathfrak{M} := (W, \{R_{I(B)}\}_{B \subseteq \mathcal{AC}}, \{R_{S(B)}\}_{B \subseteq \mathcal{AC}}, m, V)$. Then each element of the set $\mathcal{D}_B$, $B \neq \emptyset$ represents the empty set or an equivalence class with respect to the equivalence relation $R_{I(B)}$. In the latter case, in fact, $\bigwedge_{i=1}^{n}(b_i, v_i)$ represents the equivalence class of objects which take the value v_i for the attribute $b_i, i = 1, 2, \ldots, n$. More formally, we have the following proposition.

Proposition 91. *Let* $\mathfrak{M} := (W, \{R_{I(B)}\}_{B \subseteq \mathcal{AC}}, \{R_{S(B)}\}_{B \subseteq \mathcal{AC}}, m, V)$ *be an* IS-*model. Then*

$$\{[\![\alpha]\!]_{\mathfrak{M}} : \alpha \in \mathcal{D}_B\} \setminus \{\emptyset\} = \{[w]_{R_{I(B)}} : w \in W\}.$$

Proof. Let us take $[\![\alpha]\!]_{\mathfrak{M}} \neq \emptyset$ for a $\alpha \in \mathcal{D}_B$. Let α be the B-basic wff $(b_1, v_1) \wedge \cdots \wedge (b_n, v_n)$. We need to show $[\![\alpha]\!]_{\mathfrak{M}} = [w]_{R_{I(B)}}$ for some $w \in W$. Let us take a $w \in [\![\alpha]\!]_{\mathfrak{M}}$. Our claim is that $[\![\alpha]\!]_{\mathfrak{M}} = [w]_{R_{I(B)}}$. In fact,

$$\begin{aligned} w' \in [\![\alpha]\!]_{\mathfrak{M}} &\Leftrightarrow w, w' \in m(b_i, v_i) \text{ for all } i \in \{1, 2, \ldots, n\} \\ &\Leftrightarrow (w, w') \in R_{I(b_i)} \text{ for all } i \in \{1, 2, \ldots, n\} \\ &\Leftrightarrow (w, w') \in R_{I(B)} \\ &\Leftrightarrow w' \in [w]_{R_{I(B)}}. \end{aligned}$$

Next, we prove the other inclusion. So let $w \in W$. We need to find $\alpha \in \mathcal{D}_B$ such that $[\![\alpha]\!]_{\mathfrak{M}} = [w]_{R_{I(B)}}$. We note that there exists a unique $v_i \in \mathcal{VC}_{b_i} \cup \{*\}$, such that $w \in m(b_i, v_i)$, $i = 1, 2, \ldots, n$. Take α to be $(b_1, v_1) \wedge \cdots \wedge (b_n, v_n)$. As above, one can show that $[\![\alpha]\!]_{\mathfrak{M}} = [w]_{R_{I(B)}}$. □

Here we note that if models $\mathfrak{M}$ and $\mathfrak{M}'$ differ only on the valuation V, then $[\![\alpha]\!]_{\mathfrak{M}} = [\![\alpha]\!]_{\mathfrak{M}'}$ for all $\alpha \in \mathcal{D}_B$.

The next propositions show how the wffs of $\mathbb{F}$ and $\mathbb{F}_f$ may be used to express the concepts presented in Sect. 2.1 related to dependencies in data and data reduction.

Let $\mathcal{S} := (W, \mathcal{AC}, \bigcup_{a \in \mathcal{AC}} \mathcal{VC}_a, f)$ be a DIS and consider the corresponding standard DIS-structure $\mathfrak{F}_{\mathcal{S}}$.

Proposition 92. *Let $P, Q, S \subseteq \mathcal{AC}$, and p, q be distinct propositional variables. Then the following hold.*

1. *$P \Rightarrow Q$ if and only if $[I(Q)]p \rightarrow [I(P)]p$ is valid in $\mathfrak{F}_{\mathcal{S}}$, i.e. $\mathfrak{M} \models [I(Q)]p \rightarrow [I(P)]p$, for all models $\mathfrak{M}$ based on $\mathfrak{F}_{\mathcal{S}}$.*
2. *$P \not\equiv Q$ if and only if $\neg[I(\emptyset)]([I(Q)]p \rightarrow [I(P)]p) \wedge \neg[I(\emptyset)]([I(P)]q \rightarrow [I(Q)]q)$ is satisfiable in $\mathfrak{F}_{\mathcal{S}}$, i.e. there is $\mathfrak{M}$ based on $\mathfrak{F}_{\mathcal{S}}$, and $w \in W$ where the wff is satisfied.*
3. *$b \in P$ is dispensable in P if and only if $[I(P)]p \leftrightarrow [I(P \setminus \{b\})]p$ is valid in $\mathfrak{F}_{\mathcal{S}}$.*
4. *$P \subseteq_f \mathcal{AC}$ is dependent if and only if $\bigvee_{b \in P}[I(\emptyset)]([I(P)]p_b \leftrightarrow [I(P \setminus \{b\})]p_b)$ is valid in $\mathfrak{F}_{\mathcal{S}}$, where $\{p_b : b \in P\}$ is a set of distinct propositional variables.*

5. *$Q \subseteq_f P$ is a reduct of P if and only if $\bigwedge_{b \in Q}\langle I(\emptyset)\rangle\neg([I(Q)]q_b \leftrightarrow [I(Q \setminus \{b\})]q_b)$ is satisfiable in $\mathfrak{F}_S$ and $[I(Q)]p \leftrightarrow [I(P)]p$ is valid in $\mathfrak{F}_S$.*

Proof. We provide the proofs of 2 and 4.
2. Let $P \not\equiv Q$. Then $Ind(P) \not\subseteq Ind(Q)$ and $Ind(Q) \not\subseteq Ind(P)$. Suppose $(x_1, y_1) \in Ind(P)$, but $(x_1, y_1) \notin Ind(Q)$ and $(x_2, y_2) \in Ind(Q)$, but $(x_2, y_2) \notin Ind(P)$. Consider a V such that $V(p) := [x_1]_{Ind(Q)}$ and $V(q) := [x_2]_{Ind(P)}$. Then, we obtain $\mathfrak{M}, x_1 \not\models [I(Q)]p \rightarrow [I(P)]p$ and $\mathfrak{M}, x_2 \not\models [I(P)]q \rightarrow [I(Q)]q$, where $\mathfrak{M} := (\mathfrak{F}_S, V)$. Thus we obtain $\mathfrak{M}, w \models \neg[I(\emptyset)]([I(Q)]p \rightarrow [I(P)]p) \wedge \neg[I(\emptyset)]([I(P)]q \rightarrow [I(Q)]q)$ for all w.

Conversely, suppose $\mathfrak{M}, w \models \neg[I(\emptyset)]([I(Q)]p \rightarrow [I(P)]p) \wedge \neg[I(\emptyset)]([I(P)]q \rightarrow [I(Q)]q)$ for some $\mathfrak{M} := (\mathfrak{F}_S, V)$ and w. Then there exist x_1 and x_2 such that $\mathfrak{M}, x_1 \not\models [I(Q)]p \rightarrow [I(P)]p$ and $\mathfrak{M}, x_2 \not\models [I(P)]q \rightarrow [I(Q)]q$. Thus, we must have y_1, y_2 such that $(x_1, y_1) \in Ind(P)$, $(x_1, y_1) \notin Ind(Q)$, $(x_2, y_2) \in Ind(Q)$ and $(x_2, y_2) \notin Ind(P)$. Thus $P \not\equiv Q$.

4. First suppose P is dependent and consider a model $\mathfrak{M} := (\mathfrak{F}_S, V)$ and an object w. We need to show $\mathfrak{M}, w \models \bigvee_{b \in P}[I(\emptyset)]([I(P)]p_b \leftrightarrow [I(P \setminus \{b\})]p_b)$. Since P is dependent, there exists $b \in P$ such that $Ind(P) = Ind(P \setminus \{b\})$. Therefore, we have $\mathfrak{M}, w' \models [I(P)]p_b \leftrightarrow [I(P \setminus \{b\})]p_b)$ for all w' and hence $\mathfrak{M}, w \models \bigvee_{b \in P}[I(\emptyset)]([I(P)]p_b \leftrightarrow [I(P \setminus \{b\})]p_b)$.

Conversely, let $\bigvee_{b \in P}[I(\emptyset)]([I(P)]p_b \leftrightarrow [I(P \setminus \{b\})]p_b)$ be valid in $\mathfrak{F}_S$. If possible, let P not be dependent. Then $Ind(P) \subsetneq Ind(P \setminus \{b\})$, for all $b \in P$. Therefore, for all $b \in P$, there exist x_b, y_b such that $(x_b, y_b) \in Ind((P \setminus \{b\})$ but $(x_b, y_b) \notin Ind(P)$. Consider a valuation V such that $V(p_b) = [x_b]_{Ind(P)}$ and let $\mathfrak{M} := (\mathfrak{F}_S, m, V)$ for some m. Then we obtain $\mathfrak{M}, x_b \not\models [I(P)]p_b \leftrightarrow [I(P \setminus \{b\})]p_b$ and hence for all w, $\mathfrak{M}, w \not\models [I(\emptyset)][I(P)]p_b \leftrightarrow [I(P \setminus \{b\})]p_b$ for all $b \in P$. This contradicts that $\bigvee_{b \in P}[I(\emptyset)]([I(P)]p_b \leftrightarrow [I(P \setminus \{b\})]p_b)$ is valid in $\mathfrak{F}_S$. □

Proposition 93. *Let $\mathfrak{M} := (\mathfrak{F}_S, V)$ be the standard DIS-model on $\mathfrak{F}_S$, for some valuation function V. Let $P, Q, S \subseteq_f AC$. Then the following hold.*

1. *$[\![\bigvee_{\alpha \in \mathcal{D}_Q}[I(P)]\alpha]\!]_{\mathfrak{M}} = POS_P(Q)$.*
2. *$b \in P$ is Q-dispensable in P if and only if $\bigvee_{\alpha \in \mathcal{D}_Q}[I(P)]\alpha \leftrightarrow \bigvee_{\alpha \in \mathcal{D}_Q}[I(P \setminus \{b\})]\alpha$ is valid in $\mathfrak{M}$.*
3. *$b \in P$ is Q-indispensable in P if and only if $\langle I(\emptyset)\rangle\neg(\bigvee_{\alpha \in \mathcal{D}_Q}[I(P)]\alpha \leftrightarrow \bigvee_{\alpha \in \mathcal{D}_Q}[I(P \setminus \{b\})]\alpha)$ is valid in $\mathfrak{M}$.*
4. *P is Q-independent in P if and only if $\bigwedge_{b \in P}\langle I(\emptyset)\rangle\neg(\bigvee_{\alpha \in \mathcal{D}_Q}[I(P)]\alpha \leftrightarrow \bigvee_{\alpha \in \mathcal{D}_Q}[I(P \setminus \{b\})]\alpha)$ is valid in $\mathfrak{M}$.*
5. *$S \subseteq P$ is Q-reduct of P if and only if $[I(\emptyset)](\bigwedge_{b \in S}\langle I(\emptyset)\rangle\neg(\bigvee_{\alpha \in \mathcal{D}_Q}[S]\alpha \leftrightarrow \bigvee_{\alpha \in \mathcal{D}_Q}[S \setminus \{b\}]\alpha) \wedge (\bigvee_{\alpha \in \mathcal{D}_Q}[I(P)]\alpha \leftrightarrow \bigvee_{\alpha \in \mathcal{D}_Q}[S]\alpha))$ is valid in $\mathfrak{M}$.*

Proof. We provide the proofs of 1, 2 and 4.

1. $x \in [\![\bigvee_{\alpha \in \mathcal{D}_Q} [I(P)]\alpha]\!]_{\mathfrak{M}}$
   $x \in [\![[I(P)]\alpha]\!]_{\mathfrak{M}}$ for some $\alpha \in \mathcal{D}_Q$
   $\Leftrightarrow x \in \underline{([\![\alpha]\!]_{\mathfrak{M}})}_{Ind(P)}$ for some $\alpha \in \mathcal{D}_Q$ (by Observation 6)
   $\Leftrightarrow x \in \underline{X}_{Ind(P)}$ for some $X \in U/Ind(Q)$ (by Proposition 91)
   $\Leftrightarrow x \in POS_P(Q)$.
2. $POS_P(Q) = POS_{P \setminus \{b\}}(Q)$
   $\Leftrightarrow [\![\bigvee_{\alpha \in \mathcal{D}_Q} [I(P)]\alpha]\!]_{\mathfrak{M}} = [\![\bigvee_{\alpha \in \mathcal{D}_Q} [I(P \setminus \{b\})]\alpha]\!]_{\mathfrak{M}}$ (using (1))
   $\Leftrightarrow \mathfrak{M} \models \bigvee_{\alpha \in \mathcal{D}_Q} [I(P)]\alpha \leftrightarrow \bigvee_{\alpha \in \mathcal{D}_Q} [I(P \setminus \{b\})]\alpha$.
4. $b \in P$ is $Q-$ *independent* in P
   $\Leftrightarrow POS_P(Q) \neq POS_{P \setminus \{b\}}(Q)$
   $\Leftrightarrow [\![\bigvee_{\alpha \in \mathcal{D}_Q} [I(P)]\alpha]\!]_{\mathfrak{M}} \neq [\![\bigvee_{\alpha \in \mathcal{D}_Q} [I(P \setminus \{b\})]\alpha]\!]_{\mathfrak{M}}$
   $\Leftrightarrow \mathfrak{M}, x \models \neg(\bigvee_{\alpha \in \mathcal{D}_Q} [I(P)]\alpha \leftrightarrow \bigvee_{\alpha \in \mathcal{D}_Q} [I(P \setminus \{b\})]\alpha)$ for some $x \in W$
   $\Leftrightarrow \mathfrak{M} \models \langle I(\emptyset)\rangle\neg(\bigvee_{\alpha \in \mathcal{D}_Q} [I(P)]\alpha \leftrightarrow \bigvee_{\alpha \in \mathcal{D}_Q} [I(P \setminus \{b\})]\alpha)$. □

We shall show in Proposition 95 that the expressive power of $\mathbb{F}$ will not be affected even if we take $[L(B)]$, $L \in \{I, S\}$, where $B = \emptyset$ or an infinite subset of $\mathcal{AC}$, as the only modal operators in the language. But before that, let us give the following simple fact about IS-structures and similarity relations.

Proposition 94. *Let* $\mathfrak{F} := (W, \{R_{I(B)}\}_{B \subseteq \mathcal{AC}}, \{R_{S(B)}\}_{B \subseteq \mathcal{AC}}, m)$ *be an* IS *structure. Then* $(x, y) \in R_{S(B)}$ *if and only if there exist* $C \subseteq B, F \subseteq B \setminus C$ *and* $\beta \in \mathcal{D}_C$ *such that* $x, y \in [\![\beta]\!]_{\mathfrak{M}}$, $x \in m(a, *)$, $y \in m(b, *)$ *for all* $a \in F$ *and* $b \in (B \setminus C) \setminus F$.

Proof. Let $(x, y) \in R_{S(B)}$, and consider the following sets.
$C := \{a \in B : x, y \in m(a, v) \text{ for some } v \in \mathcal{VC}_a \cup \{*\}\}$;
$F := \{a \in B \setminus C : x \in m(a, *)\}$.
Suppose that $\beta \in \mathcal{D}_C$ is such that $\beta := \bigwedge_{x \in m(b,v)} (b, v)$.

We shall prove that for all $b \in (B \setminus C) \setminus F$, we have $y \in m(b, *)$. First note that since $(x, y) \in R_{S(B)} \subseteq R_{S(\{b\})}$ and $b \notin C$, so we must have either $x \in m(b, *)$ or $y \in m(b, *)$. But as $b \notin F$, $y \in m(b, *)$. Conversely, suppose the right hand side holds. We need to prove $(x, y) \in R_{S(B)}$. Let us take an arbitrary $b \in B$. It is enough to show that $(x, y) \in R_{S(\{b\})}$. But this can easily be verified by considering the three cases when b belongs to F, or C, or $(B \setminus C) \setminus F$. □

For $B \subseteq_f \mathcal{AC}$, let us denote by $(B, *)$, the wff $\bigwedge_{b \in B} (b, *)$ or $\top$ according as B is non-empty or empty respectively.

Proposition 95. *Let* $B \subseteq_f \mathcal{AC}$. *The following wffs are valid in all* IS*-models.*

(a) $[I(B)]\alpha \leftrightarrow \bigwedge_{\beta \in \mathcal{D}_B} (\beta \to [I(\emptyset)](\beta \to \alpha))$.
(b) $[S(B)]\alpha \leftrightarrow \bigwedge_{C \subseteq B} \bigwedge_{F \subseteq B \setminus C} \bigwedge_{\beta \in \mathcal{D}_C} \Big(\beta \wedge (F, *) \to$
$$[I(\emptyset)]((((B \setminus C) \setminus F, *) \wedge \beta) \to \alpha)\Big).$$

Proof. (*a*). One can prove the validity in all IS-models using Proposition 91.
(*b*). When $B = \emptyset$, the wff $[S(B)]\alpha \leftrightarrow \bigwedge_{C\subseteq B}\bigwedge_{F\subseteq B\setminus C}\bigwedge_{\beta\in\mathcal{D}_C}(\beta \wedge (F,*) \rightarrow [I(\emptyset)]((((B\setminus C)\setminus F,*)\wedge\beta) \rightarrow \alpha))$ reduces to $[S(\emptyset)]\alpha \leftrightarrow (\top \rightarrow [I(\emptyset)](\top \rightarrow \alpha))$, which is obviously valid in all IS-models. So, let us prove the validity of the wff $[S(B)]\alpha \leftrightarrow \bigwedge_{C\subseteq B}\bigwedge_{F\subseteq B\setminus C}\bigwedge_{\beta\in\mathcal{D}_C}(\beta\wedge(F,*) \rightarrow [I(\emptyset)]((((B\setminus C)\setminus F,*)\wedge\beta) \rightarrow \alpha))$ in all IS-models for $B \neq \emptyset$.

First suppose $\mathfrak{M}, w \models [S(B)]\alpha$. Let $C \subseteq B$, $F \subseteq B\setminus C$, $\beta \in \mathcal{D}_C$ and $w' \in W$ such that the following holds.
(i) $\mathfrak{M}, w \models \beta$; (ii) $\mathfrak{M}, w \models (F,*)$; (iii) $\mathfrak{M}, w' \models ((B\setminus C)\setminus F,*)$; (iv) $\mathfrak{M}, w' \models \beta$.

We need to show $\mathfrak{M}, w' \models \alpha$. But using Proposition 94 and the above conditions (i)–(iv), we obtain $(w,w') \in R_{S(B)}$, and hence the desired result.

The other direction can similarly be proved, again using Proposition 94. □

Proposition 95 shows that for every wff α, there exists a wff α' such that $\alpha \leftrightarrow \alpha'$ is valid in all IS-models, and α' does not involve any modal operator $[L(B)], L \in \{I,S\}$, where $B(\neq \emptyset) \subseteq_f \mathcal{AC}$. Here, we note that the complexity of α', denoted as $|\alpha'|$, could be very large compared to α. For instance, if α is of the form $\underbrace{[I(B)][I(B)]\ldots[I(B)]}_{n-times}\beta$, where $|\mathcal{D}_B| = m$, then $|\alpha'| > m^n|\beta|$. The situation becomes worse when α is of the form $\underbrace{[S(B)][S(B)]\ldots[S(B)]}_{n-times}\beta$.

One can think of strengthening this result by requiring α' to be such that it does not even involve the modal operator $[L(\emptyset)]$, $L \in \{I,S\}$. This is not possible as shown by the following example.

Example 21. Let $b \in \mathcal{AC}$ and $v_b^1, v_b^2 \in \mathcal{VC}_b$. For each $a\ (\neq b) \in \mathcal{AC}$, let $v_a^1, v_a^2, v_a^3 \in \mathcal{VC}_a$. Now consider the standard IS-structures $\mathfrak{F}_{\mathcal{S}}$ and $\mathfrak{F}_{\mathcal{S}'}$ generated respectively from the ISs $\mathcal{S}$ and $\mathcal{S}'$ given below (Table 8).

Let us consider the models $\mathfrak{M} := (\mathfrak{F}_{\mathcal{S}}, V)$ and $\mathfrak{M}' := (\mathfrak{F}_{\mathcal{S}'}, V)$, for any V. For $c \neq b$, we see that $\mathfrak{M}, x \models [I(b)](c, v_c^1)$, while $\mathfrak{M}', x \not\models [I(b)](c, v_c^1)$.

But one can show that for any α which does not involve *any* modal operator $[L(F)]$, $F \subseteq_f \mathcal{AC}$, $L \in \{I,S\}$,

$$\mathfrak{M}, x \models \alpha \text{ if and only if } \mathfrak{M}', x \models \alpha.$$

So the wff $[I(b)](c, v_c^1)$ cannot be logically equivalent to a wff that does not contain any modal operator $[L(F)]$ with $F \subseteq_f \mathcal{AC}$. Similarly, $[S(b)]((c, v_c^1) \vee (c, v_c^3))$,

Table 8. ISs $\mathcal{S}$ and $\mathcal{S}'$

	$\mathcal{S}$		$\mathcal{S}'$	
	$a\ (\neq b)$	b	$a\ (\neq b)$	b
x	v_a^1	v_b^1	v_a^1	v_b^1
y	v_a^2	v_b^2	v_a^3	$*$
z	v_a^3	$*$	v_a^2	v_b^1

$c \neq b$, cannot be logically equivalent to such a wff as $\mathfrak{M}, x \models [S(b)]((c, v_c^1) \vee (c, v_c^3))$ and $\mathfrak{M}', x \not\models [S(b)]((c, v_c^1) \vee (c, v_c^3))$.

One can also use a deterministic information system to prove the above fact.

We end this section by illustrating the satisfiability/validity of some wffs, the latter referring to any of the classes of IS, DIS, standard IS and standard DIS-structures.

Proposition 96.

1. *$[S(B)]p \leftrightarrow ((a_1, v_1) \wedge (a_2, v_2) \wedge \cdots (a_n, v_n))$ is satisfiable.*
2. *$\neg[I(B)]p \leftrightarrow (\bigwedge_{\alpha \in \mathcal{D}_B}(\alpha \rightarrow \langle I(\emptyset)\rangle(\alpha \wedge \neg p)))$ is valid, $B \subseteq_f \mathcal{AC}$.*
3. *$\bigwedge_{i \in \{1,2,\ldots,n\}}(b_i, v_i) \leftrightarrow [I(B)](\bigwedge_{i \in \{1,2,\ldots,n\}}(b_i, v_i))$, $B := \{b_1, b_2, \ldots, b_n\}$, is valid.*
4. *$\bigwedge_{i \in \{1,2,\ldots,n+1\}}\langle I(\emptyset)\rangle(p_i \wedge (b, v)) \rightarrow \bigvee_{i \neq j}\langle I(\emptyset)\rangle(p_i \wedge p_j)$ is satisfiable.*
5. *$[I(\emptyset)][I(B)](a, v) \leftrightarrow [I(\emptyset)](a, v)$ is valid.*
6. *$\langle I(B)\rangle\delta \rightarrow [I(B)]\delta$ is valid, where δ is a wff obtained by applying only Boolean connectives on descriptors (b, v), $b \in B$.*
7. *$[I(\emptyset)]\neg(a, *) \rightarrow ([I(a)]\alpha \leftrightarrow [S(a)]\alpha)$ is valid.*

If the wff in 1 is valid in an SIS-model $\mathfrak{M} := (\mathfrak{F}_S, V)$, then an object $x \in \underline{V(p)}_{Ind_{S'}(B)}$ in all the completions S' of S if and only if it takes the value v_i for the attribute a_i, $1 \leq i \leq n$ (cf. Proposition 2). Validity of the wff in 2 means that an object x is not a positive element of a set X with respect to an attribute set, say B, if and only if there exists an object y which takes the same attribute value as x for each attribute of B but $y \notin X$. The wff in 3 represents the fact that an object x takes the value v_i for the attributes b_i, $1 \leq i \leq n$, if and only if every object $R_{\{b_1, b_2, \ldots, b_n\}}$-related to x also takes the value v_i for the attributes b_i, $1 \leq i \leq n$. The wff in 4 is valid only in an IS-structure where $|m(b, v)| \leq n$. The wff in 6 represents the fact that any property defined using only Boolean connectives and attributes from the set B is definable with respect to $Ind(B)$. The wff in 7 says that if we have complete information about each object regarding an attribute a, then $Ind(\{a\}) = Sim(\{a\})$.

12.3 Axiomatization for the Logics of Information Systems

In this section we present axiomatic systems and prove the soundness and completeness theorems with respect to different classes of models. Let $B, C \subseteq \mathcal{AC}$, $F \in \{P : P \subseteq_i \mathcal{AC}\} \cup \{\emptyset\}$, $v, v' \in \mathcal{VC}_a \cup \{*\}$, $u \in \mathcal{VC}_a$ and $L \in \{I, S\}$.

Axiom schema:

Ax1. All axioms of classical propositional logic.
Ax2 $[L(B)](\alpha \rightarrow \beta) \rightarrow ([L(B)]\alpha \rightarrow [L(B)]\beta)$.
Ax3 $[L(F)]\alpha \rightarrow \alpha$.
Ax4 $\langle L(F)\rangle[L(F)]\alpha \rightarrow \alpha$.
Ax5 $[I(F)][I(F)]\alpha \rightarrow [I(F)]\alpha$.

Ax6 $[L(C)]\alpha \rightarrow [L(B)]\alpha$ for $C \subseteq B \subseteq \mathcal{AC}$.
Ax7 $(a, v) \rightarrow \neg(a, v')$, for $v \neq v'$.
Ax8 $\bigvee_{v \in \mathcal{VC}_a \cup \{*\}} (a, v)$.
Ax9 $(a, v) \rightarrow [I(a)](a, v)$.
Ax10 $((a, v) \wedge [I(B \cup \{a\})]\alpha) \rightarrow [I(B)]((a, v) \rightarrow \alpha)$.
Ax11 $(a, u) \rightarrow [S(a)]((a, u) \vee (a, *))$.
Ax12 $((a, u) \wedge [S(B \cup \{a\})]\alpha) \rightarrow [S(B)](((a, u) \vee (a, *)) \rightarrow \alpha)$.
Ax13 $((a, *) \wedge [S(B \cup \{a\})]\alpha) \rightarrow [S(B)]\alpha$.
Ax14 $[S(F)]\alpha \rightarrow [I(F)]\alpha$.
Ax15 $[I(\emptyset)]\alpha \rightarrow [S(\emptyset)]\alpha$.
AxC1 $\neg(a, *)$.
AxC2 $[S(F)]\alpha \leftrightarrow [I(F)]\alpha$.

Rules of inference:

$$N.\ \frac{\alpha}{[L(B)]\alpha} \qquad MP.\ \frac{\alpha \quad \alpha \rightarrow \beta}{\beta}$$

Ax2-Ax5 are the usual ($S5$-)modal axioms. So, for $F \in \{P : P \subseteq_i \mathcal{AC}\} \cup \{\emptyset\}$, $[I(F)]$ satisfies all the $S5$-axioms and $[S(F)]$ satisfies axioms T and B of modal logic. The notion of theoremhood is defined in the usual way. Let LIS and LDIS be the deductive systems consisting of the inference rules N, MP and the axioms Ax1-Ax15 and Ax1-Ax15,AxC1,AxC2 respectively. Similarly, let $\mathrm{LIS_f}$ and $\mathrm{LDIS_f}$ be the deductive systems *in the language* $\mathbb{F}_f$ consisting of the inference rules N, MP and the axioms Ax1-Ax15 and Ax1-Ax15,AxC1,AxC2 respectively. Note that for $F \subseteq_i \mathcal{AC}$, $\alpha \rightarrow \langle I(F)\rangle\alpha$ is not an axiom of the systems IS_f and DIS_f as $\langle I(F)\rangle\alpha \notin \mathbb{F}_f$. For $\Lambda \in \{\mathrm{LIS}, \mathrm{LIS_f}, \mathrm{LDIS}, \mathrm{LDIS_f}\}$, we write $\vdash_\Lambda$ to denote that α is a theorem of the deductive system Λ. Let us denote the set of all wffs of the language of the system $\Lambda \in \{\mathrm{LIS}, \mathrm{LIS_f}, \mathrm{LDIS}, \mathrm{LDIS_f}\}$ by $\mathbb{F}_\Lambda$. So, $\mathbb{F}_{\mathrm{LIS}} = \mathbb{F}_{\mathrm{LDIS}} = \mathbb{F}$ and $\mathbb{F}_{\mathrm{LIS_f}} = \mathbb{F}_{\mathrm{LDIS_f}} = \mathbb{F}_f$. We would also like to mention that for $B \subseteq_f \mathcal{AC}$, it is not necessary to write the $S5$-axioms for the operators $[I(B)]$ and the axioms T and B for $[S(B)]$, as these are theorems of each of the above defined systems.

Ax6-Ax8 correspond to (IS7), (IS2) and (IS1) respectively of Definition 70. Ax9, Ax10 and Ax6 for I establish the relationship between the indiscernibility relation and attribute, attribute-value pairs. Ax10 is the syntactic counterpart for the condition (IS8a). Note that $[I(B \cup b)]\alpha \rightarrow [I(B)]\alpha \wedge [I(b)]\alpha$ would appear to be the counterpart of the condition $R_{I(B)} \cap R_{I(b)} \subseteq R_{I(B \cup \{b\})}$, but in fact, there are IS-models in which it is not valid. Similarly, Ax11-Ax13 and Ax6 for S establish the relationship between the similarity relation and attribute, attribute-value pairs. In fact, Ax12 is the syntactic counterpart for the condition (IS8b). Ax14 and Ax15 relates the indiscernibility and similarity relation. AxC1 and AxC2 corresponds to the property (DIS1) and (DIS2) of DIS-structures respectively.

We observe that the n-agent epistemic logic $S5_n^D$ [22] with knowledge operators K_i ($i = 1, \ldots, n$) and distributed knowledge operators D_G for groups G of agents, is embeddable in LIS, LDIS, LIS$_\mathfrak{f}$ and LDIS$_\mathfrak{f}$ with $|\mathcal{AC}| \geq n$. Suppose $\mathcal{AC} := \{a_1, a_2, \ldots, a_m\}$, $m \geq n$. Then the embedding Ψ fixes propositional variables, and takes $K_i\alpha$ to $[I(a_i)]\ \Psi(\alpha)$ and $D_{\{i_1,i_2,\ldots,i_s\}}\alpha$ to $[I(\{a_{i_1}, a_{i_2}, \ldots, a_{i_s}\})]$ $\Psi(\alpha)$. On the other hand, LIS, LDIS, LIS$_\mathfrak{f}$ and LDIS$_\mathfrak{f}$ are more expressive than $S5_n^D$, having the extra feature of the descriptors. Any study of indiscernibility/similarity relations induced by information systems naturally involves descriptors, as these determine both what value an object will take for an attribute, and also the indiscernibility/similarity relation itself (shown by Ax6, Ax9-Ax13).

To illustrate the proof system, let us give a Hilbert-style proof of a LIS-wff.

Proposition 97. *For $a, b \in \mathcal{AC}$, $\emptyset \neq B \subseteq_f \mathcal{AC}$ $v \in \mathcal{VC}_a$ and $u \in \mathcal{VC}_b$, we have the following.*

Th1 $\vdash_{\text{LIS}} [I(\emptyset)]\neg(a,*) \to ([I(a)]\alpha \leftrightarrow [S(a)]\alpha)$.
Th2 $\vdash_{\text{LIS}} [I(B)]\alpha \to \alpha$.
Th3 $\vdash_{\text{LIS}} (a,v) \wedge (b,u) \leftrightarrow [I(\{a,b\})]((a,v) \wedge (b,u))$.

Proof. Th1: Let $v \in \mathcal{VC}_a$.

1. $\vdash_{\text{LIS}} (a,v) \wedge [I(a)]\alpha \to [I(\emptyset)]((a,v) \to \alpha)$ (Ax10).
2. $\vdash_{\text{LIS}} (a,v) \wedge [I(a)]\alpha \to [S(\emptyset)]((a,v) \to \alpha)$ (Ax15 and PL).
3. $\vdash_{\text{LIS}} (a,v) \wedge [I(a)]\alpha \to [S(a)]((a,v) \to \alpha)$ (Ax6 and PL).
4. $\vdash_{\text{LIS}} (a,v) \wedge [I(a)]\alpha \to ([S(a)](a,v) \to [S(a)]\alpha)$ (Ax2 and PL).
5. $\vdash_{\text{LIS}} (a,v) \wedge [S(a)](a,v) \wedge [I(a)]\alpha \to [S(a)]\alpha)$ (PL).
6. $\vdash_{\text{LIS}} (a,v) \to [S(a)]((a,v) \vee (a,*))$ (Ax11).
7. $\vdash_{\text{LIS}} ((a,v) \wedge [S(a)]\neg(a,*)) \to [S(a)](a,v)$((6) and PL).
8. $\vdash_{\text{LIS}} [I(\emptyset)]\neg(a,*) \to [S(a)]\neg(a,*)$(Ax15, Ax6 and PL).
9. $\vdash_{\text{LIS}} (a,v) \wedge [I(\emptyset)]\neg(a,*) \to [S(a)](a,v)$ ((7), (8) and PL).
10. $\vdash_{\text{LIS}} (a,v) \wedge [I(a)]\alpha \wedge [I(\emptyset)]\neg(a,*) \to [S(a)]\alpha$ ((5), (9) and PL)

Moreover, since $\vdash_{\text{LIS}} (a,*) \wedge [I(\emptyset)]\neg(a,*) \to \bot$, we obtain $\vdash_{\text{LIS}} (a,*) \wedge [I(a)]\alpha \wedge [I(\emptyset)]\neg(a,*) \to [S(a)]\alpha$. Thus, we obtain $\vdash_{\text{LIS}} (a,v) \wedge [I(a)]\alpha \wedge [I(\emptyset)]\neg(a,*) \to [S(a)]\alpha$ for all $v \in \mathcal{VC}_a \cup \{*\}$ and hence, using PL, we obtain

11. $\vdash_{\text{LIS}} (\bigvee_{v \in \mathcal{VC}_a \cup \{*\}}(a,v)) \wedge [I(a)]\alpha \wedge [I(\emptyset)]\neg(a,*) \to [S(a)]\alpha$. (Ax3)
12. $\vdash_{\text{LIS}} [I(\emptyset)]\neg(a,*) \to \neg(a,*)$ (Ax3).
13. $\vdash_{\text{LIS}} \neg(a,*) \to \bigvee_{v \in \mathcal{VC}_a \cup \{*\}}(a,v)$ (Ax8 and PL).
14. $\vdash_{\text{LIS}} [I(\emptyset)]\neg(a,*) \to \bigvee_{v \in \mathcal{VC}_a \cup \{*\}}(a,v)$ ((12), (13) and PL).
15. $\vdash_{\text{LIS}} [I(a)]\alpha \wedge [I(\emptyset)]\neg(a,*) \to [S(a)]\alpha$. ((11), (14) and PL).
16. $\vdash_{\text{LIS}} [I(\emptyset)]\neg(a,*) \to ([I(a)]\alpha \to [S(a)]\alpha)$ (PL).
17. $\vdash_{\text{LIS}} [S(a)]\alpha \to [I(a)]\alpha$ (Ax14).
18. $\vdash_{\text{LIS}} [I(\emptyset)]\neg(a,*) \to ([I(a)]\alpha \leftrightarrow [S(a)]\alpha)$ ((16), (17) and PL).

Th2: First, we prove the result for singleton B. Let $B = \{a\}$.
Let $v \in \mathcal{VC}_a \cup \{*\}$.

1. $\vdash_{\text{LIS}} (a, v) \wedge [I(a)]\alpha \rightarrow ((a, v) \rightarrow \alpha)$ (Ax10, Ax3, PL).
2. $\vdash_{\text{LIS}} (a, v) \wedge [I(a)]\alpha \rightarrow \alpha$ (1, PL).
   Since (2) holds for all $v \in \mathcal{VC}_a \cup \{*\}$, by PL, we obtain,
3. $\vdash_{\text{LIS}} \bigvee_{v \in \mathcal{VC}_a \cup \{*\}} (a, v) \wedge [I(a)]\alpha \rightarrow \alpha$.
4. $\vdash_{\text{LIS}} [I(a)]\alpha \rightarrow \alpha$ (Ax8 and PL).

Now, assuming s$\vdash_{\text{LIS}} [I(B)]\alpha \rightarrow \alpha$ and following exactly the above steps, one can prove $\vdash_{\text{LIS}} [I(B \cup \{a\})]\alpha \rightarrow \alpha$.
Th3: Let $B = \{a, b\}$.

1. $\vdash_{\text{LIS}} (a, v) \wedge (b, u) \rightarrow [I(a)](a, v) \wedge [I(b)](b, u)$ (Ax9 and PL).
2. $\vdash_{\text{LIS}} [I(a)](a, v) \wedge [I(b)](b, u) \rightarrow [I(B)](a, v) \wedge [I(B)](b, u)$ (Ax6 and PL).
3. $\vdash_{\text{LIS}} [I(B)](a, v) \wedge [I(B)](b, u) \rightarrow [I(B)]((a, v) \wedge (b, u))$ (Modal (K-)theorem).
4. $\vdash_{\text{LIS}} (a, v) \wedge (b, u) \rightarrow [I(B)]((a, v) \wedge (b, u))$ ((1), (2), (3) and PL).
5. $\vdash_{\text{LIS}} [I(B)]((a, v) \wedge (b, u)) \rightarrow (a, v) \wedge (b, u)$ (Th2).
6. $\vdash_{\text{LIS}} (a, v) \wedge (b, u) \leftrightarrow [I(B)]((a, v) \wedge (b, u))$ ((4), (5) and PL). □

It is not difficult to obtain

Theorem 29 (Soundness).

(i) For $\Lambda \in \{\text{LIS}, \text{LIS}_\text{f}\}$ and $\alpha \in \mathbb{F}_\Lambda$, if $\vdash_\Lambda \alpha$, then $\models_{IS} \alpha$ and hence $\models_{SIS} \alpha$.
(ii) For $\Lambda \in \{\text{LDIS}, \text{LDIS}_\text{f}\}$ and $\alpha \in \mathbb{F}_\Lambda$, if $\vdash_\Lambda \alpha$, then $\models_{DIS} \alpha$ and hence $\models_{SDIS} \alpha$.

Completeness. The modal operator $[I(B)]$ is very similar to the distributed knowledge operator of epistemic logic: the relation corresponding to $[I(B)]$ is given by $\bigcap_{b \in B} R_{I(b)}$. However, as clear from the previous section, the deductive systems considered here are different, because the logics have an extra feature of descriptors. As we shall see in this section, this feature, in fact, makes the proof of completeness much simpler than the completeness proofs for epistemic logic with distributed knowledge operators presented in, for instance, [1,23,102].

As in normal modal logic, we have the following result.
Let $\Lambda \in \{\text{LIS}, \text{LDIS}, \text{LIS}_\text{f}, \text{LDIS}_\text{f}\}$.

Proposition 98 (Lindenbaum's Lemma). *If Σ is a $\Lambda-$consistent set of wffs of $\mathbb{F}_\Lambda$, then there is a $\Lambda-$maximal consistent set $\Sigma^+ \subseteq \mathbb{F}_\Lambda$ such that $\Sigma \subseteq \Sigma^+$.*

We now describe the *canonical model.*

Definition 72 (Canonical model). *By the canonical model for Λ, we mean the model $\mathfrak{M}^\Lambda := (W^\Lambda, \{R^\Lambda_{I(B)}\}_{B \subseteq AC}, \{R^\Lambda_{S(B)}\}_{B \subseteq AC}, m^\Lambda, V^\Lambda)$ where*

- $W^\Lambda := \{w \subseteq \mathbb{F}_\Lambda : wis\Lambda-$ maximally consistent set $\}$,

- *for each* $B \subseteq \mathcal{AC}$ *and* $L \in \{I, S\}$, $(w, w') \in R^{\Lambda}_{L(B)}$ *if and only if for all wff* α, $[L(B)]\alpha \in w$ *implies* $\alpha \in w'$,
- $m^{\Lambda}(a, v) := \{w \in W^{\Lambda} : (a, v) \in w\}$,
- $V^{\Lambda}(p) := \{w \in W^{\Lambda} : p \in w\}$.

By giving the same argument as in the standard normal modal logic case, we obtain

Proposition 99 (Truth Lemma). *For any wff* $\beta \in \mathbb{F}_{\Lambda}$ *and* $w \in W^{\Lambda}$ *(the domain of* $\mathfrak{M}^{\Lambda}$*),* $\beta \in w$ *if and only if* $\mathfrak{M}^{\Lambda}, w \models \beta$.

Using the Truth Lemma and Proposition 98 we have

Proposition 100. *If* $\alpha \in \mathbb{F}_{\Lambda}$ *is* $\Lambda-$*consistent then there exists a* $\Lambda-$*maximal consistent set* $\Sigma \subseteq \mathbb{F}_{\Lambda}$ *such that* $\mathfrak{M}^{\Lambda}, \Sigma \models \alpha$.

The next proposition gives some properties of the canonical models.

Proposition 101. *The canonical model* $\mathfrak{M}^{\Lambda}$, $\Lambda \in \{\mathrm{LIS}, \mathrm{LIS_f}\}$ *satisfies* $(IS1)$–$(IS5)$, $(IS7)$, $(IS8a)$, $(IS8b)$, *and*

(IS6a) $R_{I(\emptyset)} = R_{S(\emptyset)}$,
(IS9a) *for* $b \in \mathcal{AC}$, *if* $(w, w') \in R_{I(b)}$, *then there exists* $v \in \mathcal{VC}_b \cup \{*\}$ *such that* $w, w' \in m(b, v)$.
(IS10a) *For* $b \in \mathcal{AC}$, *if* $(w, w') \in R_{S(b)}$, *then either of the following holds.*
(a) $w, w' \in m(b, v)$ *for some* $v \in \mathcal{VC}_b$.
(b) $w \in m(b, *)$.
(c) $w' \in m(b, *)$.

Moreover, $\mathfrak{M}^{\mathrm{LDIS}}$ *and* $\mathfrak{M}^{\mathrm{LDIS_f}}$ *also satisfy (DIS1) and (DIS2) in addition to the above conditions.*

Proof. We only prove that $\mathfrak{M}^{\mathrm{LIS}}$ satisfies (IS8a), (IS8b) and (IS9a). The rest can be done similarly.

(IS8a) Let $(w, w') \in R^{\Lambda}_{I(B)}$ and let there exist a $v \in \mathcal{VC}_b \cup \{*\}$ such that $w, w' \in m^{\Lambda}(b, v)$. Further, suppose $[I(B \cup \{b\})]\alpha \in w$. We need to prove $\alpha \in w'$. Using Ax10 and the fact that $(b, v) \wedge [I(B \cup \{b\})]\alpha \in w$, we obtain $[I(B)]((b, v) \rightarrow \alpha) \in w$. This gives $(b, v) \rightarrow \alpha \in w'$ and hence $\alpha \in w'$ as $(b, v) \in w'$.

(IS8b) Let $(w, w') \in R^{\Lambda}_{S(B)}$ and $[S(B \cup \{b\})]\alpha \in w$. We need to show $\alpha \in w'$.

Case (a): $(w, w') \in m(b, v)$ for some $v \in \mathcal{VC}_b$. Using Ax12 and the fact that $(b, v) \wedge [S(B \cup \{b\})]\alpha \in w$, we obtain $[S(B)](((b, v) \vee (b, *)) \rightarrow \alpha) \in w$. This gives $((b, v) \vee (b, *)) \rightarrow \alpha \in w'$ and hence $\alpha \in w'$ as $(b, v) \vee (b, *) \in w'$.

Case (b): $w \in m(b, *)$. This follows from Ax13.

Case (c): $w' \in m(b, *)$. From Ax8, it follows that either $(b, *) \in w$ or, $(b, v) \in w$ for some $v \in \mathcal{VC}_b$. In the case when $(b, v) \in w$ for some $v \in \mathcal{VC}_b$, we obtain $[S(B)](((b, v) \vee (b, *)) \rightarrow \alpha) \in w$ using Ax12. Thus we have $((b, v) \vee (b, *)) \rightarrow \alpha \in w'$ and hence $\alpha \in w'$ as $(b, *) \in w'$.

Similarly, if $(b,*) \in w$, then we obtain $\alpha \in w'$ using Ax13.

(IS9a) Let $(w,w') \in R^{\Lambda}_{I(b)}$. Then Ax8 guarantees the existence of a $v \in \mathcal{VC}_b \cup \{*\}$ such that $w \in m^{\Lambda}(b,v)$, i.e. $(b,v) \in w$. Then by Ax9, we obtain $[I(b)](b,v) \in w$. So $(b,v) \in w'$, i.e. $w' \in m^{\Lambda}(b,v)$.

(IS10a) Let $(w,w') \in R^{\Lambda g}_{S(b)}$. From Ax8, it follows that either $(b,*) \in w$ or $(b,v) \in w$ for some $v \in \mathcal{VC}_b$. When $(b,*) \in w$, then we obtain the desired result. So let us consider the second case. Here, by Ax11, we obtain $[S(b)]((b,v) \vee (b,*)) \in w$ and hence either $w' \in m^{\Lambda g}(b,v)$ or $w' \in m^{\Lambda g}(b,*)$. □

Note that we still have not proved $R_{I(\emptyset)} = W \times W$ and the other directions of (IS9) and (IS10). In order to incorporate these properties, we construct a new model from $\mathfrak{M}^{\Lambda}$.
Let $\mathfrak{M}^{\Lambda g} := (W^{\Lambda g}, \{R^{\Lambda g}_{I(B)}\}_{B \subseteq AC}, \{R^{\Lambda g}_{S(B)}\}_{B \subseteq AC}, m^{\Lambda g}, V^{\Lambda g})$ be the sub-model of $\mathfrak{M}^{\Lambda}$ generated by Σ using the equivalence relation $R^{\Lambda}_{I(\emptyset)}$, i.e.

- $W^{\Lambda g}$ is the equivalence class of Σ with respect to the equivalence relation $R^{\Lambda}_{I(\emptyset)}$.
- $R^{\Lambda g}_{L(B)}, m^{\Lambda g}, V^{\Lambda g}$, $L \in \{I,S\}$, are respectively the restrictions of $R^{\Lambda}_{L(B)}$, m^{Λ}, V^{Λ} to $W^{\Lambda g}$ (that is: $R^{\Lambda g}_{L(B)} := R^{\Lambda}_{L(B)} \cap (W^{\Lambda g} \times W^{\Lambda g})$, $m^{\Lambda g}(a,v) := m^{\Lambda}(a,v) \cap W^{\Lambda g}$, and $V^{\Lambda g}(p) := V^{\Lambda}(p) \cap W^{\Lambda g}$, for $p \in PV$).

We then have

Proposition 102.

(i) For $\Lambda \in \{\text{LIS}, \text{LIS}_{\text{f}}\}$, $\mathfrak{M}^{\Lambda g} := (W^{\Lambda g}, \{R^{\Lambda g}_{I(B)}\}_{B \subseteq AC}, \{R^{\Lambda g}_{S(B)}\}_{B \subseteq AC}, m^{\Lambda g}, V^{\Lambda g})$ is an IS-model.
(ii) For $\Lambda \in \{\text{LDIS}, \text{LDIS}_{\text{f}}\}$, $\mathfrak{M}^{\Lambda g} := (W^{\Lambda g}, \{R^{\Lambda g}_{I(B)}\}_{B \subseteq AC}, \{R^{\Lambda g}_{S(B)}\}_{B \subseteq AC}, m^{\Lambda g}, V^{\Lambda g})$ is a DIS-model.

Proof. We only prove (i) for $\Lambda = \text{LIS}$. The other cases can be proved in the same fashion. Clearly (IS1)–(IS8) and (IS9a), (IS10a) are satisfied. We only prove the other directions of (IS9) and (IS10). First we prove the other direction of (IS9). Let there exist $v \in \mathcal{VC}_b \cup \{*\}$ such that $w, w' \in m^{\Lambda g}(b,v)$. Let $[I(b)]\alpha \in w$. We want to show $\alpha \in w'$. Here we have $(b,v) \wedge [I(b)]\alpha \in w$ and hence from Ax10 with $B = \emptyset$, we obtain $[I(\emptyset)]((b,v) \to \alpha) \in w$. Since $wR^{\Lambda g}_{I(\emptyset)}w'$, we obtain $(b,v) \to \alpha \in w'$. This together with $(b,v) \in w'$ gives $\alpha \in w'$.

Now, we prove the other direction of (IS10). Suppose either of (a)–(c) holds. We have to show $(w,w') \in R^{\Lambda g}_{S(b)}$. Let $[S(b)]\alpha \in w$. We have to show $\alpha \in w'$. Let us first consider the case when $w, w' \in m^{\Lambda g}(b,v)$ for some $v \in \mathcal{VC}_b$. Then we have $(b,v) \wedge [S(b)]\alpha \in w$ and hence from Ax12 with $B = \emptyset$, we obtain $[S(\emptyset)](((b,v) \vee (b,*)) \to \alpha) \in w$. Since $wR^{\Lambda g}_{S(\emptyset)}w'$, we obtain $((b,v) \vee (b,*)) \to \alpha \in w'$. This together with $(b,v) \in w'$ gives $\alpha \in w'$. The other cases can be done in the same way. □

Since $R^{\Lambda}_{L(B)} \subseteq R^{\Lambda}_{I(\emptyset)}$ for all $B \subseteq \mathcal{AC}$ and $L \in \{I, S\}$, $\mathfrak{M}^{\Lambda g}$ is also a generated sub-model of $\mathfrak{M}^{\Lambda}$ with respect to $R^{\Lambda}_{L(B)}$, $L \in \{I, S\}$. An easy induction on the complexity of the wff α gives us

Proposition 103. *For each wff α and $w \in W^{\Lambda g}$, we have*

$$\mathfrak{M}^{\Lambda}, w \models \alpha \textit{ if and only if } \mathfrak{M}^{\Lambda g}, w \models \alpha.$$

Thus we obtain the following completeness theorem.

Theorem 30 (Completeness).

(i) For $\Lambda \in \{\mathrm{LIS}, \mathrm{LIS_f}\}$ and $\alpha \in \mathbb{F}_{\Lambda}$, if $\models_{IS} \alpha$, then $\vdash_{\Lambda} \alpha$.
(ii) For $\Lambda \in \{\mathrm{LDIS}, \mathrm{LDIS_f}\}$ and $\alpha \in \mathbb{F}_{\Lambda}$, if $\models_{DIS} \alpha$, then $\vdash_{\Lambda} \alpha$.

Observe that the above theorem does not give completeness with respect to the class of standard (C)IS-models. But due to Proposition 90, we have the following.

Theorem 31. *Let $\alpha \in \mathbb{F}_{\mathrm{LIS_f}} = \mathbb{F}_{\mathrm{LDIS_f}} = \mathbb{F}_f$. Then,*

(i) $\models_{SIS} \alpha$ implies $\vdash_{\mathrm{LIS_f}} \alpha$,
(ii) $\models_{SDIS} \alpha$ implies $\vdash_{\mathrm{LDIS_f}} \alpha$.

From this theorem it follows that $\mathrm{LIS_f}$ and $\mathrm{LDIS_f}$ are the logics for ISs and DISs respectively which can express properties of indiscernibility and similarity relations corresponding to any finite set of attributes.

12.4 Decidability

The decidability problem of LIS and LDIS is open. However, one obtains the answer for $\mathrm{LIS_f}$ and $\mathrm{LDIS_f}$. Let $k \in \{IS, DIS, SIS, SDIS\}$.

Theorem 32. *We can decide for a given $\alpha \in \mathbb{F}_f$, whether $\models_k \alpha$.*

Let $\mathcal{A}^* \subseteq_f \mathcal{AC}$. Consider the language $\mathbb{L}(\mathcal{A}^*)$ which is the same as $\mathbb{L}$ except that the attribute and attribute-value constants are taken from the set $\mathcal{A}^*$ and $\bigcup_{a \in \mathcal{A}^*} \mathcal{VC}_a$ respectively. The set $\mathbb{F}_{\mathcal{A}^*}$ of wffs of $\mathbb{L}(\mathcal{A}^*)$ are defined in the standard way. Let $\mathcal{D}_{\mathcal{A}^*}$ be the set of descriptors (a, v) from $\mathbb{L}(\mathcal{A}^*)$. The semantics of the logic $\mathbb{L}(\mathcal{A}^*)$ will be based on the structure of the form $\mathfrak{F} := (W, \{R_{I(B)}\}_{B \subseteq \mathcal{A}^*}, \{R_{S(B)}\}_{B \subseteq \mathcal{A}^*}, m)$, where $m : \mathcal{D}_{\mathcal{A}^*} \to 2^W$. The notions of IS/DIS/$SIS$/$SDIS$-models and structures are defined for $\mathbb{L}(\mathcal{A}^*)$ in the same way as we have done earlier.

Let $\mathfrak{M} := (W, \{R_{I(B)}\}_{B \subseteq \mathcal{AC}}, \{R_{S(B)}\}_{B \subseteq \mathcal{AC}}, m, V)$ be a model of $\mathbb{L}$. Let $\mathfrak{M}(\mathcal{A}^*)$ be the model $(W, \{R'_{I(B)}\}_{B \subseteq \mathcal{A}^*}, \{R'_{S(B)}\}_{B \subseteq \mathcal{A}^*}, m', V)$ of $\mathbb{L}(\mathcal{A}^*)$, where

- $m' := m|_{\mathcal{D}_{\mathcal{A}^*}}$,
- $R'_{L(B)} := R_{L(B)}$ for $B \subseteq \mathcal{A}^*$ and $L \in \{I, S\}$.

By an easy induction on α, we obtain

Proposition 104. *For all $\alpha \in \mathbb{F}_{\mathcal{A}^*}$ and for all $w \in W$,*

$$\mathfrak{M}, w \models \alpha \text{ if and only if } \mathfrak{M}(\mathcal{A}^*), w \models^{\mathcal{A}^*} \alpha.$$

Proposition 105. *Let $k \in \{IS, SIS, DIS, SDIS\}$.*

(i) If $\mathfrak{M}$ is a k-model, then $\mathfrak{M}(\mathcal{A}^*)$ is also a k-model.
(ii) Given a k-model $\mathfrak{M}'$ of $\mathbb{L}(\mathcal{A}^*)$, there exists a k-model $\mathfrak{M}$ such that $\mathfrak{M}' = \mathfrak{M}(\mathcal{A}^*)$.

Proof. We only prove (ii). (i) is obvious.
Let $\mathfrak{M}' := (W, \{R'_{I(B)}\}_{B \subseteq \mathcal{AC}}, \{R'_{S(B)}\}_{B \subseteq \mathcal{AC}}, m', V)$. For each $b \in \mathcal{AC} \setminus \mathcal{A}^*$, take a $v_b \in \mathcal{VC}_b$. Now consider the k-model $\mathfrak{M} := (W, \{R_{I(B)}\}_{B \subseteq \mathcal{AC}}, \{R_{S(B)}\}_{B \subseteq \mathcal{AC}}, m, V)$, where, for $L \in \{I, S\}$,

$R_{L(b)} := R'_{L(b)}$ for $b \in \mathcal{A}^*$,
$R_{L(b)} := W \times W$ for $b \notin \mathcal{A}^*$,
$R_{L(B)} := \bigcap_{b \in B} R_{L(b)}$,
$m(b, v) := m'(b, v)$ for $b \in \mathcal{A}^*$, and
for $b \notin \mathcal{A}^*$, $m(b, v)$ is $\emptyset$ or $W \times W$ according as $v \neq v_b$ or $v = v_b$.

One can easily verify that $\mathfrak{M}$ is a k-model. Moreover, $\mathfrak{M}' = \mathfrak{M}(\mathcal{A}^*)$. □

From Propositions 104 and 105, we obtain

Proposition 106. *Let $k \in \{IS, DIS, SIS, SDIS\}$. Then for all $\alpha \in \mathbb{F}_{\mathcal{A}^*}$,*

$$\models_k \alpha \text{ if and only if } \models_k^{\mathcal{A}^*} \alpha.$$

Let us consider an $\alpha \in \mathbb{F}_f$ which involves only primitive symbols, and let $sub(\alpha)$ denote the set of all sub-wffs of α. Consider the set $\mathcal{A}^*_\alpha \subseteq_f \mathcal{AC}$ defined as follows:

$\mathcal{A}^*_\alpha := \{a \in \mathcal{AC} : (a, v) \in sub(\alpha)$ for some $v \in \mathcal{VC}_a \cup \{*\}\} \cup \bigcup\{B \subseteq_f \mathcal{AC} : [L(B)]\beta \in sub(\alpha)$ for some β and $L \in \{I, S\}\}$.

Note that $\mathcal{A}^*_\alpha \subseteq_f \mathcal{AC}$ and $\alpha \in \mathbb{F}_{\mathcal{A}^*_\alpha}$. Therefore, from Proposition 106, it follows that checking the validity of the wff α in all $(S)IS/(S)DIS$-models of $\mathbb{L}$ reduces to checking its validity in all $(S)IS/(S)DIS$-models of $\mathbb{L}(\mathcal{A}^*_\alpha)$. Thus in order to prove Theorem 32, we shall prove the following result. Let $k \in \{IS, DIS, SIS, SDIS\}$.

Theorem 33. *We can decide for a given $\alpha \in \mathbb{F}_f$, whether $\models_k^{\mathcal{A}^*_\alpha} \alpha$.*

To prove this we shall use the standard filtration technique [9], with natural modifications to the definitions. In the following, we assume that Σ is a finite sub-wff closed set of $\mathbb{F}_{\mathcal{A}^*_\alpha}$-wffs.

Let $\mathfrak{M} := (W, \{R_{I(B)}\}_{B \subseteq \mathcal{A}^*_\alpha}, \{R_{S(B)}\}_{B \subseteq \mathcal{A}^*_\alpha}, m, V)$ be a model. We define a binary relation $\equiv_\Sigma$ on W as follows:

$$w \equiv_\Sigma w', \text{ if and only if for all } \beta \in \Sigma \cup \mathcal{D}_{\mathcal{A}^*_\alpha}, \ \mathfrak{M}, w \models^{\mathcal{A}^*_\alpha} \beta \text{ if and only if } \mathfrak{M}, w' \models^{\mathcal{A}^*_\alpha} \beta.$$

Definition 73 (Filtration model).

- $W^f := \{[w] : w \in W\}$, $[w]$ *is the equivalence class of* w *with respect to the equivalence relation* $\equiv_\Sigma$*;*
- $R^f_{L(B)} \subseteq W^f \times W^f$, $L \in \{I, S\}$ *is defined as:*
  $([w], [u]) \in R^f_{L(B)}$ *if and only if there exist* $w' \in [w]$ *and* $u' \in [u]$ *such that* $(w', u') \in R_{L(B)}$*;*
- $V^f : PV \to 2^{W^f}$ *is defined as:* $V^f(p) := \{[w] \in W^f : w \in V(p)\}$ *;*
- $m^f(a, v) := \{[w] : w \in m(a, v)\}$, $v \in \mathcal{VC}_a \cup \{*\}$.

Let $\mathfrak{M}^f$ *denote the model* $(W^f, \{R^f_{I(B)}\}_{B \subseteq \mathcal{A}^*_\alpha}, \{R^f_{S(B)}\}_{B \subseteq \mathcal{A}^*_\alpha}, m^f, V^f)$.

Proposition 107. *The domain of* $\mathfrak{M}^f$ *contains at most* $2^{|\Sigma \cup \mathcal{D}_{\mathcal{A}^*_\alpha}|}$ *elements.*

Proof. Consider the map $g : W^f \to 2^{\Sigma \cup \mathcal{D}_{\mathcal{A}^*_\alpha}}$ defined as,

$$g([w]) = \{\beta \in \Sigma \cup \mathcal{D}_{\mathcal{A}^*_\alpha} : \mathfrak{M}, w \models^{\mathcal{A}^*_\alpha} \beta\}.$$

Since this map is injective, the domain of $\mathfrak{M}^f$ contains at most $2^{|\Sigma \cup \mathcal{D}_{\mathcal{A}^*_\alpha}|}$ elements. □

Proposition 108.

(i) If the model $\mathfrak{M}$ *is an (S)IS model then* $\mathfrak{M}^f$ *is also an (S)IS model.*
(ii) If the model $\mathfrak{M}$ *is a (S)DIS model then* $\mathfrak{M}^f$ *is also a (S)DIS model.*

Proof. We give the proof of (i).
(IS1) and (IS2) are easy to check.
(IS6): Let $L \in \{I, S\}$. Let us consider $[w], [u] \in W^f$. Since $R_{L(\emptyset)} = W \times W$, $(w, u) \in R_{L(\emptyset)}$ and hence $([w], [u]) \in R^f_{L(\emptyset)}$.

(IS7): $([w], [u]) \in R^f_{L(B)}$
$\Rightarrow (w', u') \in R_{L(B)}$ for some $w' \in [w]$ and $u' \in [u]$
$\Rightarrow (w', u') \in R_{L(C)}$
$\Rightarrow ([w], [u]) \in R^f_{L(C)}$.

(IS8a): $([w], [u]) \in R^f_{I(B)}$ and $[w], [u] \in m^f(b, v)$ for some $v \in \mathcal{VC}_b \cup \{*\}$
$\Rightarrow (w', u') \in R_{I(B)}$ and $w', u' \in m(b, v)$ for some $w' \in [w]$ and $u' \in [u]$
$\Rightarrow (w', u') \in R_{I(B \cup \{b\})}$
$\Rightarrow ([w], [u]) \in R^f_{I(B \cup \{b\})}$.
(IS8b) can be proved in similar way.

(IS9): $([w], [u]) \in R^f_{I(b)}$
$\Leftrightarrow (w', u') \in R_{I(b)}$ for some $w' \in [w]$ and $u' \in [u]$
$\Leftrightarrow w', u' \in m(b, v)$ for some $v \in \mathcal{VC}_b \cup \{*\}$, for some $w' \in [w]$ and $u' \in [u]$
$\Leftrightarrow [w], [u] \in m^f(b, v)$ for some $v \in \mathcal{VC}_b \cup \{*\}$.
(IS10) can be proved similarly.

From (IS7) and (IS8), it follows that for $L \in \{I, S\}$, $R^f_{L(B)} = \bigcap_{b \in B} R^f_{L(b)}$ for all $B \subseteq A^*_\alpha$. Note that A^*_α is a finite set. Using these and the fact that for $b \in A^*_\alpha$, $R^f_{I(b)}$ and $R^f_{S(b)}$ are respectively equivalence and tolerance relations, we obtain (IS3) and (IS4). (IS5) follows from the fact that $R^f_{I(b)} \subseteq R^f_{S(b)}$.

In order to complete the proof, we need to show that if $\mathfrak{M} := (W, \{R_{I(B)}\}_{B \subseteq \mathcal{A}^*_\alpha}, \{R_{S(B)}\}_{B \subseteq \mathcal{A}^*_\alpha}, m, V)$ is a standard IS-model, then $\mathfrak{M}^f := (W^f, \{R^f_{I(B)}\}_{B \subseteq \mathcal{A}^*_\alpha}, \{R^f_{S(B)}\}_{B \subseteq \mathcal{A}^*_\alpha}, m^f, V^f)$ is also so.

Suppose the IS-structure $\mathfrak{F}_{\mathcal{S}} := (W, \{R_{I(B)}\}_{B \subseteq \mathcal{A}^*_\alpha}, \{R_{S(B)}\}_{B \subseteq \mathcal{A}^*_\alpha}, m)$ is generated by the IS $\mathcal{S} := (W, \mathcal{A}^*_\alpha, \bigcup_{a \in \mathcal{A}^*_\alpha} \mathcal{VC}_a \cup \{*\}, g)$. Then we have $w \in m(a, v)$ if and only if $g(w, a) = v$, $v \in \mathcal{VC}_a \cup \{*\}$, $Ind_{\mathcal{S}}(B) = R_{I(B)}$ and $Sim_{\mathcal{S}}(B) = R_{S(B)}$ for all $B \subseteq \mathcal{A}^*_\alpha$.

Note that if $[w] = [w']$, then $g(w, a) = g(w', a)$ for all $a \in \mathcal{A}^*_\alpha$. From this it follows that the mapping $g^f : W^f \times \mathcal{A}^*_\alpha \to \bigcup_{a \in \mathcal{A}^*_\alpha} \mathcal{VC}_a \cup \{*\}$ defined by

$$g^f([w], a) = v \text{ if and only if } g(w, a) = v, \;\; v \in \mathcal{VC}_a \cup \{*\}$$

is well defined.

Consider the IS $\mathcal{S}^f := (W^f, \mathcal{A}^*_\alpha, \bigcup_{a \in \mathcal{A}^*_\alpha} \mathcal{VC}_a \cup \{*\}, g^f)$.

We shall show that $(W^f, \{R^f_{I(B)}\}_{B \subseteq \mathcal{A}^*_\alpha}, \{R^f_{S(B)}\}_{B \subseteq \mathcal{A}^*_\alpha}, m^f)$ is generated by $\mathcal{S}^f$.

Here, $[w] \in m^f(a, v)$
$\Leftrightarrow w \in m(a, v)$
$\Leftrightarrow g(w, a) = v$
$\Leftrightarrow g^f([w], a) = v$.

Moreover, for $a \in \mathcal{A}^*_\alpha$, $([w], [u]) \in R^f_{I(a)}$
$\Leftrightarrow (w', u') \in R_{I(a)}$ for some $w' \in [w]$ and $u' \in [u]$
$\Leftrightarrow (w', u') \in Ind_{\mathcal{S}}(\{a\})$ for some $w' \in [w]$ and $u' \in [u]$
$\Leftrightarrow g(w', a) = g(u', a)$ for some $w' \in [w]$ and $u' \in [u]$
$\Leftrightarrow g^f([w], a) = g^f([u], a)$
$\Leftrightarrow ([w], [u]) \in Ind_{\mathcal{S}^f}(\{a\})$.
and hence $Ind_{\mathcal{S}^f}(B) = \bigcap_{a \in B} Ind_{\mathcal{S}^f}(a) = \bigcap_{a \in B} R^f_{I(a)} = R^f_{I(B)}$, ($\because B$ is finite).
Similarly, one can prove $Sim_{\mathcal{S}^f}(B) = R^f_{S(B)}$. □

Thus we have

Proposition 109 (Filtration Theorem). *For all wffs $\beta \in \Sigma \cup \mathcal{D}_{\mathcal{A}^*_\alpha}$, all models $\mathfrak{M}$, and all objects $w \in W$, $\mathfrak{M}, w \models^{\mathcal{A}^*_\alpha} \beta$ if and only if $\mathfrak{M}^f, [w] \models^{\mathcal{A}^*_\alpha} \beta$.*

Proposition 110 (Finite model property). *Let $k \in \{IS, SIS, DIS, SDIS\}$. Let $\alpha \in \mathbb{F}_{\mathcal{A}^*_\alpha}$ and Σ be the set of all sub-wffs of α. If α is satisfiable in a k-model of $\mathbb{L}(\mathcal{A}^*_\alpha)$ then it is satisfiable in a finite k-model of $\mathbb{L}(\mathcal{A}^*_\alpha)$ with at most $2^{|\Sigma \cup \mathcal{D}_{\mathcal{A}^*_\alpha}|}$ elements.*

From this proposition, we obtain Theorem 33.

12.5 Dynamic Logics for Information Systems

In Sect. 12.1 we defined the notions of information and update of information systems. In this section, we shall propose dynamic logics for information systems, where we can express these notions. This is done by enriching the language of the logics defined in Sect. 12.2 with modal operators which carry information. In order to give the satisfiability relation for these modal operators, we introduce the concept of *update of models*.

Syntax. Let us consider the language $\mathbb{L}^D$ the alphabet of which is that of the language $\mathbb{L}$ with the added symbols '$\vee$' and ' ;'. In order to define the wffs of $\mathbb{L}^D$, we need to define what we mean by an information here.

Definition 74. *The set* Inf *of* information *is the smallest set such that*

- $(\phi, a, v) \in Inf$, *where* $\phi \in \mathbb{F}$, $a \in \mathcal{AC}$ *and* $v \in \mathcal{VC}_a$,
- *if* $\sigma, \sigma' \in Inf$, *then* $\sigma;\sigma'$, $\sigma \vee \sigma' \in Inf$.

Information of the form (ϕ, a, v) will be called an *atomic information.* It signifies that the objects represented by ϕ take the value v for the attribute a. Information of the form $\sigma \vee \sigma'$ says that either σ is the case or σ'. $\sigma;\sigma'$ signifies that first we get the information σ and then σ'.
The set $\mathbb{F}^D$ of wffs of $\mathbb{L}^D$ is obtained by adding the following clause to the wff-formation rules of $\mathbb{L}$:

$$\text{if } \alpha \in \mathbb{F}^D \text{ and } \sigma \in Inf \text{ then } [\sigma]\alpha \in \mathbb{F}^D.$$

$\mathbb{F}^D_f$ will denote the set of all wffs of $\mathbb{L}^D$ which do not involve any modal operator $[L(B)]$, $L \in \{I, S\}$, where B is infinite.

Semantics. Let us first define the notion of update of a model.

Definition 75. *Let* $\mathfrak{M} := (W, \{R_{I(B)}\}_{B \subseteq \mathcal{AC}}, \{R_{S(B)}\}_{B \subseteq \mathcal{AC}}, m, V)$ *be a model. Then the model obtained by* updating $\mathfrak{M}$ *with the atomic information* $\sigma_0 = (\phi, a, v)$, *denoted by* $\mathfrak{M}^{\sigma_0}$, *is the model* $(W, \{R^{\sigma_0}_{I(B)}\}_{B \subseteq \mathcal{AC}}, \{R^{\sigma_0}_{S(B)}\}_{B \subseteq \mathcal{AC}}, m^{\sigma_0}, V^{\sigma_0})$ *where*

- $V^{\sigma_0} := V$,
- m^{σ_0} *is given as follows:*
  - $m^{\sigma_0}(a', v') := m(a', v')$ *for* $a' \in \mathcal{AC} \setminus \{a\}$ *and* $v' \in \mathcal{VC}_a \cup \{*\}$,
  - $m^{\sigma_0}(a, v') := m(a, v') \setminus [\![\phi]\!]_{\mathfrak{M}}$ *for* $v' \in (\mathcal{VC}_a \cup \{*\}) \setminus \{v\}$,
  - $m^{\sigma_0}(a, v) := m(a, v) \cup [\![\phi]\!]_{\mathfrak{M}}$,
- $R^{\sigma_0}_{L(B)}$, $L \in \{I, S\}, B \subseteq \mathcal{AC}$ *are defined as follows:*
  - *For* $a \notin B$, $R^{\sigma_0}_{L(B)} := R_{L(B)}$,
  - *If* $a \in B$, *then* $(x, y) \in R^{\sigma_0}_{I(B)}$ *if and only if either of the following holds.*
    - $x, y \notin [\![\phi]\!]_{\mathfrak{M}}$ *and* $(x, y) \in R_{I(B)}$,

  * $x \in [\![\phi]\!]_{\mathfrak{M}},\ y \notin [\![\phi]\!]_{\mathfrak{M}},\ y \in m(a,v)$ *and* $(x,y) \in R_{I(B\setminus\{a\})}$,
  * $x \notin [\![\phi]\!]_{\mathfrak{M}},\ y \in [\![\phi]\!]_{\mathfrak{M}},\ x \in m(a,v)$ *and* $(x,y) \in R_{I(B\setminus\{a\})}$,
  * $x, y \in [\![\phi]\!]_{\mathfrak{M}}$ *and* $(x,y) \in R_{I(B\setminus\{a\})}$.
- *If* $a \in B$, *then* $(x,y) \in R^{\sigma_0}_{S(B)}$ *if and only if either of the following holds.*
  * $x, y \notin [\![\phi]\!]_{\mathfrak{M}}$ *and* $(x,y) \in R_{S(B)}$,
  * $x \in [\![\phi]\!]_{\mathfrak{M}},\ y \notin [\![\phi]\!]_{\mathfrak{M}},\ y \in m(a,v) \cup m(a,*)$ *and* $(x,y) \in R_{S(B\setminus\{a\})}$,
  * $x \notin [\![\phi]\!]_{\mathfrak{M}},\ y \in [\![\phi]\!]_{\mathfrak{M}},\ x \in m(a,v) \cup m(a,*)$ *and* $(x,y) \in R_{S(B\setminus\{a\})}$,
  * $x, y \in [\![\phi]\!]_{\mathfrak{M}}$ *and* $(x,y) \in R_{S(B\setminus\{a\})}$.

Note that $R^{\sigma_0}_{L(\emptyset)} = R_{L(\emptyset)},\ L \in \{I, S\}$.

Observation 7. It is possible that $x \in m^{\sigma_0}(a,v)$ with $x \in [\![\phi]\!]_{\mathfrak{M}} \cap m(a,v')$, $v \neq v'$. But $x \notin m^{\sigma_0}(a,v')$, by definition. So the old value v' of x under a is replaced by the new value v. This was noted earlier also in Sect. 12.1.

Thus each information σ induces a relation $\mathfrak{R}_\sigma$ on the set of all models as follows. Let $R_1; R_2$ denote the composition of the two relations R_1 and R_2.

- For atomic information σ_0, $(\mathfrak{M}, \mathfrak{M}') \in \mathfrak{R}_{\sigma_0}$ if and only if $\mathfrak{M}' = \mathfrak{M}^{\sigma_0}$.
- $\mathfrak{R}_{\sigma;\sigma'} := \mathfrak{R}_\sigma; \mathfrak{R}_{\sigma'}$.
- $\mathfrak{R}_{\sigma\vee\sigma'} := \mathfrak{R}_\sigma \cup \mathfrak{R}_{\sigma'}$.

So $\mathfrak{R}_{\sigma_0}$, where σ_0 is an atomic information, behaves like a function, i.e. for each $\mathfrak{M}$, there exists a unique $\mathfrak{M}'$ such that $(\mathfrak{M}, \mathfrak{M}') \in \mathfrak{R}_{\sigma_0}$. Further, we could have information σ which is non-deterministic, i.e. update of a model with this information could result in more than one model.

Satisfiability of the wff $\alpha \in \mathbb{F}^D$ in a model $\mathfrak{M}$ at the world w is defined by extending Definition 69:

$$\mathfrak{M}, w \models [\sigma]\alpha \text{ if and only if } \mathfrak{M}', w \models \alpha, \quad \text{for all } \mathfrak{M}' \text{ such that } (\mathfrak{M}, \mathfrak{M}') \in \mathfrak{R}_\sigma.$$

At this point we need to address some natural questions. One could ask whether the update of a (standard) IS-model is also an (standard) IS-model. Moreover, how are these notions of information and information update related with the ones defined for ISs in Sect. 12.1? We will answer these questions and finally, we would like to show how the language of the dynamic extension defined above can be used to express the properties related to information and its effect on the approximation of sets. The rest of this section will deal with these issues. Recall the notations used in Definition 65. Let $p_1, p_2, \ldots$ and $x_1, x_2, \ldots$ respectively be an enumeration of the set PV of propositional variables and the set Var of variables. Let us consider the map $\Psi : \mathbb{F} \to \mathcal{T}_{\mathfrak{H}(\mathcal{AC},\mathcal{VC})}$ defined inductively, where $\mathcal{VC} := \bigcup_{a \in \mathcal{AC}} \mathcal{VC}_a$.

$$\Psi(p_i) := x_i.$$
$$\Psi((a,v)) := (a,v).$$
$$\Psi(\neg\alpha) := \sim \Psi(\alpha).$$
$$\Psi(\alpha \wedge \beta) := \Psi(\alpha) \sqcap \Psi(\beta).$$

$$\Psi([I(B)]\alpha) := I_B\Psi(\alpha).$$
$$\Psi([S(B)]\alpha) := S_B\Psi(\alpha).$$

Note that Ψ is a bijection. Moreover, for a standard $(C)IS$-model $\mathfrak{M} := (\mathfrak{F}_\mathcal{S}, V)$ with domain W and an assignment $\theta : Var \to 2^W$ such that $\theta(x_i) := V(p_i)$, we obtain

$$[\![\phi]\!]_\mathfrak{M} = \theta_\mathcal{S}(\Psi(\phi)) \text{ for all } \phi \in \mathbb{F},$$

where $\theta_\mathcal{S}$ is the extension of θ to $\mathcal{T}_{\mathfrak{H}(\mathcal{AC},\mathcal{VC})}$.

Consider now the set Inf (cf. Definition 74) and $Inf_{(\mathcal{AC},\mathcal{VC})}$ (cf. Definition 66). Using Ψ, we get the bijection $\mathfrak{C} : Inf \to Inf_{(\mathcal{AC},\mathcal{VC})}$ defined inductively as:

$\mathfrak{C}((\phi, a, v)) := (\Psi(\phi), a, v)$, for atomic (ϕ, a, v),
$\mathfrak{C}(\sigma_1 \vee \sigma_2) := \mathfrak{C}(\sigma_1) \vee \mathfrak{C}(\sigma_2)$,
$\mathfrak{C}(\sigma_1; \sigma_2) := \mathfrak{C}(\sigma_1); \mathfrak{C}(\sigma_2)$.

The next proposition answers the first question raised above. Let $k \in \{$IS, DIS, $SIS, SDISs\}$.

Proposition 111. *Let $\sigma \in Inf$ and $\mathfrak{M}, \mathfrak{M}'$ be two models such that $(\mathfrak{M}, \mathfrak{M}') \in \mathfrak{R}_\sigma$. If $\mathfrak{M}$ is a k-model, then $\mathfrak{M}'$ is also a k-model.*

Proof. The proof is by induction on the complexity of σ. □

Proposition 112 below relates the update of ISs (DISs) (cf. Definition 67) and update of IS-models (DIS-models) (cf. Definition 75).

Proposition 112. *Let $\mathfrak{F}_\mathcal{S}$ and $\mathfrak{F}_{\mathcal{S}'}$ be two standard* IS*-structures (standard* DIS*-structures) with the same domain W. Let $V : PV \to 2^W$ and $\theta : Var \to 2^W$ be such that $\theta(x_i) := V(p_i)$, for all i. Then for all $\sigma \in Inf$, we have*

$$(\mathfrak{M}, \mathfrak{M}') \in \mathfrak{R}_\sigma, if \textit{ and only } if \mathcal{S}' \in U(\mathcal{S}, \mathfrak{C}(\sigma), \theta),$$

where $\mathfrak{M} := (\mathfrak{F}_\mathcal{S}, V)$ and $\mathfrak{M}' := (\mathfrak{F}_{\mathcal{S}'}, V)$.

Proof. The proof is by induction on the complexity of σ. We only provide the proof of the basis case. The induction case can be proved using Proposition 111. Let σ be of the form (ϕ, a, v).
We first note the following two trivial facts.

(i) For different $\mathcal{K}_1$, $\mathcal{K}_2$, we obtain different $\mathfrak{F}_{\mathcal{K}_1}, \mathfrak{F}_{\mathcal{K}_2}$;
(ii) For atomic σ, $(\mathfrak{M}_1, \mathfrak{M}_2), (\mathfrak{M}_1, \mathfrak{M}_3) \in \mathfrak{R}_\sigma$ implies $\mathfrak{M}_2 = \mathfrak{M}_3$.

Let $\mathfrak{F}_\mathcal{S} := (W, \{R_{I(B)}\}_{B \subseteq \mathcal{AC}}, \{R_{S(B)}\}_{B \subseteq \mathcal{AC}}, m)$ and $\mathcal{S} := (W, \mathcal{AC}, \bigcup_{a \in \mathcal{AC}} \mathcal{VC}_a \cup \{*\}, f)$. Let $\mathcal{K} = (W, \mathcal{AC}, \bigcup_{a \in \mathcal{AC}} \mathcal{VC}_a \cup \{*\}, f_k) \in U(\mathcal{S}, \mathfrak{C}(\sigma), \theta)$. Then due to the above-mentioned facts, it is enough to show that

$$\mathfrak{F}_\mathcal{S}^\sigma = \mathfrak{F}_\mathcal{K}, \tag{33}$$

where $\mathfrak{F}_S^\sigma = (W, \{R^\sigma_{I(B)}\}_{B\subseteq\mathcal{AC}}, \{R^\sigma_{S(B)}\}_{B\subseteq\mathcal{AC}}, m^\sigma)$ (cf. Definition 75). In order to prove (33), we need to show the following.

$$m^\sigma(b,u) = \{w \in W : f_k(w,b) = u\} \tag{34}$$
$$R^\sigma_{I(B)} = Ind_{\mathcal{K}}(B) \text{ and } R^\sigma_{S(B)} = Sim_{\mathcal{K}}(B) \tag{35}$$

Let us prove (34).
First suppose $b \neq a$. Then,

$w \in m^\sigma(b,u) = m(b,u)$
$\Leftrightarrow f(w,b) = u$
$\Leftrightarrow f_{\mathcal{K}}(w,b) = u.$

Now, consider the case when $b = a$. Then, we have,

$w \in m^\sigma(a,v) = m(a,v) \cup [\![\phi]\!]_{\mathfrak{M}}$
$\Leftrightarrow f(w,a) = v$ or $w \in [\![\phi]\!]_{\mathfrak{M}} = \theta_S(\Psi(\phi))$
$\Leftrightarrow f_{\mathcal{K}}(w,a) = v.$

Moreover, for $u \neq v$, we have,

$w \in m^\sigma(a,u) = m(a,u) \setminus [\![\phi]\!]_{\mathfrak{M}}$
$\Leftrightarrow f(w,a) = u$ and $w \notin [\![\phi]\!]_{\mathfrak{M}} = \theta_S(\Psi(\phi))$
$\Leftrightarrow f_{\mathcal{K}}(w,a) = u.$

One can prove (35) similarly. □

The questions of the kind posed in Sect. 12.1 can now be looked upon as questions of satisfiability of some wffs in particular models. For instance, let us consider the standard IS-model $\mathfrak{M} := (\mathfrak{F}_{\mathcal{K}_1}, V)$, for the IIS $\mathcal{K}_1$ of the example given there. Suppose V is the assignment such that $V(p) := X$ and $V(q) := \{P4\}$. The information $\mathcal{I}_1$ and $\mathcal{I}_2$ obtained at the end of Sect. 12.1 now take the form

$\sigma_1 := (q, N, yes); ((q, T, high) \vee (q, T, very\ high)),$
$\sigma_2 := (((T, high) \vee (T, very\ high)) \wedge (N, yes), H, yes),$

Let D be the attribute set $\{T, H, N\}$. Then the questions (Q1)–(Q5) can be identified with the following.

(Q1) At which objects is the wff $[I(D)]p \wedge [\sigma_1]\neg[I(D)]p$ satisfiable?
(Q2) Do we have $\mathfrak{M}, P5 \models [\sigma_1;\sigma_2][S(D)]p$?
(Q3) Is $[\sigma_1;\sigma_2](\langle I(D)\rangle p \rightarrow [I(D)]p)$ valid in $\mathfrak{M}$?
(Q4) Are the wffs $(H, no) \wedge (N, no) \wedge p$ and $[\sigma_1;\sigma_2]((H, no) \wedge (N, no) \wedge p)$ satisfiable?
(Q5) Is $(N, no) \wedge [\sigma_1;\sigma_2](N, yes)$ satisfiable?

Observe that for giving the wff for Q2, we have used Proposition 2.

The following proposition gives some wffs involving information operators. Here satisfiability and validity referred to the class of all IS-models.

Proposition 113.

1. *Let* $\sigma := \sigma_1; \sigma_2; \ldots; \sigma_n$*, where* $\sigma_i := (\phi_i, a, v_i)$*. Then* $[S(a)]p \to [\sigma][I(a)]p$ *is valid in all models* $\mathfrak{M}$*, where* $[\![\phi_i]\!]_{\mathfrak{M}} \subseteq [\![(a, *)]\!]_{\mathfrak{M}}$*.*
2. $\neg\langle I(B)\rangle p \wedge [\sigma][I(B)]p$ *is a contradiction.*
3. $\langle I(B)\rangle p \wedge [\sigma][I(B)]p$ *is satisfiable.*
4. $[I(\emptyset)]p \to [(p, a, v)][I(\emptyset)](a, v)$ *is valid.*
5. $[I(\emptyset)]\neg(p \wedge q) \to [I(\emptyset)]\Big([(p, a, v)][q, b, u]\alpha \leftrightarrow [q, b, u][p, a, v]\alpha\Big)$ *is valid.*
6. $[I(\emptyset)]\Big(([I(B)]p \leftrightarrow [I(B)]q) \wedge (\langle I(B)\rangle p \leftrightarrow \langle I(B)\rangle q)\Big) \wedge [\sigma]\langle I(\emptyset)\rangle\neg\Big(([I(B)]p \leftrightarrow [I(B)]q) \wedge (\langle I(B)\rangle p \leftrightarrow \langle I(B)\rangle q)\Big)$ *is satisfiable, where* σ *is atomic information.*

Let us elucidate some of the above. The validity of the wff in 1 guarantees that if an object $x \in \underline{X}_{Ind_{\mathcal{S}'}(\{a\})}$ for each completion $\mathcal{S}'$ of an IS $\mathcal{S}$, then $x \in \underline{X}_{Ind_{\mathcal{K}}(\{a\})}$, where $\mathcal{K}$ is obtained from $\mathcal{S}$ by updating the missing information about a. The wff in 2 represents the fact that if an object is a negative element of a set, then no information can convert it into a positive element of the set. On the other hand, the satisfiability of the wff in 3 signifies that a possible element of a set may become a positive element of the set due to a gain in information. Validity of the wff in 5 guarantees that if we get information regarding two disjoint sets then the order in which the information is obtained will not affect the update. Satisfiability of the wff in 6 indicates that two sets that are roughly equal (i.e. having identical lower and upper approximations) may not remain so after gain in information.

In Sect. 8, different patterns of changes of partitions due to an inflow of information were considered. For instance, with the flow of information partitions may become finer; moreover, in an ideal situation, a stage may be reached where all objects become distinguishable. In the next proposition, we give wffs which characterize such patterns.

Let $\mathfrak{F} := (W, \{R_{I(B)}\}_{B \subseteq \mathcal{AC}}, \{R_{S(B)}\}_{B \subseteq \mathcal{AC}}, m)$ be an IS-structure.

Proposition 114.

1. *Let* $\alpha := [I(B)]q \to [(p, a, v)][I(B)]q$ *and* $\sigma_0 = (p, a, v)$ *be an atomic information. Then* α *is valid in* $\mathfrak{F}$ *if and only if for every model* $\mathfrak{M} := (\mathfrak{F}, V)$ *and corresponding updated model* $\mathfrak{M}^{\sigma_0}$*, we have* $R^{\sigma_0}_{I(B)} \subseteq R_{I(B)}$*.*
2. *Let* $p \neq q$*,* $\alpha := [(p, a, v)](\langle I(B)\rangle q \to [I(B)]q)$ *and* $\sigma_0 := (p, a, v)$ *be an atomic information. Then* α *is valid in* $\mathfrak{F}$ *if and only if for every model* $\mathfrak{M} := (\mathfrak{F}, V)$ *and corresponding updated model* $\mathfrak{M}^{\sigma_0}$*, we have* $xR^{\sigma_0}_{I(B)}y \Leftrightarrow x = y$*.*
3. *Let* $B \neq \emptyset$*,* $\sigma_0 := (p, a, v)$ *an atomic information and*

$$
\begin{aligned}
\alpha := \quad & \Big((p \wedge \neg(a, v)) \to [I(B \setminus \{a\})](\neg p \to \neg(a, v))\Big) \\
\wedge & \Big((\neg p \wedge (a, v)) \to [I(B \setminus \{a\})](p \to (a, v))\Big) \\
\wedge & \Big(p \to \bigwedge_{(a, v_1), v_1 \in \mathcal{VC}_a \cup \{*\}} ((a, v_1) \to [I(B \setminus \{a\})](p \to (a, v_1)))\Big)
\end{aligned}
$$

Then for every IS-*model* $\mathfrak{M} := (\mathfrak{F}, V)$,

$$\mathfrak{M} \models \alpha \text{ if and only if } R^{\sigma_0}_{I(B)} \subseteq R_{I(B)}.$$

Proof. We give only one direction for items (2) and (3).

2. If possible, let $\mathfrak{M} := (W, \{R_{I(B)}\}_{B\subseteq \mathcal{AC}}, \{R_{S(B)}\}_{B\subseteq \mathcal{AC}}, m, V))$ be an IS-model and $\mathfrak{M}^{\sigma_0} := (W, \{R^{\sigma_0}_{I(B)}\}_{B\subseteq \mathcal{AC}}, \{R^{\sigma_0}_{S(B)}\}_{B\subseteq \mathcal{AC}}, m^{\sigma_0}, V))$ be the corresponding updated IS-model with the information $\sigma_0 = (p, a, v)$ such that there exist $x, y \in W$, $x \neq y$ and $(x, y) \in R^{\sigma_0}_{I(B)}$. Consider the IS-model $\mathfrak{M}^* := (W, \{R_{I(B)}\}_{B\subseteq \mathcal{AC}}, \{R_{S(B)}\}_{B\subseteq \mathcal{AC}}, m, V^*))$, where $V^*(r) = V(r)$, $r \neq q$ and $V^*(q) = \{x\}$ and corresponding updated IS-model $(\mathfrak{M}^*)^{\sigma_0} := (W, \{R^{*\sigma_0}_{I(B)}\}_{B\subseteq \mathcal{AC}}, \{R^{*\sigma_0}_{S(B)}\}_{B\subseteq \mathcal{AC}}, m^{*\sigma_0}, V^{\sigma_0})$ with respect to the information σ_0. Since $V^*(p) = V(p)$, we have $R^{*\sigma_0}_{I(C)} = R^{\sigma_0}_{I(C)}$, for all $C \subseteq \mathcal{AC}$. Therefore, we have $(x, y) \in R^{*\sigma_0}_{I(B)}$ and hence, $\mathfrak{M}^{*\sigma_0}, x \not\models \langle I(B)\rangle q \rightarrow [I(B)]q$. This gives $\mathfrak{M}^*, x \not\models [(p, a, v)](\langle I(B)\rangle q \rightarrow [I(B)]q)$, a contradiction.

3. Let $\mathfrak{M} \models \alpha$. We shall prove that $R^{\sigma_0}{}_{I(B)} \subseteq R_{I(B)}$. If $a \notin B$, then one gets $R^{\sigma_0}{}_{I(B)} = R_{I(B)}$. So let us assume that $a \in B$ and $(x, y) \in R^{\sigma_0}{}_{I(B)}$. We need to prove that $(x, y) \in R_{I(B)}$. We only consider the case when $x \in V(p)$ and $y \notin V(p)$. The other cases can be done in the same way. Using the definition of $R^{\sigma_0}{}_{I(B)}$, one obtains $y \in m(a, v)$, $(x, y) \in R_{I(B\setminus\{a\})}$. Thus we only need to prove $(x, y) \in R_{I(a)}$ as $R_{I(B\setminus\{a\})} \cap R_{I(a)} = R_{I(B)}$. If possible, let $x \notin m(a, v)$. Since $\mathfrak{M}, x \models ((p \wedge \neg(a, v)) \rightarrow [I(B \setminus \{a\})](\neg p \rightarrow \neg(a, v)))$, we get $y \notin m(a, v)$, a contradiction. □

12.6 Axiomatization for the Dynamic Logics of Information Systems

Let DLIS and DLDIS be the deductive systems obtained by extending LIS and LDIS respectively by adding the Axioms Ax16-Ax22 and the rule $N(Inf)$ given below.

Axiom schema:

Ax16 $[(\phi, a, v)]p \leftrightarrow p$.
Ax17 $[(\phi, a, v)]\neg\alpha \leftrightarrow \neg[(\phi, a, v)]\alpha$.
Ax18 $[(\phi, a, v)](\alpha \rightarrow \beta) \leftrightarrow ([(\phi, a, v)]\alpha \rightarrow [(\phi, a, v)]\beta)$.
Ax19(a) $[(\phi, a, v)](a', v') \leftrightarrow (a', v')$, $a' \in \mathcal{AC} \setminus \{a\}$ and $v' \in \mathcal{VC} \cup \{*\}$.
Ax19(b) $[(\phi, a, v)](a, v') \leftrightarrow (\neg\phi \wedge (a, v'))$, $v' \in (\mathcal{VC} \cup \{*\}) \setminus \{v\}$.
Ax19(c) $[(\phi, a, v)](a, v) \leftrightarrow (\neg\phi \rightarrow (a, v))$.
Ax20(a) $[(\phi, a, v)][L(B)]\alpha \leftrightarrow [L(B)]\alpha$, $a \notin B$, $L \in \{I, S\}$.
Ax20(b) For $a \in B$,

$$\begin{aligned}[(\phi, a, v)][I(B)]\alpha \leftrightarrow \Big(&\big(\phi \rightarrow [I(B \setminus \{a\})](\phi \rightarrow [(\phi, a, v)]\alpha)\big) \\ &\wedge \big(\phi \rightarrow [I(B \setminus \{a\})](\neg\phi \wedge (a, v) \rightarrow [(\phi, a, v)]\alpha)\big) \\ &\wedge \big((\neg\phi \wedge (a, v)) \rightarrow [I(B \setminus \{a\})](\phi \rightarrow [(\phi, a, v)]\alpha)\big) \\ &\wedge \big(\neg\phi \rightarrow [I(B)](\neg\phi \rightarrow [(\phi, a, v)]\alpha)\big) \Big).\end{aligned}$$

Ax20(c) For $a \in B$,

$$[(\phi, a, v)][S(B)]\alpha \leftrightarrow \Big(\big(\phi \to [S(B \setminus \{a\})](\phi \to [(\phi, a, v)]\alpha)\big) \\ \wedge \big(\phi \to [S(B \setminus \{a\})](\neg\phi \wedge ((a, v) \vee (a, *)) \to [(\phi, a, v)]\alpha)\big) \\ \wedge \big((\neg\phi \wedge ((a, v) \vee (a, *))) \to [S(B \setminus \{a\})](\phi \to [(\phi, a, v)]\alpha)\big) \\ \wedge \big(\neg\phi \to [S(B)](\neg\phi \to [(\phi, a, v)]\alpha)\big) \Big).$$

Ax21 $[\sigma; \sigma']\alpha \leftrightarrow [\sigma][\sigma']\alpha$.
Ax22 $[\sigma \vee \sigma']\alpha \leftrightarrow [\sigma]\alpha \wedge [\sigma']\alpha$.

Rules of inference:

$$N(Inf). \qquad \frac{\alpha}{[(\phi, a, v)]\alpha}$$

for every atomic information (ϕ, a, v).

The axioms listed above may be further simplified, but we have given them in this form just to connect with Definition 75. Once again we note that $\mathbb{F} \subseteq \mathbb{F}^D$. Let *DLIS*$_\mathrm{f}$ and DLDIS$_\mathrm{f}$ be the deductive systems *in the language* $\mathbb{F}_f^D$ obtained from LIS$_\mathrm{f}$ and LDIS$_\mathrm{f}$ respectively by adding the Axioms Ax16-Ax22 and the rule $N(Inf)$. We shall write $\vdash_{\mathrm{LIS}}^D \alpha$, $\vdash_{\mathrm{LDIS}}^D \alpha$, $\vdash_{\mathrm{LIS_f}}^D \alpha$ and $\vdash_{\mathrm{LDIS_f}}^D \alpha$ according as α is a theorem of the system DLIS, DLDIS, DLIS$_\mathrm{f}$ and DLDIS$_\mathrm{f}$ respectively. Let us denote the set of all wffs of the language of the systems DLIS, DLDIS, DLIS$_\mathrm{f}$ and DLDIS$_\mathrm{f}$ respectively by $\mathbb{F}_{\mathrm{LIS}}^D$, $\mathbb{F}_{\mathrm{LDIS}}^D$, $\mathbb{F}_{\mathrm{LIS_f}}^D$ and $\mathbb{F}_{\mathrm{LDIS_f}}^D$. Note that $\mathbb{F}_{\mathrm{LIS}}^D = \mathbb{F}_{\mathrm{LDIS}}^D = \mathbb{F}^D$ and $\mathbb{F}_{\mathrm{LIS_f}}^D = \mathbb{F}_{\mathrm{LDIS_f}}^D = \mathbb{F}_f^D$. Next we prove the soundness theorem.

Theorem 34 (Soundness).

(i) For $\Lambda \in \{\mathrm{LIS}, \mathrm{LIS_f}\}$ *and* $\alpha \in \mathbb{F}_\Lambda^D$*, if* $\vdash_\Lambda^D \alpha$*, then* $\models_{IS} \alpha$ *and hence* $\models_{SIS} \alpha$.
(ii) For $\Lambda \in \{\mathrm{LDIS}, \mathrm{LDIS_f}\}$ *and* $\alpha \in \mathbb{F}_\Lambda^D$*, if* $\vdash_\Lambda^D \alpha$*, then* $\models_{DIS} \alpha$ *and hence* $\models_{SDIS} \alpha$.

Proof. We only prove the validity of Ax20(b) in all IS-models.
Let $\mathfrak{M}, w \models [(\phi, a, v)][I(B)]\alpha$. Then

$$\mathfrak{M}^{\sigma_0}, w \models [I(B)]\alpha, \ \sigma_0 = (\phi, a, v). \tag{36}$$

We need to prove

(i) $\mathfrak{M}, w \models \phi \to [I(B \setminus \{a\})](\phi \to [(\phi, a, v)]\alpha)$,
(ii) $\mathfrak{M}, w \models \phi \to [I(B \setminus \{a\})](\neg\phi \wedge (a, v) \to [(\phi, a, v)]\alpha)$,
(iii) $\mathfrak{M}, w \models (\neg\phi \wedge (a, v)) \to [I(B \setminus \{a\})](\phi \to [(\phi, a, v)]\alpha)$,
(iv) $\mathfrak{M}, w \models \neg\phi \to [I(B)](\neg\phi \to [(\phi, a, v)]\alpha)$.

(i) Let $\mathfrak{M}, w \models \phi$, $(w, w') \in R_{I(B \setminus \{a\})}$ and $\mathfrak{M}, w' \models \phi$. This gives $(w, w') \in R^{\sigma_0}_{I(B)}$. Therefore, by (36), we obtain $\mathfrak{M}^{\sigma_0}, w' \models \alpha$, i.e. $\mathfrak{M}, w' \models [(\phi, a, v)]\alpha$.

The other cases can similarly be proved.

Now suppose (i)–(iv) hold. We will show $\mathfrak{M}, w \models [(\phi, a, v)][I(B)]\alpha$. Let $(w, w') \in R^{\sigma_0}_{I(B)}$. We need to prove $\mathfrak{M}^{\sigma_0}, w' \models \alpha$.

Here we have only four possibilities as given in Definition 75. First let $w \in [\![\phi]\!]_{\mathfrak{M}}$, $w' \notin [\![\phi]\!]_{\mathfrak{M}}$, $w' \in m(a, v)$ and $(w, w') \in R_{I(B \setminus \{a\})}$. So using (ii), we obtain the desired result. All the other cases can be proved in the same manner.

Using Proposition 111, it is shown that the rules of inference preserve validity. □

Completeness. We follow the standard technique of dynamic epistemic logic to prove the completeness theorem. Let $\Lambda \in \{\text{LIS}, \text{LDIS}, \text{LIS}_\text{f}, \text{LDIS}_\text{f}\}$. We first show that for every $\alpha \in \mathbb{F}^D_\Lambda$, there exists $\alpha^* \in \mathbb{F}_\Lambda$ such that $\vdash^D_\Lambda \alpha \leftrightarrow \alpha^*$. So in that case the completeness of the system Λ will give us the desired completeness result. Let us proceed step by step with the following lemmas.

Lemma 1. *If* $\vdash^D_\Lambda \alpha \leftrightarrow \alpha'$ *and* $\vdash^D_\Lambda \beta \leftrightarrow \beta'$, *then* $\neg\alpha, \alpha \wedge \beta, [I(B)]\alpha, [S(B)]\alpha, [(\phi, a, v)]\alpha$ *are respectively logically equivalent to* $\neg\alpha', \alpha' \wedge \beta', [I(B)]\alpha', [S(B)]\alpha'$ *and* $[(\phi, a, v)]\alpha'$.

Proof. We only consider the case $[(\phi, a, v)]\alpha$. In fact, this follows from Ax18 and inference rule $N(Inf)$. □

By this lemma and by using the axioms, we obtain

Lemma 2. *If* $\vdash^D_\Lambda \beta \leftrightarrow \beta'$ *and* β *is a sub-wff of* α, *then* $\vdash^D_\Lambda \alpha \leftrightarrow \alpha'$ *where* α' *is obtained from* α *by substituting some occurrences of* β *by* β'.

An easy induction on the complexity of the wff α gives us

Lemma 3. *For all* $\alpha \in \mathbb{F}^D_\Lambda$ *and information* (ϕ, a, v), *there exists* $\alpha^* \in \mathbb{F}_\Lambda$ *such that* $\vdash^D_\Lambda [(\phi, a, v)]\alpha \leftrightarrow \alpha^*$.

Lemma 4. *For all* $\alpha \in \mathbb{F}^D_\Lambda$ *and information* σ, *there exists* $\alpha^* \in \mathbb{F}_\Lambda$ *such that* $\vdash^D_\Lambda [\sigma]\alpha \leftrightarrow \alpha^*$.

Finally, we get

Proposition 115. *For all* $\alpha \in \mathfrak{L}^D_\Lambda$, *there exists* $\alpha^* \in \mathbb{F}_\Lambda$ *such that* $\vdash^D_\Lambda \alpha \leftrightarrow \alpha^*$.

Theorem 35 (Completeness).

(i) *For* $\Lambda \in \{\text{LIS}, \text{LIS}_\text{f}\}$ *and* $\alpha \in \mathbb{F}^D_\Lambda$, *if* $\models_{IS} \alpha$, *then* $\vdash^D_\Lambda \alpha$.
(ii) *For* $\Lambda \in \{\text{LDIS}, \text{LDIS}_\text{f}\}$ *and* $\alpha \in \mathbb{F}^D_\Lambda$, *if* $\models_{DIS} \alpha$, *then* $\vdash^D_\Lambda \alpha$.
(iii) *For* $\alpha \in \mathbb{F}^D_{\text{LIS}_\text{f}} = \mathbb{F}^D_f$, *if* $\models_{SIS} \alpha$, *then* $\vdash^D_{\text{LIS}_\text{f}} \alpha$.
(iv) *For* $\alpha \in \mathbb{F}^D_{\text{LDIS}_\text{f}} = \mathbb{F}^D_f$, *if* $\models_{SDIS} \alpha$, *then* $\vdash^D_{\text{LDIS}_\text{f}} \alpha$.

Proof. We only prove (i). The rest can be done in the same way. Given $\models_{IS} \alpha$. Proposition 115 guarantees the existence of a $\alpha^* \in \mathbb{F}_\Lambda$ such that $\vdash^D_\Lambda \alpha \leftrightarrow \alpha^*$. Using soundness we obtain $\models_{IS} \alpha^*$ and completeness (Theorem 30) gives $\vdash_\Lambda \alpha^*$. Hence we have $\vdash^D_\Lambda \alpha$. □

From Proposition 35 (items (iii) and (iv)), it follows that the fragments $\mathbb{F}^D_{IS_f}$ and $\mathbb{F}^D_{DIS_f}$ are the dynamic logics for ISs and DISs respectively.

We end this section by noting the relation between the dynamic logic DLDIS$_f$ and dynamic information systems (DIS) (cf. Definition 7). DLDIS$_f$ can study a scenario where the attribute values of the objects are changing with time. This is also the situation in the case of the DIS. As mentioned in Sect. 8.1, in the case of DISs, we have a number of DISs $\mathcal{K}_t$ corresponding to each time point t with the same set of objects, attributes, attribute values. The link with updates may be presented as follows. Suppose $t < t'$ and the DISs $\mathcal{K}_t$ and $\mathcal{K}_{t'}$ differ for only objects belonging to the sets $W_i,\ i = 1, 2, \ldots, n$ and the difference is for the attributes a_i, $i = 1, 2, \ldots, n$. Further, suppose that in $\mathcal{K}_t$, objects belonging to the set W_i take the value u_i for the attribute a_i. Then one can say that the $SDIS$-model $\mathfrak{M} := (\mathfrak{F}_{\mathcal{K}_t}, V)$ based on the $SDIS$-structure generated from $\mathcal{K}_t$, is obtained from the $SDIS$-model $\mathfrak{M}' := (\mathfrak{F}_{\mathcal{K}_{t'}}, V)$ due to an update with respect to the information $\bigwedge_{i=1}^n (p_i, a_i, u_i)$, where $V(p_i) = W_i$.

12.7 A Comparison with Dynamic Epistemic Logics

In this section, we compare the features of the dynamic logics defined in Sect. 12.5 with dynamic epistemic logics.

In public announcement logic PAL [104], the updated model is obtained by restricting the original model to some subset of the domain. More complex epistemic actions like the ones discussed in [4,31], may result in the refinement of accessibility relations while the domain of the model remains unchanged, and they may even result in the enlargement of the domain of the model. Contrary to these, in the case of dynamic logics for information systems defined in Sect. 12.5, the domain of the model remains unchanged whatever information is provided. Relations can change in any manner. Two objects not related earlier may become related after the update. However, we may have some situation where change in the relation follows some special pattern such as refinement.

We do not have the equivalence of $[\phi]K_a\psi$ and $K_a\psi$ in PAL, which makes it clear that the public announcement has an effect on the knowledge of the agents. Similarly, in the dynamic epistemic logic considered in [4], where more complicated actions were considered, we do not have the validity of the wff $[\alpha]\Box_A\phi \leftrightarrow \Box_A\phi$. In the dynamic epistemic logic considered in [31], we have the validity of the wff $[U_B\pi]\Box_a\phi \leftrightarrow \Box_a\phi$, provided $a \notin B$. This is due to the fact that the information is provided to the agents belonging to the group B and as the agent $a \notin B$, its information state remains unchanged.

Similar is the situation in dynamic logics for information systems. We have the validity of the wff $[\sigma][I(B)]\phi \leftrightarrow [I(B)]\alpha$ and $[\sigma]\langle I(B)\rangle\phi \leftrightarrow \langle I(B)\rangle\phi$, provided σ does not carry information regarding any attributes present in B. Thus it follows that the lower and upper approximations of the sets with respect to only those attributes will be affected about which information is provided.

In dynamic epistemic logics, usually an action or flow of information does not change the value of the atomic propositions. For instance, in PAL, the validity of the wff $[\phi]p \leftrightarrow (\phi \rightarrow p)$ represents this fact. Similarly, the wffs $[\alpha]p \leftrightarrow pre(\alpha) \rightarrow p$ and $[U_B\pi]p \leftrightarrow p$ are valid in the dynamic epistemic logic considered in [4] and [31] respectively. In [103], an action called 'public assignment' expressed as $[p := \phi]$, is considered which *only* affects the atomic information p. The update of the epistemic model $\mathfrak{M} := (U, \{R_n\}_{n \in Ag}, V)$ with respect to the public assignment $[p := \phi]$ gives the model $(U, \{R_n\}_{n \in Ag}, V')$, where $V'(q) = V(q)$ for $q \neq p$ and $V'(p) = [\![\phi]\!]_{\mathfrak{M}}$. Note that relations remain unchanged in this case.

In dynamic logics for information systems, we have two kinds of atomic wffs: the propositional variables and descriptors. Although we have the validity of $[\sigma]p \leftrightarrow p$, we may have information σ and descriptor (a, v) such that the wff $[\sigma](a, v) \leftrightarrow (a, v)$ is not valid. This shows that atomic facts may change due to flow of information. We also note that the update due to public assignment is different from one due to the actions like (ϕ, a, v). Actually, in this case, not only is the assignment to the atomic wff (a, v) affected, but unlike public assignment, the assignments to other atomic wffs as well as the relations may get affected. This is because relations are connected with the attribute, attribute value pairs. Note that $[\![\phi]\!]_{\mathfrak{M}} \subseteq m'(a, v)$, but unlike public assignment, we do not necessarily have $[\![\phi]\!]_{\mathfrak{M}} = m'(a, v)$.

The information or action may be a total or partial function in dynamic epistemic logics. For example, the public announcement action $[\phi]$ in PAL is a partial function as we have the validity of $\langle\phi\rangle\psi \rightarrow [\phi]\psi$, but $\langle\phi\rangle\top$ is not valid. On the other hand, in the dynamic epistemic logic considered in [4], the action $[U_B\pi]$ is a total function which follows from the validity of the wff $\neg[U_B\pi]\psi \rightarrow [U_B\pi]\neg\psi$. There are some actions which are non-deterministic. In fact, the only source of non-determinism is the non-deterministic action operator $\cup$.

Similar is the situation in dynamic logics for information systems. The information (ϕ, a, v) is a total function, as the wff $\neg[(\phi, a, v)]\alpha \leftrightarrow [(\phi, a, v)]\neg\alpha$ is valid. The information operator $\vee$ is the only source of non-determinism.

13 Multi-agent Scenario: Revisited

In this section, we shall extend our study to the multi-agent scenario. For the purpose, let us first extend the notion of (deterministic/incomplete) information systems. Consider a set $Ag := \{1, 2, \ldots, n\}$ of agents.

Definition 76. *A tuple* $\mathcal{S} := (U, \mathcal{A}, \{Val_a\}_{a \in \mathcal{A}} \cup \{*\}, \{f_i\}_{i \in Ag})$ *is called a* multi-agent information system *(MIS), where* $U, \mathcal{A}, Val_a, *$ *are as in Definition 9, and for each* $i \in Ag$, $f_i : U \times \mathcal{A} \rightarrow \bigcup\{Val_a : a \in \mathcal{A}\} \cup \{*\}$ *is such that* $f_i(x, a) \in Val_a \cup \{*\}$.

An information system which satisfies $f_i(x, a) = *$ *for some* $x \in U$, $a \in \mathcal{A}$ *and* $i \in Ag$, *will be called a* multi-agent incomplete information system *(MIIS).*

On the other hand, if $f_i(x,a) \neq *$ for all $x \in U$, $a \in \mathcal{A}$ and all $i \in Ag$, then it will be called a multi-agent deterministic information system *(MDIS).*

For each $B \subseteq \mathcal{A}$ and agent i, one then obtains $Ind_{\mathcal{S}}(i,B)$ and $Sim_{\mathcal{S}}(i,B)$.

Definition 77.

- $(x,y) \in Ind_{\mathcal{S}}(i,B)$ *if and only if* $f_i(x,a) = f_i(y,a)$ *for all* $a \in B$.
- $(x,y) \in Sim_{\mathcal{S}}(i,B)$ *if and only if* $f_i(x,a) = f_i(y,a)$, *or* $f_i(x,a) = *$, *or* $f_i(y,a) = *$, *for all* $a \in B$.

In Sect. 13.1, we shall present logics for MISs and MDISs and as before, we extend them to define corresponding dynamic logics in Sect. 13.2. Just as in the single agent case, information could be provided to an individual agent or a group of agents from outside. Moreover, in this multi-agent situation, an agent may update her information with help from agents inside the system as well: she may borrow information of another agent. So, the dynamic logics proposed in Sect. 13.2 can express flow of information from outside the system as well as transfer of information between the agents. All the results that hold for the single agent case can be carried over to the multi-agent case and so we omit the detail.

The content of this section is based on the article [48].

13.1 The Logics for MIS and MDIS

Let us consider the language $\mathbb{L}_M$ to be the same as the language of the logics for IS and DIS except that it also has a set $Ag := \{1, 2, \ldots, n\}$ of agents. In order to incorporate agents, the descriptors and modalities are modified to be of the form (i,a,v) and $[L(i,B)]$ respectively, where $i \in Ag$, $a \in \mathcal{AC}$, $v \in \mathcal{VC}_a \cup \{*\}$, $B \subseteq \mathcal{AC}$ and $L \in \{I, S\}$. So the wffs of $\mathbb{L}_M$ are defined recursively as:

$$\top \mid \bot \mid (i,a,v) \mid p \mid \neg\alpha \mid \alpha \wedge \beta \mid [L(i,B)]\alpha.$$

Let $\mathcal{D}_M$ be the set of descriptors. We use the earlier notation $\mathbb{F}$ and $\mathbb{F}_f$ respectively for the set of wffs of $\mathbb{L}_M$ and the set of wffs which do not involve any modal operator $[L(i,B)]$, where B is an infinite subset of $\mathcal{AC}$.

Semantics. The semantics of $\mathbb{L}_M$ is based on the notions of MIS and $MDIS$ structures, defined as follows.

Definition 78. *By a MIS-structure ($MDIS$-structure), we mean a tuple* $\mathfrak{F} := (W, \{R_{I(i,B)}\}_{i \in Ag, B \subseteq \mathcal{AC}}, \{R_{S(i,B)}\}_{i \in Ag,\ B \subseteq \mathcal{AC}}, m)$, *where* $m : \mathcal{D}_M \to 2^W$ *is such that for each* $i \in Ag$, $(W, \{R_{I(i,B)}\}_{B \subseteq \mathcal{AC}}, \{R_{S(i,B)}\}_{B \subseteq \mathcal{AC}}, m^i)$ *is an* IS*-structure (*DIS*-structure),* m^i *being defined as* $m^i(a,v) := m(i,a,v)$.

Let us call a model $\mathfrak{M} := (\mathfrak{F}, V)$, $V : PV \to 2^W$, a MIS-model or a $MDIS$-model according as $\mathfrak{F}$ is a MIS or a $MDIS$ structure.

The satisfiability of a wff $\alpha \in \mathbb{F}$ in a model $\mathfrak{M} := (\mathfrak{F}, V)$ at an object $w \in W$ is defined in the standard way. For instance,

$\mathfrak{M}, w \models (i, a, v)$ if and only if $w \in m(i, a, v)$, for $(i, a, v) \in \mathcal{D}_M$.
$\mathfrak{M}, w \models [L(i, B)]\alpha$, $L \in \{I, S\}$, if and only if for all w' in W with $(w, w') \in R_{L(i,B)}$, $\mathfrak{M}, w' \models \alpha$.

We define the notions of standard MIS-models ($SMIS$-models) and standard $MDIS$-models ($SMDIS$-models) in the line of standard IS and DIS models. Let us write $\models_k \alpha$ if α is valid in the class of all k-models, $k \in$ {MIS, SMIS, MDIS, SMDIS}.

Connection with Strong/Weak Approximations. In the case of MISs, the notions of strong/weak lower and upper approximations (cf. Definition 19) can be defined for the indiscernibility and similarity relations corresponding to a set of agents and a set of attributes. For instance, for the set $P \subseteq Ag$ and $B \subseteq \mathcal{AC}$, we obtain strong lower approximation for indiscernibility as the set $\bigcap_{i \in P} \underline{X}_{Ind(i,B)}$. Since, here we are dealing with a fixed and finite set of agents, the language given above can express these notions of strong/weak approximations using the disjuncts and conjuncts. For instance, the wff $\bigwedge_{i \in Ag}[I(i, B)]p$ gives the strong lower approximation of the set represented by p for the indiscernibility relation.

Axiomatization and Decidability. Let LMIS be the logic consisting of the axioms AxM1-AxM15 along with the inference rules MP and N given below. Further, suppose LMDIS is the extension of LMIS with the axioms AXCM1 and AxCM2.

Let $B, C \subseteq \mathcal{AC}$, $F \in \{P : P \subseteq_i \mathcal{AC}\} \cup \{\emptyset\}$, $v, v' \in \mathcal{VC}_a \cup \{*\}$, $u \in \mathcal{VC}_a$ and $L \in \{I, S\}$.

Axiom schema:

AxM1. All axioms of classical propositional logic.
AxM2 $[L(i, B)](\alpha \to \beta) \to ([L(i, B)]\alpha \to [L(i, B)]\beta)$.
AxM3 $[L(i, F)]\alpha \to \alpha$.
AxM4 $\langle L(i, F)\rangle[L(i, F)]\alpha \to \alpha$.
AxM5 $[I(i, F)][I(i, F)]\alpha \to [I(i, F)]\alpha$.
AxM6 $[L(i, C)]\alpha \to [L(i, B)]\alpha$ for $C \subseteq B \subseteq \mathcal{AC}$.
AxM7 $(i, a, v) \to \neg(i, a, v')$, for $v \neq v'$.
AxM8 $\bigvee_{v \in \mathcal{VC}_a \cup \{*\}}(i, a, v)$.
AxM9 $(i, a, v) \to [I(i, a)](i, a, v)$.
AxM10 $((i, a, v) \wedge [I(i, B \cup \{a\})]\alpha) \to [I(i, B)]((i, a, v) \to \alpha)$.
AxM11 $(i, a, u) \to [S(i, a)]((i, a, u) \vee (i, a, *))$.
AxM12 $((i, a, u) \wedge [S(i, B \cup \{a\})]\alpha) \to [S(i, B)](((i, a, u) \vee (i, a, *)) \to \alpha)$.
AxM13 $((i, a, *) \wedge [S(i, B \cup \{a\})]\alpha) \to [S(i, B)]\alpha$.
AxM14 $[S(i, F)]\alpha \to [I(i, F)]\alpha$.
AxM15 $[I(i, \emptyset)]\alpha \to [S(i, \emptyset)]\alpha$.

AxM16 $[I(i, \emptyset)]\alpha \rightarrow [L(j, B]\alpha$.
AxCM1 $\neg(i, a, *)$.
AxCM2 $[S(i, F)]\alpha \leftrightarrow [I(i, F)]\alpha$.

Rules of inference:

$$N. \ \frac{\alpha}{[L(i, B)]\alpha} \qquad MP. \ \frac{\alpha \quad \alpha \rightarrow \beta}{\beta}$$

One can easily identify the axioms AxM1-AxM15, AxCM1,AxCM2 with the Ax1-Ax15, AxC1,AxC2 respectively given in Sect. 12.3. Axiom AxM16 states that all the relations in the system are contained in the relations $R_{I(i,\emptyset)}$ for each $i \in Ag$. In order to obtain the completeness theorem, we use the generated submodel as in the case of logics for information systems (cf. Sect. 12.3), and there we will require this fact.

Let LMIS_f and LMDIS_f be the deductive systems LMIS and LMDIS respectively restricted to the language $\mathbb{F}_f$. Moreover, let us denote the set of all wffs of the language of the system Λ, $\Lambda \in \{\text{LMIS}, \text{LMDIS}, \text{LMIS}_\text{f}, \text{LMDIS}_\text{f}\}$, by $\mathbb{F}_\Lambda$. Note that $\mathbb{F}_{\text{LMIS}} = \mathbb{F}_{\text{LMDIS}} = \mathbb{F}$ and $\mathbb{F}_{\text{LMIS}_\text{f}} = \mathbb{F}_{\text{LMDIS}_\text{f}} = \mathbb{F}_f$.

Following exactly the same technique as in Sect. 12.3, one can obtain the following soundness and completeness theorems.

Theorem 36 (Soundness).

(i) For $\Lambda \in \{\text{LMIS}, \text{LMIS}_\text{f}\}$ *and* $\alpha \in \mathbb{F}_\Lambda$, *if* $\vdash_\Lambda \alpha$, *then* $\models_{MIS} \alpha$ *and hence* $\models_{SMIS} \alpha$.
(ii) For $\Lambda \in \{\text{LMDIS}, \text{LMDIS}_\text{f}\}$ *and* $\alpha \in \mathbb{F}_\Lambda$, *if* $\vdash_\Lambda \alpha$, *then* $\models_{MDIS} \alpha$ *and hence* $\models_{SMDIS} \alpha$.

Theorem 37 (Completeness).

(i) For $\Lambda \in \{\text{LMIS}, \text{LMIS}_\text{f}\}$ *and* $\alpha \in \mathbb{F}_\Lambda$, *if* $\models_{MIS} \alpha$, *then* $\vdash_\Lambda \alpha$.
(ii) For $\alpha \in \mathbb{F}_{\text{LMIS}_\text{f}}$, *if* $\models_{SMIS} \alpha$, *then* $\vdash_{\text{LMIS}_\text{f}} \alpha$.
(iii) For $\Lambda \in \{\text{LMDIS}, \text{LMDIS}_\text{f}\}$ *and* $\alpha \in \mathbb{F}_\Lambda$, *if* $\models_{MDIS} \alpha$, *then* $\vdash_\Lambda \alpha$.
(iv) For $\alpha \in \mathbb{F}_{\text{LMDIS}_\text{f}}$, *if* $\models_{SMDIS} \alpha$, *then* $\vdash_{\text{LMDIS}_\text{f}} \alpha$.

Note that the above theorem does not give the completeness of LMIS with respect to the class of standard MIS-structures. However, we obtain the completeness of LMIS_f with respect to the class of standard MIS-structures (item (ii)) and hence LMIS_f is a logic for multi-agent information systems. Similarly, LMDIS_f is a logic for multi-agent deterministic information systems. These logics can express the properties of indiscernibility and similarity relations (cf. Definition 77), corresponding to any finite set of attributes.

We also have the following decidability result. The proof is similar to Theorem 32. Let $k \in \{MIS, MDIS, SMIS, SMDIS\}$.

Theorem 38. *We can decide for a given* $\alpha \in \mathbb{F}_f$, *whether* $\models_k \alpha$.

13.2 Dynamic Logics for MIS and MDIS

In this section, we shall define dynamic logics for MIS and $MDIS$ by extending the logics of Sect. 13.1. As mentioned earlier, here information may not necessarily be for individual agents but may be for a group of agents. We would also like to capture the flow of information between the agents. Thus we modify the set Inf of information (cf. Definition 74) as follows.

Definition 79. *The set Inf of* information *is the smallest set such that*

- $(\phi, P, a, v), (\phi, P, j, a) \in Inf$, *where* $\phi \in \mathbb{F}$, $a \in \mathcal{AC}, v \in \mathcal{VC}_a$, $P \subseteq Ag$ *and* $j \in Ag$,
- *if* $\sigma, \sigma' \in Inf$, *then* $\sigma;\sigma'$, $\sigma \vee \sigma' \in Inf$.

An information of the form $(\phi, \{i\}, a, v)$ and $(\phi, \{i\}, j, a)$, simply written as (ϕ, i, a, v) and (ϕ, i, j, a), will be called an *atomic information.* (ϕ, P, a, v) signifies that information is obtained by the agents $i \in P$ according to which the objects represented by ϕ take the value v for the attribute a. On the other hand, (ϕ, P, j, a) signifies that the agents belonging to the set P replace the information about the objects represented by ϕ regarding the attribute a with the information the agent j has about these objects regarding the same attribute a. Thus $((i, a, *), i, j, a)$ says that the agent i adopts the information regarding attribute a that agent j has about any object lying in the interpretation of $(i, a, *)$ (i.e. an object for which i has no information for attribute a).

The set $\mathbb{F}^D$ of wffs is then obtained by extending the wff-formation rules of $\mathbb{L}_M$ with the clause:

$$\text{if } \sigma \in Inf,\ \alpha \in \mathbb{F}^D \text{ then } [\sigma]\alpha \in \mathbb{F}^D.$$

Semantics. Our next task is to define the notion of update of the models. Note that we need to define updates with respect to both the types of atomic information, viz. (ϕ, i, a, v) and (ϕ, i, j, a).

Let $\mathfrak{M} := (W, \{R_{I(i,B)}\}_{i \in Ag, B \subseteq \mathcal{AC}}, \{R_{S(i,B)}\}_{i \in Ag,\ B \subseteq \mathcal{AC}}, m)$ be a model and consider an atomic information σ_0. The model obtained by updating $\mathfrak{M}$ with σ_0, denoted as $\mathfrak{M}^{\sigma_0}$, is given by Definitions 80 and 81, depending on the type of atomic information σ_0.

Definition 80. *Let* $\sigma_0 := (\phi, i, a, v)$.

Then $\mathfrak{M}^{\sigma_0} := (W, \{R^{\sigma_0}_{I(i,B)}\}_{i \in Ag, B \subseteq \mathcal{AC}}, \{R^{\sigma_0}_{S(i,B)}\}_{i \in Ag,\ B \subseteq \mathcal{AC}}, m^{\sigma_0}, V)$, *where*

- m^{σ_0} *is given as follows:*
  - $m^{\sigma_0}(j, b, u) := m(j, b, u)$ *for* $j \neq i$ *or* $b \neq a$,
  - $m^{\sigma_0}(i, a, v') := m(i, a, v') \setminus [\![\phi]\!]_{\mathfrak{M}}$ *for* $v \neq v'$,
  - $m^{\sigma_0}(i, a, v) := m(i, a, v) \cup [\![\phi]\!]_{\mathfrak{M}}$.
- $R^{\sigma_0}_{L(j,B)}$, $L \in \{I, S\}$ *are defined as follows:*
  - *For* $a \notin B$ *or* $j \neq i$, $R^{\sigma_0}_{L(j,B)} := R_{L(j,B)}$, $L \in \{I, S\}$.
  - *If* $a \in B$, *then* $(x, y) \in R^{\sigma_0}_{I(i,B)}$ *if and only if either of the following holds.*

  * $x, y \notin [\![\phi]\!]_{\mathfrak{M}}$ *and* $(x, y) \in R_{I(i,B)}$,
  * $x \in [\![\phi]\!]_{\mathfrak{M}}$, $y \notin [\![\phi]\!]_{\mathfrak{M}}$, $y \in m(i, a, v)$ *and* $(x, y) \in R_{I(i,B\setminus\{a\})}$,
  * $x \notin [\![\phi]\!]_{\mathfrak{M}}$, $y \in [\![\phi]\!]_{\mathfrak{M}}$, $x \in m(i, a, v)$ *and* $(x, y) \in R_{I(i,B\setminus\{a\})}$,
  * $x, y \in [\![\phi]\!]_{\mathfrak{M}}$ *and* $(x, y) \in R_{I(i,B\setminus\{a\})}$.
- *If* $a \in B$, *then* $(x, y) \in R^{\sigma_0}_{S(i,B)}$ *if and only if either of the following holds.*
  * $x, y \notin [\![\phi]\!]_{\mathfrak{M}}$ *and* $(x, y) \in R_{S(i,B)}$,
  * $x \in [\![\phi]\!]_{\mathfrak{M}}$, $y \notin [\![\phi]\!]_{\mathfrak{M}}$, $y \in m(i, a, v) \cup m(i, a, *)$ *and* $(x, y) \in R_{S(i,B\setminus\{a\})}$,
  * $x \notin [\![\phi]\!]_{\mathfrak{M}}$, $y \in [\![\phi]\!]_{\mathfrak{M}}$, $x \in m(i, a, v) \cup m(i, a, *)$ *and* $(x, y) \in R_{S(i,B\setminus\{a\})}$,
  * $x, y \in [\![\phi]\!]_{\mathfrak{M}}$ *and* $(x, y) \in R_{S(i,B\setminus\{a\})}$.

Definition 81. *Let* $\sigma_0 := (\phi, i, j, a)$.

Then $\mathfrak{M}^{\sigma_0} := (W, \{R^{\sigma_0}_{I(i,B)}\}_{i\in Ag, B\subseteq AC}, \{R^{\sigma_0}_{S(i,B)}\}_{i\in Ag,\ B\subseteq AC}, m^{\sigma_0}, V)$, *where*

- m^{σ_0} *is given as follows:*
  - $m^{\sigma_0}(k, b, u) := m(k, b, u)$ *for* $k \neq i$ *or* $b \neq a$,
  - $m^{\sigma_0}(i, a, v) := (m(j, a, v) \cap [\![\phi]\!]_{\mathfrak{M}}) \cup (m(i, a, v) \cap [\![\neg\phi]\!]_{\mathfrak{M}})$.
- $R^{\sigma_0}_{L(j,B)}$, $L \in \{I, S\}$ *are defined as follows:*
  - *For* $a \notin B$ *or* $j \neq i$, $R^{\sigma_0}_{L(j,B)} := R_{L(j,B)}$, $L \in \{I, S\}$.
  - *If* $a \in B$, *then* $(x, y) \in R^{\sigma_0}_{I(i,B)}$ *if and only if either of the following holds.*
    * $x, y \notin [\![\phi]\!]_{\mathfrak{M}}$ *and* $(x, y) \in R_{I(i,B)}$,
    * $x \in [\![\phi]\!]_{\mathfrak{M}}$, $y \notin [\![\phi]\!]_{\mathfrak{M}}$, $(x, y) \in R_{I(i,B\setminus\{a\})}$ *and* $x \in m(j, a, v), y \in m(i, a, v)$ *for some* v,
    * $x \notin [\![\phi]\!]_{\mathfrak{M}}$, $y \in [\![\phi]\!]_{\mathfrak{M}}$, $(x, y) \in R_{I(i,B\setminus\{a\})}$ *and* $x \in m(i, a, v), y \in m(j, a, v)$ *for some* v,
    * $x, y \in [\![\phi]\!]_{\mathfrak{M}}$ *and* $(x, y) \in R_{I(j,\{a\})} \cap R_{I(i,B\setminus\{a\})}$.
  - *If* $a \in B$, *then* $(x, y) \in R^{\sigma_0}_{S(i,B)}$ *if and only if either of the following holds.*
    * $x, y \notin [\![\phi]\!]_{\mathfrak{M}}$ *and* $(x, y) \in R_{S(i,B)}$,
    * $x \in [\![\phi]\!]_{\mathfrak{M}}$, $y \notin [\![\phi]\!]_{\mathfrak{M}}$, $(x, y) \in R_{S(i,B\setminus\{a\})}$ *and (i)* $y \in m(i, a, *)$, *or (ii)* $x \in m(j, a, *)$, *or(iii)* $y \in m(i, a, v)$ *and* $x \in m(j, a, v)$ *for some* $v \in \mathcal{VC}_a$,
    * $x \notin [\![\phi]\!]_{\mathfrak{M}}$, $y \in [\![\phi]\!]_{\mathfrak{M}}$, $(x, y) \in R_{S(i,B\setminus\{a\})}$ *and (i)* $x \in m(i, a, *)$, *or (ii)* $y \in m(j, a, *)$, *or(iii)* $x \in m(i, a, v)$ *and* $y \in m(j, a, v)$ *for some* $v \in \mathcal{VC}_a$,
    * $x, y \in [\![\phi]\!]_{\mathfrak{M}}$ *and* $(x, y) \in R_{S(j,\{a\})} \cap R_{S(i,B\setminus\{a\})}$.

Each information $\sigma \in Inf$ induces a relation $\mathfrak{R}_\sigma$ on the set of all models. As before, $R_1; R_2$ denotes the composition of the two relations R_1 and R_2.

- for atomic information $\sigma_0 := (\phi, i, a, v)$ or $\sigma_0 := (\phi, i, j, a)$, $(\mathfrak{M}, \mathfrak{M}') \in \mathfrak{R}_{\sigma_0}$ if and only if $\mathfrak{M}' = \mathfrak{M}^{\sigma_0}$,
- for $\sigma := (\phi, P, a, v)$, $\mathfrak{R}_\sigma := \mathfrak{R}_{\sigma_1}; \mathfrak{R}_{\sigma_2}; \cdots ; \mathfrak{R}_{\sigma_m}$, where $P := \{i_1, i_2, \ldots, i_m\}$ and $\sigma_k := (\phi, i_k, a, v)$,
- for $\sigma := (\phi, P, j, a)$, $\mathfrak{R}_\sigma := \mathfrak{R}_{\sigma_1}; \mathfrak{R}_{\sigma_2}; \cdots ; \mathfrak{R}_{\sigma_m}$, where $P := \{i_1, i_2, \ldots, i_m\}$ and $\sigma_k := (\phi, i_k, j, a)$,

- $\mathfrak{R}_{\sigma;\sigma'} := \mathfrak{R}_\sigma ; \mathfrak{R}'_{\sigma'}$,
- $\mathfrak{R}_{\sigma\vee\sigma'} := \mathfrak{R}_\sigma \cup \mathfrak{R}'_{\sigma}$.

For $\mathfrak{R}_{\sigma_i}$ and $\mathfrak{R}_{\sigma_j}$, observe that $\mathfrak{R}_{\sigma_i};\mathfrak{R}_{\sigma_j} = \mathfrak{R}_{\sigma_j};\mathfrak{R}_{\sigma_i}$.

Let $k \in \{MIS, MDIS, SMIS, SMDIS\}$. We obtain the following similar to Proposition 111.

Proposition 116. *Let $\sigma \in Inf$ and $\mathfrak{M}, \mathfrak{M}'$ be two models such that $(\mathfrak{M}, \mathfrak{M}') \in \mathfrak{R}_\sigma$. If $\mathfrak{M}$ is a k-model, then $\mathfrak{M}'$ is also a k-model.*

Finally, we define the satisfiability of the wff $\alpha \in \mathbb{F}^D$ in a model $\mathfrak{M}$ at the world w by extending the definition of satisfiability relation for the wffs of $\mathbb{F}$ with the clause:

$\mathfrak{M}, w \models [\sigma]\alpha$ if and only if $\mathfrak{M}', w \models \alpha$, for all $\mathfrak{M}'$ such that $(\mathfrak{M}, \mathfrak{M}') \in \mathfrak{R}_\sigma$.

The following example shows how the language defined above can be used to express the information flow between the agents.

Example 22. We modify the IIS $\mathcal{K}_1$ of Sect. 12.1 (cf. Table 6) and turn it into a MIIS $\mathcal{S}_1$ given in Table 9 below. We now have three agents 1, 2, 3 in the picture. Information I_1 and I_2 given in the context of $\mathcal{K}_1$ could now be meant for particular agents of the system. These could be modified as:

Table 9. MIS $\mathcal{S}_1$

	1			2			3		
	T	H	N	T	H	N	T	H	N
P1	High	*	No	High	*	No	High	No	No
P2	Very high	*	Yes	No	*	No	Very high	Yes	Yes
P3	*	No	No	*	No	No	No	*	No
P4	No	No	No	*	Yes	Yes	Very high	*	Yes
P5	No	Yes	No	*	Yes	No	No	Yes	*
P6	High	No	Yes	*	No	Yes	High	No	*

I_3: Agent 1 has got the information that her prior report regarding the temperature and nausea of the patient P4 is not correct. In fact, P4 has nausea and temperature (very high or high).

I_4: All agents have got the information that if a patient has high or very high temperature and nausea, then she must have headache.

So, information I_3 is for the agent 1 and information I_4 is for all the agents of the system.

In addition to getting information from outside, agents may also update their information by using the information of other agents. For instance, the following action can take place.

I_5: Agent 2 fills her information gap by borrowing from agent 1 and then to fill the remaining gap she borrows from agent 3.

I_6: Agents 1 and 2 replace their information about the temperature of the patients with the information of the agent 3.

Let $\mathfrak{F}_{\mathcal{S}_1}$ be the standard MIS-structure generated by $\mathcal{S}_1$. Consider a standard MIS-model $\mathfrak{M} := (\mathfrak{F}_{\mathcal{S}_1}, V)$, where V is a valuation function such that $V(p) := \{P4\}$. Then the information $I_3 - I_6$ can be expressed by $\sigma_3 - \sigma_6$ respectively in the language defined above as follows.

$\sigma_3 := (p, 1, N, yes) \; ; \; ((p, 1, T, high) \cup (p, 1, T, very\ high))$,
$\sigma_4 := \sigma_4^1; \sigma_4^2; \sigma_4^3$, where
$\sigma_4^i := (\phi_i, i, H, yes)$, ϕ_i is $((i, T, high) \vee (i, T, very\ high)) \wedge (i, N, yes)$, $i \in \{1, 2, 3\}$.
$\sigma_5 := \sigma_5^1; \sigma_5^3$, where
$\sigma_5^i := ((2, T, *), 2, i, T) \; ; \; ((2, H, *), 2, i, H) \; ; \; ((2, N, *), 2, i, N)$, $i \in \{1, 3\}$.
$\sigma_6 := (\top, \{1, 2\}, 3, T)$, where $\top$ is the logical constant for true.

Note that in σ_4, the ordering of occurrences of σ_4^1, σ_4^2 and σ_4^3 will not matter. So, for example, one can take σ_4 as $\sigma_4^2; \sigma_4^1; \sigma_4^3$. On the other hand, in σ_5, the ordering matters – information $\sigma_5^1; \sigma_5^3$ and $\sigma_5^3; \sigma_5^1$ have different effects.

In the above example, one may be interested to see if agent 2 has complete information about the attributes after updating with I_5. This enquiry corresponds to checking the satisfiability of the wff $[\sigma_5]((2, T, *) \vee (2, H, *) \vee (2, N, *))$ in the model $\mathfrak{M}$.

Finally, we give the following reduction axioms, which lead us to the axiomatization of the dynamic logics corresponding to the above semantics. Let σ_0 be an atomic information, $L \in \{I, S\}$ and p denote a propositional variable.

Reduction axioms

AxM17. $[\sigma_0]p \leftrightarrow p$.
AxM18. $[\sigma_0]\neg\alpha \leftrightarrow \neg[\sigma_0]\alpha$.
AxM19. $[\sigma_0](\alpha \rightarrow \beta) \leftrightarrow ([\sigma_0]\alpha \rightarrow [\sigma_0]\beta)$.
AxM20(a). $[(\phi, i, a, v)](k, b, v') \leftrightarrow (k, b, v')$, for $k \neq i$, or $b \neq a$.
AxM20(b). $[(\phi, i, a, v)](i, a, v') \leftrightarrow (\neg\phi \wedge (i, a, v'))$, $v' \in (\mathcal{VC} \cup \{*\}) \setminus \{v\}$.
AxM20(c). $[(\phi, i, a, v)](i, a, v) \leftrightarrow (\neg\phi \rightarrow (i, a, v))$.
AxM21(a). $[(\phi, i, a, v)][L(k, B)]\alpha \leftrightarrow [L(k, B)]\alpha$, where $a \notin B$ or $i \neq k$.
AxM21(b). For $a \in B$,

$$[(\phi, i, a, v)][I(i, B)]\alpha \leftrightarrow \Big(\big(\phi \rightarrow [I(i, B \setminus \{a\})](\phi \rightarrow [(\phi, i, a, v)]\alpha)\big)$$
$$\wedge \big(\phi \rightarrow [I(i, B \setminus \{a\})](\neg\phi \wedge (i, a, v) \rightarrow [(\phi, i, a, v)]\alpha)\big)$$
$$\wedge \big((\neg\phi \wedge (i, a, v)) \rightarrow [I(i, B \setminus \{a\})](\phi \rightarrow [(\phi, i, a, v)]\alpha)\big)$$
$$\wedge \big(\neg\phi \rightarrow [I(i, B)](\neg\phi \rightarrow [(\phi, i, a, v)]\alpha)\big) \Big).$$

AxM21(c). For $a \in B$,

$$[(\phi,i,a,v)][S(i,B)]\alpha \leftrightarrow \Big(\big(\phi \rightarrow [S(i,B\setminus\{a\})](\phi \rightarrow [(\phi,i,a,v)]\alpha)\big)$$
$$\wedge\big(\phi \rightarrow [S(i,B\setminus\{a\})](\neg\phi \wedge ((i,a,v)\wedge(i,a,*)) \rightarrow [(\phi,i,a,v)]\alpha)\big)$$
$$\wedge\big((\neg\phi \wedge ((i,a,v)\vee(i,a,*))) \rightarrow [S(i,B\setminus\{a\})](\phi \rightarrow [(\phi,i,a,v)]\alpha)\big)$$
$$\wedge\big(\neg\phi \rightarrow [S(i,B)](\neg\phi \rightarrow [(\phi,i,a,v)]\alpha)\big) \Big).$$

AxM22(a). $[(\phi,i,j,a)](k,b,v) \leftrightarrow (k,b,v)$, for $k \neq i$, or $b \neq a$.
AxM22(b). $[(\phi,i,j,a)](i,a,v) \leftrightarrow \Big(((j,a,v)\wedge\phi) \vee ((i,a,v)\wedge\neg\phi)\Big)$.
AxM23(a). $[(\phi,i,j,a)][L(k,B)]\alpha \leftrightarrow [L(k,B)]\alpha$, where $a \notin B$ or $i \neq k$.
AxM23(b). For $a \in B$,

$$[(\phi,i,j,a)][I(i,B)]\alpha \leftrightarrow \Big((\neg\phi \rightarrow [I(i,B)](\neg\phi \rightarrow [(\phi,i,j,a)]\alpha))$$
$$\wedge(\phi \rightarrow \bigwedge_{v\in\mathcal{VC}_a\cup\{*\}} ((j,a,v) \rightarrow [I(i,B\setminus\{a\})](((i,a,v)\wedge\neg\phi) \rightarrow [(\phi,i,j,a)]\alpha)))$$
$$\wedge(\neg\phi \rightarrow \bigwedge_{v\in\mathcal{VC}_a\cup\{*\}} ((i,a,v) \rightarrow [I(i,B\setminus\{a\})](((j,a,v)\wedge\phi) \rightarrow [(\phi,i,j,a)]\alpha)))$$
$$\wedge(\phi \rightarrow \bigwedge_{v\in\mathcal{VC}_a\cup\{*\}} ((j,a,v) \rightarrow [I(i,B\setminus\{a\})]((j,a,v)\wedge\phi) \rightarrow [(\phi,i,j,a)]\alpha))\Big).$$

AxM23(c). For $a \in B$,

$$[(\phi,i,j,a)][S(i,B)]\alpha \leftrightarrow \Big((\neg\phi \rightarrow [S(i,B)](\neg\phi \rightarrow [(\phi,i,j,a)]\alpha))$$
$$\wedge(\phi \rightarrow \bigwedge_{v\in\mathcal{VC}_a\cup\{*\}} ((j,a,v) \rightarrow [S(i,B\setminus\{a\})]((((i,a,v)\vee(i,a,*))\wedge\neg\phi) \rightarrow$$
$$[(\phi,i,j,a)]\alpha)))$$
$$\wedge(\neg\phi\wedge(i,a,*) \rightarrow [S(i,B\setminus\{a\})](\phi,i,j,a)\alpha)$$
$$\wedge(\neg\phi\wedge\neg(i,a,*) \rightarrow \bigwedge_{v\in\mathcal{VC}_a\cup\{*\}} ((i,a,v) \rightarrow [S(i,B\setminus\{a\})]((j,a,v)\wedge\phi \rightarrow$$
$$[(\phi,i,j,a)]\alpha)))$$
$$\wedge(\phi\wedge(j,a,*) \rightarrow [S(i,B\setminus\{a\})][(\phi,i,j,a)]\alpha)$$
$$\wedge[S(i,B\setminus\{a\})](\phi\wedge(j,a,*) \rightarrow [(\phi,i,j,a)]\alpha)$$
$$\wedge(\phi \rightarrow \bigwedge_{v\in\mathcal{VC}_a} ((j,a,v) \rightarrow [S(i,B\setminus\{a\})]((j,a,v)\wedge\phi \rightarrow [(\phi,i,j,a)]\alpha)))\Big).$$

AxM24(a). For $P := \{i_1, i_2, \ldots, i_m\}$,
$[(\phi,P,a,v)]\alpha \leftrightarrow [\sigma_1][\sigma_2]\cdots[\sigma_m]\alpha$, where $\sigma_k := (\phi,i_k,a,v)$.
AxM24(b). For $P := \{i_1, i_2, \ldots, i_m\}$,
$[(\phi,P,j,a)]\alpha \leftrightarrow [\sigma_1][\sigma_2]\cdots[\sigma_m]\alpha$, where $\sigma_k := (\phi,i_k,j,a)$.

AxM25. $[\sigma;\sigma']\alpha \leftrightarrow [\sigma][\sigma']\alpha$.
AxM26. $[\sigma \vee \sigma']\alpha \leftrightarrow [\sigma]\alpha \wedge [\sigma']\alpha$.

The soundness of the above axioms can be proved in the usual way. Axioms AxM17-AxM21, AxM25,AxM26 can be identified with the reduction axioms Ax16-Ax22 for the single agent case. In the multi-agent case, we also have operators for capturing the flow of information to a group of agents instead of just one. Moreover, we also have operators for interaction between the agents, that is, operators which correspond to a situation where an agent updates her knowledge base by borrowing information from another agent of the system. Axioms AxM22-AxM24 gives the reduction axioms for these operators.

14 Summary and Future Directions

In this final section, we sum up the main work that has been attempted with varying degrees of success. We also mention some issues coming out from the article which need further work. We reiterate that the terms "agent" and "source" are used synonymously

Rough set theory is studied in the following three situations.

1. Information arrives from different agents.
2. Information evolves with time.
3. A combination of (1) and (2).

1. **Multi-agent situation** (cf. Sects. 3–7).
   - The notion of multiple-source approximation system with distributed knowledge base ($MSAS^D$) is used to represent the multi-agent situation, where we have equivalence relations representing the knowledge base of individual agents as well as groups of agents. (cf. Sect. 3).
   - Notions of strong/weak lower and upper approximations are proposed for $MSAS^D$, which are obtained as a generalization of Pawlak's lower and upper approximations. Properties of these notions are studied. In particular, it is determined how strong/weak lower and upper approximations are related with the (i) lower, upper approximations with respect to distributed knowledge base of groups of agents and (ii) approximations in tolerance approximation spaces. (cf. Sect. 3.1).
   - Notions of strong, weak, lower and upper definable set based on strong/weak lower and upper approximations are defined and their properties are investigated. (cf. Sect. 3.2).
   - In order to express how much the information provided by the $MSAS^D$ depends on that of an agent or a group of agents, the notion of dependency is introduced and properties are investigated. (cf. Sect. 3.3).
   - Logic $LMSAS^D$ for $MSAS^D$ is proposed. (cf. Sects. 4.1–4.3 for syntax and semantics).
     - It is shown how the language of $LMSAS^D$ can be used to reason about the properties of rough sets in multiple-source situation. (cf. Sect. 4.4).

- Relationship with some known logics are given. In particular, it is shown that $S5_n^D$ and KTB are embeddable in LMSAS^{D}. (cf. Sect. 4.5).
- A sound and complete deductive system for LMSAS^{D} is given. (cf. Sect. 5).
- Decidability with respect to different classes of interpretations is proved. Moreover, decidability of a fragment is also proved with respect to the class of interpretations based on MSAS^{D}. (cf. Sect. 6).
- An algebraic semantics of LMSAS^{D} is given and the corresponding soundness and completeness theorems are proved. (cf. Sect. 7).

2. **Information varying with time** (cf. Sects. 8–12).
   - The notion of dynamic I space, $I \in \{K, K4, T, B, S4, KTB, KB4, S5\}$, is considered which represents the knowledge base, or the information that we have about the domain at different time points. (cf. Sect. 8).
   - The dynamic $S5$ spaces, also called dynamic approximation spaces (DASs), are focused upon, where the knowledge base is represented by an equivalence relation. Different types of DASs are defined depending on different patterns of flow of information. (cf. Sect. 8.1).
   - Logics for dynamic I spaces are obtained which can express the approximations of sets relative to time. (cf. Sect. 8.2).
     - Characterizing wffs for dynamic I spaces and different types of DASs are given.(cf. Sect. 8.2).
     - Comparisons are made between the semantics of these logics and that of related modal systems. (cf. Sect. 9).
     - Tableau-based proof procedures for the logics are proposed and the corresponding soundness and completeness theorems are proved. (cf. Sect. 10).
     - Some decidability results are discussed. (cf. Sect. 11).
   - It is observed that the logics proposed in Sect. 8 can study the effect of flow of information on the approximations of sets, but the flow of information itself does not appear in the picture. So, we move towards a dynamic logic which can also talk about information and information updates.
     - Logics LIS$_{\mathrm{f}}$ and LDIS$_{\mathrm{f}}$ respectively for the information systems and deterministic information systems are proposed. These logics have some desirable properties crucial for our purpose. (cf. Sect. 12.2).
     - It is shown that LDIS$_{\mathrm{f}}$ can express notions related to dependencies in data and data reduction. (cf. Sect. 12.2).
     - Deductive systems are proposed and corresponding soundness and completeness theorems are proved. (cf. Sect. 12.3).
     - Decidability of LIS$_{\mathrm{f}}$ and LDIS$_{\mathrm{f}}$ is proved. (cf. Sect. 12.4)).
     - A formalization of the notion of information and information update for information systems is presented. (cf. Sect. 12.1).
     - LIS$_{\mathrm{f}}$ and LDIS$_{\mathrm{f}}$ are extended to define dynamic logics where this update of information can be captured. (cf. Sect. 12.5).

- Reduction axioms are provided which lead us to decidability and sound and complete deductive systems for these dynamic extensions. (cf. Sect. 12.6).
- These dynamic logics are contrasted with different dynamic epistemic logics. (cf. Sect. 12.7).

3. **Combination of situations (1) and (2)** (cf. Sect. 13).
   - Information now could be meant for a group of agents instead of just one agent. Moreover, the flow of information between agents, that is, the situation when an agent borrows information from another agent of the system is also captured. (cf. Sect. 13.2).
   - In the line of the single agent case, dynamic logics are defined. Axiomatization, decidability can be proved following the technique of the single agent case. (cf. Sects. 13.1–13.2).

14.1 Directions for Further Work

There are some issues, both from the rough set and logic perspectives, that need investigation. We briefly sketch these issues.

Since the proposal of strong/weak approximations in [45], a lot of work has been done on the multi-granulation rough set model (MGRS) consisting of a number of (equivalence) relations on the same domain (cf. e.g. [57,64,66,67, 86–88,94,107–109]). These have contributed both on the theoretical framework as well as to the applicability of MGRS. We would like to mention here that the notions of the optimistic and pessimistic approximations proposed by Qian et al. [86,87] on MGRS are actually the same as the notions of strong and weak approximations respectively, introduced by us in [45]. The article [57] also proposed approximation operators for a multiple source situation, although it considers the knowledge base of an individual source to be given by a tolerance relation instead of equivalence. These approximation operators are defined taking into account only the information that sources have about the objects, but it does not consider the view of the sources regarding membership of the objects. That is, only the knowledge base R of the sources are used. On the contrary, the notions of strong/weak approximations are based on a totally different principle: one takes into account whether the source considers an object as a positive, negative or boundary element of the set. So, in that case, one uses the approximations $\underline{X}_R$ and $\overline{X}_R$ corresponding to the knowledge base R of each source of the system.

Apart from approximations of sets, the notions of definability of sets and dependency of information on a group of agents are investigated here in the multi-agent case. There are some other important rough set notions such as membership function, reduction in knowledge base, dependency in knowledge etc. which need to be studied in a multi-agent setting. In this direction, one can find some initial work on membership functions in [51]. Another direction of research would be the lattice structure arising from the concepts defined on MSAS^{D}. The article [51] presents an investigation on this which needs to be pursued further.

Our study of the multi-agent situation is based on Pawlak's rough set model which considers equivalence relations. As in the case of DASs, one may generalize the MSAS^{D}s by considering relations other than equivalence. It is not difficult to extend the scheme of the current work to some such generalizations of MSAS^{D}s. However, there are some other cases which need a fresh study in the multi-agent setting, as these may give rise to a situation which does not occur in our current set up. For instance, covering and neighbourhood based approximation spaces admit a situation where one agent considers an object to be a positive element of a set, but another considers that object to be a negative element of the same set. It seems worth investigating, how the basic ideas of this work may be extended to these formalisms.

The current work on the multi-agent scenario is based on the assumption that each agent is equally preferred while deciding the membership of an object. But one may incorporate the notion of preference order on the set of agents. Notions such as approximations of sets or membership functions would then depend on the knowledge base of the agents of the system as well as on the position of the agents in the hierarchy giving the preference of agents. A first step in this direction may be found in [50], where different notions of approximations based on the preference order of the agents are proposed. The strong/weak lower and upper approximations are obtained as a special case of these notions. The logic for this multi-agent rough set model based on preference order is not yet investigated, but it appears that LMSAS^{D} can be suitably extended to serve the purpose. Similarly, the notions of multi-agent information system (MIS) and multi-agent deterministic information system (MDIS) (cf. Definition 76) can be generalized by incorporating the notion of preference order on the set of agents. Moreover, in the line of [50], these generalized MIS and $MDIS$ would generate different generalized notions of approximations of sets. Thus, a natural question would be about a dynamic logic for these generalizations of MIS and MDIS which can study the effect of flow of information on the *generalized notions* of approximations of sets.

One may compare, in the context of the logic $\underline{\mathcal{F}}_w$ for weak lower and strong upper approximations (cf. Sect. 4.5), the work of Murakami in [74]. A sound and complete deductive system for a propositional modal logic that has a global modal operator and a modal operator ∇ is presented in [74]. A model $\mathfrak{M}$ is based on a structure of the form (U, Π), where Π is a set of partitions on U. The truth condition for ∇ is given as:

$$\mathfrak{M}, w \models \nabla\alpha \text{ if and only if there is a } P \in \Pi \text{ such that } [w]_P = [\![\alpha]\!]_{\mathfrak{M}}. \qquad (\nabla)$$

No restriction is imposed on the cardinality of Π, and the main results do not appear to be affected even if we restrict the cardinality of Π to be at most $|\mathbb{N}|$. Observe that if we replace equality by inclusion '$\subseteq$' in (∇), we get:

$$\mathfrak{M}, w \models \nabla\alpha \text{ if and only if there is a } P \in \Pi \text{ such that } w \in \underline{[\![\alpha]\!]_{\mathfrak{M}}}_{R^P},$$

where R^P is the equivalence relation corresponding to the partition P. The system with this modified truth condition on ∇ may thus be identified with the extension of $\underline{\mathcal{F}}_w$ with the global modal operator. It may be of interest to check if some variant of the technique used in [74], helps to obtain a complete axiomatization of this extension.

There are some issues with LMSAS^{D} and the logics for dynamic I spaces which are still not answered. For instance, decidability is still open in both cases. In case of LMSAS^{D} and the logics $L(T, I)$, we have found decidable fragments. However, a satisfactory decidable fragment for the logics $L(T, I)$ is yet to be obtained. It appears that the deductive system of the logics for dynamic I spaces will consist of the axioms of the modal system I for $\Box$ and the usual axioms of linear temporal logic [71] with modifications to account for finiteness both in the past and future. However, the corresponding completeness theorem is yet to be proved. Moreover, one can investigate some interesting variant/extensions of these logics. For instance, instead of linear finite time line, one may consider a branching time structure. Similarly, one may add descriptors to the language to see the changes in attribute values of the objects with time.

Another pending issue is the complexity analysis of the decidability results presented in this article.

A nice property of Hilbert-style axiomatization of modal systems is that, one can switch from one system to another by just adding or removing axioms. This property is missing in the proposed tableau procedures for the logics of dynamic I spaces. In fact, the ν-rule depends on the accessibility relation on the set of N-prefixes which varies with the system. Moreover, due to the ν-rule, one needs to keep track of all N-prefixes occurring in the branch. In [73], single step tableaux (SST) for modal logics are presented which have the property of Hilbert-style axiomatization as mentioned above. Moreover, SST modifies Fitting's modal tableau [28] by replacing the ν rule such that one is not required to keep track of previously occurring prefixes. In the line of [73], one can propose SST for $\mathcal{L}$ (the language of logics for dynamic I spaces) as well. We also need to modify the ν-rule. For instance, for the dynamic $K4$ spaces, one can replace the ν-rule with the following rules.

$$\frac{\sigma\nu}{\sigma t\nu_0} \qquad\qquad \frac{\sigma\nu}{\sigma t\nu}$$

It is not difficult to see the soundness theorem. Independent proofs of completeness and decidability are not yet investigated. However, using Theorem 17, one can prove the completeness theorem by showing that the ν-rule is deducible here. Suppose $(\sigma, \sigma') \in R_n^{K4}$, $char(\sigma) = char(\sigma') = n$ and σv occurs on a branch θ. We need to show that $\sigma'\nu_0$ can be introduced in θ using the rules mentioned above. In fact, by giving arguments similar to Proposition 67 which was done for $S5$, we obtain σ' to be of the form $\sigma l_1 l_2 \cdots l_r$. Now, using the second rule above $r - 1$ times successively starting from $\sigma\nu$, we obtain $\sigma l_1 l_2 \cdots l_{r-1}\nu \in \theta$. Then applying the first rule on $\sigma l_1 l_2 \cdots l_{r-1}\nu$, we obtain $\sigma l_1 l_2 \cdots l_{r-} l_r \nu_0 \in \theta$, that is $\sigma'\nu_0 \in \theta$.

Following Orłowska's proposal of information structures, we can generalize the notion of MSAS^{D} to define structures of the form $(U, \{R_{(P,B)}\}_{P\subseteq_f N, B\subseteq \mathcal{A}})$, where we now require $R_{(P,B)} = \bigcap_{i\in P}\bigcap_{a\in B} R_{(\{i\},\{a\})}$. The relation $R_{(P,B)}$ represents the distributed knowledge base of the group P of agents with respect to the attribute set B. An obvious modification of LMSAS^{D} will give us a logic for such a structure. Moreover, it does not appear to be a difficult exercise to get a deductive system for such a logic. One may use the technique of copying to obtain the completeness theorem. The situation is not so simple when we also want to bring the descriptors into the picture. In fact, in such a case, our interest would be on a logic for structures of the form

$$\mathfrak{F} := (W, \{R_{I(P,B)}\}_{P\subseteq Ag, B\subseteq \mathcal{AC}}, \{R_{S(P,B)}\}_{P\subseteq Ag,\ B\subseteq \mathcal{AC}}, m),$$

where $\mathfrak{F} := (W, \{R_{I(i,B)}\}_{i\in Ag, B\subseteq \mathcal{AC}}, \{R_{S(i,B)}\}_{i\in Ag,\ B\subseteq \mathcal{AC}}, m)$ is a *MIS*-structure (*MDIS*-structure) and $R_{L(P,B)} = \bigcap_{i\in P}\bigcap_{a\in B} R_{L(i,\{a\})}$, $L \in \{I, S\}$. A logic for such a structure can be given by enriching the language of the logic LMIS_{f} (LMDIS_{f}) with the modal operators $[L(P, B)]$, which will play the role of the syntactic counterpart of the relations $R_{L(P,B)}$. It is also not difficult to extend the axiom schema given in Sect. 13.1 to this case. But the proof of the completeness theorem does not seem to be an easy task. Copying will not work here due to the presence of descriptors in the language.

In Sect. 12, logics are proposed for information systems with a *finite set* of attribute values Val_a for each attribute a, as is the case in usual practical problems. One may think of removing this restriction on the cardinality of Val_a. Observe that without this restriction axiom Ax8 would become an infinitary wff. It is not difficult to see that each attribute a can be viewed as a predicate P_a over the set of attributes. So, $P_a(v)$ would indicate that the object takes the value v for the attribute a. Thus, instead of working with the descriptors, one may use these predicates. This formalism would also work in the case when Val_a is infinite for some (or, all) a. We refer to [56] for details on this line of research.

Another important direction is to carry out our study for non-deterministic information systems (NIS). The uncertainty in NIS about attribute values of the objects may be removed or reduced with time due to flow of information. One may look for a dynamic logic capturing this aspect of NIS following the line of investigation of dynamic logics for information systems presented here. In this direction, a first step would be to obtain a logic for NIS with the features similar to the logic of DIS proposed in Sect. 12.2. The logics proposed in [43] seems to serve this purpose.

In literature one can find several works on the study of rough sets under a situation where knowledge base evolves with time (cf. e.g. [16,17,19,69,70,105]). In this article, our proposal of dynamic logics is motivated from the formalism of the notion of information and information update presented in Sect. 12.1. Depending on the applications and type of information systems, one may need other notions of information and corresponding update. It would be worth investigating if the scheme of the current work in obtaining a dynamic logic would then be applicable. At this point, we would like to add here that the dynamic logics of Sect. 12

cannot capture information asking for the introduction or deletion of objects to or from the domain. With time, we may get information about new objects, and introduction of these objects might affect the concept approximations. Moreover, sometimes, we may wish to view only a part of the domain instead of the whole, while taking decisions. For instance, at some point, there may be insufficient information about certain objects with respect to some attributes, and we may wish, at that point, to ignore those objects while making decisions. As evident from the work in [41,70,93,99], any study of rough set theory in the perspective of update cannot be complete without touching these aspects of updates. The update logic proposed in [56] overcomes these limitations.

In order to capture the situation where information regarding the attribute-values of the objects are not precise but given in terms of probability, [44] proposed the notion of probabilistic information system (PIS). Notions of distinguishability relations and corresponding approximation operators for PISs are proposed and studied there. It is shown that the DISs, IISs and NISs are all special instances of PISs. Moreover, the approximation operators defined on DIS (relative to indiscernibility), IISs and NISs (relative to similarity relations) all originate from a single approximation operator defined on PISs. It appears to us that the multi-agent situation can be captured through PISs as well. Formulating a logic for PIS is yet another pending work.

Acknowledgements. I would like to thank Prof. Mohua Banerjee, my thesis supervisor, for her many suggestions and constant support during this research. In every sense, this thesis would have never been finished without her support.

References

1. Balbiani, P.: Axiomatization of logics based on Kripke models with relative accessibility relations. In: Orłowska, E. (ed.) Incomplete Information: Rough Set Analysis, pp. 553–578. Physica Verlag, Heidelberg, New York (1998)
2. Balbiani, P., Orłowska, E.: A hierarchy of modal logics with relative accessibility relations. J. Appl. Non-Class. Logics **9**(2–3), 303–328 (1999)
3. Baldoni, M., Giordano, L., Martelli, A.: A tableau calculus for multimodal logics and some (un)decidability results. In: de Swart, H. (ed.) Proceedings of TABLEAUX 1998. LNAI, vol. 1397, pp. 44–59. Springer, Heidelberg (1998)
4. Baltag, A., Moss, S., Solecki, S.: The logic of public announcements, common knowledge, and private suspicions. Technical report SEN-R9922, CWI, Amsterdam (1999)
5. Banerjee, M., Chakraborty, M.K.: Rough consequence, rough algebra. In: Ziarko, W.P. (ed.) RSKD 1993. WC, pp. 196–207. Springer, London (1994)
6. Banerjee, M., Khan, M.A.: Propositional logics from rough set theory. In: Peters, J.F., Skowron, A., Düntsch, I., Grzymała-Busse, J.W., Orłowska, E., Polkowski, L. (eds.) Transactions on Rough Sets VI. LNCS, vol. 4374, pp. 1–25. Springer, Heidelberg (2007)

7. Banerjee, M., Khan, M.A.: Rough set theory: a temporal logic view. In: Chakraborty, M.K., Löwe, B., Mitra, M.N., Sarukkai, S. (eds.) Studies in Logic. Proceedings of Logic, Navya-Nyāya and Applications: Homage to Bimal Krishna Matilal, Kolkata, India, vol. 15, pp. 1–20. College Publications, London (2008). An earlier version of this paper was presented and published in the pre-proceedings of the conference, January 2007
8. Bennett, B., Dixon, C., Fisher, M., Hustadt, U., Franconi, E., de Rijke, M.: Combinations of modal logics. Artif. Intell. Rev. **17**(1), 1–20 (2002)
9. Blackburn, P., de Rijke, M., Venema, Y.: Modal Logic. Cambridge University Press, Cambridge (2001)
10. Bonanno, G.: Axiomatic characterization of the AGM theory of belief revision in a temporal logic. Artif. Intell. **171**, 144–160 (2007)
11. Bonanno, G.: A sound and complete temporal logic for belief revision. In: Dégremont, C., Keiff, L., Rückert, H. (eds.) Dialogues, Logics and Other Strange Things: Essays in Honour of Shahid Rahman, pp. 67–80. College Publications, London (2008)
12. Bull, R.A.: On modal logic with propositional quantifiers. J. Symb. Log. **34**(2), 257–263 (1969)
13. Cerrito, S., Mayer, M.C., Praud, S.: First order linear temporal logic over finite time structures. In: Ganzinger, H., McAllester, D., Voronkov, A. (eds.) LPAR 1999. LNCS (LNAI), vol. 1705, pp. 62–76. Springer, Heidelberg (1999)
14. Chang, C.C., Keisler, H.J.: Model Theory. Elsevier Science Publications B.V., Amsterdam (1992)
15. Chellas, B.F.: Modal Logic. Cambridge University Press, Cambridge (1980)
16. Chen, H., Li, T., Qiao, S., Ruan, D.: A rough set based dynamic maintenance approach for approximations in coarsening and refining attribute values. Int. J. Intell. Syst. **25**, 1005–1026 (2010)
17. David, C.: Temporal dynamics in information tables. Fundamenta Informaticae **115**(1), 57–74 (2012)
18. Demri, S., Orłowska, E.: Incomplete Information: Structure, Inference, Complexity. Springer, Heidelberg (2002)
19. Deng, D., Huang, H.-K.: Dynamic reduction based on rough sets in incomplete decision systems. In: Yao, J.T., Lingras, P., Wu, W.-Z., Szczuka, M.S., Cercone, N.J., Ślęzak, D. (eds.) RSKT 2007. LNCS (LNAI), vol. 4481, pp. 76–83. Springer, Heidelberg (2007)
20. Dixon, C., Fisher, M.: Tableaux for temporal logics of knowledge: synchronous systems of perfect recall or no learning. In: Proceedings of Tenth International Symposium on Temporal Representation and Reasoning and the Fourth International Conference on Temporal Logic (TIME-ICTL 2003), pp. 62–71. IEEE Computer Society Press (2003)
21. Dubois, D., Prade, H.: Rough fuzzy sets and fuzzy rough sets. Int. J. Gen. Syst. **17**, 191–200 (1990)
22. Fagin, R., Halpern, J.Y., Moses, Y., Vardi, M.Y.: Reasoning About Knowledge. The MIT Press, Cambridge (1995)
23. Fagin, R., Halpern, J.Y., Vardi, M.Y.: What can machines know? On the properties of knowledge in distributed systems. J. ACM **39**, 328–376 (1992)
24. Fan, T.F., Hu, W.C., Liau, C.J.: Decision logics for knowledge representation in data mining. In: COMPSAC 2001: Proceedings of the 25th International Computer Software and Applications Conference on Invigorating Software Development, p. 626. IEEE Computer Society, Washington, DC (2001)

25. Fan, T.-F., Liu, D.-R., Tzeng, G.-H.: Arrow decision logic. In: Ślęzak, D., Wang, G., Szczuka, M.S., Düntsch, I., Yao, Y. (eds.) RSFDGrC 2005. LNCS (LNAI), vol. 3641, pp. 651–659. Springer, Heidelberg (2005)
26. Farinas Del Cerro, L., Orłowska, E.: DAL – a logic for data analysis. Theoret. Comput. Sci. **36**, 251–264 (1997)
27. Finger, M., Gabbay, D.M.: Adding a temporal dimension to a logic system. J. Log. Lang. Inf. **1**, 203–233 (1992)
28. Fitting, M.: Proof Methods for Modal and Intuitionistic Logics. D. Reidel Publishing Co., Dordrecht (1983)
29. Gabbay, D.M., Shehtman, V.B.: Product of modal logics, part 1. Log. J. IGPL **6**(1), 73–146 (1998)
30. Gargov, G.: Two completeness theorems in the logic for data analysis. Technical report 581, Institute of Computer Science, Polish Academy of Sciences, Warsaw (1986)
31. Gerbrandy, J.: Bisimulations on planet Kripke. Ph.D. thesis, ILLC, University of Amsterdam (1999)
32. Governatori, G.: Labelled tableaux for multi-modal logics. In: Baumgartner, P., Hähnle, R., Posegga, J. (eds.) TABLEAUX 1995. LNCS (LNAI), vol. 918, pp. 79–94. Springer, Heidelberg (1995)
33. Grzymała-Busse, J.W., Rzasa, W.: Local, global approximations for incomplete data. Trans. Rough Sets **VIII**, 21–34 (2008)
34. Halpern, J.Y., Moses, Y.: Knowledge and common knowledge in a distributed environment. J. ACM **37**(3), 549–587 (1990)
35. Halpern, J.Y., van der Meyden, R., Vardi, M.Y.: Complete axiomatizations for reasoning about knowledge and time. SIAM J. Comput. **33**(3), 674–703 (2004)
36. He, X., Xu, L., Shen, W.: Dynamic information system and its rough set model based on time sequence. In: Proceedings of the IEEE International Conference on Granular Computing, pp. 542–545, May 2006
37. Hodkinson, I.: Monodic packed fragment with equality is decidable. Stud. Logica **72**(2), 185–197 (2002)
38. Hodkinson, I., Kontchakov, R., Kurucz, A., Wolter, F., Zakharyaschev, M.: On the computational complexity of decidable fragments of first-order linear temporal logics. In: TIME 2003, Proceedings of the 10th International Symposium on Temporal Representation and Reasoning, pp. 91–98 (2003)
39. Hodkinson, I., Wolter, F., Zakharyaschev, M.: Decidable fragments of first-order temporal logics. Ann. Pure Appl. Logic **106**(1–3), 85–134 (2000)
40. Hughes, G.E., Cresswell, M.J.: A New Introduction to Modal Logic. Routledge, London (1996)
41. Jerzy, B., Słowiński, R.: Incremental induction of decision rules from dominance-based rough approximations. Electron. Notes Theor. Comput. Sci. **82**, 40–51 (2003)
42. Khan, M.A.: Multiple-source approximation systems, evolving information systems and corresponding logics: a study in rough set theory. Ph.D. thesis, Indian Institute of Technology Kanpur (2010)
43. Khan, M.A.: A modal logic for non-deterministic information systems. In: Banerjee, M., Krishna, S.N. (eds.) ICLA 2015. LNCS, vol. 8923, pp. 119–131. Springer, Heidelberg (2015)
44. Khan, M.A.: A probabilistic approach to information system and rough set theory. In: Chakraborty, M.K., Skowron, A., Maiti, M., Kar, S. (eds.) Facets of Uncertainties and Applications. Springer Proceedings in Mathematics and Statistics, vol. 125, pp. 113–123. Springer, Heidelberg (2015). (chapter 9)

45. Khan, M.A., Banerjee, M.: Formal reasoning with rough sets in multiple-source approximation systems. Int. J. Approx. Reason. **49**(2), 466–477 (2008)
46. Khan, M.A., Banerjee, M.: Multiple-source approximation systems: membership functions and indiscernibility. In: Wang, G., Li, T., Grzymala-Busse, J.W., Miao, D., Skowron, A., Yao, Y. (eds.) RSKT 2008. LNCS (LNAI), vol. 5009, pp. 80–87. Springer, Heidelberg (2008)
47. Khan, M.A., Banerjee, M.: An algebraic semantics for the logic of multiple-source approximation systems. In: Sakai, H., Chakraborty, M.K., Hassanien, A.E., Ślęzak, D., Zhu, W. (eds.) RSFDGrC 2009. LNCS, vol. 5908, pp. 69–76. Springer, Heidelberg (2009)
48. Khan, M.A., Banerjee, M.: A dynamic logic for multi-agent partial knowledge information systems. In: Proceedings of the Workshop on Logical Methods for Social Concepts (LMSC 2009) at ESSLLI 2009, Bordeaux, France (2009). http://www.irit.fr/Andreas.Herzig/Esslli09/
49. Khan, M.A., Banerjee, M.: A logic for complete information systems. In: Sossai, C., Chemello, G. (eds.) ECSQARU 2009. LNCS, vol. 5590, pp. 829–840. Springer, Heidelberg (2009)
50. Khan, M.A., Banerjee, M.: A preference-based multiple-source rough set model. In: Szczuka, M., Kryszkiewicz, M., Ramanna, S., Jensen, R., Hu, Q. (eds.) RSCTC 2010. LNCS, vol. 6086, pp. 247–256. Springer, Heidelberg (2010)
51. Khan, M.A., Banerjee, M.: A study of multiple-source approximation systems. In: Peters, J.F., Skowron, A., Słowiński, R., Lingras, P., Miao, D., Tsumoto, S. (eds.) Transactions on Rough Sets XII. LNCS, vol. 6190, pp. 46–75. Springer, Heidelberg (2010)
52. Khan, M.A., Banerjee, M.: A logic for multiple-source approximation systems with distributed knowledge base. J. Philos. Log. **40**(5), 663–692 (2011)
53. Khan, M.A., Banerjee, M.: Logics for information systems and their dynamic extensions. ACM Trans. Comput. Log. **12**(4), 29 (2011)
54. Khan, M.A., Banerjee, M.: Logics for some dynamic spaces-I. J. Log. Comput. **25**(3), 827–856 (2015)
55. Khan, M.A., Banerjee, M.: Logics for some dynamic spaces-II. J. Log. Comput. **25**(3), 857–878 (2015)
56. Khan, M.A., Banerjee, M., Rieke, R.: An update logic for information systems. Int. J. Approx. Reason. **55**, 436–456 (2014)
57. Khan, M.A., Ma, M.: A modal logic for multiple-source tolerance approximation spaces. In: Banerjee, M., Seth, A. (eds.) Logic and Its Applications. LNCS, vol. 6521, pp. 124–136. Springer, Heidelberg (2011)
58. Komorowski, J., Pawlak, Z., Polkowski, L., Skowron, A.: Rough sets: a tutorial. In: Pal, S.K., Skowron, A. (eds.) Rough Fuzzy Hybridization: A New Trend in Decision-Making, pp. 3–98. Springer, Singapore (1999)
59. Kripke, S.A.: A completeness theorem in modal logic. J. Symb. Log. **24**(1), 1–14 (1959)
60. Kryszkiewicz, M.: Rough set approach to incomplete information systems. Inf. Sci. **112**, 39–49 (1998)
61. Kryszkiewicz, M.: Rules in incomplete information systems. Inf. Sci. **113**, 271–292 (1999)
62. Ladner, R., Reif, J.: The logic of distributed protocols. In: Halpern, J.Y. (ed.) Proceedings of the 1986 Conference on Theoretical Aspects of Reasoning About Knowledge, pp. 207–222. Morgan Kaufmann Publishers Inc., San Francisco (1986)

63. Lehmann, D.J.: Knowledge, common knowledge and related puzzles (extended summary). In: PODC 1984: Proceedings of the Third Annual ACM Symposium on Principles of Distributed Computing, pp. 62–67. ACM, New York (1984)
64. Liang, J.Y., Wang, F., Dang, C.Y., Qian, Y.H.: An efficient rough feature selsction algorithm with a multi-granulation view. Int. J. Approx. Reason. **53**(7), 1080–1093 (2012)
65. Liau, C.J.: An overview of rough set semantics for modal and quantifier logics. Intl. J. Uncertain. Fuzziness Knowl.-Based Syst. **8**(1), 93–118 (2000)
66. Lin, G.P., Liang, Y.H., Qian, J.Y.: Multigranulation rough sets: from partition to covering. Inf. Sci. **241**, 101–118 (2013)
67. Lin, G.P., Qian, Y.H., Li, J.J.: NMGRS: neighborhood-based multigranulation rough sets. Int. J. Approx. Reason. **53**(7), 1080–1093 (2012)
68. Lin, T.Y., Yao, Y.Y.: Neighborhoods system: measure, probability and belief functions. In: Proceedings of the 4th International Workshop on Rough Sets and Fuzzy Sets and Machine Discovery, pp. 202–207, November 1996
69. Liu, D., Li, T., Zhang, J.: A rough set-based incremental approach for learning knowledge in dynamic incomplete information systems. Int. J. Approx. Reason. **55**, 1764–1786 (2014)
70. Liu, D., Li, T.R., Ruan, D., Zou, W.L.: An incremental approach for inducing knowledge from dynamic information systems. Fundamenta Informaticae **94**(2), 245–260 (2009)
71. Manna, Z., Pnueli, A.: The Temporal Logic of Reactive and Concurrent Systems Specification. Springer, Heidelberg (1991)
72. Margaris, A.: First Order Mathematical Logic. Blaisdell Publishing Company, London (1966)
73. Massacci, F.: Single step tableaux for modal logics. J. Autom. Reason. **24**, 319–364 (2000)
74. Murakami, Y.: Modal logic of partitions. Ph.D. thesis, University Graduate School, Indiana University (2005)
75. Orłowska, E.: Dynamic information system. Fundamenta Informaticae **5**, 101–118 (1982)
76. Orłowska, E.: Logic of indiscernibility relations. In: Goos, G., Hartmanis, J. (eds.) Computation Theory. LNCS, vol. 208, pp. 177–186. Springer, Heidelberg (1985)
77. Orłowska, E.: Logic of nondeterministic information. Stud. Logica **1**, 91–100 (1985)
78. Orłowska, E.: Kripke semantics for knowledge representation logics. Studia Logica **XLIX**, 255–272 (1990)
79. Orłowska, E.: Many-valuedness and uncertainty. Intl. J. Mult. Valued Log. **4**, 207–277 (1999)
80. Orłowska, E., Pawlak, Z.: Expressive power of knowledge representation systems. Int. J. Man Mach. Stud. **20**(5), 485–500 (1984)
81. Orłowska, E., Pawlak, Z.: Representation of nondeterministic information. Theoret. Comput. Sci. **29**, 27–39 (1984)
82. Pagliani, P.: Pretopologies and dynamic spaces. Fundamenta Informaticae **59**(2–3), 221–239 (2004)
83. Pawlak, Z.: Rough sets. Intl. J. Comput. Inf. Sci. **11**(5), 341–356 (1982)
84. Pawlak, Z.: Rough Sets. Theoretical Aspects of Reasoning About Data. Kluwer Academic Publishers, Dordrecht (1991)
85. Pomykała, J.A.: Approximation, similarity and rough constructions. ILLC prepublication series for computation and complexity theory CT-93-07, University of Amsterdam (1993)

86. Qian, Y.H., Liang, J.Y., Yao, Y.Y., Dang, C.Y.: Incomplete multigranulation rough set. IEEE Trans. Syst. Man Cybern. Part A **40**(2), 420–431 (2010)
87. Qian, Y.H., Liang, J.Y., Yao, Y.Y., Dang, C.Y.: MGRS: a multi-granulation rough set. Inf. Sci. **180**, 949–970 (2010)
88. Qian, Y.H., Zhang, H., Sang, Y.L., Liang, J.Y.: Multigranulation decision-theoretic rough sets. Int. J. Approx. Reason. **55**(1), 225–237 (2014)
89. Rasiowa, H.: An Algebraic Approach to Non-classical Logics. North-Holland Publishing Company, Amsterdam (1974)
90. Rasiowa, H., Marek, W.: On reaching consensus by groups of intelligent agents. In: Ras, Z.W. (ed.), Proceedings of the International Symposium on Methodologies, Intelligent Systems ISMIS 1989, Charlotte, NC, pp. 234–243. North-Holland, New York (1989)
91. Rasiowa, H., Sikorski, R.: A proof of the completeness theorem of godel. Fundamenta Informaticae **37**, 193–203 (1950)
92. Rauszer, C.M.: Rough logic for multiagent systems. In: Masuch, M., Polos, L. (eds.) Knowledge Representation and Reasoning Under Uncertainty. LNCS, vol. 808, pp. 161–181. Springer, Heidelberg (1994)
93. Shan, N., Ziarko, W.: Data-based acquisition and incremental modification of classification rules. Comput. Intell. **11**, 357–370 (1995)
94. She, Y.H., He, X.L.: On the structure of the multigranulation rough set model. Knowl.-Based Syst. **36**, 81–92 (2012)
95. Skowron, A., Stepaniuk, J.: Tolerance approximation spaces. Fundamenta Informaticae **27**, 245–253 (1996)
96. Ślęzak, D., Ziarko, W.: The investigation of the Bayesian rough set model. Int. J. Approx. Reason. **40**, 81–91 (2005)
97. Stefanowski, J., Tsoukiàs, A.: On the extension of rough sets under incomplete information. In: Zhong, N., Skowron, A., Ohsuga, S. (eds.) RSFDGrC 1999. LNCS (LNAI), vol. 1711, pp. 73–82. Springer, Heidelberg (1999)
98. Tanaka, Y., Ono, H.: Rasiowa-Sikorski lemma and Kripke completeness of predicate and infinitary modal logics. In: Zakharyaschev, Z., Segerberg, K., de Rijke, M., Wansing, H. (eds.) Advances in Modal Logic, vol. 2, pp. 401–419. CSLI Publications, Stanford (2001)
99. Tong, L., An, L.: Incremental learning of decision rules based on rough set theory. In: IWCICA 2002: Proceedings of the 4th World Congress on Intelligent Control and Automation, pp. 420–425 (2002)
100. Vakarelov, D.: Abstract characterization of some knowledge representation systems and the logic NIL of nondeterministic information. In: Jorrand, P., Sgurev, V. (eds.) Artificial Intelligence II, pp. 255–260. North-Holland, Amsterdam (1987)
101. Vakarelov, D.: Modal logics for knowledge representation systems. Theoret. Comput. Sci. **90**, 433–456 (1991)
102. van der Hoek, W., Meyer, J.J.: Making some issues of implicit knowledge explicit. Found. Comput. Sci. **3**, 193–223 (1992)
103. van Ditmarsch, H., van der Hoek, W., Kooi, B.: Dynamic epistemic logic with assignment. In: Dignum, F., Dignum, V., Koenig, S., Kraus, S., Singh, M., Wooldridge, M. (eds.) AAMAS 2005: Proceedings of the Fourth International Joint Conference on Autonomous Agents and Multiagent Systems, pp. 141–148. ACM, New York (2005)
104. van Ditmarsch, H., van der Hoek, W., Kooi, B.: Dynamic Epistemic Logic. Springer, Berlin (2007)

105. Wojna, A.: Constraint based incremental learning of classification rules. In: Ziarko, W.P., Yao, Y. (eds.) RSCTC 2000. LNCS (LNAI), vol. 2005, p. 428. Springer, Heidelberg (2001)
106. Wooldridge, M., Dixon, C., Fisher, M.: A tableau-based proof method for temporal logics of knowledge and belief. J. Appl. Non-Class. Log. **8**(3), 225–258 (1998)
107. Wu, W.Z., Leung, Y.: Theory and applications of granular labeled partitions in multi-scale decision tables. Inf. Sci. **181**, 3878–3897 (2011)
108. Xu, W.H., Wang, Q.R., Zhang, X.T.: Multi-granulation fuzzy rough sets in a fuzzy tolerance approximation space. Int. J. Fuzzy Syst. **13**(4), 246–259 (2011)
109. Yang, X.B., Qian, Y.H., Yang, J.Y.: Hierarchical structures on multigranulation spaces. J. Comput. Sci. Technol. **27**(6), 1169–1183 (2012)
110. Yao, Y.Y., Liu, Q.: A generalized decision logic in interval-set-valued information tables. In: Zhong, N., Skowron, A., Ohsuga, S. (eds.) RSFDGrC 1999. LNCS (LNAI), vol. 1711, pp. 285–293. Springer, Heidelberg (1999)
111. Ziarko, W.: Variable precision rough set model. J. Comput. Syst. Sci. **46**, 39–59 (1993)

Author Index

Printed in the United States
By Bookmasters

Printed in the United States
By Bookmasters